FRONTIERS IN AUTISM RESEARCH

New Horizons for Diagnosis and Treatment

FRONTIERS IN AUTISM RESEARCH

New Horizons for Diagnosis and Treatment

Editor

Valerie W Hu

The George Washington University
School of Medicine and Health Sciences, USA

World Scientific

NEW JERSEY • LONDON • SINGAPORE • BEIJING • SHANGHAI • HONG KONG • TAIPEI • CHENNAI

Published by

World Scientific Publishing Co. Pte. Ltd.

5 Toh Tuck Link, Singapore 596224

USA office: 27 Warren Street, Suite 401-402, Hackensack, NJ 07601

UK office: 57 Shelton Street, Covent Garden, London WC2H 9HE

Library of Congress Cataloging-in-Publication Data

Frontiers in autism research : new horizons for diagnosis and treatment / [edited by] Valerie W. Hu, The George Washington University, School of Medicine and Health Sciences, USA.

pages cm

Includes bibliographical references and index.

ISBN 978-9814602150 (hardcover : alk. paper)

1. Autism spectrum disorders. 2. Autism spectrum disorders--Treatment. I. Hu, Valerie W., editor of compilation.

RC553.A88F76 2014

616.85'882--dc23

2014005502

British Library Cataloguing-in-Publication Data

A catalogue record for this book is available from the British Library.

In-house Editor: Darilyn Yap

Typeset by Stallion Press
Email: enquiries@stallionpress.com

Printed in Singapore by B & Jo Enterprise Pte Ltd

Dedication

To my husband, Dave, for his love and support,

to my son, Matt, for inspiring me,

and to the families of individuals with ASD

who share the journey, commitment, and quest

for a better future for their loved ones.

Contents

Preface		xi
About the Editor		xv
List of Contributors		xvii
Dissecting the Genetic Architecture of ASD		1
Chapter 1	Phenotype Definition: A Cornerstone of Autism Research, Diagnosis and Therapy *Valerie W. Hu*	3
Chapter 2	From Molecular Pathways to ASD Therapy: Insights from Syndromic Forms of Autism *Laura Nicholls, Radhika Ramadas and Irina Voineagu*	23
Chapter 3	Language Impairment in Autism Spectrum Disorders *Natalia Rakhlin and Elena L. Grigorenko*	47
Chapter 4	Whole Genome Sequencing in Autism: Clinical Translation *Ryan K.C. Yuen and Stephen W. Scherer*	69
Chapter 5	The Impact of Integrative Unconventional Data Analysis Approaches on Advancing Autism Genetics Research *Zohreh Talebizadeh and Ayten Shah*	99

Chapter 6 Construction of a Genetic Classifier for ASD Using Gene Pathway Analysis 119
Gursharan Chana, Renee Testa, Piers Gillett, Daniel Williams, Chad A. Bousman, Daniela Zantomio, Ian P. Everall, Christos Pantelis and Efstratios Skafidas

Gene Dysregulation in ASD: From Transcription to Translation 145

Chapter 7 Genome-Wide Expression Studies of Blood and Lymphoblastoid Cell Lines in Autism Spectrum Disorders 147
Boryana Stamova and Frank R. Sharp

Chapter 8 Searching in the "Dark": Non-coding RNA as a New Avenue of Autism Research 175
Tewarit Sarachana and Valerie W. Hu

Chapter 9 Targeting Noncoding RNA for Treatment of Autism Spectrum Disorders 203
Jessica DeWitt and Daniel B. Campbell

Chapter 10 A-to-I RNA Editing in Autism Spectrum Disorder 229
Alal Eran, Isaac S. Kohane and Louis M. Kunkel

Chapter 11 Translational Control of Autism and Fragile-X Syndrome 249
Christos G. Gkogkas, Argel Aguilar-Valles, Arkady Khoutorsky and Nahum Sonenberg

Epigenetic, Environmental, and Physiological Contributions to ASD 277

Chapter 12 Epigenetics in Autism 279
Matthew Ginsberg and Marvin Natowicz

Chapter 13 The Epigenetics of Autism — Running Beyond the Bases 303
Simon G. Gregory

Chapter 14 Genes and Environment in Autism Spectrum Disorders: An Integrated Perspective 335
Kelly M. Bakulski, Alison B. Singer and M. Daniele Fallin

Chapter 15 The Potential Brain Drain from Environmental Exposures on the Methylome and Genome Across Generations 375
Janine M. LaSalle and Keith W. Dunaway

Chapter 16 Oxidative Stress and Mitochondrial Dysfunction in ASDs 407
Feng Gu, Ved Chauhan and Abha Chauhan

Chapter 17 Maternal Autoantibodies in Autism Spectrum Disorder 429
Karen L. Jones and Judy Van de Water

Chapter 18 Why is Autism More Common in Males? 451
Simon Baron-Cohen, Michael V. Lombardo, Bonnie Auyeung, Emma Ashwin, Bhismadev Chakrabarti and Rebecca Knickmeyer

Moving Towards Personalized Treatment of ASD and Lifespan Issues 471

Chapter 19 Future Directions in Psychopharmacology of Autism Spectrum Disorder 473
Deepali Mankad and Evdokia Anagnostou

Chapter 20 Medical Co-Morbidities in Autism: Clues to Underlying Biological Mechanisms and/or Diagnostic Subtypes? 497
Margaret L. Bauman

Chapter 21	Translational Implications of a Whole-Body Approach to Brain Health in Autism: How Transduction between Metabolism and Electrophysiology Points to Mechanisms for Neuroplasticity *Martha R. Herbert*	515
Chapter 22	Achieving Optimal Outcomes in Autism: Treating Potentially Reversible Conditions Associated with Autism Spectrum Disorder *Richard E. Frye and Daniel A. Rossignol*	557
Chapter 23	Transcranial Magnetic Stimulation: Application in Autism Treatment *Manuel F. Casanova and Estate M. Sokhadze*	583
Chapter 24	Music Therapy: Personalized Interventions for Individuals with Autism Spectrum Disorder *Petra Kern*	607
Chapter 25	The Contributions of Art Therapy in Treatment, Assessment, and Research with People Who Have Autism Spectrum Disorders *Donna Betts, Rebecca Harmer and Gabriela Schmulevich*	627
Chapter 26	Shifting Paradigms: An Examination of Our Understanding of Adult Autism *Rhonda J. Greenhaw*	655
Index		683

Preface

Autism spectrum disorder (ASD) has rapidly become one of the most prevalent neurodevelopmental disorders affecting both children and adults. Despite the rapid rise in the number of cases reported and the high heritability associated with ASD, there is still very little understanding of its etiology and lifelong progression as well as the many factors, both intrinsic and environmental, that may increase risk for ASD. Currently, there are neither validated biomarkers for diagnostic screening nor pharmacologic therapies specific for ASD. Diagnosis is primarily based upon reports of functional and behavioral abnormalities, with medical treatment taking a mostly trial-and-error approach, since the pathology affecting most individuals with ASD is unknown. This uncertainty is in part due to the significant clinical and phenotypic heterogeneity associated with ASD. Yet, progress is being made in identifying genes, metabolic pathways, and cellular/tissue functions disrupted by autism, and there is evidence that targeted treatment of some of these genetically-defined conditions or dysfunctional pathways has led to improved outcomes for subgroups of individuals on the spectrum.

This book focuses on the emerging and expanding areas of research on ASD and their potential to lead to better diagnosis and more effective therapies. These patient-focused research areas include innovative and integrative approaches to genetic/genomic analyses and investigations of epigenetic contributions, including the role of noncoding RNAs, DNA methylation, alternative splicing, RNA editing, and faulty translation in gene regulation and expression, metabolic and immune dysfunction, co-morbidities, as well as hormonal and gene-environment interactions that may increase risk for ASD.

Advances in targeted or personalized therapies as well as novel and alternative interventions, such as transcranial magnetic stimulation, music and art therapy, are also presented. In addition, current research on issues involving adults with ASD, a population that has been sorely neglected in the field of autism, is discussed with the goal of improving outcomes and quality of life for individuals with ASD throughout their lifespan. The emergent and unifying themes across the chapters include: (1) the need to reduce the heterogeneity in the ASD population under study in order to better define pathological conditions that can be targeted for therapy in specific subgroups, (2) the importance of large-scale gene regulation which can be impacted by both dysfunctional genes and environment, and (3) the need to take a whole-body systems approach to treating individuals on the spectrum, while respecting individual differences and capitalizing on abilities.

Within each chapter, experts review cutting-edge research as well as provide their perspective on the future of research in their respective areas, including the challenges involved and the types of studies or advances that are necessary to move the field forward to achieve predicted translational (that is, practical) goals. These unique forward-looking features are designed to provide focus and "food for thought" for future studies on autism, while the executive summaries succinctly highlight the major take-home messages of each chapter. In exploring the frontiers of autism research, it is also important to keep in mind the sage advice of Sir William Bragg (Nobel Laureate in Physics, 1915): *"The important thing in science is not so much to obtain new facts as to discover new ways of thinking about them."*

Since many of the chapters are projecting research towards novel therapeutic strategies and other practical interventions, it is anticipated that this book will be of value not only to established investigators and students engaged in autism research, but also to the general population, especially families of individuals affected by ASD and their professional caregivers.

In closing, I would like to thank all of the individuals who contributed their valuable time, expertise, and effort to writing the chapters for this book, as well as all of the individuals at World Scientific who have helped to make this project a reality. Last but not least, I would like to thank my son, Matthew Hu-Smith, for providing the inspiration and motivation for my own research on autism.

Valerie W. Hu, Ph.D.

About the Editor

Valerie W. Hu

Valerie W. Hu, Ph.D., is a Professor of Biochemistry and Molecular Medicine at The George Washington University School of Medicine and Health Sciences in Washington, DC as well as the mother of a son with an autism spectrum disorder (ASD). Dr. Hu was trained as a chemist, with a Ph.D. in Chemistry from the California Institute of Technology and a B.S. in Chemistry from the University of Hawaii. She has a long research history in cross-disciplinary studies focused on protein structure-function relationships and membrane-protein interactions. In late 2004, because of her personal interest in ASD, she redirected her research focus towards autism. Dr. Hu has since become a leader in the application of multi-disciplinary, integrative genomics approaches to ASD which involve the integration of large-scale data from gene expression, behavioral, genetic, and epigenetic analyses. The primary goals of her research are biomarker-based diagnosis and personalized treatment of autism spectrum disorders. Attaining these goals will require a better understanding of the underlying pathobiology giving rise to different manifestations of autism. This pathobiology will be revealed in part through the identification of altered genes, dysfunctional pathways, and aberrant gene regulatory mechanisms specific to the different subtypes of ASD, coupled with the identification of environmental risk factors for ASD. Dr. Hu hopes that the autism research embodied and envisioned in this book will help shed light on these processes and lead towards earlier and more definitive diagnosis, evidence-based treatment options, and ultimately more enriched lives for all individuals with ASD across their lifespan.

List of Contributors

Argel Aguilar-Valles

Department of Biochemistry & Goodman Cancer Research Centre, McGill University, Montreal, QC, H3A 1A3, Canada; Department of Neuroscience, Université de Montréal, Montreal, QC, H3C 3J7, Canada

Evdokia Anagnostou

Bloorview Research Institute, University of Toronto, Toronto, Ontario, Canada

Emma Ashwin

Autism Research Centre, Department of Psychiatry, University of Cambridge, UK

Bonnie Auyeung

Autism Research Centre, Department of Psychiatry, University of Cambridge, UK

Kelly M. Bakulski

Department of Epidemiology, Johns Hopkins Bloomberg School of Public Health and Wendy Klag Center for Autism and Developmental Disabilities, Johns Hopkins Bloomberg School of Public Health, Baltimore, MD 21205; and Center for Excellence in Genomic Science, Johns Hopkins University, Baltimore, MD 21205, USA

Simon Baron-Cohen

Autism Research Centre, Department of Psychiatry, University of Cambridge, UK

Margaret L. Bauman

Department of Anatomy and Neurobiology, Boston University School of Medicine, Boston, MA 02118, USA

Donna Betts

Art Therapy Program, The George Washington University, Alexandria, VA 22314, USA

Chad A. Bousman

Department of Psychiatry, The University of Melbourne, Parkville, Victoria, Australia

Daniel B. Campbell

Zilkha Neurogenetic Institute, University of Southern California; Department of Psychiatry and the Behavioral Sciences, University of Southern California Keck School of Medicine, Los Angeles, CA 90089, USA

Manuel F. Casanova

Department of Psychiatry and Behavioral Sciences, School of Medicine, University of Louisville, Louisville, KY 40202, USA

Bhismadev Chakrabarti

Autism Research Centre, Department of Psychiatry, University of Cambridge, UK; Centre for Integrative Neuroscience and Neurodynamics, School of Psychology and Clinical Language Sciences, University of Reading, UK

Gursharan Chana

Centre for Neural Engineering, Victoria Research Laboratory, National Information and Communications Technology Australia, Department of Electrical and Electronic Engineering, and Department of Psychiatry, The University of Melbourne, Parkville, Victoria, Australia

Abha Chauhan

NYS Institute for Basic Research in Developmental Disabilities, Staten Island, NY 10314, USA

Ved Chauhan

NYS Institute for Basic Research in Developmental Disabilities, Staten Island, NY 10314, USA

Jessica DeWitt

Program in Biological and Biomedical Sciences and the Zilkha Neurogenetic Institute, University of Southern California, Los Angeles, CA 90089, USA

Keith W. Dunaway

Medical Microbiology and Immunology, Integrative Genetics and Genomics graduate group, M.I.N.D. Institute, Genome Center, University of California at Davis, Sacramento, CA 95817, USA

Alal Eran

Center for Biomedical Informatics and Department of Pediatrics, Harvard Medical School, Boston, MA 02115; Children's Hospital Informatics Program, Boston Children's Hospital, Boston, MA 02115, USA

Ian P. Everall

Department of Psychiatry, The University of Melbourne, Parkville, Victoria, Australia

M. Daniele Fallin

Department of Epidemiology, Wendy Klag Center for Autism and Developmental Disabilities, Department of Biostatistics, and Department of Mental Health, Johns Hopkins Bloomberg School of Public Health, Baltimore, MD 21205; Center for Excellence in Genomic Science, Johns Hopkins University, Baltimore, MD 21205, USA

Richard E. Frye

Arkansas Children's Hospital Research Institute, Department of Pediatrics, University of Arkansas for Medical Sciences, Little Rock AR 72202, USA

Piers Gillett

Centre for Neural Engineering, Victoria Research Laboratory, National Information and Communications Technology Australia, Department of Electrical and Electronic Engineering, and Department of Psychiatry, The University of Melbourne, Parkville, Victoria, Australia

Matthew Ginsberg

Cleveland Clinic Lerner College of Medicine, Cleveland Clinic, Cleveland, OH 44195, USA

Christos G. Gkogkas

Department of Biochemistry & Goodman Cancer Research Centre, McGill University, Montreal, QC, H3A 1A3, Canada; Current affliation: Patrick Wild Centre and Centre for Integrative Physiology, University of Edinburgh, Hugh Robson Building, George Square, EH8 9XD, Edinburgh, UK

Rhonda J. Greenhaw

Office of Collaborative Programs and Hussman Center for Adults with Autism, Towson University College of Health Professions, Towson, MD 21252, USA

Simon G. Gregory

Duke Molecular Physiology Institute, Duke University Medical Center, Durham, NC 27710, USA

Elena L. Grigorenko

Yale Child Study Center, Yale University, New Haven, CT 06520, USA; Moscow State University for Psychology and Pedagogy, Moscow, Russian Federation

Feng Gu

NYS Institute for Basic Research in Developmental Disabilities, Staten Island, NY 10314, USA

Rebecca Harmer

Capital Region Children's Center, Washington, DC 20006, USA

Martha Herbert

TRANSCEND Research Program, Massachusetts General Hospital; Harvard Medical School; Higher Synthesis Health, Boston, MA 02129, USA

Valerie W. Hu

Department of Biochemistry and Molecular Medicine, The George Washington University School of Medicine and Health Sciences, Washington, DC 20037, USA

Karen L. Jones

Department of Rheumatology, Allergy, and Clinical Immunology, and the M.I.N.D. Institute, University of California, Davis, Sacramento, CA 95817, USA

Petra Kern

Music Therapy Consulting, Santa Barbara, CA 93103, USA; Marylhurst University, Portland, OR 97036, USA; University of Louisville, Louisville, KY 40202, USA

Arkady Khoutorsky

Department of Biochemistry & Goodman Cancer Research Centre, McGill University, Montreal, QC, H3A 1A3, Canada

Rebecca Knickmeyer

Autism Research Centre, Department of Psychiatry, University of Cambridge, UK; Department of Psychiatry, University of North Carolina–Chapel Hill, Chapel Hill, NC 27599, USA; Neurosciences Hospital, Chapel Hill, NC 27514, USA

Isaac S. Kohane

Center for Biomedical Informatics and Department of Pediatrics, Harvard Medical School, Boston, MA 02115; Children's Hospital Informatics Program, Boston Children's Hospital, Boston, MA 02115, USA

Louis M. Kunkel

Division of Genetics and Genomics and the The Manton Center for Orphan Disease Research, Boston Children's Hospital, Boston, MA 02115; Department of Pediatrics and Department of Genetics, Harvard Medical School, Boston, MA 02115, USA

Janine M. LaSalle

Medical Microbiology and Immunology, Integrative Genetics and Genomics graduate group, M.I.N.D. Institute, Genome Center, University of California at Davis, Sacramento, CA 95817, USA

Michael V. Lombardo

Autism Research Centre, Department of Psychiatry, University of Cambridge, UK

Deepali Mankad

Bloorview Research Institute, University of Toronto, Toronto, Ontario, Canada

Marvin Natowicz

Cleveland Clinic Lerner College of Medicine, and Pathology & Laboratory Medicine, Genomic Medicine, Neurological and Pediatrics Institutes, Cleveland Clinic, Cleveland, OH 44195, USA

Laura Nicholls

School of Biotechnology and Biomolecular Sciences, University of New South Wales, Kensington, Sydney NSM 2052, Australia

Christos Pantelis

Department of Psychiatry and Melbourne Neuropsychiatry Centre, Department of Psychiatry, The University of Melbourne & Melbourne Health, Parkville, Victoria, Australia

Natalia Rakhlin

Yale Child Study Center, Yale University, New Haven, CT 06520, USA

Radhika Ramadas

School of Biotechnology and Biomolecular Sciences, University of New South Wales, Kensington, Sydney NSM 2052, Australia

Daniel A. Rossignol

Rossignol Medical Center, Irvine, CA 92618, USA

Tewarit Sarachana

Department of Biochemistry and Molecular Medicine, The George Washington University School of Medicine and Health Sciences, Washington, DC 20037, USA; Current address: Department of Clinical Chemistry, Faculty of Allied Health Sciences, Chulalongkorn University, Bangkok, Thailand

Stephen W. Scherer

Program in Genetics and Genome Biology, The Centre for Applied Genomics, The Hospital for Sick Children, Toronto, Ontario M5G 0A4 Canada; McLaughlin Centre, University of Toronto, Toronto, Ontario M5S 1A1, Canada

Gabriela Schmulevich

Art Therapy Program, The George Washington University, Alexandria, VA 22314, USA

Ayten Shah

Children's Mercy Hospital, Kansas City, MO 64108, USA

Frank R. Sharp

Department of Neurology and the M.I.N.D. Institute, University of California at Davis, Sacramento, CA 95817, USA

Alison B. Singer

Department of Epidemiology and Wendy Klag Center for Autism and Developmental Disabilities, Johns Hopkins Bloomberg School of Public Health, Baltimore, MD 21205, USA

Efstratios Skafidas

Centre for Neural Engineering, Victoria Research Laboratory, National Information and Communications Technology Australia, Dept. of Electrical and Electronic Engineering, and Department of Psychiatry, The University of Melbourne, Parkville, Victoria, Australia

Estate M. Sokhadze

Department of Psychiatry and Behavioral Sciences, School of Medicine, University of Louisville, Louisville, KY 40202, USA

Nahum Sonenberg

Department of Biochemistry & Goodman Cancer Research Centre, McGill University, Montreal, QC, H3A 1A3, Canada

Boryana Stamova

Department of Neurology and the M.I.N.D. Institute, University of California at Davis, Sacramento, CA 95817, USA

Zohreh Talebizadeh

Children's Mercy Hospital, Kansas City, MO 64108; University of Missouri-Kansas City School of Medicine, Kansas City, MO 64108, USA

Renee Testa

Melbourne Neuropsychiatry Centre, Department of Psychiatry, The University of Melbourne & Melbourne Health, Parkville, Victoria, Australia; Department of Psychology, Monash University, Clayton, Victoria, Australia

Judy Van de Water

Department of Rheumatology, Allergy, and Clinical Immunology, and the M.I.N.D. Institute, University of California, Davis, Sacramento, CA 95817, USA

Irina Voineagu

School of Biotechnology and Biomolecular Sciences, University of New South Wales, Kensington, Sydney NSM 2052, Australia

Daniel Williams

Centre for Neural Engineering and Department of Psychiatry, The University of Melbourne, Parkville, Victoria, Australia

Ryan K.C. Yuen

McLaughlin Centre and The Centre for Applied Genomics, Hospital for Sick Children and University of Toronto, Peter Gilgan Centre for Research and Learning, Toronto, Ontario M5S 1A1, Canada

Daniela Zantomio

Department of Haematology, Austin Health, Heidelberg, Victoria, Australia

Dissecting the Genetic Architecture of ASD

Chapter 1

Phenotype Definition: A Cornerstone of Autism Research, Diagnosis and Therapy

Valerie W. Hu*

Abstract

Autism spectrum disorder (ASD) refers to a group of neurodevelopmental disorders that impact the ability of individuals to engage in reciprocal social interactions and communication, and that are often associated with stereotyped, repetitive behaviors as well as restricted interests. The variable degree to which individuals exhibit the primary symptoms of ASD results in considerable clinical heterogeneity, typically ascribed to both genetic and biological heterogeneity, which includes metabolic disturbances as well as gene-environment interactions. Such heterogeneity poses a significant challenge to all studies in which individuals on the spectrum are treated as a single case group in comparison to a neurotypical control group. In this chapter, I will review studies in which phenotype definition has been used to reduce heterogeneity among individuals in specific study groups, resulting in the increased ability to detect genetic variants and gene expression signatures as well as biomarkers associated with subphenotypes of ASD.

Keywords: Autism spectrum disorder; Heterogeneity; Subphenotype; Categorization; Dimensional analysis; Gene expression; Genetics; Biomarkers; Subtype-specific therapeutic targets.

*Correspondence to: Department of Biochemistry and Molecular Medicine, The George Washington University, School of Medicine and Health Sciences, 2300 Eye St., NW, Washington, DC 20037. Email: valhu@gwu.edu

Introduction

Autism spectrum disorder (ASD) refers to a highly heritable neurodevelopmental syndrome defined by a group of symptoms that, according to the latest clinical diagnostic criteria published in the Diagnostic and Statistical Manual of Mental Disorders, 5th edition[1] are marked by: (1) "Persistent deficits in social communication and social interaction across multiple contexts…" and by (2) "Restricted, repetitive patterns of behavior, interests, or activities…". Within these two broadly defined functional and behavioral domains are a myriad of symptoms that may or may not include difficulties with language production and usage to varying degrees, failure to comprehend or use nonverbal communication (i.e., gestures), difficulty coping with various sensory stimuli, self-injury, aggression, abnormal gait, hypotonia, and other physiological symptoms, such as hyperventilation and fainting. Interestingly, "savant skills", which refer to abilities far above the norm in a specific area (such as art, music, memory, or computational skills) are often exhibited by individuals with ASD with a frequency that is estimated to be ten times that within the general population. In short, the term ASD describes a highly heterogeneous syndrome that presents a significant challenge for geneticists and biologists who are seeking to establish a genetic or molecular "cause" for this disorder.

At present, there are neither definitive genetic markers nor molecules that are unequivocal for the diagnosis of "idiopathic" (of unknown cause) ASD, despite the identification of specific genes that are causal for some syndromes such as fragile X, Rett syndrome, and tuberous sclerosis, that associate to various degrees with ASD, and rare genetic variants that together account for less than 20% of ASD.[2] Although there are ongoing pilot clinical studies that are investigating the efficacy of drugs developed for these genetically-defined forms of ASD such as fragile X[3] on idiopathic ASD, there are no specific therapies developed explicitly for idiopathic ASD or for the core symptoms of ASD. In order to increase our understanding of the pathogenesis and biology of ASD as well as to facilitate progress in diagnosis and treatment of the majority of individuals with ASD, it is imperative that we incorporate phenotype definition into genetic and other biological analyses of ASD.[4] In this chapter, I will review examples of how individuals with idiopathic ASD may be divided into phenotypic subgroups (or subphenotypes), and then describe how we have used this approach to identify biological phenotypes which suggest novel subtype-dependent therapeutic targets and to demonstrate the potential for identifying subtype-specific biomarkers for diagnosis of ASD.

Heterogeneity: Categorical and Dimensional Approaches to Stratification of ASD

The behavioral/clinical heterogeneity in 11 individual cases of autism was noted in the very first report by Leo Kanner in 1943,[5] who not only coined the term "autistic disturbances" in reference to these cases but also provided a description of the "pathognomonic" features that distinguished these cases from everything else he had previously seen in the clinic. These "autistic" features included: (1) the inability to relate to other people and ordinary situations, described as an "extreme autistic aloneness", (2) difficulty with pragmatic language, (3) literalness with respect to both expressive and receptive language, (4) excellent rote memory, (5) an overpowering need not to be disturbed, (6) aversion to intrusions of any kind, including sensory, (7) obsessive desire for sameness, and (8) limitations in or lack of spontaneity. Since this first description of "autism" and because of early twin and family studies that revealed significant heritability in autism,[6–8] there have been numerous attempts to identify genetic mutations or variants that may be responsible for the etiology of ASD.[9–11] However, with a few exceptions, many of these studies have failed to replicate the findings of earlier studies or on different cohorts. The failure to replicate genetic findings has been attributed primarily to the heterogeneity of the ASD population, which recent whole exome sequencing has demonstrated to be even more heterogeneous than previously imagined, with a current estimate of autism candidate genes (identified on the basis of rare variants) to number in the hundreds.[12–15] Whole-genome sequencing that includes the non-coding regions of DNA as well as protein coding regions, while providing orders of magnitude more information on rare variants than whole exome sequencing, further increases the complexity of the genetics of ASD.[16] Only relatively recently, have there been efforts to decrease the heterogeneity of ASD for genetic analyses.

Two general approaches have been used to reduce the heterogeneity of ASD cases for genetic and other analyses: one involves the categorization of probands into subgroups based on attributes such as savant skills, compulsions, repetitive behaviors, or course of the disorder (e.g., regression vs. early onset without regression),[17–20] while the other uses dimensional stratification along a specific symptom domain or quantitative trait, such as verbal and nonverbal abilities and social responsiveness, which can be scored for severity.[19,21–23] Both of these strategies have proven useful in increasing the sensitivity to detect genetic variants that would otherwise be missed. Dimensional stratification also allows for the inclusion of non-autistic relatives who display a given, but subclinical, endophenotype[24] (or trait) which would increase the power of associating a genetic variant with a specific trait, particularly in linkage analyses. Such a study design has been used recently in an extended pedigree analysis to examine genetic loci associated with the broad autism phenotype (BAP), which is primarily focused on personality traits and pragmatic language deficits.[25,26] In general, the results of this study support the hypothesis that incorporating endophenotypes into genetic analyses can reveal previously undetected linkage regions that associate with different components of the BAP, such as repetitive/ritualistic behavior, social functioning, pragmatic language, and anxiety. However, a locus that previously showed linkage with language phenotypes (namely 7q) in an analysis that included both probands and their nondiagnosed parents[27] was not replicated in this study, although another language-related locus at 13q was found to be associated with anxiety and social communication in the later study. Another recent linkage study[28] that included language and behavioral phenotypes of all family members in the analysis identified two novel loci, 15q23-26 and 16p12, for language phenotypes, two loci for social behaviors at 14q and 15q, and another locus for repetitive behaviors at 13q, which is distinct from the repetitive behavior loci (6p and 19p) from the Piven *et al.* study.[25] While informative with respect to the familial and genetic architecture of ASD and endophenotypes associated with ASD, linkage analyses alone do not provide much insight into the molecular pathways that may contribute to the BAP or ASD.

Incorporation of ASD Phenotypes into Transcriptome Analyses

To obtain a better understanding of the specific genes and metabolic/signaling pathways that underlie the pathobiology of ASD, we undertook a series of genome-wide gene expression (i.e., transcriptome) profiling studies starting first with monozygotic twins discordant for diagnosis of ASD,[29] then pairs of affected-unaffected siblings,[30] and finally unrelated individuals with ASD in comparison to unaffected age- and sex-matched controls.[31] While the experimental designs using twins or siblings took advantage of identical or similar genotypes to reduce genetic heterogeneity, we thought it necessary to subgroup unrelated individuals into more similar phenotypic subgroups to reduce clinical heterogeneity (which may in part be due to genetic heterogeneity) among the cases studied. To accomplish this, we developed a novel method of subgrouping individuals based on cluster analyses[32] of severity scores taken from the Autism Diagnostic Interview-Revised diagnostic instrument,[33] a phenotypically-rich parent/caregiver questionnaire which is not only considered the "gold standard" for behavioral diagnosis of ASD, but also uniquely and consistently available for all diagnosed individuals within the Autism Genetic Resources Exchange (AGRE) from which we derived lymphoblastoid cell lines (LCL) for our expression analyses.[31] Another advantage of the ADI-R diagnostic was the availability of quantitative data on severity across a multitude of symptoms and functional deficits exhibited by individuals with ASD which allowed for both phenotypic subgrouping as well as quantitative trait analyses, as will be described in a later section on a meta-analysis of pre-existing genotype data from genome-wide association studies (GWAS).

Briefly, multivariate cluster analyses (hierarchical clustering, principal components analysis, and K-means cluster analyses) were performed using 123 severity scores on 63 items (the higher number of scores reflecting "current" and "ever" scores on the same items) across virtually all functional domains that are impacted by ASD.[32] These included spoken language and nonverbal communication, social and play skills, stereotypies, repetitive and ritualistic behaviors, circumscribed interests, self-injurious as well as aggressive behaviors, physical sensitivities and mannerisms, and savant skills. By incorporating all of these behaviors into our subgrouping analysis, we hoped to identify subgroups of individuals that reflected as a whole the constellation of symptoms that constitute ASD (but to varying degrees of severity), rather than focus on a single domain, such as spoken language. The initial result of this subphenotyping strategy which was applied to ADI-R scores from 1351 individuals with ASD is shown in Figure 1. By inspecting the items associated with the highest severity scores within the subgroups, we identified four subgroups distinguished by: (1) severe language impairment (spoken language); (2) higher frequency of savant skills, but of otherwise moderate severity across other items; (3) mild severity, as reflected by a score of 0 (i.e., normal) on a large number of items; and (4) intermediate severity, but without severe language impairment or noticeable savant skills. We next performed expression profiling of LCLs from 26–31 individuals from three of the four subgroups and compared the expression data against that from LCLs of 29 controls.[31] Figure 2A shows that there are both qualitative and quantitative differences in gene expression across the three subgroups relative to controls, while Figure 2B shows that the three subgroups studied are distinct with unique differentially expressed genes (relative to controls) in each subgroup that reflect impact on different sets of functions, but with some overlapping genes, notably a set

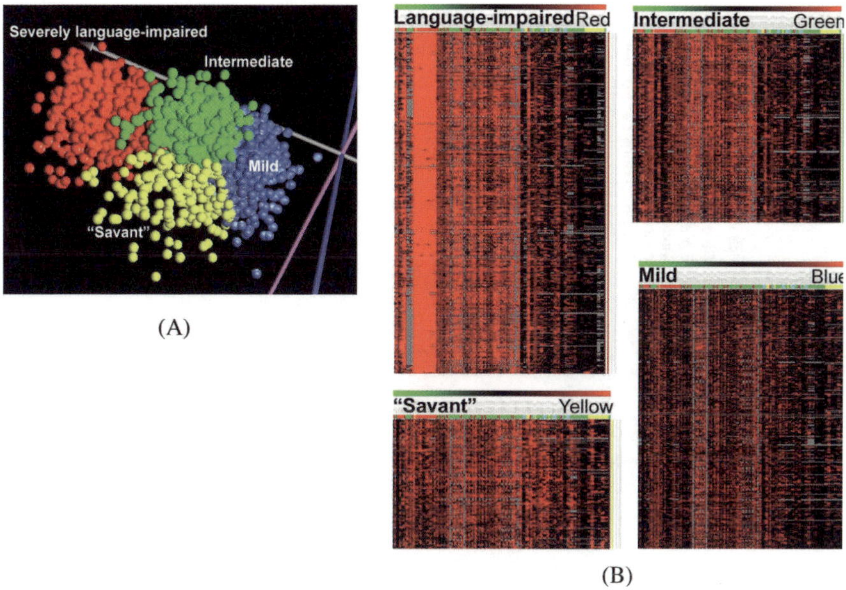

Figure 1. Results of cluster analyses of ADI-R scores. (A) Principal components analysis (PCA). Each point in the figure represents an individual whose position on the graph is determined by 123 item scores from the ADI-R diagnostic instrument. The color assigned to the individuals reflects the colors assigned to the different phenotypic subgroups resulting from the K-means cluster analysis (see Fig. 1B). (B) K-means cluster analysis with K = 4. Individuals are represented by rows in the clusters while ADI-R items are represented by columns. The intensity of red reflects degree of severity (from 1 to 3) on a particular item while black indicates a "normal" score of 0. Figures are from Hu *et al.*[30] based on cluster analyses of ADI-R scores as described by Hu and Steinberg.[32]

of 20 long noncoding transcripts of unknown function that are shared by all three subgroups. The significance of some of these expression differences with regard to subgroup-specific therapeutic approaches will be discussed later.

Subtype-dependent Class Prediction Analyses for Biomarker Identification

One of the goals of our gene expression studies on cells derived from peripheral tissues (i.e., blood) was the identification of

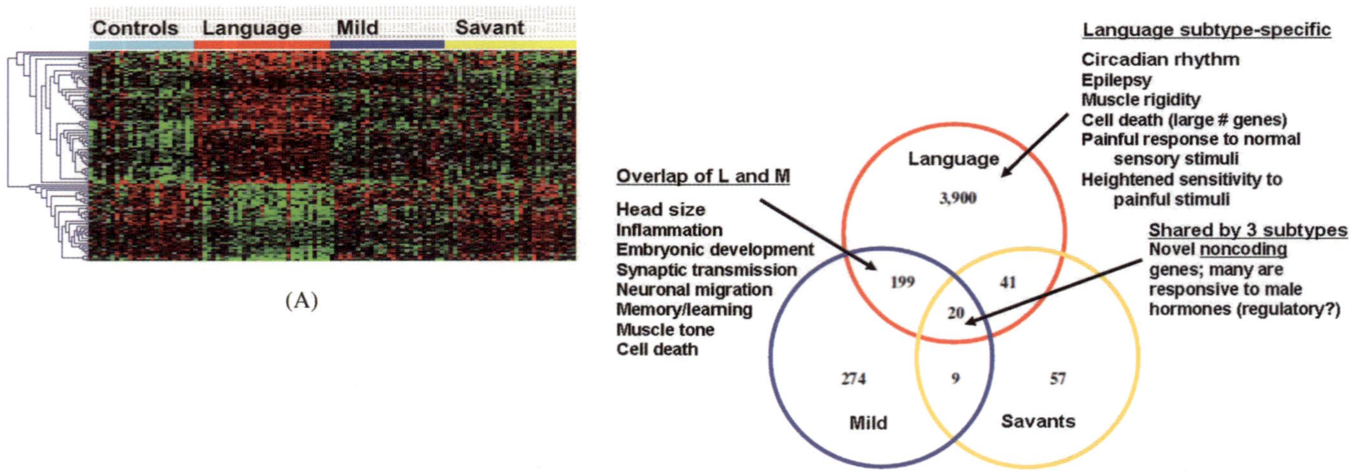

Figure 2. Results of gene expression profiling of 3 of the subphenotypes of ASD and controls. Colored bars above the heat map of gene expression (A) indicate the subgroups of individuals with ASD and controls. Rows: genes; Columns: individuals. Green: reduced expression relative to a universal human RNA standard; Red: increased expression. (B) Venn diagram of the number of differentially expressed genes in each of three subtypes of ASD relative to the controls as well as their associated functions. Figures are adapted from Hu *et al.*[31]

signatures of differentially expressed genes that may be useful as biomarkers for diagnostic screening. Thus, based on the subgroup-dependent expression profiles identified in the studies described above, we performed class prediction analyses to identify the most robust differentially expressed genes that may distinguish cases from controls.[34] These studies identified 18–29 subgroup-dependent genes (transcripts) that were able to distinguish controls from cases that were assigned to the different phenotypic subgroups with greater than 93% accuracy (Table 1). Importantly, the specificities and sensitivities of each set of subgroup-specific genes were also greater than 90%. Interestingly, when all the cases were combined into one case group, the cases could still be separated from the controls, but with only ~81% accuracy, a sensitivity of 91% and a specificity of 61%. Using quantitative nuclease protection assays, we confirmed the ability of a subset of genes (14 out of 29) from the language subgroup to separate 22 of the original cases from 22 of the original controls with 78% sensitivity and 80% specificity. We further validated this set of genes on a new cohort of cases and controls, achieving ~90% sensitivity but only 75% specificity. Despite the need to increase the size of the validation group and to optimize the selection of genes for a biomarker screen coupled with the use of primary peripheral tissues, our study illustrates the potential for the utilization of gene expression signatures for diagnosis of ASD, and further demonstrates the increased sensitivity

Table 1. Results of class prediction analyses which identified transcripts (genes) that could distinguish between cases and controls with reasonably high levels of accuracy, sensitivity and specificity.[34] It is clear that phenotype definition of the cases improves upon all three of these class prediction parameters. L: Language-impaired subtype; M: Mild; S: Savant; A: All (combined cases); C: Controls

Case-Control	% Accuracy	(# Genes)	% Sensitivity	% Specificity
L vs C	93.3	(29)	96.6	90.3
M vs C	94.5	(26)	96.0	93.3
S vs C	94.0	(18)	96.6	90.5
A vs C	81.8	(74)	91.2	61.1

and specificity that can be achieved by incorporating subtypes into the class prediction analyses.

Value of ADI-R-derived Phenotypic Subgroups in Genetic Analyses

To examine whether or not the phenotypic subgrouping protocol based on ADI-R cluster analyses might improve upon the ability to identify ASD-associated common single nucleotide polymorphisms (SNPs) from genome-wide association analyses (GWA), we performed a meta-analysis[35] of genotype data from a previously published GWA study[36] which included over 550,000 SNPs from over 4000 individuals with ASD and over 7000 controls. However, because we could obtain ADI-R scoresheets, necessary for our cluster analyses, for only 1867 affected individuals included in the former study, we limited our analyses to these cases and included 2438 controls from the former study. The ADI-R data were then used in two ways: (1) to generate quantitative trait distribution profiles across the case group by summing item scores on specific traits, which included spoken language, nonverbal communication, play skills, social development, and insistence on sameness, with multiple item scores contributing to each trait for each individual; (2) to divide the individuals with ASD into phenotypic subgroups by cluster analyses of 123 ADI-R scores as previously described.[32] Our meta-analysis of the pre-existing GWA data was done in three stages. The first stage involved quantitative trait association analyses to select SNPs that were nominally associated ($P \leq 10^{-5}$) with each of the five traits, which reduced the total number of SNPs to 167 unique SNPs, which we postulated to be more likely functionally relevant due to their trait association. The second phase of the study involved subtype-dependent case-control association analyses for each set of trait-associated SNPs, while the third phase involved case-control association analyses with the combined set of Bonferroni-significant subtype-dependent trait-associated SNPs using the four subtypes as independent case groups as well as a combined case group of 1867 individuals. The net result of this multi-stage QTL-subphenotype association analysis of pre-existing genotype data was the identification of 18 novel and highly significant SNPs.[35] Ten of these SNPs were shared between subtypes, but with distinctly different odds ratios relative to controls (that is, predisposing in one subtype and protective in the other), confirming the genetic heterogeneity among the subtypes at these loci. Notably, no significant SNPs were identified when all individuals with ASD were combined into one case group, thus demonstrating that heterogeneity reduction, and not just increased numbers of individuals, leads to increased power in genetic analyses. Interestingly, all of the SNPs were in noncoding regions of genes that are in functionally relevant pathways with respect to ASD, or in CNV regions previously associated with ASD.[37] The pathways included axonogenesis, neurogenesis, long-term synaptic potentiation and steroid metabolism, the latter of which we had also identified by gene expression profiling of LCL from sib pairs.[30] It should be mentioned that applying this phenotypic subgrouping approach to genetic linkage analyses of ASD in pedigrees coupled with novel stratification to remove intra-family heterogeneity also resulted in significant improvement in the LOD scores for specific loci,[38] which is discussed in Chapter 5 by Talebizadeh and Shah. As described in the next section, some of the genes implicated by the meta-analysis of GWAS data, as well as several of those revealed by subtype-dependent gene expression profiling, suggest novel targets for therapy as well as possible pharmacogenomic applications.

Value of Phenotype-Definition for Targeted Therapies and Pharmacogenomics

As mentioned earlier, some of the subphenotype-associated differentially expressed genes may suggest novel approaches to subgroup-specific therapies. For example, among 15 circadian rhythm-related genes that were found to be differentially expressed only in the severely language-impaired subtype of ASD are two genes, *AA-NAT* and *DPYD*, whose deficiency suggests novel therapeutic approaches. Because AA-NAT is the rate-limiting enzyme in the biosynthesis of melatonin, an endogenous sleep-inducing hormone, a documented deficiency of this enzyme (or melatonin) in a particular individual suggests that melatonin supplementation may improve not only the sleep disturbances that are frequently exhibited in individuals with severe language impairment, but also neurological functions, inasmuch as circadian rhythm has been linked to synaptic regulation.[39] Indeed, a recent review of the literature on the use of melatonin for sleep disorders in ASD indicates that providing melatonin supplementation to individuals with ASD produced improvements in sleep parameters (i.e., duration, onset latency, and fewer nighttime awakenings) with some indications of improvement in daytime behaviors as well.[40]

Dihydropyrimidine dehydrogenase (DPYD), on the other hand, is an enzyme involved in the breakdown of nucleic acid (from DNA) which results in the production of beta-alanine, an amino acid which has been shown to be an inhibitory neurotransmitter that binds both GABA and glycine receptors at the respective agonist binding sites, resulting in inhibitory neurotransmission.[41,42] The reduction in this neurotransmitter caused by DPYD deficiency is therefore expected to lead to decreased inhibitory signaling at the synapse, which has been postulated to be a central component of the neuropathology giving rise to ASD.[43] Interestingly, DPYD deficiency is strongly associated with epilepsy, mental and motor retardation, as well as ASD.[41] Thus, a novel approach to treating ASD individuals with this enzyme deficiency might be to use beta-alanine (a nutritional supplement that is brain-available) to counteract the deficiency. As beta-alanine is also postulated to be an "endogenous anticonvulsant molecule",[44] it may also be useful in the treatment of epilepsy (a co-morbidity in a sizable fraction of individuals with ASD), as suggested by Weaver and colleagues, although not without ethnicity-dependent side effects.[45] From a pharmacogenomics perspective, an individual with ASD presenting with DPYD deficiency may also be a good candidate for anticonvulsant medications, which are often prescribed for ASD on a trial-and-error basis with no biological rationale.

Similarly, biological insights generated from subtype-dependent genetic analyses can also inform pharmacogenomics and targeted therapies. As an example, two SNPs in an intron of *HTR4* (a serotonin receptor) were found to be highly associated with the severe-language phenotype of ASD,[35] for which we also found decreased expression of *HTR4*.[31] Interestingly, HTR4 deficiency in the gut is associated with digestive disorders[46] which are frequently observed in individuals with ASD, particularly in the more severely affected individuals, possibly in relation to circadian dysfunction. Two chemokine genes, *CCL20* and *CCL25*, implicated by our subtype-dependent association study, are also linked to gastrointestinal (GI) disease,[47] and thus serve as additional targets for treatment of individuals with ASD who present with comorbid GI disorders. In fact, the initial discovery of genetic variants in the *MET* oncogene in relation to ASD was based on genetic analysis of the subgroup of individuals with ASD and comorbid GI disorders.[48] As will be discussed in Chapter 21 by Herbert, taking a whole-body approach to treatment of ASD may have a significant impact on neurological functioning as well.

Future Directions

In the preceding sections, I have summarized some of the studies from our laboratory that utilized cluster analyses of ADI-R scores to subgroup ASD cases for gene expression, genetics, and class prediction analyses which collectively demonstrate the value of heterogeneity reduction to associate both behavioral phenotypes and potential biomarkers with biological phenotypes (as represented by distinct gene expression signatures) and to increase the signal-to-noise in GWAS. As the latest genomic technologies now provide the ability to sequence the whole human genome (see Chapter 4 by Yuen and Scherer), it becomes even more important to use phenotype definition to extract biologically meaningful information from such large-scale studies that, because of the sheer magnitude of data collected, are associated with a lot of "noise". The same situation applies also to genome-wide studies of the epigenome and other 'omes (e.g., metabolome, microbiome), as well as to clinical trials, as it is apparent that no single medication/treatment is efficacious for all individuals with ASD. Indeed, there are many ways to achieve reduction of heterogeneity that are likely to facilitate the association of a specific phenotype of ASD with genetic/biological differences from the norm, and to elucidate the genes, pathways, and pathogenic mechanisms that may contribute to a particular subtype of ASD. Aside from the previously mentioned use of basic behavioral phenotypes such as deficits in language, nonverbal communication, or social responsiveness as well as the presence of savant skills or repetitive behaviors to reduce heterogeneity for genetic analyses, additional informative phenotypes might be derived through metabolic analyses (see Chapter 22 by Frye and Rossignol), association with mitochondrial dysfunction (Chapter 16 by Gu *et al.*), the presence and/or nature of maternal autoantibodies (Chapter 17 by Jones and Van de Water), comorbidities (Chapter 20 by Bauman) or comorbidity clusters,[49] degree of language impairment (Chapter 3 by Rakhlin and Grigorenko), functional magnetic resonance imaging (fMRI),[50] cluster analysis of principal pathogenetic components,[51] and even by well-documented environmental exposures, as discussed in Chapter 14 by Bakulski *et al.*

In addition to providing a better understanding of the many factors contributing to the pathogenesis and pathobiology of ASD, the *anticipated translational outcomes* from this "Divide and Conquer" approach to overcoming the heterogeneity of ASD are: (1) the *development of subtype-dependent biomarkers for diagnosis* that may better inform pharmacogenomics with regard to ASD, and (2) the *identification of subtype-specific genes/molecular targets* that can lead to the development of novel therapeutic options for specific subgroups of individuals with ASD based on sound biological rationales, thereby *advancing the treatment of ASD towards personalized medicine.*

Executive Summary

- The heterogeneity of ASD poses significant challenges to research aimed at identifying the underlying genetic and/or molecular bases (or correlates) of autism.

- Dividing individuals with ASD into subgroups according to differences in behavioral symptoms, comorbidities, or metabolic and immune abnormalities is anticipated to be useful in identifying different pathogenic mechanisms operating within a given subgroup.
- Moving forward, more attention must be paid to phenotype definition in studying ASD in order to harness the power of large-scale 'omics' technologies to reveal subgroup-specific genes as well as metabolic and signaling pathways that may provide novel targets for diagnosis as well as therapeutics.

Acknowledgements

The work that is described from Dr. Hu's laboratory was supported in part by NIMH grant R21 MH073393 (VWH) and in part by a grant from the GWU McCormick Genomics Center (VWH).

References

1. American Psychiatric Association: *Diagnostic and Statistical Manual of Mental Disorders*, 5th edition: Arlington, VA: American Psychiatric Association; 2013.
2. Betancur C: Etiological heterogeneity in autism spectrum disorders: more than 100 genetic and genomic disorders and still counting. *Brain Res* 2011, **1380**:42–77.
3. Jacquemont S, Berry-Kravis E, Hagerman R, von Raison F, Gasparini F, Apostol G, Ufer M, Des Portes V, Gomez-Mancilla B: The challenges of clinical trials in fragile X syndrome. *Psychopharmacology* (*Berl*) 2013, :1–14.
4. Hu V: From Genes to Environment: Using integrative genomics to build a "systems level" understanding of autism spectrum disorders. *Child Dev* 2013, **84**(1):89–103.
5. Kanner L: Autistic disturbances of affective contact. *Nervous Child* 1943, **2**: 217–250.
6. Folstein S, Rutter M: Infantile autism: a genetic study of 21 twin pairs. *J Child Psychol Psychiatry* 1977, **18**(4):297–321.
7. Bailey A, Le Couteur A, Gottesman I, Bolton P, Simonoff E, Yuzda E, Rutter M: Autism as a strongly genetic disorder: evidence from a British twin study. *Psychol Med* 1995, **25**(1):63–77.
8. Hallmayer J, Glasson EJ, Bower C, Petterson B, Croen L, Grether J, Risch N: On the twin risk in autism. *Am J Hum Genet* 2002, **71**(4):941–946.
9. Grigorenko EL: Pathogenesis of autism: a patchwork of genetic causes. *Future Neurol* 2009, **4**(5):591–599.

10. Freitag CM, Staal W, Klauck SM, Duketis E, Waltes R: Genetics of autistic disorders: review and clinical implications. *Eur Child Adolesc Psychiatry* 2010, **19**(3):169–178.
11. Murdoch JD, State MW: Recent developments in the genetics of autism spectrum disorders. *Curr Op Genet Dev* 2013, **23**(3):310–315.
12. Iossifov I, Ronemus M, Levy D, Wang Z, Hakker I, Rosenbaum J, Yamrom B, Lee Y-, Narzisi G, Leotta A, Kendall J, Grabowska E, Ma B, Marks S, Rodgers L, Stepansky A, Troge J, Andrews P, Bekritsky M, Pradhan K, Ghiban E, Kramer M, Parla J, Demeter R, Fulton L, Fulton R, Magrini V, Ye K, Darnell J, Darnell R, Mardis E, Wilson R, Schatz M, McCombie W, Wigler M: *De Novo* Gene Disruptions in Children on the Autistic Spectrum. *Neuron* 2012, **74**(2):285–299.
13. Neale BM, Kou Y, Liu L, Ma'Ayan A, Samocha KE, Sabo A, Lin C-, Stevens C, Wang L-, Makarov V, Polak P, Yoon S, Maguire J, Crawford EL, Campbell NG, Geller ET, Valladares O, Schafer C, Liu H, Zhao T, Cai G, Lihm J, Dannenfelser R, Jabado O, Peralta Z, Nagaswamy U, Muzny D, Reid JG, Newsham I, Wu Y, Lewis L, Han Y, Voight BF, Lim E, Rossin E, Kirby A, Flannick J, Fromer M, Shakir K, Fennell T, Garimella K, Banks E, Poplin R, Gabriel S, Depristo M, Wimbish JR, Boone BE, Levy SE, Betancur C, Sunyaev S, Boerwinkle E, Buxbaum JD, Cook Jr. EH, Devlin B, Gibbs RA, Roeder K, Schellenberg GD, Sutcliffe JS, Daly MJ: Patterns and rates of exonic *de novo* mutations in autism spectrum disorders. *Nature* 2012, **484**(7397):242–245.
14. O'Roak BJ, Vives L, Girirajan S, Karakoc E, Krumm N, Coe BP, Levy R, Ko A, Lee C, Smith JD, Turner EH, Stanaway IB, Vernot B, Malig M, Baker C, Reilly B, Akey JM, Borenstein E, Rieder MJ, Nickerson DA, Bernier R, Shendure J, Eichler EE: Sporadic autism exomes reveal a highly interconnected protein network of *de novo* mutations. *Nature* 2012, **484**(7397):246–250.
15. Sanders SJ, Murtha MT, Gupta AR, Murdoch JD, Raubeson MJ, Willsey AJ, Ercan-Sencicek AG, Dilullo NM, Parikshak NN, Stein JL, Walker MF, Ober GT, Teran NA, Song Y, El-Fishawy P, Murtha RC, Choi M, Overton JD, Bjornson RD, Carriero NJ, Meyer KA, Bilguvar K, Mane SM, Šestan N, Lifton RP, Günel M, Roeder K, Geschwind DH, Devlin B, State MW: *De novo* mutations revealed by whole-exome sequencing are strongly associated with autism. *Nature* 2012, **484**(7397):237–241.
16. Jiang Y-, Yuen RKC, Jin X, Wang M, Chen N, Wu X, Ju J, Mei J, Shi Y, He M, Wang G, Liang J, Wang Z, Cao D, Carter MT, Chrysler C, Drmic IE, Howe JL, Lau L, Marshall CR, Merico D, Nalpathamkalam T, Thiruvahindrapuram B, Thompson A, Uddin M, Walker S, Luo J, Anagnostou E, Zwaigenbaum L, Ring RH, Wang J, Lajonchere C, Wang J, Shih A, Szatmari P, Yang H, Dawson G, Li Y, Scherer SW: Detection of clinically relevant genetic variants in autism spectrum disorder by whole-genome sequencing. *Am J Hum Genet* 2013, **93**(2):249–263.

17. Nurmi EL, Dowd M, Tadevosyan-Leyfer O, Haines JL, Folstein SE, Sutcliffe JS: Exploratory subsetting of autism families based on savant skills improves evidence of genetic linkage to 15q11-q13. *J Am Acad Child Adolesc Psychiatry* 2003, **42**(7):856–863.
18. Cannon D, Miller J, Robison R, Villalobos M, Wahmhoff N, Allen-Brady K, McMahon W, Coon H: Genome-wide linkage analyses of two repetitive behavior phenotypes in Utah pedigrees with autism spectrum disorders. *Mol Autism* 2010, **1**(1).
19. Coon H, Villalobos M, Robison R, Camp N, Cannon D, Allen-Brady K, Miller J, McMahon W: Genome-wide linkage using the Social Responsiveness Scale in Utah autism pedigrees. *Mol Autism* 2010, **1**(1).
20. Gregg JP, Lit L, Baron CA, Hertz-Picciotto I, Walker W, Davis RA, Croen LA, Ozonoff S, Hansen R, Pessah IN, Sharp FR: Gene expression changes in children with autism. *Genomics* 2008, **91**(1):22–29.
21. Duvall JA, Lu A, Cantor RM, Todd RD, Constantino JN, Geschwind DH: A quantitative trait locus analysis of social responsiveness in multiplex autism families. *Am J Psychiatry* 2007, **164**(4):656–662.
22. Alarcon M, Yonan AL, Gilliam TC, Cantor RM, Geschwind DH: Quantitative genome scan and Ordered-Subsets Analysis of autism endophenotypes support language QTLs. *Mol Psychiatry* 2005, **10**(8):747–757.
23. Chen GK, Kono N, Geschwind DH, Cantor RM: Quantitative trait locus analysis of nonverbal communication in autism spectrum disorder. *Mol Psychiatry* 2006, **11**(2):214–220.
24. Gottesman II, Gould TD: The endophenotype concept in psychiatry: etymology and strategic intentions. *Am J Psychiatry* 2003, **160**(4):636–645.
25. Piven J, Vieland VJ, Parlier M, Thompson A, O'Conner I, Woodbury-Smith M, Huang Y, Walters KA, Fernandez B, Szatmari P: A molecular genetic study of autism and related phenotypes in extended pedigrees. *J Neurodev Disord* 2013, **5**:30.
26. Piven J: The broad autism phenotype: A complementary strategy for molecular genetic studies of autism. *Am J Med Genet* 2001, **105**:34–35.
27. Bradford Y, Haines J, Hutcheson H, Gardiner M, Braun T, Sheffield V, Cassavant T, Huang W, Wang K, Vieland V, Folstein S, Santangelo S, Piven J: Incorporating language phenotypes strengthens evidence of linkage to autism. *Am J Med Genet Neuropsychiatric Genet* 2001, **105**(6):539–547.
28. Bartlett CW, Hou L, Flax JF, Hare A, Cheong SY, Fermano Z, Zimmerman-Bier B, Cartwright C, Azaro MA, Buyske S, Brzustowicz LM: A genome scan for loci shared by autism spectrum disorder and language impairment. *Am J Psychiatry* 2013, (Oct. 30).
29. Hu VW, Frank BC, Heine S, Lee NH, Quackenbush J: Gene expression profiling of lymphoblastoid cell lines from monozygotic twins discordant in

severity of autism reveals differential regulation of neurologically relevant genes. *BMC Genomics* 2006, **7**:118.

30. Hu VW, Nguyen A, Kim KS, Steinberg ME, Sarachana T, Scully MA, Soldin SJ, Luu T, Lee NH: Gene expression profiling of lymphoblasts from autistic and nonaffected sib pairs: altered pathways in neuronal development and steroid biosynthesis. *PLoS One* 2009, **4**(6):e5775.
31. Hu VW, Sarachana T, Kim KS, Nguyen A, Kulkarni S, Steinberg ME, Luu T, Lai Y, Lee NH: Gene expression profiling differentiates autism case-controls and phenotypic variants of autism spectrum disorders: evidence for circadian rhythm dysfunction in severe autism. *Autism Res* 2009, **2**(2):78–97.
32. Hu VW, Steinberg ME: Novel clustering of items from the Autism Diagnostic Interview-Revised to define phenotypes within autism spectrum disorders. *Autism Res* 2009, **2**(2):67–77.
33. Lord C, Rutter M, Couteur AL: Autism diagnostic interview-revised: a revised version of a diagnostic interview for caregivers of individuals with possible pervasive developmental disorders. *J Autism Dev Disord* 1994, **24**(5):659–685.
34. Hu VW, Lai Y: Developing a predictive gene classifier for autism spectrum disorders based upon differential gene expression profiles of phenotypic subgroups. *N Am J Med Sci* 2013, **6**(3):107–116.
35. Hu VW, Addington A, Hyman A: Novel autism subtype-dependent genetic variants are revealed by quantitative trait and subphenotype association analyses of published GWAS data. *PLoS ONE* 2011, **6**(4).
36. Wang K, Zhang H, Ma D, Bucan M, Glessner JT, Abrahams BS, Salyakina D, Imielinski M, Bradfield JP, Sleiman PMA, Kim CE, Hou C, Frackelton E, Chiavacci R, Takahashi N, Sakurai T, Rappaport E, Lajonchere CM, Munson J, Estes A, Korvatska O, Piven J, Sonnenblick LI, Alvarez Retuerto AI, Herman EI, Dong H, Hutman T, Sigman M, Ozonoff S, Klin A, Owley T, Sweeney JA, Brune CW, Cantor RM, Bernier R, Gilbert JR, Cuccaro ML, McMahon WM, Miller J, State MW, Wassink TH, Coon H, Levy SE, Schultz RT, Nurnberger JI, Haines JL, Sutcliffe JS, Cook EH, Minshew NJ, Buxbaum JD, Dawson G, Grant SFA, Geschwind DH, Pericak-Vance MA, Schellenberg GD, Hakonarson H: Common genetic variants on 5p14.1 associate with autism spectrum disorders. *Nature* 2009, **459**(7246):528–533.
37. Pinto D, Pagnamenta AT, Klei L, Anney R, Merico D, Regan R, Conroy J, Magalhaes TR, Correia C, Abrahams BS, Almeida J, Bacchelli E, Bader GD, Bailey AJ, Baird G, Battaglia A, Berney T, Bolshakova N, Bölte S, Bolton PF, Bourgeron T, Brennan S, Brian J, Bryson SE, Carson AR, Casallo G, Casey J, Chung BHY, Cochrane L, Corsello C, Crawford EL, Crossett A, Cytrynbaum C, Dawson G, De Jonge M, Delorme R, Drmic I, Duketis E, Duque F, Estes A, Farrar P, Fernandez BA, Folstein SE, Fombonne E, Freitag CM, Gilbert J, Gillberg C, Glessner JT, Goldberg J, Green A, Green J, Guter SJ, Hakonarson

H, Heron EA, Hill M, Holt R, Howe JL, Hughes G, Hus V, Igliozzi R, Kim C, Klauck SM, Kolevzon A, Korvatska O, Kustanovich V, Lajonchere CM, Lamb JA, Laskawiec M, Leboyer M, Le Couteur A, Leventhal BL, Lionel AC, Liu X-, Lord C, Lotspeich L, Lund SC, Maestrini E, Mahoney W, Mantoulan C, Marshall CR, McConachie H, McDougle CJ, McGrath J, McMahon WM, Merikangas A, Migita O, Minshew NJ, Mirza GK, Munson J, Nelson SF, Noakes C, Noor A, Nygren G, Oliveira G, Papanikolaou K, Parr JR, Parrini B, Paton T, Pickles A, Pilorge M, Piven J, Ponting CP, Posey DJ, Poustka A, Poustka F, Prasad A, Ragoussis J, Renshaw K, Rickaby J, Roberts W, Roeder K, Roge B, Rutter ML, Bierut LJ, Rice JP, Salt J, Sansom K, Sato D, Segurado R, Sequeira AF, Senman L, Shah N, Sheffield VC, Soorya L, Sousa I, Stein O, Sykes N, Stoppioni V, Strawbridge C, Tancredi R, Tansey K, Thiruvahindrapduram B, Thompson AP, Thomson S, Tryfon A, Tsiantis J, Van Engeland H, Vincent JB, Volkmar F, Wallace S, Wang K, Wang Z, Wassink TH, Webber C, Weksberg R, Wing K, Wittemeyer K, Wood S, Wu J, Yaspan BL, Zurawiecki D, Zwaigenbaum L, Buxbaum JD, Cantor RM, Cook EH, Coon H, Cuccaro ML, Devlin B, Ennis S, Gallagher L, Geschwind DH, Gill M, Haines JL, Hallmayer J, Miller J, Monaco AP, Nurnberger Jr JI, Paterson AD, Pericak-Vance MA, Schellenberg GD, Szatmari P, Vicente AM, Vieland VJ, Wijsman EM, Scherer SW, Sutcliffe JS, Betancur C: Functional impact of global rare copy number variation in autism spectrum disorders. *Nature* 2010, **466**(7304):368–372.

38. Talebizadeh Z, Arking DE, Hu VW: A novel stratification method in linkage studies to address inter- and intra-family heterogeneity in autism. *PLoS ONE* 2013, **8**(6):e67569.
39. Bourgeron T: The possible interplay of synaptic and clock genes in autism spectrum disorders. *Cold Spring Harb Symp Quant Biol* 2007, **72**:645–654.
40. Rossignol DA, Frye RE: Melatonin in autism spectrum disorders. *Curr Clin Pharmacol* 2013 (Sep 20).
41. Van Kuilenburg ABP, Vreken P, Abeling NGGM, Bakker HD, Meinsma R, Van Lenthe H, De Abreu RA, Smeitink JAM, Kayserili H, Apak MY, Christensen E, Holopainen I, Pulkki K, Riva D, Botteon G, Holme E, Tulinius M, Kleijer WJ, Beemer FA, Duran M, Niezen-Koning KE, Smit GPA, Jakobs C, Smit LME, Moog U, Spaapen LJM, Van Gennip AH: Genotype and phenotype in patients with dihydropyrimidine dehydrogenase deficiency. *Hum Genet* 1999, **104**(1):1–9.
42. Tiedje KE, Stevens K, Barnes S, Weaver DF: ß-Alanine as a small molecule neurotransmitter. *Neurochem Int* 2010, **57**(3):177–188.
43. Hussman JP: Suppressed GABAergic inhibition as a common factor in suspected etiologies of autism. *J Autism Dev Disord* 2001, **31**(2):247–248.

44. Tan CYK, Wainman D, Weaver DF: N-, a-, and ß-substituted 3-aminopropionic acids: design, syntheses and antiseizure activities. *Bioorganic Med Chem* 2003, **11**(1):113–121.
45. MacPhee S, Weaver IN, Weaver DF: An evaluation of interindividual responses to the orally administered neurotransmitter beta-alanine. *J Amino Acids* 2013, **2013**:Article ID: 429847.
46. Sikander A, Rana SV, Prasad KK: Role of serotonin in gastrointestinal motility and irritable bowel syndrome. *Clinica Chimica Acta* 2009, **403**(1–2):47–55.
47. Nishimura M, Kuboi Y, Muramoto K, Kawano T, Imai T: Chemokines as novel therapeutic targets for inflammatory bowel disease. In *Annals of the New York Academy of Sciences.* Edited by Anonymous 2009:350–356. [Anonymous , *vol 1173.*]
48. Campbell DB, Sutcliffe JS, Ebert PJ, Militerni R, Bravaccio C, Trillo S, Elia M, Schneider C, Melmed R, Sacco R, Persico AM, Levitt P: A genetic variant that disrupts MET transcription is associated with autism. *Proc Natl Acad Sci U S A* 2006, **103**(45):16834–16839.
49. Doshi-Velez F, Ge Y, Kohane I: Comorbidity clusters in autism spectrum disorders: an electronic health record time-series analysis. *Pediatrics* 2013 (Dec 9).
50. Piggot J, Shirinyan D, Shemmassian S, Vazirian S, Alarcón M: Neural systems approaches to the neurogenetics of autism spectrum disorders. *Neuroscience* 2009, **164**(1):247–256.
51. Sacco R, Lenti C, Saccani M, Curatolo P, Manzi B, Bravaccio C, Persico AM: Cluster analysis of autistic patients based on principal pathogenetic components. *Autism Res* 2012, **5**(2):137–147.

Chapter 2

From Molecular Pathways to ASD Therapy: Insights from Syndromic Forms of Autism

Laura Nicholls, Radhika Ramadas and Irina Voineagu*

Abstract

Autism spectrum disorders (ASD) are highly heritable, yet genetically and clinically very heterogeneous. Currently, ASD treatment relies heavily on early behavioural intervention, which is partially effective but not curative, and specific pharmacological therapy for ASD is not yet available. Syndromic forms of ASD are genetic syndromes with a high incidence of autistic symptoms, which offer a unique opportunity for understanding the neurobiological mechanisms of ASD. Here we discuss the current progress in characterizing the molecular and neurobiological mechanisms underlying fragile X syndrome, Rett syndrome and tuberous sclerosis, as well as novel therapies developed for these syndromic forms of ASD. We further review recent evidence for common mechanisms underpinning syndromic and idiopathic ASD, and discuss the possibility of employing pharmacological agents developed for syndromic ASD to effectively treat idiopathic ASD.

Keywords: Rett syndrome; Fragile X syndrome; Tuberous sclerosis; Syndromic autism; Autism genetics.

*Correspondence to: School of Biotechnology and Biomolecular Sciences, University of NSW, Kensington, Sydney NSW 2052 Australia. Email: i.voineagu@unsw.edu.au

Introduction

Autism spectrum disorders (ASD) are a group of neurodevelopmental conditions defined by variable deficits in three core domains: language, reciprocal social interactions, and repetitive behaviours. The clinical and genetic landscape of the disorder is highly heterogeneous: each ASD case varies in the severity of impairment across behavioural, and cognitive dimensions, and despite a strong genetic liability, known heritable causes account for less than 20% of all cases, with none of these individually contributing to more than 2% of cases.[1]

Under the Diagnostic and Statistical Manual of Mental Disorders (DSM-IV)[2] ASD encompass autistic disorder, Asperger syndrome and pervasive developmental disorder – not otherwise specified (PDD-NOS). At the most severe end of the spectrum, autistic disorder requires deficits in the three core domains, with onset before the age of three. Asperger syndrome is defined by impaired social communication and restrictive-repetitive behavior, yet normal language, while PDD-NOS requires deficits in at least one core domain, and exclusion from the diagnosis of autistic disorder, Asperger syndrome or other pervasive developmental disorders. Further, there is significant co-occurrence of ASD with other clinical dysfunctions such as motor and sensory abnormalities,[3,4] epilepsy,[5] and intellectual disability, particularly in those with autistic disorder.[6]

Like many psychiatric disorders, ASD diagnosis and categorization is challenging due to a reliance on observed behaviour and cognitive function, without validation by biomarkers. As a result of such intrinsic diagnostic ambiguity, the assignment of an individual to the autistic disorder, Asperger syndrome or PDD-NOS subtypes varies between clinicians despite significant congruence in how symptoms are catalogued.[7,8] Further, few studies have identified robust differences between subtypes[9–11] and assignment to ASD categories has proven a poor predictor of clinical outcomes.[12,13] It has been suggested that inadequate diagnostic criteria is partially to blame for the recently observed increase in ASD cases.[14] At the implementation of the DSM-IV, autism was still considered a rare disorder, affecting 0.1% of the population,[15] while the prevalence of ASD has now risen to approximately 1.1%, and the increase has largely been attributed to the evolution of diagnostic criteria to embrace a broader conception of ASD, and a heightened awareness in the general population.[16] Nevertheless, a true increase due to genetic and environmental factors cannot and have not been ruled out (see Ref. [17] for a review).

As a result of these diagnostic sensitivity and specificity issues, in the latest edition of the DSM (DSM-V), ASD were targeted for an overhaul of their taxonomic structure. Under DSM-V criteria, there would be a single diagnosis of ASD, without the existing subtypes, based on impairments in two core domains of social-communication and restrictive-repetitive behaviour. To fulfill the ASD diagnosis, all social and communicative deficits and at least two repetitive-restrictive behaviours, interests or activities must manifest at an early age. Although it has been stated that this more dimensional approach to ASD diagnosis can preserve the integrity of ASD diagnoses and allow subtyping based on more meaningful distinctions (reviewed in Ref. [18]), the full effects of implemented DSM-V criteria remain to be evaluated.

Currently, the only effective therapy in ASD is early behavioral intervention,[19] which aims to improve language abilities, communication skills and alleviate repetitive and self-injurious behaviors. The pharmacological therapy used in ASD is entirely non-specific and includes antipsychotics, antidepressant and anxiolytics, which are minimally effective.[20] On the other hand, the development of specific pharmacological therapy entirely depends on the elucidation of the molecular mechanisms underlying ASD.

Much of our current understanding of the molecular underpinnings of autism comes from the large body of research on autism genetics. Although the clinical heterogeneity of ASD is mirrored by an equally daunting genetic heterogeneity,[21] several rare monogenic syndromes show high penetrance of autistic symptoms. Fragile X syndrome, Rett syndrome and tuberous sclerosis patients fulfill ASD diagnostic criteria in more than 20% of cases.[1] These syndromic forms of ASD are particularly informative as they allow the investigation of specific molecular mechanisms as well as the development of animal models and targeted drug screening. Consequently, targeted therapy is advancing much faster for syndromic ASD, and some therapeutic approaches are already undergoing clinical trials.[22]

In this chapter we discuss recent insights into the molecular basis of ASD from several syndromic forms of autism, and discuss whether convergent pathways may underlie syndromic and idiopathic ASD, with a focus on the possible role of convergent pathways in developing effective therapy.

Genetics of Idiopathic ASD

Although initially conceived as a psychogenic condition attributed to an aloof style of parenting,[23] the hereditary basis of ASD became apparent from early disease concordance studies showing ASD concordance rates of 70–90% between monozygotic twins, 6–10 % between dizygotic twins,[24] and 3–26% in siblings.[25–27] Recurrence risk for siblings is three times greater if the sibling is male, giving credence to the argument for reduced penetrance in females, and two-fold if there is more than one affected older sibling.[27] Further evidence for the role of genetic burden in ASD comes from the observation that unaffected family members of ASD cases often have mild autistic-like phenotypes.[28] This phenomenon is most pronounced in multiplex families, where it is unlikely that the condition arose as a result of *de novo* mutation.[29]

Given the high heritability of autism, significant research efforts have been invested in identifying the genetic changes contributing to ASD risk. Here we give a brief overview of the current understanding of the genetic factors implicated in idiopathic ASD (comprehensively reviewed in Refs. [21, 30]).

Common genetic variation

The contribution of common genetic variation to ASD risk has been evaluated by several large-scale genome-wide association studies

(GWAS) in recent years.[31] A study by Wang *et al.* identified six SNPs at genome-wide significance ($p < 5 \times 10^{-8}$) on chromosome 5p14 between *CDH9* and *CDH10*.[32] More recently, it has been shown that the ASD-associated SNPs were transcribed as part of a non-coding transcript antisense to a moesin pseudogene.[33] A second GWAS identified an ASD-associated SNP within an intron of *MACROD2*.[31] A third study by Weiss *et al.* identified an ASD-associated SNP on chromosome 5p15 between *SEMA5A* and *TAS2R1*.[34] Notably each GWAS failed to replicate the findings of the others, which suggests that the effect sizes of common genetic variation to ASD risk are too small to detect with the current cohort sizes (often less than 2000 ASD cases). Nevertheless, recent meta-analyses of the above GWAS data have demonstrated that common variants of very small effect sizes additively account for 40–60% of ASD heritability.[35]

Thus, it is currently believed that each patient with idiopathic ASD carries a combination of common genetic variants with low-effect sizes and rare genetic variants with larger effect sizes but none of which are necessary or sufficient to cause the disease.[36]

Rare genetic variation

Abnormalities in chromosome structure were the first rare mutations associated with ASD, with karyotypic rearrangements visible in approximately 6–7% of ASD cases.[1] Maternally inherited duplications in the 15q11-15q13 region, associated with Angelman syndrome, are particularly common and occur in 1–2% of ASD cases.[1] Large structural aberrations implicated in ASD have been observed in all chromosomes, yet causal relationships have been difficult to determine.[37]

The role of rare variants in ASD risk has most recently and fruitfully been explored through large-scale copy number variation (CNV) analyses and exome sequencing, where exomes are the protein-coding regions of the genome. Microarray-based interrogations of CNVs associated with ASD have uncovered several recurrent themes.[38–43] First, each study confirmed that inherited CNVs are equally common in cases as in controls. Second, the prevalence of *de novo* CNV is 3–5% higher in probands than controls.[38–43] Third, the CNVs detected in probands tend to be larger and contain more genes.[38–43] Network and set enrichment analysis of CNV data found the ASD-associated CNVs to converge on the ubiquination and neuronal cell adhesion pathways,[44] as well as synaptogenesis and axon guidance.[45]

Next generation sequencing has facilitated a deeper scrutiny of rare *de novo* mutations in autism, notably in four recent studies which undertook large-scale sequencing and analysis of the exomes of ASD parents-child trios.[46–49] Proband exomes were compared with those of both parents and an unaffected sibling, where possible, to provide a clear indication of *de novo* mutations with the potential to impact disease state. It was found that probands had a higher proportion of non-synonymous and nonsense mutations than controls,[49] and that the rates of frame-shift, splice-site and *de novo* mutations were higher in the exonic regions of ASD cases.[47] Notably, the latter study also showed an association between several genes affected by rare *de novo* mutation and the fragile X mental retardation protein. Therefore, in addition to emphasizing the increased frequency of rare variants predicted to alter protein structure in ASD, the studies gave further support to the convergence of these mutations within pathways regulating neural plasticity.

Syndromic ASD

Despite the wealth of knowledge informing our current understanding of ASD genetic aetiology, the majority of ASD cases remain without an identified cause.[1] The monogenic syndromic forms of autism such as Rett syndrome, fragile X syndrome and tuberous sclerosis offer a unique opportunity to elucidate the underlying biology of ASD. For example, gene expression could be contrasted between (1) mutation carriers and controls, to elucidate mutation-specific features; (2) mutation carriers with and without ASD, to determine ASD-specific features; (3) ASD cases from different syndromic backgrounds, to identify features common to

ASD.[1] As such, the study of syndromic forms of ASD provides an efficient and elegant approach for determining any prevailing molecular pathology across the spectrum.

Rett syndrome

Rett syndrome (RTT) is an X-linked neurodevelopmental disorder with high ASD penetrance, primarily affecting girls at a frequency of 1/10 000 live births.[50] Following 6–18 months of normal development, children with RTT gradually lose speech and purposeful hand use, and develop microcephaly, seizures, ataxia, intellectual disability and repetitive hand stereotypies, with 50% also expressing autism.[51] Further motor deterioration occurs later in adolescence, characterised by scoliosis, pronounced dystonia and occasionally the loss of the ability to walk.[52] Greater than 95% of typical RTT is caused by *de novo* mutations in *MECP2,* which encodes the transcriptional modulator methyl-CpG-binding protein 2.[50,53] Neuropathological studies have documented severely impaired synaptogenesis and decreased dendritic arborisations in children with RTT,[54] although the degree of impairment varies by brain region.[55]

MeCP2 belongs to the methyl-CpG-binding domain (MBD) family of proteins. While the MBD domain binds to symmetrically methylated CpG sites,[56,57] MeCP2 also possesses a nuclear localization signal, a transcriptional repression domain which recruits histone deacetylases involved in gene silencing,[58] and a C-terminal domain of undetermined yet important function.[59] MeCP2 is abundantly expressed in the CNS where it functions both as a transcriptional activator and repressor. However, cataloguing MeCP2 target genes in the human brain in a genome-wide manner has proven difficult and much of our knowledge of MeCP2 activity remains based on studies of knock-out or transgenic MeCP2 mouse models.[60] One major study thus identified approximately 400 genes subject to MeCP2 repression, and a further 2000 genes which appear to be targets of MeCP2 activation.[61] Many of the genes implicated as targets of MeCP2 regulation are involved in neurite growth and synaptogenesis, such as *A2BP1*, which regulates splicing of neuronal genes and has previously been linked to autism susceptibility.[62] A few small, array-based studies have investigated transcriptional variation between the brains of RTT cases and controls and found that RTT brains display comparative downregulation of presynaptic, neuron-specific, and neurodevelopmental genes, and upregulation of glia-specific genes and genes typically involved in neurodegeneration.[63]

Fragile X syndrome

Fragile X syndrome (FXS) is the most common inherited form of intellectual disability and is also one of the leading causes of autism; approximately 21% of children diagnosed with FXS are subsequently diagnosed with ASD.[64] FXS arises from the silencing of the fragile X mental retardation protein (FMRP, encoded by *FMR1*) due to expanded CGG repeats or, less commonly from missense mutations.[65–67] FMRP is an RNA-binding protein which associates with primarily synaptic mRNA through any of three RNA-binding domains[68] and appears to repress translation by associating with the ribosome to block both initiation and elongation.[69,70] FMRP is greatly enriched in dendritic spines and has been shown to regulate several families of synaptic and other autism-related proteins.[71] Hence, loss of functional FMRP is thought to elicit large downstream effects on synaptic function and plasticity.[72]

FMR1 overlaps a fragile site on the X chromosome, is composed of 17 exons subject to substantial alternative splicing, and has a CGG repeat of variable length in the 5′ untranslated region.[73] Based on the CGG repeat length, *FMR1* alleles can be classified as: normal, intermediate, premutation and full mutation. Normal *FMR1* alleles contain 5–44 repeats, most commonly 29 or 30. The intermediate allele carries 45–54 repeats, and is unstable but not pathogenic.[74] The premutation allele carries between 54 and 200 repeats and is meiotically unstable, especially when maternally transmitted. Full mutations, or alleles with greater than 200 repeats, are always derived from maternally transmitted premutations, and the risk of expansion to a full mutation is directly proportional to the length of the maternal premutation.[75] Expansion into the full mutation elicits epigenetic silencing of *FMR1*[66,75] causing FXS; methylation of the gene occurs in embryogenesis and triggers further histone modification, reinforcing the strength of gene repression.[76,77].

Although the exact mechanism by which FMRP functions remains to be fully elucidated, the Group 1 metabotropic glutamate receptor (mGluR) theory of FXS provides a compelling explanation for the neurologic and psychiatric aspects of the syndrome.[78] mGluR-dependent long term depression (mGluR-LTD), an activity-dependent, transient disruption of synaptic transmission, plays a key role in synaptic plasticity and is triggered by local translation of postsynaptic, dendritically-localised mRNAs in response to synaptic stimulation. The elevated local protein synthesis causes internalization of α-amino-3-hydroxyl-4-isoxazole propionic acid receptors (AMPAR); endocytosis of AMPAR is essential to mGluR-LTD.[79] FMRP is thought to help regulate the process by normally repressing mRNA translation unless mGluR is activated. Consequently, in the absence of FMRP, constitutive protein expression in dendrites appears to dysregulate AMPAR internalisation and hence mGluR-LTD.[78]

Several convergent lines of evidence support this explanation of FXS. *FMR1* knockout (KO) mice have enhanced mGluR-LTD in the hippocampus, and show increased mGluR-LTD activity even without new protein synthesis.[79] Dephosphorylation of FMRP by protein phosphatase 2A (PP2A) occurs shortly after mGluR activation and coincides with the translation of FMRP target messages, such as PSD-95.[80] Fluorescent *in situ* hybridisation shows FMRP-regulated mRNAs localise to dendrites of cortical and hippocampal neurons in both wild-type and *FMR1* KO mice, yet stimulus-induced translation is impaired in KO mice.[81] Soon after a transient spike in FMRP translation in response to mGluR stimulation, the protein is ubiquinated and degraded, yet preventing the degradation of FMRP with proteasome inhibitors suppresses mGluR-LTD.[82]

Knockdown of *FMR1* in cultured hippocampal cells results in an increased basal rate of AMPAR internalization and a reduced response to synaptic stimulation.[83] Hence, intellectual disability in FXS is thought to be a result of both the constitutively active mGluR-LTD and an inability to further increase synaptic protein synthesis in response to stimuli.[84]

Mammalian target of rapamycin (mTOR) is another key translational regulator located downstream of mGluR signalling (see Fig. 3 in Ref. [85]). mTOR activation influences synaptic plasticity through a localised signalling cascade regulating translation of synaptic proteins.[86] *FMR1* KO mice display enhanced hippocampal mTOR signalling, possibly as a result of elevated PIKE-S synthesis. PIKE-S is a target of FMRP regulation and is involved in the intermediate signalling steps between mGluR stimulation and mTOR activation.

Recently, high-throughput sequencing of genes involved in the mGluR signalling pathway showed significant enrichment of rare, potentially damaging *de novo* mutations in idiopathic ASD cases compared with controls.[87] Taken together, these studies validate the roles of synaptic dysfunction and neuronal signalling in both FXS and ASD, and further the argument that pathophysiological mechanisms convergent within these pathways are common to both disorders.

Tuberous sclerosis

Tuberous sclerosis complex (TSC) is a dominantly inherited, multisystem genetic disorder with a prevalence of approximately one case per 11,000 live births.[88] TSC is characterised by the appearance of hamartomas (tumour-like nodules) in the skin and, in 90% of cases, the CNS.[89] Epilepsy and intellectual disability have very high comorbidity with the condition (80–90% and 44–60% of cases, respectively[90]) while ASD co-occurs in up to 50% of TSC cases.[91] TSC is caused by mutations in either *TSC1* or *TSC2*, genes which encode the proteins hamartin (TSC1) and tuberin (TSC2). Hamartin and tuberin form a complex involved in the inhibition of mTOR and hence, in TSC, mTOR is hyperactive due to the absence of a functional TSC1/2 complex. Although a large number of different mutations in the tuberous sclerosis proteins have been catalogued, most result in protein truncations and occur in *TSC2*.[92] Like FMRP, mTOR is involved in translational control and its uninhibited activity corresponds to an increase in protein synthesis and cell growth.[93]

In addition to playing a broad role in the regulation of translation and cell proliferation, the TSC1/2 complex appears to specifically influence axonal, dendritic and synaptic maturation and function. Axon formation is mediated by extracellular signalling and intrinsic, asymmetric polarization.[94] The TSC2/1 complex appears enriched within all neurites. Growth factors such as isulin-like growth factor-1 (IGF-1) and brain derived growth factor (BDNF) activate PI3K and Akt, which subsequently phosphorylate and deactivate TSC2.[95] Local inactivation of the TSC1/2 complex within a nascent axon allows its specific propagation through mTOR signalling while growth of the remaining neurites is inhibited.[96] Further, inactivation of TSC1/2 and the subsequent activation of mTOR causes heightened synthesis of the SAD kinases,[96] which are key regulators of neuronal polarization.[96,97] Finally, expression of a functional TSC1/2 complex seems essential for proper axon myelination.[98,99]

Loss of functional TSC1/2 also affects dendritic spine formation and structure, and dendrite arborisation. The dendrites of *TSC1* KO mice display increased spine length and head width and reduced spine density.[100] Cultured hippocampal *TSC1* KO mouse neurons share this dendritic morphology and also appear to possess enhanced AMPAR activation.[101] In addition, and contrary to the silencing of FMRP associated with FXS, loss of either TSC1 or TSC2 in the mouse hippocampus abolishes mGluR-LTD.[101,102] Inhibition of mTOR in *TSC2*$^{+/-}$ mice rescues both their aberrant protein synthesis and mGluR-LTD.[102] Further, brief treatment of adult *TSC2*$^{+/-}$ mice with rapamycin (an mTOR inhibitor) reverses cognitive deficits.[103] Both FXS and TSC implicate hyperactive synaptic mTOR signalling as pathogenic.

Emerging Pharmacological Therapies for Syndromic ASD

One of the most exciting discoveries over the last decade of research on syndromic forms of ASD was that the behavioral and neurological phenotypes of several mouse models could be reversed in young and adult mice, indicating that the neurodevelopmental insult is not permanent, and thus targeted pharmacological therapies may be effective in human patients. Restoring *Mecp2* expression in adult *Mecp2* deficient mice reversed the neurological and behavioral deficits.[104] Equally remarkable, mGluR5 agonists rescued the phenotype in *Tsc* mice, while mGluR5 antagonists were able to reverse pathological changes in *Fmr1* mice.[83,104]

Based on the accumulating evidence that increased signalling activity via mGluR5 is involved in the pathophysiology of fragile X syndrome, clinical trials have already been initiated to test the efficiency of mGluR5 inhibitors in fragile X syndrome patients. Fenobam, a negative allosteric modulator of mGluR5 has been tested in an open-label study of 12 fragile X patients.[105] This study reported moderate reversal of pre-pulse inhibition deficits, and further randomized double-blind studies of mGluR5 inhibitors are underway.

The results of the first randomized double-blind study to be completed have shown that significant treatment effects could be observed selectively in patients with full *FMR1* promoter methylation and absence of *FMR1* expression in blood, while patients with partial *FMR1* methylation showed no improvement with mGluR5 inhibitor treatment.[106]

Evidence for Convergent Pathways between Syndromic and Idiopathic ASD

Given that pharmacological therapies are more rapidly emerging for syndromic forms of ASD, it becomes important to elucidate whether there are common mechanisms underlying the different

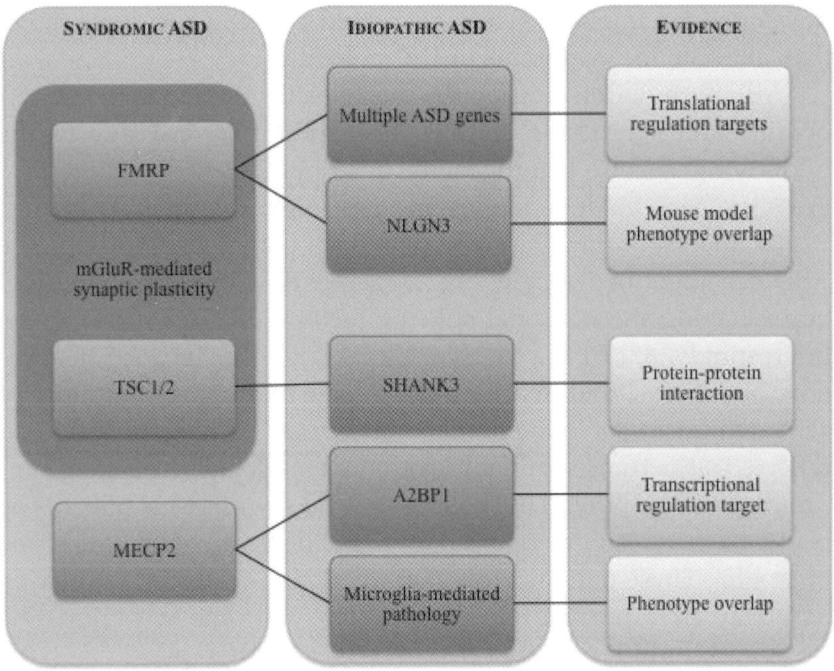

Figure 1. Molecular and cellular mechanisms linking syndromic and idiopathic ASD.

forms of syndromic ASD as well as mechanistic overlaps between syndromic and idiopathic ASD. Recent progress addressing this question is schematically summarized in Figure 1.

As mentioned above, both fragile X syndrome and tuberous sclerosis are associated with impaired mGluR5-mediated synaptic plasticity and protein synthesis. However, while fragile X mice show hyperactivation of mGluR5 signaling, TSC1 or TSC2 deficiencies abolish mGluR5-dependent synaptic plasticity. Thus while both syndromes involve the same signalling pathways, they do so in opposite manners, yet leading to similar autistic behavioural changes. This important insight from animal models indicates that (1) synaptic plasticity is finely tuned, and changes in either direction can trigger pathological phenotypes and (2) using pharmacotherapy developed for fragile X syndrome would possibly aggravate rather than alleviate the symptoms in TSC patients.

Although understanding the mechanistic overlap between idiopathic and syndromic ASD is less straightforward due to the underlying genetic heterogeneity of idiopathic ASD, recent progress has been made in identifying possible convergent pathways.

By analysing transcriptome data from patients with idiopathic ASD we identified a network of genes downregulated in ASD brain, having the neuronal splicing factor *A2BP1* as a hub gene. Interestingly, *A2BP1* is a transcriptional target of *MECP2*.[61] In addition, patients with idiopathic autism have an increased number of activated microglia in the brain parenchyma,[107] a phenomenon also reflected at a transcriptional level.[108] Remarkably, in a Rett syndrome mouse model,[109] the behavioral and neurological phenotype could be reversed by repopulation of the brain with wild-type microglia. These data indicate that microglial-mediated pathology may play an important role in both Rett syndrome and idiopathic ASD, and could represent a fruitful target for therapy.

A recent study used a yeast two-hybrid system of a human cDNA library to interrogate protein-protein interactions between proteins encoded by genes implicated in both syndromic and idiopathic ASD.[110] This study showed that *SHANK3*, a gene implicated in idiopathic ASD interacts with *TSC1*, and that the two proteins share 21 protein binding partners, suggesting that common molecular changes may be involved in TSC patients and ASD patients carrying *SHANK3* mutations.

An investigation of FMRP targets in a human embryonic cell line, HEK293, identified 93 genes previously implicated in ASD as targets of FMRP.[111] Conversely, a recent exome sequencing study found that genes harbouring *de novo* mutations in ASD probands are enriched for FMRP regulatory targets.[47] Further, a *NLGN3* knock-out mouse model of idiopathic ASD has been shown to exhibit mGluR-dependent synaptic plasticity changes similar to those observed in fragile X mice.[112]

Although not providing a definitive answer, these studies suggest that idiopathic ASD patients carrying certain genetic variants (e.g. *SHANK3* or *NLGN3* mutations) could benefit from specific pharmacological treatment designed for syndromic ASD.

Future Directions

Targeted pharmacological therapy for syndromic ASD is starting to appear within reach, thus offering a unique opportunity for potentially effective treatment for patients with idiopathic ASD, provided that the underlying biological mechanisms are shared. Thus, further research aiming to delineate the commonalities between syndromic and idiopathic ASD would be highly relevant.

As discussed in this chapter, phenotypic overlaps between mouse models of genes implicated in idiopathic ASD and syndromic ASD are very encouraging and more such studies are likely to emerge in the near future. However, it should be noted that the phenotypic manifestation, including drug responses, depends heavily on the genomic context, and thus the results from mouse studies may not be easily translatable to predict treatment responses in ASD patients.

An alternative avenue is the identification of specific endophenotypes, either neurophysiological markers or biological markers, that are characteristic of syndromic ASD and a subset of idiopathic ASD. A recent study compared electroencephalographic (EEG) connectivity in idiopathic ASD and TSC with the aim to identify common neurobiological mechanisms.[113] This type of data could potentially identify the subset of ASD patients likely to benefit from treatments designed for syndromic ASD.

Applying novel therapies developed for syndromic ASD to a randomly selected set of patients with idiopathic ASD carries the risk of potential adverse effects (as discussed above for fragile X syndrome and TSC), and may be falsely deemed ineffective, as a result of the heterogeneity characteristic of idiopathic ASD. It thus remains crucial to identify subsets of ASD patients that are responsive to specific pharmacological therapies, and it is expected that personalized medicine would bring fruitful contributions to such endeavors.

Executive Summary

- Idiopathic ASD is a highly heritable but genetically heterogeneous condition. The disease heterogeneity has hampered the development of targeted pharmacological therapy despite intense research efforts.
- Fragile X syndrome, Rett syndrome and tuberous sclerosis are genetic syndromes with high incidence of autism.
- The molecular mechanisms underlying syndromic ASD have been extensively characterized in animal models and cell culture systems. As a result, the development of targeted pharmacological therapies is progressing faster for syndromic ASD than for idiopathic ASD.
- A novel class of pharmacological agents for ASD, mGluR5 antagonists, are currently undergoing clinical trials for the treatment of fragile X syndrome.
- Several common molecular and neurobiological mechanisms underlying syndromic and idiopathic ASD have been recently characterized (see Figure 1), offering hope that targeted therapy may be effective for patients with idiopathic ASD.

References

1. Abrahams B, Geschwind D: Advances in autism genetics: on the threshold of a new neurobiology. *Nat Rev Genet* 2008, **9**(5):341–355.
2. *Diagnostic and Statistical Manual of Mental Disorders: DSM-IV*, 4th edn. Washington, D.C: American Psychiatric Association; 1994.
3. Leekam S, Nieto C, Libby S, Wing L, Gould J: Describing the sensory abnormalities of children and adults with autism. *J Autism Dev Disord* 2007, **37**(5): 894–910.
4. Ming X, Brimacombe M, Wagner G: Prevalence of motor impairment in autism spectrum disorders. *Brain Dev* 2007, **29**(9):565–570.
5. Tuchman R, Rapin I: Epilepsy in autism. *Lancet Neurol* 2002, **1**(6):352–358.
6. Gillberg C, Billstedt E: Autism and Asperger syndrome: coexistence with other clinical disorders. *Acta Psychiat Scand* 2000, **102**(5):321–330.
7. Walker D, Thompson A, Zwaigenbaum L, Goldberg J, Bryson S, Mahoney W, Strawbridge C, Szatmari P: Specifying PDD-NOS: a comparison of PDD-NOS, Asperger syndrome, and autism. *J Am Acad Child Adolesc Psychiatry* 2004, **43**(2):172–180.

8. Lord C, Petkova E, Hus V, Gan W, Lu F, Martin DM, Ousley O, Guy L, Bernier R, Gerdts J *et al.*: A multisite study of the clinical diagnosis of different autism spectrum disorders. *Arch Gen Psychiatry* 2012, **69**(3):306–313.
9. Macintosh K, Dissanayake C: Annotation: The similarities and differences between autistic disorder and Asperger's disorder: a review of the empirical evidence. *J Child Psychol Psychiatry* 2004, **45**(3):421–434.
10. Witwer A, Lecavalier L: Examining the validity of autism spectrum disorder subtypes. *J Autism Dev Disord* 2008, **38**(9):1611–1624.
11. Snow A, Lecavalier L: Comparing autism, PDD-NOS, and other developmental disabilities on parent-reported behavior problems: little evidence for ASD subtype validity. *J Autism Dev Disord* 2011, **41**(3):302–310.
12. Szatmari P, Bryson S, Boyle M, Streiner D, Duku E: Predictors of outcome among high functioning children with autism and Asperger syndrome. *J Child Psychol Psychiatry* 2003, **44**(4):520–528.
13. Szatmari P, Bryson S, Duku E, Vaccarella L, Zwaigenbaum L, Bennett T, Boyle M: Similar developmental trajectories in autism and Asperger syndrome: from early childhood to adolescence. *J Child Psychol Psychiatry* 2009, **50**(12):1459–1467.
14. Fombonne E: Epidemiological surveys of autism and other pervasive developmental disorders: an update. *J Autism Dev Disord* 2003, **33**(4):365–382.
15. Bryson S, Clark B, Smith I: First report of a Canadian epidemiological study of autistic syndromes. *J Child Psychol Psychiatry* 1988, **29**(4):433–445.
16. Geschwind D: Advances in autism. *Ann Rev Med* 2009, **60**:367–380.
17. Hertz-Picciotto I, Delwiche L: The rise in autism and the role of age at diagnosis. *Epidemiology (Cambridge, Mass)* 2009, **20**(1):84–90.
18. Grzadzinski R, Huerta M, Lord C: DSM-5 and autism spectrum disorders (ASDs): an opportunity for identifying ASD subtypes. *Mol Autism* 2013, **4**(1):12.
19. Rogers SJ, Vismara LA: Evidence-based comprehensive treatments for early autism. *J Clin Child Adolesc Psychol* 2008, **37**(1):8–38.
20. Dove D, Warren Z, McPheeters ML, Taylor JL, Sathe NA, Veenstra-VanderWeele J: Medications for adolescents and young adults with autism spectrum disorders: a systematic review. *Pediatrics* 2012, **130**(4):717–726.
21. State M, Levitt P: The conundrums of understanding genetic risks for autism spectrum disorders. *Nat Neurosci* 2011, **14**(12):1499–1506.
22. Smith DG, Ehlers MD: Mining and modeling human genetics for autism therapeutics. *Curr Opin Neurobiol* 2012, **22**(5):902–910.
23. Kanner L: Autistic disturbances of affective contact. *Nervous Child* 1943, **2**(3):217–250.
24. Bailey A, Le Couteur A, Gottesman I, Bolton P, Simonoff E, Yuzda E, Rutter M: Autism as a strongly genetic disorder: evidence from a British twin study. *Psychol Med* 1995, **25**(1):63–77.

25. Ronald A, Hoekstra R: Autism spectrum disorders and autistic traits: a decade of new twin studies. *Am J Med Genet Neuropsychiatr genet* 2011, **156B**(3): 255–274.
26. Berg J, Geschwind D: Autism genetics: searching for specificity and convergence. *Genome Biol* 2012, **13**(7):247.
27. Ozonoff S, Young G, Carter A, Messinger D, Yirmiya N, Zwaigenbaum L, Bryson S, Carver L, Constantino J, Dobkins K *et al.*: Recurrence risk for autism spectrum disorders: a Baby Siblings Research Consortium study. *Pediatrics* 2011, **128**(3):95.
28. Bishop D, Maybery M, Maley A, Wong D, Hill W, Hallmayer J: Using self-report to identify the broad phenotype in parents of children with autistic spectrum disorders: a study using the Autism-Spectrum Quotient. *J Child Psychol Psychiatry* 2004, **45**(8):1431–1436.
29. Sasson N, Lam K, Childress D, Parlier M, Daniels J, Piven J: The broad autism phenotype questionnaire: prevalence and diagnostic classification. *Autism Res* 2013, **6**(2):134–143.
30. Murdoch JD, State MW: Recent developments in the genetics of autism spectrum disorders. *Curr Opin Genet Dev* 2013.
31. Anney R, Klei L, Pinto D, Regan R, Conroy J, Magalhaes T, Correia C, Abrahams B, Sykes N, Pagnamenta A *et al.*: A genome-wide scan for common alleles affecting risk for autism. *Hum Mol Genet* 2010, **19**(20):4072–4082.
32. Wang K, Zhang H, Ma D, Bucan M, Glessner J, Abrahams B, Salyakina D, Imielinski M, Bradfield J, Sleiman P *et al.*: Common genetic variants on 5p14.1 associate with autism spectrum disorders. *Nature* 2009, **459**(7246):528–533.
33. Kerin T, Ramanathan A, Rivas K, Grepo N, Coetzee GA, Campbell DB: A noncoding RNA antisense to moesin at 5p14.1 in autism. *Sci Transl Med* 2012, **4**(128):128ra140.
34. Weiss L, Arking D, Gene Discovery Project of Johns H, the Autism C, Daly M, Chakravarti A: A genome-wide linkage and association scan reveals novel loci for autism. *Nature* 2009, **461**(7265):802–808.
35. Anney R, Klei L, Pinto D, Almeida J, Bacchelli E, Baird G, Bolshakova N, Bölte S, Bolton P, Bourgeron T *et al.*: Individual common variants exert weak effects on the risk for autism spectrum disorderspi. *Hum Mol Genet* 2012, **21**(21):4781–4792.
36. State MW: The genetics of child psychiatric disorders: focus on autism and Tourette syndrome. *Neuron* 2010, **68**(2):254–269.
37. Devlin B, Scherer S: Genetic architecture in autism spectrum disorder. *Curr Opin Genet Dev* 2012, **22**(3):229–237.
38. Autism Genome Project C, Szatmari P, Paterson A, Zwaigenbaum L, Roberts W, Brian J, Liu X-Q, Vincent J, Skaug J, Thompson A *et al.*: Mapping autism risk loci using genetic linkage and chromosomal rearrangements. *Nat Genet* 2007, **39**(3):319–328.

39. Levy D, Ronemus M, Yamrom B, Lee Y-h, Leotta A, Kendall J, Marks S, Lakshmi B, Pai D, Ye K *et al.*: Rare de novo and transmitted copy-number variation in autistic spectrum disorders. *Neuron* 2011, **70**(5):886–897.
40. Marshall CR, Noor A, Vincent JB, Lionel AC, Feuk L, Skaug J, Shago M, Moessner R, Pinto D, Ren Y *et al.*: Structural variation of chromosomes in autism spectrum disorder. *Am J Hum Genet* 2008, **82**(2):477–488.
41. Pinto D, Pagnamenta A, Klei L, Anney R, Merico D, Regan R, Conroy J, Magalhaes T, Correia C, Abrahams B *et al.*: Functional impact of global rare copy number variation in autism spectrum disorders. *Nature* 2010, **466**(7304):368–372.
42. Sanders S, Ercan-Sencicek A, Hus V, Luo R, Murtha M, Moreno-De-Luca D, Chu S, Moreau M, Gupta A, Thomson S *et al.*: Multiple recurrent de novo CNVs, including duplications of the 7q11.23 Williams syndrome region, are strongly associated with autism. *Neuron* 2011, **70**(5):863–885.
43. Sebat J, Lakshmi B, Malhotra D, Troge J, Lese-Martin C, Walsh T, Yamrom B, Yoon S, Krasnitz A, Kendall J *et al.*: Strong association of de novo copy number mutations with autism. *Science (New York, NY)* 2007, **316**(5823):445–449.
44. Glessner J, Wang K, Cai G, Korvatska O, Kim C, Wood S, Zhang H, Estes A, Brune C, Bradfield J *et al.*: Autism genome-wide copy number variation reveals ubiquitin and neuronal genes. *Nature* 2009, **459**(7246):569–573.
45. Gilman S, Iossifov I, Levy D, Ronemus M, Wigler M, Vitkup D: Rare de novo variants associated with autism implicate a large functional network of genes involved in formation and function of synapses. *Neuron* 2011, **70**(5):898–907.
46. O'Roak BJ, Vives L, Girirajan S, Karakoc E, Krumm N, Coe BP, Levy R, Ko A, Lee C, Smith JD *et al.*: Sporadic autism exomes reveal a highly interconnected protein network of de novo mutations. *Nature* 2012, **485**(7397):246–250.
47. Iossifov I, Ronemus M, Levy D, Wang Z, Hakker I, Rosenbaum J, Yamrom B, Lee Y-H, Narzisi G, Leotta A *et al.*: De novo gene disruptions in children on the autistic spectrum. *Neuron* 2012, **74**(2):285–299.
48. Neale B, Kou Y, Liu L, Ma'ayan A, Samocha K, Sabo A, Lin C-F, Stevens C, Wang L-S, Makarov V *et al.*: Patterns and rates of exonic de novo mutations in autism spectrum disorders. *Nature* 2012, **485**(7397):242–245.
49. Sanders SJ, Murtha MT, Gupta AR, Murdoch JD, Raubeson MJ, Willsey AJ, Ercan-Sencicek AG, DiLullo NM, Parikshak NN, Stein JL *et al.*: De novo mutations revealed by whole-exome sequencing are strongly associated with autism. *Nature* 2012, **485**(7397):237–241.
50. Laurvick C, de Klerk N, Bower C, Christodoulou J, Ravine D, Ellaway C, Williamson S, Leonard H: Rett syndrome in Australia: a review of the epidemiology. *J Pediatr* 2006, **148**(3):347–352.
51. Wulffaert J, Van Berckelaer-Onnes IA, Scholte EM: Autistic disorder symptoms in Rett syndrome. *Autism* 2009, **13**(6):567–581.

52. Hagberg B, Aicardi J, Dias K, Ramos O: A progressive syndrome of autism, dementia, ataxia, and loss of purposeful hand use in girls: Rett's syndrome: report of 35 cases. *Ann Neurol* 1983, **14**(4):471–479.
53. Amir R, Van den Veyver I, Wan M, Tran C, Francke U, Zoghbi H: Rett syndrome is caused by mutations in X-linked MECP2, encoding methyl-CpG-binding protein 2. *Nat Genet* 1999, **23**(2):185–188.
54. Kaufmann W, Taylor C, Hohmann C, Sanwal I, Naidu S: Abnormalities in neuronal maturation in Rett syndrome neocortex: preliminary molecular correlates. *Eur Child Adolesc Psychiatry* 1997, **6** (Suppl 1):75–77.
55. Armstrong D, Dunn JK, Antalffy B, Trivedi R: Selective dendritic alterations in the cortex of Rett syndrome. *J Neuropathol Exp Neurol* 1995, **54**(2):195–201.
56. Lewis J, Meehan R, Henzel W, Maurer-Fogy I, Jeppesen P, Klein F, Bird A: Purification, sequence, and cellular localization of a novel chromosomal protein that binds to methylated DNA. *Cell* 1992, **69**(6):905–914.
57. Richard M, Joe DL, Adrian PB: Characterization of MeCP2, a vertebrate DNA binding protein with affinity for methylated DNA. *Nucleic Acids Res* 1992, **20**.
58. Nan X, Ng H, Johnson C, Laherty C, Turner B, Eisenman R, Bird A: Transcriptional repression by the methyl-CpG-binding protein MeCP2 involves a histone deacetylase complex. *Nature* 1998, **393**(6683):386–389.
59. Neul J, Fang P, Barrish J, Lane J, Caeg E, Smith E, Zoghbi H, Percy A, Glaze D: Specific mutations in methyl-CpG-binding protein 2 confer different severity in Rett syndrome. *Neurology* 2008, **70**(16):1313–1321.
60. Samaco R, Neul J: Complexities of Rett syndrome and MeCP2. *J Neurosci* 2011, **31**(22):7951–7959.
61. Chahrour M, Jung S, Shaw C, Zhou X, Wong S, Qin J, Zoghbi H: MeCP2, a key contributor to neurological disease, activates and represses transcription. *Science (New York, NY)* 2008, **320**(5880):1224–1229.
62. Martin C, Duvall J, Ilkin Y, Simon J, Arreaza M, Wilkes K, Alvarez-Retuerto A, Whichello A, Powell C, Rao K *et al.*: Cytogenetic and molecular characterization of A2BP1/FOX1 as a candidate gene for autism. *Am J Med Genet B Neuropsychiatr Genet* 2007, **144B**(7):869–876.
63. Colantuoni C, Jeon O, Hyder K, Chenchik A, Khimani A, Narayanan V, Hoffman E, Kaufmann W, Naidu S, Pevsner J: Gene expression profiling in postmortem Rett Syndrome brain: differential gene expression and patient classification. *Neurobiol Dis* 2001, **8**(5):847–865.
64. Hatton D, Sideris J, Skinner M, Mankowski J, Bailey D, Roberts J, Mirrett P: Autistic behavior in children with fragile X syndrome: prevalence, stability, and the impact of FMRP. *Am J Med Genet A* 2006, **140A**(17):1804–1813.
65. Collins SC, Bray SM, Suhl JA, Cutler DJ, Coffee B, Zwick ME, Warren ST: Identification of novel FMR1 variants by massively parallel sequencing in developmentally delayed males. *Am J Med Genet A* 2010, **152**(10):2512–2520.

66. Pieretti M, Zhang F, Fu Y, Warren S, Oostra B, Caskey C, Nelson D: Absence of expression of the FMR-1 gene in fragile X syndrome. *Cell* 1991.
67. Verkerk AJMH, Pieretti M, Sutcliffe JS, Fu Y-H, Kuhl D, Pizzuti A, Reiner O, Richards S, Victoria MF, Zhang F: Identification of a gene (FMR-1) containing a CGG repeat coincident with a breakpoint cluster region exhibiting length variation in fragile X syndrome. *Cell* 1991, **65**(5):905–914.
68. Ashley C, Wilkinson K, Reines D, Warren S: FMR1 protein: conserved RNP family domains and selective RNA binding. *Science* 1993, **262**.
69. Feng Y, Gutekunst C, Eberhart D, Yi H, Warren S, Hersch S: Fragile X mental retardation protein: nucleocytoplasmic shuttling and association with somatodendritic ribosomes. *J Neurosci* 1997, **17**(5):1539–1547.
70. Bhakar A, Dölen G, Bear M: The pathophysiology of fragile X (and what it teaches us about synapses). *Ann Rev Neurosci* 2012, **35**:417–443.
71. Brown V, Jin P, Ceman S, Darnell J, O'Donnell W, Tenenbaum S, Jin X, Feng Y, Wilkinson K, Keene J *et al.*: Microarray identification of FMRP-associated brain mRNAs and altered mRNA translational profiles in fragile X syndrome. *Cell* 2001, **107**(4):477–487.
72. Kindler S, Kreienkamp H-J: The role of the postsynaptic density in the pathology of the fragile x syndrome. *Results Probl Cell Differ* 2012, **54**:61–80.
73. Nelson D, Orr H, Warren S: The unstable repeats — three evolving faces of neurological disease. *Neuron* 2013, **77**(5):825–843.
74. Zhong N, Ju W, Pietrofesa J, Wang D, Dobkin C, Brown W: Fragile X "gray zone" alleles: AGG patterns, expansion risks, and associated haplotypes. *Am J Med Genet* 1996, **64**(2):261–265.
75. Fu Y, Kuhl D, Pizzuti A, Pieretti M, Sutcliffe J, Richards S, Verkerk A, Holden J, Fenwick R, Warren S: Variation of the CGG repeat at the fragile X site results in genetic instability: resolution of the Sherman paradox. *Cell* 1991, **67**(6): 1047–1058.
76. Sutcliffe J, Nelson D, Zhang F, Pieretti M, Caskey C, Saxe D, Warren S: DNA methylation represses FMR-1 transcription in fragile X syndrome. *Hum Mol Genet* 1992, **1**(6):397–400.
77. Coffee B, Zhang F, Ceman S, Warren S, Reines D: Histone modifications depict an aberrantly heterochromatinized FMR1 gene in fragile x syndrome. *Am J Hum Genet* 2002, **71**(4):923–932.
78. Bear M, Huber K, Warren S: The mGluR theory of fragile X mental retardation. *Trends Neurosci* 2004, **27**(7):370–377.
79. Huber K, Roder J, Bear M: Chemical induction of mGluR5- and protein synthesis — dependent long-term depression in hippocampal area CA1. *J Neurophysiol* 2001, **86**(1):321–325.
80. Narayanan U, Nalavadi V, Nakamoto M, Pallas D, Ceman S, Bassell G, Warren S: FMRP phosphorylation reveals an immediate-early signaling

pathway triggered by group I mGluR and mediated by PP2A. *J Neurosci* 2007, **27**(52):14349–14357.

81. Muddashetty R, Kelić S, Gross C, Xu M, Bassell G: Dysregulated metabotropic glutamate receptor-dependent translation of AMPA receptor and postsynaptic density-95 mRNAs at synapses in a mouse model of fragile X syndrome. *J Neurosci* 2007, **27**(20):5338–5348.
82. Hou L, Antion M, Hu D, Spencer C, Paylor R, Klann E: Dynamic translational and proteasomal regulation of fragile X mental retardation protein controls mGluR-dependent long-term depression. *Neuron* 2006, **51**(4): 441–454.
83. Nakamoto M, Nalavadi V, Epstein M, Narayanan U, Bassell G, Warren S: Fragile X mental retardation protein deficiency leads to excessive mGluR5-dependent internalization of AMPA receptors. *Proc Natl Acad Sci USA* 2007, **104**(39):15537–15542.
84. Santoro M, Bray S, Warren S: Molecular mechanisms of fragile X syndrome: a twenty-year perspective. *Ann Rev Pathol* 2012, **7**:219–245.
85. Tang S, Schuman E: Protein synthesis in the dendrite. *Philos Trans R Soc Lon B Biol Sci* 2002, **357**(1420):521–529.
86. Hou L, Klann E: Activation of the phosphoinositide 3-kinase-Akt-mammalian target of rapamycin signaling pathway is required for metabotropic glutamate receptor-dependent long-term depression. *J Neurosci* 2004, **24**(28): 6352–6361.
87. Kelleher R, Geigenmüller U, Hovhannisyan H, Trautman E, Pinard R, Rathmell B, Carpenter R, Margulies D: High-throughput sequencing of mGluR signaling pathway genes reveals enrichment of rare variants in autism. *PloS One* 2012, **7**(4).
88. O'Callaghan F, Shiell A, Osborne J, Martyn C: Prevalence of tuberous sclerosis estimated by capture-recapture analysis. *Lancet* 1998, **351**(9114):1490.
89. Sahin M: Targeted treatment trials for tuberous sclerosis and autism: no longer a dream. *Curr Opin Neurobiol* 2012, **22**(5):895–901.
90. Crino P, Nathanson K, Henske E: The tuberous sclerosis complex. *New Eng J Med* 2006, **355**(13):1345–1356.
91. Jeste S, Sahin M, Bolton P, Ploubidis G, Humphrey A: Characterization of autism in young children with tuberous sclerosis complex. *J Child Neurol* 2008, **23**(5):520–525.
92. Sancak O, Nellist M, Goedbloed M, Elfferich P, Wouters C, Maat-Kievit A, Zonnenberg B, Verhoef S, Halley D, van den Ouweland A: Mutational analysis of the TSC1 and TSC2 genes in a diagnostic setting: genotype — phenotype correlations and comparison of diagnostic DNA techniques in Tuberous Sclerosis Complex. *Eur J Hum Genet* 2005, **13**(6):731–741.
93. Wullschleger S, Loewith R, Hall M: TOR signaling in growth and metabolism. *Cell* 2006, **124**(3):471–484.

94. Arimura N, Kaibuchi K: Neuronal polarity: from extracellular signals to intracellular mechanisms. *Nat Rev Neurosci* 2007, **8**(3):194–205.
95. Manning B, Tee A, Logsdon M, Blenis J, Cantley L: Identification of the tuberous sclerosis complex-2 tumor suppressor gene product tuberin as a target of the phosphoinositide 3-kinase/akt pathway. *Mol Cell* 2002, **10**(1):151–162.
96. Choi Y-J, Di Nardo A, Kramvis I, Meikle L, Kwiatkowski D, Sahin M, He X: Tuberous sclerosis complex proteins control axon formation. *Genes Dev* 2008, **22**(18):2485–2495.
97. Kishi M, Pan YA, Crump JG, Sanes JR: Mammalian SAD kinases are required for neuronal polarization. *Science* 2005, **307**(5711):929–932.
98. Ridler K, Bullmore E, De Vries P, Suckling J, Barker G, Meara S, Williams S, Bolton P: Widespread anatomical abnormalities of grey and white matter structure in tuberous sclerosis. *Psychol Med* 2001, **31**(8):1437–1446.
99. Meikle L, Talos D, Onda H, Pollizzi K, Rotenberg A, Sahin M, Jensen F, Kwiatkowski D: A mouse model of tuberous sclerosis: neuronal loss of Tsc1 causes dysplastic and ectopic neurons, reduced myelination, seizure activity, and limited survival. *J Neurosci* 2007, **27**(21):5546–5558.
100. Meikle L, Pollizzi K, Egnor A, Kramvis I, Lane H, Sahin M, Kwiatkowski D: Response of a neuronal model of tuberous sclerosis to mammalian target of rapamycin (mTOR) inhibitors: effects on mTORC1 and Akt signaling lead to improved survival and function. *J Neurosci* 2008, **28**(21):5422–5432.
101. Tavazoie S, Alvarez V, Ridenour D, Kwiatkowski D, Sabatini B: Regulation of neuronal morphology and function by the tumor suppressors Tsc1 and Tsc2. *Nat Neurosci* 2005, **8**(12):1727–1734.
102. Auerbach B, Osterweil E, Bear M: Mutations causing syndromic autism define an axis of synaptic pathophysiology. *Nature* 2011, **480**(7375):63–68.
103. Ehninger D, Han S, Shilyansky C, Zhou Y, Li W, Kwiatkowski D, Ramesh V, Silva A: Reversal of learning deficits in a $Tsc2^{+/-}$ mouse model of tuberous sclerosis. *Nat Med* 2008, **14**(8):843–848.
104. Guy J, Gan J, Selfridge J, Cobb S, Bird A: Reversal of neurological defects in a mouse model of Rett syndrome. *Science* 2007, **315**(5815):1143–1147.
105. Berry-Kravis E, Hessl D, Coffey S, Hervey C, Schneider A, Yuhas J, Hutchison J, Snape M, Tranfaglia M, Nguyen DV *et al.*: A pilot open label, single dose trial of fenobam in adults with fragile X syndrome. *J Med Genet* 2009, **46**(4):266–271.
106. Jacquemont S, Curie A, des Portes V, Torrioli MG, Berry-Kravis E, Hagerman RJ, Ramos FJ, Cornish K, He Y, Paulding C *et al.*: Epigenetic modification of the FMR1 gene in fragile X syndrome is associated with differential response to the mGluR5 antagonist AFQ056. *Sci Transl Med* 2011, **3**(64):64ra61.
107. Suzuki K, Sugihara G, Ouchi Y, Nakamura K, Futatsubashi M, Takebayashi K, Yoshihara Y, Omata K, Matsumoto K, Tsuchiya KJ *et al.*: Microglial activation in young adults with autism spectrum disorder. *JAMA Psychiatry* 2013, **70**(1):49–58.

108. Voineagu I, Wang X, Johnston P, Lowe J, Tian Y, Horvath S, Mill J, Cantor R, Blencowe B, Geschwind D: Transcriptomic analysis of autistic brain reveals convergent molecular pathology. *Nature* 2011, **474**(7351):380–384.
109. Derecki NC, Cronk JC, Lu Z, Xu E, Abbott SB, Guyenet PG, Kipnis J: Wild-type microglia arrest pathology in a mouse model of Rett syndrome. *Nature* 2012, **484**(7392):105–109.
110. Sakai Y, Shaw CA, Dawson BC, Dugas DV, Al-Mohtaseb Z, Hill DE, Zoghbi HY: Protein interactome reveals converging molecular pathways among autism disorders. *Sci Transl Med* 2011, **3**(86):86ra49.
111. Ascano M, Mukherjee N, Bandaru P, Miller J, Nusbaum J, Corcoran D, Langlois C, Munschauer M, Dewell S, Hafner M *et al.*: FMRP targets distinct mRNA sequence elements to regulate protein expression. *Nature* 2012, **492**(7429):382–386.
112. Baudouin SJ, Gaudias J, Gerharz S, Hatstatt L, Zhou K, Punnakkal P, Tanaka KF, Spooren W, Hen R, De Zeeuw CI *et al.*: Shared synaptic pathophysiology in syndromic and nonsyndromic rodent models of autism. *Science* 2012, **338**(6103):128–132.
113. Peters JM, Taquet M, Vega C, Jeste SS, Fernandez IS, Tan J, Nelson CA, 3rd, Sahin M, Warfield SK: Brain functional networks in syndromic and non-syndromic autism: a graph theoretical study of EEG connectivity. *BMC Med* 2013, **11**:54.

Chapter 3

Language Impairment in Autism Spectrum Disorders

Natalia Rakhlin and Elena L. Grigorenko*

Abstract

Studying shared functional dimensions of complex, neurodevelopmental phenotypes may be useful to developing an understanding of their shared etiologies. We apply this dimensional approach to the study of autism spectrum disorders (ASD) and language disorders (LD), and summarize the research to date that relates language impairment with ASD. Although the current picture is still mixed, this approach has the potential to shed light on the biological correlates and/or causes of ASD, and thus contribute to the development of effective clinical interventions.

Keywords: Dimensional classification of developmental disorders; Heterogeneity; Autism spectrum disorder; Developmental language disorder; Quantitative and Molecular genetics.

Introduction

In recent years, the approach to neurodevelopmental disorders, following a similar movement in general psychiatry, has started to shift from a categorical one, exemplified by the classification systems of the DSM-IV and the ICD-10, to a dimensional one. The recognized shortcomings of categorical classifications are related

*Correspondence to: Child Study Center, Yale University, 230 South Frontage Road, New Haven, CT 06519-1124. Email: elena.grigorenko@yale.edu

to their questionable validity as a nosology based on established clusters of clinical symptoms, rather than on known causal mechanisms, and their consequent limited utility for identifying the most effective treatments.[1,2] The new dimensional approach is aimed at incorporating multiple levels of analysis, derived from genetics, neuroscience, and the cognitive and behavioral sciences, to characterize the individual dimensions of function across these multiple levels of analysis, thus linking behavioral symptoms with disruptions of specific cognitive processes, captured at the level of neuroanatomy and/or neurophysiology and the structure and/or function of the genome. The dimensional approach assumes that traits observed in the phenotype of a complex disorder are separable from each other, such that individual traits that are part of the phenotypes of multiple disorders account for observed comorbidity.[3] This approach is motivated by the realization of the inherent phenotypic and genetic heterogeneity of complex disorders and the non-linearity of the mappings between genes and behavior, which render categorical approaches ineffective. Understanding the biological mechanisms underlying such disorders may require dissecting phenotypes into theoretically motivated dimensional constructs[4] and studying these constructs in cross-syndrome cohorts, phenotypically homogeneous with respect to the trait in question, rather than heterogeneous cohorts who share the same classification label.

Language research in autism spectrum disorders (ASD) is a topic particularly well-suited to studies using the dimensional approach, as (1) there is a great heterogeneity of phenotypes observed among individuals with ASD; and (2) there is another clinical category, namely Language Disorder (LD), although currently mutually exclusive with ASD for clinical diagnostic purposes, with a phenotype (specifically, impairment in language acquisition and functioning) that has long been thought to be related to ASD. Although the original description of autism did not include language impairments,[5] some early researchers conceptualized ASD and LD as forms of the same cognitive liability of different severity.[6] Subsequently, substantive differences with respect to language characteristics between children with ASD and LD have been acknowledged. First, the most consistently found type of language impairment in ASD (observed in children of all levels of functioning) concerns the use of language in social situations (i.e., pragmatic aspects of communication) rather than structural aspects of language, a reverse of the pattern found in LD, even though structural language may be also impaired in individuals with ASD and pragmatics in individuals with LD. In addition, there are other ASD-specific language-related features, e.g., "autistic regression" or commonly observed loss of communication skills in the second year of life following seemingly normal development,[7,8] aberrant vocal phenomena, such as echolalia, and semantic abnormalities, such as neologisms and pronoun reversal.

Despite these differences, a number of studies showed multiple similarities between language profiles of children with LD and some children with ASD, e.g., delays in reaching early language milestones, morphosyntactic errors (such as omission of verb tense markers), and difficulties with repeating pseudowords. These similarities were interpreted as evidence for a subtype of ASD, i.e., ASD with language impairment, the so-called ALI,[9] or that ASD should be viewed as "LD Plus".[10] On the other hand, it was also proposed that ASD and LD may overlap only superficially, even if the two do not share etiology.[11] The issue of how the two diagnostic categories, ASD and LD, are phenomenologically related, however, has become less important, as the field has developed an understanding of the multifactorial/dimensional nature of complex disorders, such as ASD and LD, and shifted to studying these dimensions at various levels with an assumption that either one or multiple dimensions may be affected in any given individual.

With respect to ASD, it had been proposed that the classic autistic behavioral traits are separable ("fractionable"), such that each one can be studied independently at the levels of behavioral symptoms, their cognitive underpinnings and etiology.[4] Understanding language impairment as a behavioral dimension of function relevant to both ASD and LD while differentiating primary language impairments from those secondary to social impairment is a way of discovering productive endophenotypes (intermediate phenotypes). Such discoveries would help overcome the challenges presented by the behavioral and genetic heterogeneity of ASD and aid the detection of the genetic variants associated with the specific facets of these conditions.

The partial overlap of symptoms in ASD and LD was argued to support the dimensional approach based on the following lines of evidence: (1) the partial overlap in language phenotypes among individuals with ASD and LD; (2) increased risk of language impairment and of subclinical language and language-related deficits in parents and siblings of probands with ASD, as well as increased risk of ASD-like symptoms in relatives of children with LD; and (3) molecular-genetic findings of shared genetic risk factors for ASD and LD. This review is aimed at summarizing the major findings in each of these areas of research to present an overview of the current understanding of the relationship between language impairment and the broad ASD phenotype. A deeper understanding of this relationship has important implications for the search for biological correlates/causes of ASD and the design of corresponding clinical interventions.

Language Impairments in ASD

Sources of phenotypic heterogeneity

A large majority of studies of language functioning in ASD have focused on the product of language acquisition, i.e., the level of phonological, lexical, grammatical or pragmatic development in groups of children with ASD, reported to be highly variable in all of the domains, but one — consistently impaired pragmatics or social use of language.[12] Relatively few studies have examined the processes or mechanisms of language acquisition in children with ASD. A considerable variability in the levels of functioning found among individuals with ASD presents a challenge for such research.

A substantial proportion of individuals with ASD (~25%–50%) are functionally nonverbal. Another large sub-group consists of individuals with no clinical levels of language abnormalities (previously identified as high functioning autism or Asperger's syndrome). The remaining group displays highly variable language functioning, with the severity of symptoms often correlating with the degree of intellectual disability.[13] Consequently, literature contains divergent findings with respect to the linguistic profiles of individuals with ASD, depending on whether the study samples from "language normal" (who at best typically have only subtle deficits in formal aspects of language, such as phonology or grammar), "language impaired" (typically displaying structural language deficits of varying magnitudes), or mixed (unselected) language ability individuals.[14] This contributes to the heterogeneity of findings reported in the literature.[11]

Another source of heterogeneity of language profiles in ASD is the dynamics of language development within the context of broader ASD symptomatology.[15] In other words, there is substantial variability in the magnitude and quality of language deficits in general (and in structural language in particular) depending on the severity of the ASD symptomatology, as well as the age of the participants in the study sample. It appears that a substantial number (up to 63%)[15] of children with ASD under the age of six exhibit mixed receptive-expressive LD, including deficits in grammar and speech sound production, with a certain number showing subsequent reduction in speech-language deficits to subclinical levels over time, and this reduction is conditioned on the overall level of functioning.[16]

In addition, different trajectories of language development are observed in children with ASD who experience so-called "autistic regression." Whereas children with a subsequent diagnosis of ASD are consistently reported to exhibit language delays,[17] in a substantial number of cases (15%–49%), instead of a delay, a rather abrupt loss of previously normally acquired skills (words and phrased speech) is observed, typically between 14 and 24 months of age.[18] This phenomenon, the cause of which remains unknown, is highly specific to ASD. Its consequences for subsequent language acquisition diverge depending on whether the regression takes place prior to or after the emergence of phrased speech. Thus, language and IQ outcomes for those children who experienced regression after gaining phrased speech appear poor, while those who experienced regression prior to the emergence of phrased speech generally do not differ from children with ASD who did not experience regression.[19]

The seeming paradox of both language delays and language loss being part of the clinical picture of ASD may be explained if one views language delays and language loss as two separate lines of developmental vulnerability. Autistic regression may be a much more wide-spread phenomenon than it appears because in a great number of cases it is masked by language delay, as children who experience substantial language delay have no overt language during the second year of life for the regression to be detectable.[20] This view is consistent with language delays and autistic regression being related to distinct etiological factors, with the former playing a role in both ASD and LD, with the latter being specific to ASD and related to the broad ASD phenotype.

Types of language deficits in ASD

Despite the variability of language profiles in individuals with ASD, certain consistent findings have emerged. The most consistent impairment associated with an ASD diagnosis is in social use of language (pragmatics); however, deficits in structural language (vocabulary and grammar) and speech sound production are also found, with frank deficits in these areas typically found in lower functioning children, and only subtle deficits of this type found in high functioning children.

Impaired social communication, i.e., pragmatics (broadly defined), is observed even among individuals with otherwise adequate language. Thus, children with ASD have difficulty initiating conversations, following conversational conventions, such as reciprocity (turn-taking), topic maintenance (responding to the topic introduced into the conversation by adding relevant information and introducing new relevant topics), and information management (providing the appropriate amount of information in a conversation depending on the listener's needs).[21] They have also been shown to underperform on narrative tasks,[22] and to have difficulty comprehending figurative language.[23] Their language has been generally described as pedantic, overly formal, and overly literal. These characteristics are among the most consistent behavioral markers of ASD.

However, pragmatic difficulties are not specific to ASD and are found in children with LD.[24,25] Furthermore, a primary pragmatic disorder, a category added to the DSM-5 (Social Communication Disorder), includes children with similar social communication difficulties as children with ASD, but without the presence of other ASD symptoms, i.e., stereotypical behaviors and difficulties in social interaction that go beyond verbal communication.[26,27] Thus, although pragmatic difficulties are most closely associated with ASD, and likely related to the broader social impairment in these children, they can also occur separately from other autistic traits. With respect to clinically significant structural language impairments (in the lexicon and grammar, i.e., phonological, morphological, syntactic, and semantic structure), the literature is mixed, even though recent informal reviews maintain that such deficits are widespread.[12,14,28] Below, we will briefly review major findings on lexical, morphosyntactic and phonological skills.

With respect to vocabulary, like children with LD, many children with ASD present with marked delays in the onset and acquisition of vocabulary.[29,30] Moreover, delayed expressive and receptive language development is one of the most sensitive early identification markers[31] and accurate predictors of subsequent adaptive behavior.[32,33] Some studies have reported that children with ASD, unlike children with LD, in general do not underperform on standardized vocabulary tests.[34] However, children with ASD with syntactic deficits, similar to peers with LD, were shown to have sparse lexicons characterized by partial word knowledge and immature knowledge of word-to-word relationships.[35] A similar behavioral overlap between a subgroup of children with ASD (those with concomitant grammatical deficits) and children with LD was demonstrated in a study that looked at comprehension of noun-noun compounds. In particular, children with poorer vocabularies who performed more poorly on non-word repetition tasks had greater difficulties regardless of their specific diagnoses.[36] This overlap points to the syntax-lexicon interface as a dimension of language that can be part of both ASD and LD phenotypes.

However, specifics of this overlap or the precise characterization of lexical semantic development of children with ASD are challenging to capture in detail. Thus, it has been shown that children with ASD, similar to TD children, exhibit the noun bias,[37] can represent meaning of both concrete and abstract nouns,[38] and show no evidence of errors in the lexical semantics of terms for natural categories, artifacts, and body parts,[39] all suggesting that lexical acquisition mechanisms are generally intact in ASD. However, abnormalities in lexical processing have also been observed, including deficits in lexical and/or category organization.[40–43] Of note is that these deficits are highly varied, including lack of the shape bias (i.e., the default assumption that a label for a novel object refers to its shape) at 24-months of age,[40] a deficit in the semantic processing of action verbs among adults with ASD,[44] an abnormal neurophysiological response to semantically incongruent words during an event-related potential task,[45] and comprehension of terms referring to emotional states.[46] Moreover, rather than using the typical "Speaker's Direction of Gaze" strategy when selecting a referent for a novel word (i.e., assuming that the referent for the novel word corresponds to the object the speaker is looking at when uttering the novel word), children with ASD used their own focus of attention, thus frequently mislabeling objects.[47] Children with ASD have been shown to have lower sensitivity to the social information encoded in gaze cues using eye-tracking.[48,49] This suggests that a social impairment (e.g., in social orientation, joint attention) may prevent children with ASD from attending to social cues in word learning, which may interfere with the acquisition of the meanings of words.[50] Yet, children with ASD have been shown to have an intact ability to use syntactic cues in word learning[51] and to exhibit normal neural responses to auditory stimuli, such as tones and speech sounds,[52] in contrast to children with LD who exhibit abnormality in these domains.[53] These contrasts suggest two different pathways leading to lexical deficits in the two disorders and a possibility that either one or both pathways may be impaired in a given child.

With respect to syntactic and morphosyntactic skills, the most prominent feature of language impairment in LD, findings in ASD have been mixed. Some studies reported no morphosyntactic errors in spontaneous speech,[54] while others reported high rates of omissions of tense marking in a language-impaired subgroup of children with ASD[55] and reduced expressive and receptive grammatical indicators.[34] High-functioning children with ASD matched on IQ and receptive vocabulary with TD age-peers have been shown to have a lower mean utterance length in free play sessions and reduced sensitivity to grammaticality violations on tense and aspect verb marking, although the reported effects were small.[56] Thus, it appears that while pronounced syntactic and morphosyntactic deficits are not an emblematic feature of the ASD phenotype, syntactically sparse expressive language, possibly reflecting slower language development, may be a wide-spread feature, with some individuals with ASD showing grammatical deficits not unlike those found in children with LD.

Finally, phonology, particularly expressive phonology, has been reported as intact in ASD in some studies,[15,34] but found to contain at least subtle abnormalities in others.[57,58] A study that used a cluster analysis approach identified four distinct linguistic profiles in 7–9-year-old children with an ASD diagnosis,[59] in two of which phonology was impaired and in two of which it was not (with or without co-occurring comprehension deficits). Thus, it appears that phonological deficits (from mild to profound), which are frequently reported for young children with LD,[60] similarly to grammar impairments, are a relatively common, but not universal feature of the clinical picture of ASD. There is some evidence that phonological deficits tend to cluster with grammatical deficits, while semantic deficits typically co-occur with pragmatic ones.[12]

Language skills and the diagnosis of ASD

It has been established that a child's language skills and the degree of language delay relative to age expectations are highly informative for the diagnosis, course, and treatment of ASD,[61] and that specific diagnostic features are closely related to the level of language skills.[62] However, language delay is not part of the latest ASD primary diagnostic criteria because, as discussed above, it is not always present nor is it specific to ASD. The adjusted definition of autism in the DSM-5 reflects the ambivalence of the field with respect to the language dimension as a domain inherently affected by the disorder. Previously, autism was defined as a triad of core impairments, with impaired communication being one of the triad, along with reciprocal social interaction and restricted repetitive behavior. In the DSM-5, only the latter two domains have been retained, with social communication problems incorporated into the domain of social interaction and a specification of language impairment that may be added. On the other hand, it is clear that language impairment is an integral part of ASD, as the severity table provided in the DSM-5 incorporates degrees of language impairment, such that an instance of the highest level of ASD severity (3) is described as "a person with few words of intelligible speech"; a Level 2 as "a person who speaks simple sentences"; and Level 1 as a person "who is able to speak in full sentences and engages in communication."

Family Studies of ASD and LD Overlap

Family studies of communication difficulties in relatives of probands with ASD and LD strongly suggest a genetic basis of the language deficits in these disorders. One strategy to find the shared etiology of LD and ASD has been to examine close relatives of probands with one diagnosis for the traits associated with the other diagnosis. Such studies demonstrated higher than expected by chance incidence of language and language-related learning deficits, such as in reading and spelling, in family members of probands with ASD.[63–65] Also, a significant number of occurrences of subclinical levels of autistic-like features, including communication and language deficits, was reported in first-degree relatives of probands with ASD, the so-called broad autism phenotype or BAP.[66,67]

Pragmatics is the most consistently reported language-related component of BAP, with parents of children with ASD scoring lower than controls.[68] Studies that specifically targeted structural language typically compared the language skills of parents of children with ASD with those of parents of children with LD. All reported a significantly lower performance in the latter group and interpreted these findings as evidence for a non-shared genetic etiology between ASD and LD.[69] A recent study did not follow such a comparative method, but instead examined the frequency of language impairments in extended families that included at least one member diagnosed with LD and one diagnosed with ASD.[70] The results indicated that heritability was higher after controlling for ASD on (1) overall language ability and (2) pragmatic skills. These results were interpreted as inconsistent with a simple additive-genetics model of heritability, but rather as consistent with non-additive genetic effects contributing to ASD; that is, they support a gene-gene interaction model, which would account for a shared genetic etiology of ASD and LD, while still allowing for the phenotypic differences that give rise to the different diagnoses.

Molecular Genetic Studies of ASD/LD Overlap

A number of molecular genetic studies have found susceptibility markers for ASD and LD, as well as contributions of *de novo* and inherited copy-number variation (CVNs) to ASD.[71] These studies strongly suggest that both disorders are complex and polygenic. In the search for the pathways from genes to phenotypes, an important question has been whether distinct aspects of ASDs have independent genetic bases or, alternatively, whether they are associated with largely overlapping risk factors. In support of the independent heritability of ASD-related domains of impairment, linkage studies have identified distinct loci associated with endophenotypes related to linguistic and non-linguistic impairments.[72] There is also some evidence for the etiological overlap of ASD and LD in specific families of genes (e.g., semaphorin genes[73–75]) or genomic locations (e.g., 17p11[76,77]).

Moreover, a plausible functional link between ASD and LD has been proposed. First, a series of papers identified *CNTNAP2* as a gene associated with an increased risk of familial ASD.[78–80] Subsequently, this gene was reported to be associated with LD,[81] suggesting a genetic mechanism of the phenotypic overlap between the two disorders. *CNTNAP2,* expressed in the developing human cortex and known to encode a neurexin, a presynaptic protein involved in neuronal adhesion, has also been found to be associated with communication ability at age 2 in the general population.[82] Furthermore, *CNTNAP2* is down-regulated by *FOXP2,* a gene previously linked to a rare monogenic communication disorder in an extended family.[83] It is also intriguing that the phenotype used in the Vernes and colleagues' study[81] was the repetition of nonsense words, a clinical marker of LD used in other studies that have identified other candidate genes for LD, and a skill shown (albeit inconsistently) to be impaired in at least subgroups of children with ASD.[34]

Conclusions

There are two distinct domains in language functioning identified as part of the communicative impairments found in ASD and LD: (1) structural language, which includes lexicon and grammar (phonology or sound patterns, morphology or word formation patterns, syntax, and compositional semantics) and (2) pragmatics. While LD is most prominently associated with deficits in the first, ASD is associated with deficits in the second domain, although both diagnoses include children with deficits in both domains. In addition, speech sound production (articulation) can be impaired in children with LD and subgroups of children with ASD, frequently overlapping with structural language impairments in the latter.

The pattern of behavioral and genetic findings suggests that the heterogeneous profiles of language deficits in children with ASD may stem from two distinct etiological sources. One is related to broader deficits in social cognition, leading to a preponderance of pragmatic deficits in ASD and related to autism-specific lexical-sematic deficits due to the child's weakness in following social cues. The other, present only in some individuals with ASD, is an additional source of vulnerability that may be directly responsible for deficits in structural language. Thus, some communication impairments in ASD, such as idiosyncratic word usage, impaired comprehension, pragmatic deficits and general difficulties with the social use of language, are likely an outcome of certain specific factors leading to emergence of the autism-typical traits (i.e., social impairment). Others, such as deficits in formal aspects of language, on the other hand, likely result from factors that generally impair receptive and expressive language in the cases of both ASD and LD. These factors may specifically affect language development or may be "generalist" factors, exerting a generally impairing effect on child development. In addition, there may be sources of vulnerability affecting development that are not specifically linked with social or language deficits, e.g., leading to a general language and cognitive delay found in both ASD and LD. Thus, the genetic model of both disorders is undoubtedly very complex, as it involves multiple genes, gene-by-gene and, likely, gene-by-environment interactions in determining disorder phenotypes. To complicate the picture further, genes may act pleiotropically leading to deficits in a number of domains.

An additional source of phenotypic complexity comes from the interactive nature of language domains, which, although often viewed as autonomous cognitive systems with distinct supporting neuro-circuits, can influence each other functionally (i.e., impairment in one may affect the acquisition of skills and/or a diminished performance in another). Thus, in a child with socially based communication impairment, especially of greater severity, the acquisition of skills in other language domains may be compromised; conversely, in a child with LD whose primary deficit is in structural language, pragmatic skills may be compromised leading to cases of phenotypic similarities without (or in addition to) meaningful etiological overlap.

Future Directions

The task of dissecting the heterogeneity in both the manifestation and etiology of ASD and LD in particular and developmental disorders in general remains a primary task of ASD research today. The facet of the field represented in this review covers suggestive evidence for and against a shared etiology of ASD and LD. To resolve these contradictions in the literature, the field has to engage in more studies where samples are constructed not based on diagnostic categories, but based on dimensional deficiencies. Although uncommon today, such studies might be extremely helpful in disentangling the mass of contradictions in which the field currently appears to be immersed.

Executive Summary

- Today's literature on ASD is voluminous and impressive; yet, there are plenty of problems with its replicability, at least partially due to the essential heterogeneity of ASD.
- One important dimension of the clinical presentation of ASD, also a source of its heterogeneity, is the level of language functioning.
- In itself, language impairment (whether an independent clinical condition or a facet of another developmental disorder, such as ASD) is a heterogeneous outcome of many underlying (a)typical developmental issues that may be manifested individually or collectively.
- When dimensions of the behavioral manifestation of ASD and LD are considered together, there is evidence of an overlap between the disorders both in terms of their psychological characteristics and in terms of their underlying etiological mechanisms.

Understanding the dimensional structure of ASD and exploring the etiology of these dimensions across multiple developmental disorders is an important task for the future.

Acknowledgements

This report was supported by NIH DC7665 and MH81756, but does not necessarily represent the positions of the NIH as grantees are encouraged to express their professional judgment. We thank Mei Tan of Yale University for her editorial help.

References

1. Levy Y, Ebstein RP: Research review: crossing syndrome boundaries in the search for brain endophenotypes. *J Child Psychol Psyc* 2009, **50**(6): 657–668.
2. Insel T: Transforming Diagnosis. In: *Director's Blog*. National Institute of Mental Health; 2013.
3. Grigorenko EL: Rethinking disorders of spoken and written language: generating workable hypotheses. *J Dev Behav Pediatr* 2007, **28**(6):478–486.
4. Happe F, Ronald A: The 'fractionable autism triad': A review of evidence from behavioural, genetic, cognitive and neural research. *Neuropsychol Rev* 2008, **18**(4):287–304.
5. Kanner L, Eisenberg L: Early infantile autism, 1943–1955. *Psychiatric Res Report* 1957, (7):55–65.
6. Cantwell D, Baker L, Rutter M: A comparative study of infantile autism and specific developmental receptive language disorder–IV. Analysis of syntax and language function. *J Child Psychol Psyc* 1978, **19**(4):351–362.
7. Goldberg WA, Osann K, Filipek PA, Laulhere T, Jarvis K, Modahl C, Flodman P, Spence MA: Language and other regression: assessment and timing. *J Autism Dev Disord* 2003, **33**(6):607–616.
8. Lord C, Shulman C, DiLavore P: Regression and word loss in autistic spectrum disorders. *J Child Psychol Psyc* 2004, **45**(5):936–955.
9. Tager-Flusberg H: Defining language phenotypes in autism. *Clin Neurosci Res* 2006, **6**(3–4):219–224.
10. Bishop DVM: Overlaps between autism and language impairment: Phenomimicry or shared etiology? *Behav Genet* 2010, **40**(5):618–629.
11. Tomblin B: Co-morbidity of autism and SLI: kinds, kin and complexity. *Int J Lang Commun Disord* 2011, **46**(2):127–137.

12. Groen WB, Zwiers MP, van der Gaag RJ, Buitelaar JK: The phenotype and neural correlates of language in autism: an integrative review. *Neurosci Biobehav R* 2008, **32**(8):1416–1425.
13. Luyster R, Kadlec MB, Carter A, Tager-Flusberg H: Language assessment and development in toddlers with autism spectrum disorders. *J Autism Dev Disord* 2008, **38**(8):1426–1438.
14. Boucher J: Research review: structural language in autistic spectrum disorder — characteristics and causes. *J Child Psychol Psyc* 2012, **53**(3):219–233.
15. Rapin I, Dunn M: Update on the language disorders of individuals on the autistic spectrum. *Brain Dev-Jpn* 2003, **25**(3):166–172.
16. Tek S, Mesite L, Fein D, Naigles L: Longitudinal analyses of expressive language development reveal two distinct language profiles among young children with Autism Spectrum Disorders. *J Autism Dev Disord* 2013.
17. Lord C, Paul R: Language and communication in autism. In: *Handbook of Autism and Pervasive Developmental Disorders.* Edited by Cohen DJ, Volkmar FR, 2 edn. New York: Wiley; 1997: 195–225.
18. Bernabei P, Cerquiglini A, Cortesi F, D'Ardia C: Regression versus no regression in the autistic disorder: developmental trajectories. *J Autism Dev Disord* 2007, **37**(3):580–588.
19. Pickles A, Simonoff E, Conti-Ramsden G, Falcaro M, Simkin Z, Charman T, Chandler S, Loucas T, Baird G: Loss of language in early development of autism and specific language impairment. *J Child Psychol Psyc* 2009, **50**(7):843–852.
20. Pickles A, Simonoff E, Conti-Ramsden G, Falcaro M, Simkin Z, Charman T, Chandler S, Loucas T, Baird G: Loss of language in early development of autism and specific language impairment. *J Child Psychol Psyc* 2009, **50**: 843–852.
21. Paul R, Orlovski SM, Marcinko HC, Volkmar F: Conversational behaviors in youth with high-functioning ASD and asperger syndrome. *J Autism Dev Disord* 2009, **39**(1):115–125.
22. Manolitsi M, Botting N: Language abilities in children with autism and language impairment: using narrative as a additional source of clinical information. *Child Lang Teach The* 2011, **27**(1):39–55.
23. Happe F: Communicative competence and theory of mind in autism: a test of relevance theory. *Cognition* 1993, **48**(2):101–119.
24. Botting N, Conti-Ramsden G: The role of language, social cognition, and social skill in the functional social outcomes of young adolescents with and without a history of SLI. *Brit J Dev Psychol* 2008, **26**:281–300.
25. Rakhlin N, Kornilov S, Palejev D, Koposov R, Chang J, Grigorenko EL: The language phenotype of a small grographically isolated Russian-speaking population: implications for genetic and clinical studies of Developmental Language Disorder. *Appl Psycholinguist* 2012, advanced on-line publication.

26. Gibson J, Adams C, Lockton E, Green J: Social communication disorder outside autism? A diagnostic classification approach to delineating pragmatic language impairment, high functioning autism and specific language impairment. *J Child Psychol Psyc* 2013, **54**(11):1186–1197.
27. Botting N, Conti-Ramsden G: Autism, primary pragmatic difficulties, and specific language impairment: can we distinguish them using psycholinguistic markers? *Dev Med and Child Neurol* 2003, **45**(8):515–524.
28. Eigsti IM, de Marchena AB, Schuh JM, Kelley E: Language acquisition in autism spectrum disorders: a developmental review. *Res Autism Spect Dis* 2011, **5**(2):681–691.
29. Luyster R, Lopez K, Lord C: Characterizing communicative development in children referred for Autism Spectrum Disorders using the MacArthur-Bates Communicative Development Inventory (CDI). *J Child Lang* 2007, **34**(3): 623–654.
30. Charman T, Drew A, Baird C, Baird G: Measuring early language development in preschool children with autism spectrum disorder using the MacArthur Communicative Development Inventory (Infant Form). *J Child Lang* 2003, **30**(1):213–236.
31. Zwaigenbaum L, Bryson S, Rogers T, Roberts W, Brian J, Szatmari P: Behavioral manifestations of autism in the first year of life. *Int J Dev Neurosci* 2005, **23**(2–3):143–152.
32. Szatmari P, Bryson SE, Boyle MH, Streiner DL, Duku E: Predictors of outcome among high functioning children with autism and Asperger syndrome. *J Child Psychol Psychiatry* 2003, **44**(4):520–528.
33. Rogers SJ, Dilalla DL: Age of symptom onset in young-children with Pervasive Developmental Disorders. *J Am Acad Child Psy* 1990, **29**(6): 863–872.
34. Kjelgaard MM, Tager-Flusberg H: An investigation of language impairment in autism: implications for genetic subgroups. *Lang Cognitive Proc* 2001, **16**(2–3):287–308.
35. McGregor KK, Berns AJ, Owen AJ, Michels SA, Duff D, Bahnsen AJ, Lloyd M: Associations between syntax and the lexicon among children with or without ASD and language impairment. *J Autism Dev Disord* 2012, **42**(1):35–47.
36. Riches NG, Loucas T, Baird G, Charman T, Simonoff E: Interpretation of compound nouns by adolescents with specific language impairment and autism spectrum disorders: an investigation of phenotypic overlap. *Int J Speech-Lang Pa* 2012, **14**(4):307–317.
37. Swensen LD, Kelley E, Fein D, Naigles LR: Processes of language acquisition in children with autism: evidence from preferential looking. *Child Dev* 2007, **78**(2):542–557.
38. Eskes GA, Bryson SE, McCormick TA: Comprehension of concrete and abstract words in autistic children. *J Autism Dev Disord* 1990, **20**(1):61–73.

39. Perkins MR, Dobbinson S, Boucher J, Bol S, Bloom P: Lexical knowledge and lexical use in autism. *J Autism Dev Disord* 2006, **36**(6):795–805.
40. Tek S, Jaffery G, Fein D, Naigles LR: Do children with Autism Spectrum Disorders show a shape bias in word learning? *Autism Res* 2008, **1**(4):208–222.
41. Dunn MA, Bates JC: Developmental change in neutral processing of words by children with autism. *J Autism Dev Disord* 2005, **35**(3):361–376.
42. Church BA, Krauss MS, Lopata C, Toomey JA, Thomeer ML, Coutinho MV, Volker MA, Mercado E: Atypical categorization in children with high-functioning autism spectrum disorder. *Psychon B Rev* 2010, **17**(6):862–868.
43. Kamio Y, Robins D, Kelley E, Swainson B, Fein D: Atypical lexical/semantic processing in high-functioning autism spectrum disorders without early language delay. *J Autism Dev Disord* 2007, **37**(6):1116–1122.
44. Moseley RL, Mohr B, Lombardo MV, Baron-Cohen S, Hauk O, Pulvermuller F: Brain and behavioral correlates of action semantic deficits in autism. *Front Hum Neurosci* 2013, **7**:725.
45. McCleery JP, Ceponiene R, Burner KM, Townsend J, Kinnear M, Schreibman L: Neural correlates of verbal and nonverbal semantic integration in children with autism spectrum disorders. *J Child Psychol Psyc* 2010, **51**(3):277–286.
46. Tager-Flusberg H, Sullivan K: Attributing mental states to story characters — a comparison of narratives produced by autistic and mentally-retarded individuals. *Appl Psycholinguist* 1995, **16**(3):241–256.
47. Walton KM, Ingersoll BR: Expressive and receptive fast-mapping in children with autism spectrum disorders and typical development: the influence of orienting cues. *Res Autism Spect Dis* 2013, **7**(6):687–698.
48. Norbury CF, Griffiths H, Nation K: Sound before meaning: word learning in autistic disorders. *Neuropsychologia* 2010, **48**(14):4012–4019.
49. McDuffie AS, Yoder PJ, Stone WL: Fast-mapping in young children with autism spectrum disorders. *First Language* 2006, **26**:421–438.
50. Charman T: Why is joint attention a pivotal skill in autism? *Philos T Roy Soc B* 2003, **358**(1430):315–324.
51. Shulman C, Guberman A: Acquisition of verb meaning through syntactic cues: a comparison of children with autism, children with specific language impairment (SLI) and children with typical language development (TLD). *J Child Lang* 2007, **34**(2):411–423.
52. Ceponiene R, Lepisto T, Shestakova A, Vanhala R, Alku P, Naatanen R, Yaguchi K: Speech-sound-selective auditory impairment in children with autism: they can perceive but do not attend. *Proc Natl Acad Sci USA* 2003, **100**(9):5567–5572.
53. Evans JL, Selinger C, Pollak SD: P300 as a measure of processing capacity in auditory and visual domains in specific language impairment. *Brain Res* 2011, **1389**:93–102.
54. Eigsti IM, Bennetto L, Dadlani MB: Beyond pragmatics: Morphosyntactic development in autism. *J Autism Dev Disord* 2007, **37**(6):1007–1023.

55. Roberts JA, Rice ML, Tager-Flusberg H: Tense marking in children with autism. *Appl Psycholinguist* 2004, **25**(3):429–448.
56. Eigsti IM, Bennetto L: Grammaticality judgments in autism: deviance or delay. *J Child Lang* 2009, **36**(5):999–1021.
57. Cleland J, Gibbon FE, Peppe SJE, O'Hare A, Rutherford M: Phonetic and phonological errors in children with high functioning autism and Asperger syndrome. *Int J Speech-Lang Pa* 2010, **12**(1):69–76.
58. Paul R, Bianchi N, Augustyn A, Klin A, Volkmar FR: Production of syllable stress in speakers with autism spectrum disorders. *Res Autism Spect Dis* 2008, **2**(1):110–124.
59. Rapin I, Dunn MA, Allen DA, Stevens MC, Fein D: Subtypes of Language Disorders in School-Age Children With Autism. *Dev Neuropsychol* 2009, **34**(1):66–84.
60. Roberts JA, Rescorla L, Giroux J, Stevens L: Phonological skills of children with specific expressive language impairment (SLI-E): Outcome at age 3. *J Speech Lang Hear R* 1998, **41**(2):374–384.
61. Anderson DK, Maye MP, Lord C: Changes in maladaptive behaviors from midchildhood to young adulthood in Autism Spectrum Disorder. *Am J Intellect Dev Disabil* 2011, **116**(5):381–397.
62. Gotham K, Risi S, Pickles A, Lord C: The autism diagnostic observation schedule: revised algorithms for improved diagnostic validity. *J Autism Dev Disord* 2007, **37**(4):613–627.
63. Fombonne E, Bolton P, Prior J, Jordan H, Rutter M: A family study of autism: cognitive patterns and levels in parents and siblings. *J Child Psychol Psyc* 1997, **38**(6):667–683.
64. Piven J, Palmer P: Cognitive deficits in parents from multiple-incidence autism families. *J Child Psychol Psyc* 1997, **38**(8):1011–1021.
65. Bolton P, Macdonald H, Pickles A, Rios P, Goode S, Crowson M, Bailey A, Rutter M: A case-control family history study of autism. *J Child Psychol Psyc* 1994, **35**(5):877–900.
66. Sasson NJ, Lam KS, Parlier M, Daniels JL, Piven J: Autism and the broad autism phenotype: familial patterns and intergenerational transmission. *J Neurodev Disord* 2013, **5**(1):11.
67. Bernier R, Gerdts J, Munson J, Dawson G, Estes A: Evidence for broader autism phenotype characteristics in parents from multiple-incidence autism families. *Autism Res* 2012, **5**(1):13–20.
68. Whitehouse AJ, Coon H, Miller J, Salisbury B, Bishop DVM: Narrowing the broader autism phenotype A study using the Communication Checklist — Adult Version **(CC-A)**. *Autism* 2010, **14**(6):559–574.
69. Lindgren KA, Folstein SE, Tomblin JB, Tager-Flusberg H: Language and reading abilities of children with Autism Spectrum Disorders and Specific Language Impairment and their first-degree relatives. *Autism Res* 2009, **2**(1):22–38.

70. Bartlett CW, Flax JF, Fermano Z, Hare A, Hou L, Petrill SA, Buyske S, Brzustowicz LM: Gene x gene interaction in shared etiology of autism and specific language impairment. *Biol Psychiatry* 2012, **72**(8):692–699.
71. Abrahams BS, Geschwind DH: Advances in autism genetics: on the threshold of a new neurobiology (vol 9, pg 341, 2008). *Nat Rev Genet* 2008, **9**(6).
72. Chen GK, Kono N, Geschwind DH, Cantor RM: Quantitative trait locus analysis of nonverbal communication in autism spectrum disorder. *Mol Psychiatr* 2006, **11**(2):214–220.
73. Calderon de Anda F, Rosario AL, Durak O, Tran T, Graff J, Meletis K, Rei D, Soda T, Madabhushi R, Ginty DD *et al.*: Autism spectrum disorder susceptibility gene TAOK2 affects basal dendrite formation in the neocortex. *Nat Neurosci* 2012, **15**:1022–1031.
74. Weiss LA, Arking DE, The Gene Discovery Project of Johns Hopkins & the Autism Consortium: A genome-wide linkage and association scan reveals novel loci for autism. *Nature* 2009, **461**:802–808.
75. Ercan-Sencicek AG, Davis Wright NR, Sanders SS, Oakman N, Valdes L, Bakkaloglu B, Doyle N, Yrigollen CM, Morgan TM, Grigorenko EL: A balanced t(10;15) translocation in a male patient with developmental language disorder. *Eur J Med Genet* 2012, **55**:128–131.
76. Potocki L, Bi W, Treadwell-Deering D, Carvalho CMB, Eifert A, Friedman EM, Glaze D, Krull K, Lee JA, Lewis RA *et al.*: Characterization of Potocki-Lupski Syndrome (dup(17)(p11.2p11.2)) and delineation of a dosage-sensitive critical interval that can convey an autism phenotype. *Am J Hum Genet* 2007, **80**:633–649.
77. Ercan-Sencicek AG, Davis Wrigh NR, Frost SJ, Fulbright e RK, Felsenfeld S, Hart L, Landi N, Mencl WE, Sanders SJ, Pugh KR *et al.*: Searching for Potocki–Lupski syndrome phenotype: A patient with language impairment and no autism. *Brain Dev* 2012.
78. Alarcon M, Abrahams BS, Stone JL, Duvall JA, Perederiy JV, Bomar JM, Sebat J, Wigler M, Martin CL, Ledbetter DH *et al.*: Linkage, association, and gene-expression analyses identify CNTNAP2 as an autism-susceptibility gene. *Am J Hum Genet* 2008, **82**(1):150–159.
79. Arking DE, Cutler DJ, Brune CW, Teslovich TM, West K, Ikeda M, Rea A, Guy M, Lin S, Cook EH *et al.*: A common genetic variant in the neurexin superfamily member CNTNAP2 increases familial risk of autism. *Am J Hum Genet* 2008, **82**(1):160–164.
80. Bakkaloglu B, O'Roak BJ, Louvi A, Gupta AR, Abelson JE, Morgan TM, Chawarska K, Klin A, Ercan-Sencicek AG, Stillman AA *et al.*: Molecular cytogenetic analysis and resequencing of Contactin Associated Protein-Like 2 in autism spectrum disorders. *Am J Hum Genet* 2008, **82**(1):165–173.
81. Vernes SC, Newbury DF, Abrahams BS, Winchester L, Nicod J, Groszer M, Alarcon M, Oliver PL, Davies KE, Geschwind DH *et al.*: A functional genetic

link between distinct developmental language disorders. *New Engl J Med* 2008, **359**(22):2337–2345.

82. Whitehouse AJ, Bishop DVM, Ang QW, Pennell CE, Fisher SE: CNTNAP2 variants affect early language development in the general population. *Genes Brain Behav* 2011, **10**(4):451–456.
83. Fisher SE, Vargha-Khadem F, Watkins KE, Monaco AP, Pembrey ME: Localisation of a gene implicated in a severe speech and language disorder. *Nat Genet* 1998, **18**(2):168–170.

Chapter 4

Whole Genome Sequencing in Autism: Clinical Translation

Ryan K.C. Yuen and Stephen W. Scherer*

Abstract

Autism Spectrum Disorder (ASD) exhibits high heritability and familial clustering. Chromosomal abnormalities, copy number variations (CNVs), smaller insertion/deletions (indels), single nucleotide variants (SNVs) and combinations thereof, have been described as etiologic factors in ASD. In fact, >100 ASD risk genes are now known with several hundred others estimated to exist. Whole-genome sequencing (WGS) holds promise as a tool for identifying all classes of genetic variation in a single experiment, which should enable the identification of the remainder of ASD risk genes. Our pilot research using WGS and thorough bioinformatics analyses to examine both *de novo* and rare inherited mutations has indeed led to an increased yield in the identification of etiologic genetic variants in ASD and its accompanying clinical syndromes. These early successes indicate that WGS will be applied to help direct personalized approaches to clinical management in many individuals and families affected by ASD.

Keywords: Whole genome sequencing; *De novo* mutations; Inherited deleterious mutations; Genetic heterogeneity; Copy number variation; Clinical translation; Medical management; Clinical diagnosis.

*Correspondence to: McLaughlin Centre and The Centre for Applied Genomics, Hospital for Sick Children and University of Toronto, 13th Floor, Peter Gilgan Centre for Research and Learning, 686 Bay Street, Toronto, Ontario M5S 1A1, Canada. Emails: RKCY: ryan.yuen@sickkids.ca, SWS: stephen.scherer@sickkids.ca

Introduction

Autism spectrum disorder (ASD) is a lifelong developmental condition affecting ~1% of the population. It is characterized by atypical development of social skills and communication, and the presence of repetitive behaviors and restricted interests. Onset is usually before the age of three and its phenotypes are highly heterogeneous. They vary from mild to severe levels of impairment, and the cognitive capabilities can also range from above average to intellectual disability. To add to this complexity, ASD is often accompanied by anxiety, intellectual disability, sleep problems, seizures, and gastrointestinal or other medical complications.[1]

Family studies indicate a significant genetic basis for ASD susceptibility. The heritability of ASD is estimated to be as high as 90% (range is 37–90%), making it one of the neuropsychiatric disorders most strongly influenced by genetics.[2,3] Decades of genetic research have led to the identification of numerous genes and syndromes associated with ASD.[4] Despite its extensive genetic heterogeneity, however, accumulating evidence converges on dysregulation of synaptic pathways.[5–7] Advances in knowledge are largely attributed to the rapidly evolving technologies, such as the development of microarray and high-throughput sequencing, which allow the capture of a previously unrealized wide-spectrum of variants in the genome. In particular, whole genome sequencing (WGS) has recently been shown to be highly effective for detection of clinically relevant genetic mutations in ASD.[8] Further understanding of the genetic causes of ASD promises to inform diagnosis/prognosis, medical management, familial recurrence risk assessment, and to facilitate drug intervention trials by stratification based on genetic pathway profiles.[1,9]

In this chapter, we will discuss (1) the understanding of ASD genetics from both pre- and post-high-throughput sequencing eras, (2) the strategies of analyzing and characterizing the genetic variants identified by WGS, (3) the implications of ASD WGS on diagnosis and treatment, and (4) the future directions of autism clinical genetic research using WGS.

Understanding of ASD Genetics from Pre- and Post-high-throughput Sequencing Eras

The genetic basis of ASD stems from the early family studies that indicate siblings of affected probands have 2–9% recurrence rate.[10] Epidemiological studies of ASD concordance in same-sex twins have also consistently favored a genetic explanation, with heritability ranging from 37% to greater than 90% under multifactorial threshold models.[2,3] Individuals with ASDs are increasingly presenting for clinical genetics evaluation with 10% having an identifiable genetic condition, such as Fragile X syndrome, and these causative mutations are detectable by sequencing or quantitative analysis of targeted genes.[11]

Genetic architecture of ASD — Knowledge before the advent of high-throughput sequencing

Linkage studies using single nucleotide polymorphisms (SNPs) initially mapped some candidate ASD risk genes which, in some instances, could apparently be replicated in different cohorts.[12–14] However, fine mapping under the peaks has generally not been successful in uncovering ASD genes, likely due to a high degree of locus heterogeneity. Large-scale genome-wide association studies (GWAS) have been performed, but so far the data indicate that few, if any, common variants have a substantial impact on ASD risk. It has been concluded that common variants may act as risk (or modifying) factors for ASD, but very large sample sizes (>10,000) will be required to detect such common variants.[15] In a very large meta-analysis, a recent study has identified common SNP variants at the *ASTN2* locus to be significantly associated with ASD,[16] or perhaps with component phenotypes of ASD, such as anxiety and or speech and language attributes.

There is now overwhelming evidence that rare (<1% minor allele frequency) and *de novo* variants, such as copy number variations (CNVs) affecting specific genes and loci contribute to the etiology of ASD (Table 1). Regarding rare CNVs, microarray studies

Table 1. Summary of main genetic findings on ASD

Approach	Detectable Genetic Variants	Clinically Relevant Variants Identified	Main Findings	Ref.
Targeted sequencing	Targeted SNVs and indels (coding/noncoding)	Known/novel variants in targeted genes	• ~10% of ASD cases have an identifiable genetic condition	10
Microarray	Common SNPs, Rare/*de novo* CNVs	Common SNPs	• Common variants may act as risk (or modifying) factors for ASD	15
		Rare/*de novo* CNVs	• Contribute to 5~10% of the ASD risk • Complex ASD cases have large CNVs • CNVs coincide with other neuropsychiatric disorders • Functional enrichment of neuronal synaptic and development genes	5–7, 10
Whole exome sequencing (WES)	SNVs, indels, CNVs (coding)	*De novo* SNVs and indels	• Contribute to ~6% of the ASD risk • 2- to 4-fold increase in *de novo* nonsense variants among ASD individuals • New ASD risk genes identified • Hundreds of genes involved in the ASD risk	24–27

(Continued)

Table 1. *(Continued)*

Approach	Detectable Genetic Variants	Clinically Relevant Variants Identified	Main Findings	Ref.
		Rare inherited SNVs	• Contribute to ~5% of the ASD risk • 1.5- to 2-fold increase of complete knockout of genes among ASD individuals • New ASD risk genes identified with biallelic mutations • Rare mutations scattered across hundreds of ASD risk genes	30–33
Whole genome sequencing (WGS)	SNVs, indels, CNVs (coding/noncoding)	*De novo*/rare inherited SNVs, indels and CNVs	• WGS has good coverage at known ASD genes • Clinically relevant variants can be found in as many as 50% of ASD cases • *De novo* mutation rate increases with paternal age • Mutation hotspots are enriched in some known ASD genes	8, 42
Targeted high-throughput sequencing	Targeted SNVs, indels, CNVs (coding/noncoding)	*De novo* SNVs and indels	• Each ASD risk gene contributed <1% frequency of ASD cases	29
		Breakpoint mapping at CNV junctions	• New ASD risk genes identified	34

have interrogated large ASD cohorts (summarized in Ref. [10]), revealing the following properties: (1) depending on the ascertainment criteria of samples, CNVs explain 5~10% of idiopathic ASD cases, but every risk locus or gene so far occurs with less than 1% frequency, (2) ASD subjects with multiple medical problems and congenital anomalies often carry one (and sometimes more than one) large *de novo* and/or inherited CNVs affecting many genes, (3) some *de novo* or inherited CNVs coincide with known medical genomic disorders, whereas others involve genes associated with ASD, intellectual disability or other neuropsychiatric conditions, and (4) there is an enrichment for CNVs that affect neuronal synaptic and development genes.[5–7,17]

These data also indicate that ASD individuals may have multiple genetic risk factors. For example, some probands with more than one potentially relevant CNV and families in which the CNV is carried by two siblings, one affected and one unaffected, indicate that establishing causation for a rare CNV can be difficult.[18–20] Some studies have shown more definitively that multiple genetic lesions in a single individual are likely a necessary liability to ASD.[21–23] A recent study identified over 103 disease genes and 44 genomic loci among individuals with some form of ASD.[4] Most of these genes have been causally implicated in intellectual disability, indicating that these two neuropsychiatric disorders often share a common genetic basis.[5] A genetic overlap between ASD and epilepsy, as well as with other neuropsychiatric diseases, is also apparent in many cases.

Genetic discoveries from high-throughput sequencing

A new approach to identify ASD risk genes includes performing whole exome sequencing (WES) of human DNA to search for *de novo* mutations (largely concentrating on single nucleotide mutations)[24–27] (Table 1). WES interrogates the coding regions, which account for ~1% of the whole genome. Reports from such projects have indicated that (1) there is a two- to four-fold increase in *de novo* nonsense variants among ASD individuals over their unaffected siblings, (2) the *de novo* SNVs contribute ~6% of the ASD risk, (3) some new

ASD risk genes (such as *CHD8, DYRK1A, KATNAL2, POGZ* and *SCN2A*) can be identified, and (4) there may be hundreds of genes involved in the ASD risk.[28,29]

In addition to the *de novo* mutations, inherited mutations were also examined in several other studies using WES technology (Table 1). These studies suggested that (1) rare biallelic mutations can be identified in known and/or novel ASD genes (such as *UBE3B, AMT, PEX7, SYNE1, VPS13B, PAH* and *POMGNT1*) by homozygosity mapping strategy in consanguineous or non-consanguineous families,[30,31] (2) rare loss-of-function (LoF) inherited mutations in autosomal genes and X-linked genes increase by 2- and 1.5-fold, respectively, in ASD cases in comparison to the controls,[32] (3) rare complete knockouts of genes by inherited LoF mutations contribute to an additional 5% of ASD cases,[32] and (4) rare mutations are scattered across hundreds of risk genes and, therefore, identification of more novel ASD genes using genetic association tests requires over thousands of samples.[33]

Other strategies have also been used for detection of potentially clinically relevant variants for ASD (Table 1). For example, Talkowski *et al.* has taken the advantage of base-pair resolution of high-throughput sequencing to map the breakpoints of previously identified large-scale balanced chromosomal abnormalities, which led to the identification of 33 putative loci/genes that are associated with ASD and other psychiatric disorders.[34] Another example is the use of serial filters against publicly available population databases to prioritize for extremely rare mutations in the ASD subjects. This strategy has been applied in two studies using whole exome[35] and WGS,[36] respectively. Such a method is particularly useful when dealing with small sample size without sequencing a comparable control population. We also note that there are also data that intergenic[37] and intragenic[38] risk loci, as well as non-coding RNAs[39] can also be involved in ASD susceptibility.

There has been significant progress in identifying new ASD risk variants as each new technology of increasing resolution (cytogenetics, microarrays, exome sequencing) has become available. Even so, when combining all of the data from these complementary approaches, the picture remains incomplete; the etiological basis of only about 20% of cases of ASD, at most, can be explained to date.[10] High heritability estimates and family studies re-confirm strong genetic contributions and suggest that more genetic variants remain to be discovered and used in disease association studies.[40,41]

Strategies of Analyzing the Genetic Variants Identified by WGS

WGS holds promise as a tool for identifying all classes of genetic variation in a single experiment, which should enable the identification of the remainder of ASD risk variants and genes. We have developed a comprehensive workflow for variant detection using WGS (Figure 1) and to assess the genomic and medical aspects of

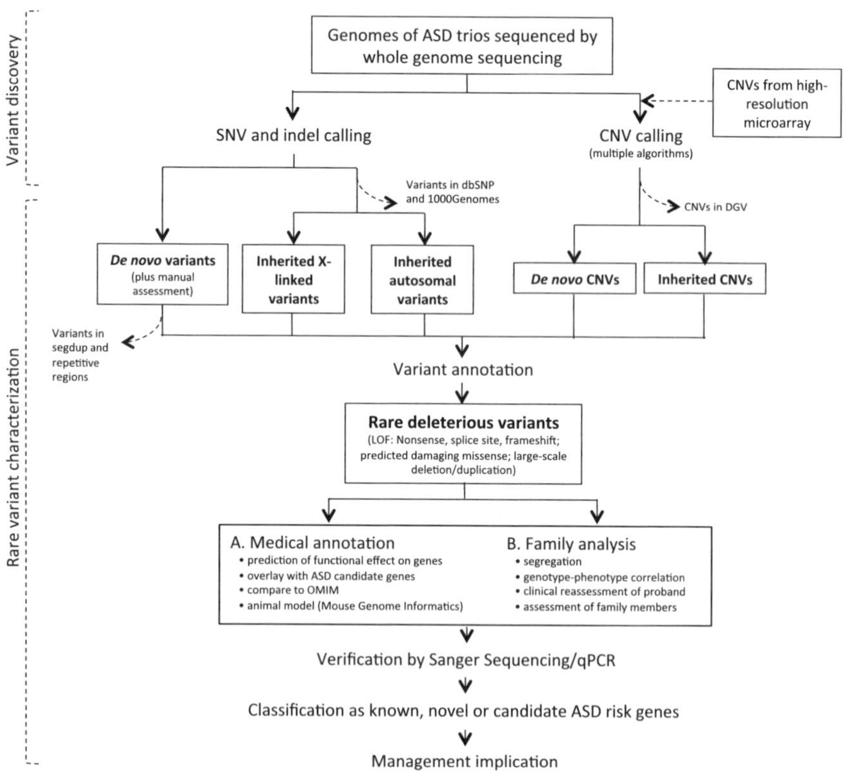

Figure 1. Workflow for detection and classification of medically relevant genetic variants in autism spectrum disorder using whole genome sequencing.

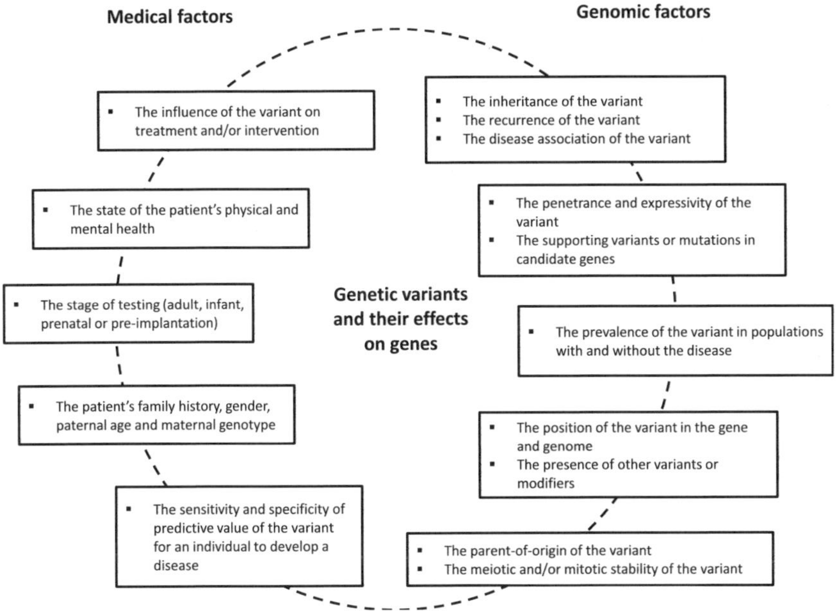

Figure 2. Medical and genomic factors to be considered during the interpretation of whole genome sequencing data. Modified from Figure 2 of Ref. [70].

each variant's relevance to ASD (Figure 2). Applying this workflow to a cohort of 32 unrelated ASD families, in a pilot study, we were able to detect clinically relevant variants to ASD or associated medical complications in 50% of the cases examined even though only focusing on the genic regions.[8] Our approach relies on developing a thorough bioinformatic pipeline allowing robust variant detection and annotation with respect to ASD. We also attempted to maximally utilize all relevant public databases on allelic frequency, as well as functional genetic information to better enable characterizing the underlying genetic variants contributing to ASD risk. Here, we have expanded this workflow to the identification and characterization of CNVs. We believe that this framework should capture the most useful information from WGS and best facilitate the translation of such information for clinical diagnosis and medical managements. There have also been other papers published recently using WGS in ASD,[36,42] but so far their design has yielded a less comprehensive dataset for follow-up studies.

Genetic variant discovery

In order to detect both *de novo* and inherited mutations, genomic DNA from the index case (also called proband) and his/her parents (i.e., a trios) and, wherever possible other family members, should be analyzed. High-quality DNA can typically be extracted from blood and sequenced to a depth of at least 30 times genomic coverage using the pair-end library approach.[8] DNA from lymphoblastoid cell lines may be used, but caution is needed given that genetic changes can be introduced through cell line creation or culturing; such changes would mistakenly appear as *de novo* variants.[43] These artificial somatic variants can potentially be eliminated or reduced using filtering procedures.[44] Independent laboratory validation of all variants (such as Sanger sequencing for SNVs and qPCR for CNVs) is needed especially in a clinical setting.[45]

Currently, Illumina and Complete Genomics are the two main companies that provide platform technologies capable of performing robust WGS. Other companies are also attempting to develop more affordable and higher-throughput sequencing platforms (e.g. Ion Proton from Life Technologies, Oxford Nanopore). Most of the high-throughput sequencing technologies generate short reads (ranging from 35–100 base pairs), which pose difficulties when attempting *de novo* DNA sequence assembly. Thereafter, sequencing reads have to be aligned, or "mapped", to the human genome reference sequence (at time of writing called version hg19) using bioinformatics tools such as the Burrows-Wheeler Aligner (BWA). Alignment tools should be selected based on the sequencing platform used. The available tools and their applications are reviewed elsewhere.[46] Local realignment and quality recalibration allows detection of single nucleotide variants (SNVs) and small insertion/deletions (indel; 1–100 base pair in size). A commonly used variant detection tool is Genome Analysis Toolkit (GATK), which also allows removal of duplicated reads. The effects (e.g., missense, nonsense or frameshift mutations, *etc.*) and classifications (e.g., in exonic, intronic, or intergenic regions) of variants across the genome can be annotated by a program such as ANNOVAR.[8]

De Novo SNVs and indels

De novo mutations are new mutations that presumably arise in a parent's germline tissue. However, detection of *de novo* mutations from high-throughput sequencing can be challenging because of the technical artifacts generated during sequencing, alignment or variant calling. Some methods have been developed to distinguish *de novo* mutations from false positive *de novo* variants. These methods range from machine-learning-based[42] to statistics-based programs.[47] In general, a potential *de novo* mutation is identified if there is a heterozygous genotype in the offspring and homozygous reference genotypes in both parents. However, this approach only applies to the autosomal regions in the genome, and additional work is needed to cover chromosome X in males (see below). Similar filtering criteria can be applied for *de novo* indel detection; the proband is heterozygous for the indel call, with the parents both being homozygous at the same position in the reference genome. However, in our experience *de novo* indels found in population databases and simple repeat regions are prone to be false-positives, so indels reported in databases such as Database SNP (dbSNP) and located in simple repeat regions are typically filtered out (Fig. 1).

Manual assessment of potential de novo exonic variants

Standard automated detection methods for detecting *de novo* variants typically only consider *de novo* variants that are homozygous in parents and heterozygous in the proband, which is not relevant for finding variants in males on the X chromosomes in males. In our experience, to accomplish this, manual curation is required to find X-linked hemizygous variants in male subjects that are not present at the same position on the X chromosome of their respective mothers (i.e. *de novo* X-linked mutations in male probands). In addition, some variants can be missed by the detection methods described above. This is particularly true for indels, which still tend to be problematic for detection under short-read high-throughput sequencing.[40] Therefore, a list of potential *de novo* exonic SNVs and indels should be prioritized from the filtered list in automated detection and validated using targeted Sanger sequencing.

CNV detection by WGS

Although clinical microarrays for CNV detection are now routinely used as a 'standard of care' in ASD,[48] the current state of technology used in diagnostic labs typically only allows detection of CNVs >100kb. WGS provides an opportunity to detect genome-wide CNVs including the smaller indels at base-pair resolution. Currently, there are four methods used for extracting CNV calls from sequencing data.[41,49] These include applying the principles of (1) "assembly-based" comparison of the query genome to the reference genome, (2) "read-pair" where end sequences of query reads are compared to the reference to look for insertions or deletions of intervening sequence, (3) "split-read" comparisons where a query sequence is intersected by a CNV breakpoint, and (4) "read-depth" method where the copy number is inferred based on deviation from expectation in a diploid reference genome. Assembly-based methods are often not applicable for the current technologies because the short-sequence reads generated by most next-generation sequencing technologies make it difficult for *de novo* assembly (especially at repetitive regions). The remaining three methods have strengths and weaknesses. For example, read-pair allows detection of unbalanced structural rearrangements such as translocation and inversion, but the size for detection is limited by the fragment size of the read pairs. Split-read provides precise breakpoint estimation, but only detects small indels (typically <100bp depending on the read length). The read-depth method can detect larger CNVs, but is more prone to be biased by GC-content and alignment quality (more detailed reviews can be found in Ref. [50]). Therefore, a combination of detection strategies is recommended (Figure 1). A high-resolution microarray can also be run in parallel for CNV detection; the same data can also be used for the evaluation of specificity and sensitivity of CNVs detected from WGS. Since it is difficult to align reads at repetitive regions of the genome, often leading to false-positive detection, we suggest removing those segments from analysis (Figure 1).

Rare variant characterization

De novo variants are generally considered more deleterious than inherited variants since they have been subjected less to evolutionary constraint; as a result they tend to be weighted in prioritization schemes. Heritability data, however, also support a significant role for autosomal inherited factors. Genome-wide studies of inherited autosomal mutations in ASD have so far mainly been limited to the model of recessive inheritance, but given the expanding number of new rare variants now known to exist in the human population.[51–53] the possibility of accumulation of deleterious variants under a haploinsufficiency model should also be considered. Likewise, given the ~4:1 male to female gender bias in ASD and the recent progress in finding X-linked forms of ASD and intellectual disability,[32,54,55] X-linked mutations in males inherited from their unaffected mothers also need to be considered.

When assessing inherited variants, the population frequency of those variants can serve as a filter to differentiate putative deleterious alterations (which would be rare in frequency) from the more likely benign events (more common in frequency). Population variant databases, such as dbSNP or the 1000 genomes project, are available for such comparisons. Data from the exome variant server hosted by NHLBI (NHLBI GO Exome Sequencing Project) can also be useful with the caveat that many subjects in this cohort have developmental disorders and therefore may not be ideal as controls. For our WGS analyses, we define a rare variant as one not being present in the dbSNP and 1000 genomes project databases.

Some others use a frequency cutoff (such as with <1% minor allele frequency), which increases the number of potential candidate variants to follow-up on.[36]

Rare deleterious variants including loss-of-function (LoF) mutations (nonsense, splice site and frameshift mutations) and damaging missense mutations are first prioritized in our analysis pipeline. As the name implies, the impact of LoF events to the gene is somewhat certain. To the contrary, the functional effect of missense mutations may not be so clear, and the *in silico* prediction tools used to predict functional impact are not always correct.[56] To increase sensitivity, we always use at least two prediction algorithms and only characterize those calls arising by both programs (SIFT and PolyPhen-2 are the most commonly used). In the end, we suggest considering only those predicted damaging missense mutations that are identified in known ASD risk genes or pathways (detailed below).

Medical classification

To evaluate the medical relevance of selected genetic variants to ASD, we developed the following assessment protocol outlined in Figure 1 (see also Figure 2): (1) to predict whether the variants are likely to have a deleterious effect on splicing of the gene or functional properties of the protein product; (2) to assess the frequency of predicted deleterious variants in a population database; (3) to compare genes affected to known ASD candidate genes; (4) to assess segregation in families and relation to risk genes using the Online Mendelian Inheritance in Man (OMIM) database, and (5) to assess functional relevance by comparison to the Mouse Genome Informatics (MGI) phenotype database. Steps (1) and (2) have been discussed in the earlier sections and here we elaborate on steps (3) to (5).

A regularly updated list of expert-annotated ASD risk genes is used in our analyses.[4] Other potential ASD candidate loci/genes can be found in online databases such as Autism Chromosome Rearrangement Database, AutDB, AutismKB, Autism Genetic Database and gene database from the Simons Foundation Autism Research Initiative. Function of genes that carry rare deleterious mutations in the documented neurodevelopmental/behavioral phenotypes of human or mouse can be examined from the human phenotype ontology (HPO) and the Mouse Genome Informatics (MGI) databases. The mode of inheritance for the genes can also be assessed by the information provided in available online resources. For human phenotypes, the numerical annotation of the OMIM entry can be used to determine the mode of inheritance. An OMIM ID beginning with 1 means autosomal dominant, 2 means autosomal recessive, and 3 indicates X-linked. For mouse phenotypes, the mode of inheritance can be inferred based on the genotype associated with the phenotype of interest: autosomal heterozygous genotypes were inferred to be dominant, autosomal homozygous genotypes were inferred to be recessive, and genes with human orthologs on the X chromosome outside of the pseudo-autosomal regions were inferred to be X-linked.

After thorough evaluation using the algorithm of analysis outlined above, the genes carrying deleterious mutations are classified as: (1) known, (2) novel or (3) candidate ASD risk genes (Figure 1). In our approach, ASD subjects carrying mutations in ASD risk genes are then classified for follow-up clinical assessments with an emphasis on clinical genetics, and possible enrollment in drug treatment and trials. Lastly, the implicated genes are prioritized for literature searches, since the gene involved may be actively studied. Details of the decision-making process regarding medical classification have been discussed in Ref. [57].

Implications of ASD Whole Genome Sequencing on Diagnosis and Treatment

In our initial published pilot study, we have applied the described workflow to analyze the WGS data in a group of families having at least one ASD child.[8] Here, we summarize the findings in some cases with additional details, and how the data can be used in the management of families.

Revealing clinical and genetic heterogeneity of ASD

Multiple lines of evidence demonstrate the importance of complex genetic factors in ASD (chromosomal abnormalities, CNV, indel, SNV) (reviewed in Ref. [10]). At the same time, careful clinical characterization has revealed ASD to be extremely heterogeneous, ranging from milder Asperger disorder to more complex forms associated with profound intellectual disability, epilepsy and other medical problems. We, as well as others, have sought *de novo* variants for a causative role, but given the well-established role of inherited factors in ASD, we also emphasize that rare autosomal and X-linked inherited variants should also be studied in families. Following this design, we ascertained that variants in 50% of the studied families were of potential significance, either because of their established association with clinical syndromes found among individuals with ASD or their link with ASD and/or related disabilities, such as intellectual disability, or medical co-morbidities, such as seizures. Some of the detected mutations could reasonably be considered pathogenic and/or have possible clinical management or genetic counseling implications for the proband or family members. For others, their role in disease needs to be closely followed in the medical literature.

Facilitating decision-making on medical management

A deleterious mutation in the Duchenne Muscular Dystrophy (*DMD*) gene in a boy with ASD alerted clinicians that he be examined for dystrophinopathy, which is associated with progressive muscle weakness and cardiomyopathy. Recent findings provide mounting evidence that DMD is a *bona fide* ASD risk gene.[58,59] In another unrelated proband, WGS detected a *de novo* mutation in *KAL1*, an X-linked gene that accounts for ~10% of Kallmann syndrome[60] and, while perhaps not directly related to ASD, general medical management can include hormone treatment to initiate puberty. In another family, the detection of a deleterious mutation in *CHD7* in one proband and his father indicates that they be assessed for CHARGE syndrome.[61,62] *CHD7* mutations are also found in rare cases of Kallmann syndrome, which, like CHARGE syndrome, can feature sensory impairment and hypogonadism.[63] Perhaps most noteworthy with respect to ASD, is that ~30% of CHARGE individuals are also on the spectrum. Families carrying mutations in *SCN2A*, *CACNA1C* and *EHMT1* should be assessed for features of epilepsy, Timothy syndrome, and Kleefstra syndrome, respectively. These individuals, along with the sibling carrier of the *VIP* mutation who reportedly exhibited symptoms of ASD, should also undergo full assessment for ASD and/or related neurodevelopmental disorders and be treated accordingly.

Providing potential pharmacological clues

Genetic variants reported in the study also highlighted potential molecular targets for pharmacologic manipulation and opened the way for personalized therapeutics in ASD. For example, personal mutations in the *CAPRIN1/AFF2* and *SCN2A* genes identify carriers as potential candidates for drug trials, using allosteric modulators of mGluR5/GABAB or GABAA receptors, which have ameliorated autism-like symptoms in mice.[64–66] Mutation of *KCNQ2* identifies a candidate pathway for treatment by Kv7 channel openers, such as retigabine.[67] The *VIP* variant highlights a potential drug target pathway, in that the family of receptors that mediate its effects (VPAC1 and VPAC2), along with related peptides like pituitary adenylate cyclase-activating polypeptide (PACAP), are considered drug targets in neurodegenerative and inflammatory diseases.[68]

Increasing diagnostic yield with improved genomic coverage

With respect to sequencing, we have detected a higher *de novo* mutation rate using our approach than was previously found in

ASD exome projects,[24–27] which is likely due to the more uniform capturing of the entire coding regions afforded by WGS. We also note that WES would not have captured 11% of the *de novo* and clinically relevant mutations found in the study, although this may depend on the capturing method used and the coverage achieved. As discussed in detail in our original paper, WGS demonstrated a more uniform coverage, while WES had more variability with a distribution of coverage.[8] An even higher extent of coverage difference was observed on the X chromosome with at least 17.5% more annotated exons (5.7% more coding) covered by WGS than by WES. Such coverage difference was even more pronounced when considering regions such as gene splice sites. These findings were supported by a recent study that reported that WES could not detect 11% of the variants detected by WGS.[69]

Future Directions of Autism Clinical Genetic Research using Whole Genome Sequencing

WGS will enable the identification of more genetic variants than other technologies used for confirmatory and predictive diagnosis. In Figure 2, we have modified and updated our thoughts from our earlier work on CNVs[70] to now include all genetic variants and how to assess this new data with respect to 'genomic' and 'medical' questions of relevance to ASD. Initial applications will most likely occur in families having a history of ASD since early identification can inform diagnosis and shorten the diagnostic path, thus allowing children to benefit more fully from early interventions.[1] In families with a child bearing a *de novo* mutation in an ASD susceptibility gene, recurrence risk might be similar to that of the general population, but gonadal mosaicism need to also be considered as we have demonstrated previously.[8,12,18] The presence of rare inherited variants in ASD candidate genes, as well as X-linked mutations in males might confer increased risk that would lead to a recommendation of increased developmental monitoring to allow earlier identification.

Resolving incomplete penetrance and variable expressivity of variants

The genetic variants selected for analysis from this workflow are extremely rare as they are not present in thousands of non-ASD individuals. Since these mutations were private to the families, genotype-phenotype correlation could only be performed in the context of the corresponding families, particularly for the inherited mutations. All autosomal inherited deleterious mutations we discuss here reside in ASD candidate genes. For some heterozygous mutations, the established mechanism of action can be haploinsufficiency. Thus, such genes were likely to be expressed in a dominant manner, in some cases possibly with incomplete penetrance and/or variable expressivity. Although each individual is a carrier of hundreds of deleterious variants, many have been previously documented to be recessive and therefore not known to be pathogenic. A recent study of rare deleterious variants in the genomes of 179 individuals revealed that very few known pathogenic variants were found in genes associated with dominant disorders after applying very stringent filters; instead they were associated with either late-onset or no overt disease phenotype,[71] suggesting that rare deleterious mutations found in genes with known dominant disorders are rare. Indeed, some families in our study showed a correlation of phenotype with the associated known outcome of the variants, but the reason why some carriers were unaffected remains to be determined. Further population genetic and functional studies of the variants, and reassessments of the carriers, are needed to address this issue.

Improving the quality of CNV calling from whole genome sequencing

WGS covers the entire genome so by definition it should provide improved genome coverage for the identification of genetic variants.[72] However, calling larger indels and CNVs from WGS data remains a challenge.[50] Using a read-depth based approach for CNV detection, we found that as many as 88% of the calls could be false-positive. In our experience, current high-throughput sequencing alone does not yield a complete variation catalog and, in fact, different approaches can produce different results.[41] In one of our other studies examining the performance of CNV calling from WGS by multiple algorithms, we found that detection of losses is associated with high sensitivity and a decent confirmation rate (69.5%), and that split-read and paired-end mapping can accurately pinpoint a variant's size.[40] However, characterization of chromosomal gains is still challenging, with decreases in detection in the medium to large size range, mostly at repetitive regions of DNA. Due to difficulty in alignments, short reads are not amenable to capturing variants at interspersed and tandem-repeat loci. One solution may be to generate multiple libraries of various insert lengths; while small libraries with tight size distribution are better at calling small variants and localizing breakpoints, large libraries excel at finding large variants. With technological advancement, longer and more accurate reads can be obtained, which will enable more accurate alignment and variation-discovery. As we test other existing/new and improved annotation algorithms, the yet undiscovered indels and CNVs in both genic and non-genic intervals will also become available for genotype and phenotype studies in ASD families.

Analyzing non-coding regions in the genome

While the non-coding regions constitute nearly 99% of the genome, most sequencing studies so far have only focused on coding regions of the genome. In those few instances where attempts have been made to analyze non-coding variants in WGS data,[42] only *de novo* variants were investigated and little emphasis was placed on interpreting their effects on the studied subjects. Interpreting the clinical significance of genetic variants found in non-coding regions remains difficult since this vast region of the genome remains poorly characterized. However, recent findings from ENCODE projects indicates that 80% of the non-coding regions indeed have a function.[73] Moreover, some genomic investigations have identified mutations in individuals with ASD in non-genic segments of DNA[37,38] and non-coding RNAs.[39,74] Some of these identified variants may act through altering the sequence of essential elements for regulating the proper expression of nearby genes. Bioinformatic tools have been developed to predict some putative regulatory sites, such as transcriptional binding motifs.[75] However, transcriptional binding factors can be tissue-specific and might differ between genomic regions depending on the context. Therefore, high-throughput functional assays are needed to evaluate and validate the effects of the DNA alternations in the transcriptional binding factors. Future improvement of technology and reduction of sequencing cost should increase the number of whole-genomes sequenced and provide a wealth of variants for cross-comparison, which will largely facilitate the overall interpretation of WGS data.

Conclusion

WGS has already proven useful in ASD research studies, and the diagnostic yield and clinical utility should increase as more undetected structural genetic variants are discovered and characterized. WGS data reinforce the growing number of observations that significant clinical and genetic heterogeneity, and pleiotropy exist in ASD. With more development, WGS promises to be the method of choice for DNA diagnostic testing in ASD, and interpretation of suspected etiologic genetic variants will become easier as more genomes from families with autism and population controls are sequenced. We anticipate that much of the new WGS data will initially come from research studies soon followed by clinical diagnostic sequencing, playing out in the same way clinical microarrays have taken a hold in autism assessment.

Executive Summary

- ASD has a high genetic basis with up to 90% heritability.
- Many ASD risk genes can be identified through advanced genomic technologies.

- Each ASD risk gene only contributes less than 1% of the ASD cases.
- There is an enrichment of neuronal synaptic and development functions from the ASD risk genes.
- Clinically relevant variants can be found in as many as 50% of ASD cases using WGS.
- Thorough evaluation of the WGS data can improve the clinical diagnosis of ASD cases.
- Best practice should involve the investigation of both *de novo* and inherited variants across a complete size spectrum of genetic variation.
- Medical interpretation of WGS data can be assisted by assessing the clinical information from the family and using the online genetic and/or phenotypic databases.
- Clinical and genetic heterogeneity of ASD supports the application of personalized medicine.
- WGS facilitates decision making on medical management including genetic counseling and treatment options.
- WGS improves genomic coverage with the potential of detecting all genetic variants relevant to ASD.
- Many clinically relevant variants exhibit incomplete penetrance and variable expressivity, which should be addressed in the future studies.
- Current CNV detection for WGS is still sub-optimal. Efforts should be made on improving CNV calling strategies.
- New strategies should be developed to characterize the role of variants in the non-coding regions of the genome.

Acknowledgements

R.K.C.Y. holds an Autism Speaks Postdoctoral Fellowship in Translational Research. S.W.S. holds the GlaxoSmithKline-CIHR Chair in Genome Sciences at the University of Toronto and The Hospital for Sick Children and Director of The Centre for Applied Genomics and University of Toronto McLaughlin Centre.

References

1. Scherer SW, Dawson G: Risk factors for autism: translating genomic discoveries into diagnostics. *Hum Genet* 2011, **130**(1):123–148.
2. Bailey A, Le Couteur A, Gottesman I, Bolton P, Simonoff E, Yuzda E, Rutter M: Autism as a strongly genetic disorder: evidence from a British twin study. *Psychol Med* 1995, **25**(1):63–77.
3. Hallmayer J, Cleveland S, Torres A, Phillips J, Cohen B, Torigoe T, Miller J, Fedele A, Collins J, Smith K *et al.*: Genetic heritability and shared environmental factors among twin pairs with autism. *Arch Gen Psychiatry* 2011, **68**(11):1095–1102.
4. Betancur C: Etiological heterogeneity in autism spectrum disorders: more than 100 genetic and genomic disorders and still counting. *Brain Res* 2011, **1380**:42–77.
5. Pinto D, Pagnamenta AT, Klei L, Anney R, Merico D, Regan R, Conroy J, Magalhaes TR, Correia C, Abrahams BS *et al.*: Functional impact of global rare copy number variation in autism spectrum disorders. *Nature* 2010, **466**(7304):368–372.
6. Gilman SR, Iossifov I, Levy D, Ronemus M, Wigler M, Vitkup D: Rare de novo variants associated with autism implicate a large functional network of genes involved in formation and function of synapses. *Neuron* 2011, **70**(5):898–907.
7. Sanders SJ, Ercan-Sencicek AG, Hus V, Luo R, Murtha MT, Moreno-De-Luca D, Chu SH, Moreau MP, Gupta AR, Thomson SA *et al.*: Multiple recurrent de novo CNVs, including duplications of the 7q11.23 Williams syndrome region, are strongly associated with autism. *Neuron* 2011, **70**(5):863–885.
8. Jiang YH, Yuen RK, Jin X, Wang M, Chen N, Wu X, Ju J, Mei J, Shi Y, He M *et al.*: Detection of Clinically Relevant Genetic Variants in Autism Spectrum Disorder by Whole-Genome Sequencing. *Am J Hum Genet* 2013.
9. Shen Y, Dies KA, Holm IA, Bridgemohan C, Sobeih MM, Caronna EB, Miller KJ, Frazier JA, Silverstein I, Picker J *et al.*: Clinical genetic testing for patients with autism spectrum disorders. *Pediatrics* 2010, **125**(4):e727–735.
10. Devlin B, Scherer SW: Genetic architecture in autism spectrum disorder. *Curr Opin Genet Dev* 2012, **22**(3):229–237.
11. Miles JH, McCathren RB, Stichter J, Shinawi M: Autism Spectrum Disorders. In: *GeneReviews*. Edited by Pagon RA, Adam MP, Bird TD, Dolan CR, Fong CT, Stephens K. Seattle (WA); 1993.
12. Szatmari P, Paterson AD, Zwaigenbaum L, Roberts W, Brian J, Liu XQ, Vincent JB, Skaug JL, Thompson AP, Senman L *et al.*: Mapping autism risk loci using genetic linkage and chromosomal rearrangements. *Nat Genet* 2007, **39**(3):319–328.
13. Weiss LA, Arking DE, Daly MJ, Chakravarti A: A genome-wide linkage and association scan reveals novel loci for autism. *Nature* 2009, **461**(7265):802–808.

14. Pagnamenta AT, Khan H, Walker S, Gerrelli D, Wing K, Bonaglia MC, Giorda R, Berney T, Mani E, Molteni M *et al.*: Rare familial 16q21 microdeletions under a linkage peak implicate cadherin 8 (CDH8) in susceptibility to autism and learning disability. *J Med Genet* 2011, **48**(1):48–54.
15. Anney R, Klei L, Pinto D, Almeida J, Bacchelli E, Baird G, Bolshakova N, Bolte S, Bolton PF, Bourgeron T *et al.*: Individual common variants exert weak effects on the risk for autism spectrum disorderspi. *Hum Mol Genet* 2012, **21**(21):4781–4792.
16. Anney R, Ripke S, Arking D, Cook E, Robinson E, Anttila V, Gill M, Santangelo S, Sutcliffe J, Daly M *et al.*: Meta-analysis of European Ancestry Individuals with Autism Spectrum Disorder reveals Strong Association 3′ of the Astrotactin 2 (ASTN2) Gene Locus on Chromosome 9. In: *World Congress of Psychiatric Genetics: Oct 2013; Boston, MA*; 2013.
17. Marshall CR, Scherer SW: Detection and characterization of copy number variation in autism spectrum disorder. *Methods Mol Biol* 2012, **838**:115–135.
18. Fernandez BA, Roberts W, Chung B, Weksberg R, Meyn S, Szatmari P, Joseph-George AM, Mackay S, Whitten K, Noble B *et al.*: Phenotypic spectrum associated with de novo and inherited deletions and duplications at 16p11.2 in individuals ascertained for diagnosis of autism spectrum disorder. *J Med Genet* 2010, **47**(3):195–203.
19. Marshall CR, Noor A, Vincent JB, Lionel AC, Feuk L, Skaug J, Shago M, Moessner R, Pinto D, Ren Y *et al.*: Structural variation of chromosomes in autism spectrum disorder. *Am J Hum Genet* 2008, **82**(2):477–488.
20. Prasad A, Merico D, Thiruvahindrapuram B, Wei J, Lionel AC, Sato D, Rickaby J, Lu C, Szatmari P, Roberts W *et al.*: A discovery resource of rare copy number variations in individuals with autism spectrum disorder. *G3 (Bethesda)* 2012, **2**(12):1665–1685.
21. Girirajan S, Rosenfeld JA, Coe BP, Parikh S, Friedman N, Goldstein A, Filipink RA, McConnell JS, Angle B, Meschino WS *et al.*: Phenotypic heterogeneity of genomic disorders and rare copy-number variants. *N Engl J Med* 2012, **367**(14):1321–1331.
22. Girirajan S, Rosenfeld JA, Cooper GM, Antonacci F, Siswara P, Itsara A, Vives L, Walsh T, McCarthy SE, Baker C *et al.*: A recurrent 16p12.1 microdeletion supports a two-hit model for severe developmental delay. *Nat Genet* 2010, **42**(3):203–209.
23. Leblond CS, Heinrich J, Delorme R, Proepper C, Betancur C, Huguet G, Konyukh M, Chaste P, Ey E, Rastam M *et al.*: Genetic and functional analyses of SHANK2 mutations suggest a multiple hit model of autism spectrum disorders. *PLoS Genet* 2012, **8**(2):e1002521.
24. Sanders SJ, Murtha MT, Gupta AR, Murdoch JD, Raubeson MJ, Willsey AJ, Ercan-Sencicek AG, DiLullo NM, Parikshak NN, Stein JL *et al.*: De novo

mutations revealed by whole-exome sequencing are strongly associated with autism. *Nature* 2012, **485**(7397):237–241.

25. O'Roak BJ, Vives L, Girirajan S, Karakoc E, Krumm N, Coe BP, Levy R, Ko A, Lee C, Smith JD *et al.*: Sporadic autism exomes reveal a highly interconnected protein network of de novo mutations. *Nature* 2012, **485**(7397):246–250.
26. Neale BM, Kou Y, Liu L, Ma'ayan A, Samocha KE, Sabo A, Lin CF, Stevens C, Wang LS, Makarov V *et al.*: Patterns and rates of exonic de novo mutations in autism spectrum disorders. *Nature* 2012, **485**(7397):242–245.
27. Iossifov I, Ronemus M, Levy D, Wang Z, Hakker I, Rosenbaum J, Yamrom B, Lee YH, Narzisi G, Leotta A *et al.*: De novo gene disruptions in children on the autistic spectrum. *Neuron* 2012, **74**(2):285–299.
28. Buxbaum JD, Daly MJ, Devlin B, Lehner T, Roeder K, State MW: The Autism Sequencing Consortium: Large-Scale, High-Throughput Sequencing in Autism Spectrum Disorders. *Neuron* 2012, **76**(6):1052–1056.
29. O'Roak BJ, Vives L, Fu W, Egertson JD, Stanaway IB, Phelps IG, Carvill G, Kumar A, Lee C, Ankenman K *et al.*: Multiplex targeted sequencing identifies recurrently mutated genes in autism spectrum disorders. *Science* 2012, **338**(6114):1619–1622.
30. Yu TW, Chahrour MH, Coulter ME, Jiralerspong S, Okamura-Ikeda K, Ataman B, Schmitz-Abe K, Harmin DA, Adli M, Malik AN *et al.*: Using whole-exome sequencing to identify inherited causes of autism. *Neuron* 2013, **77**(2):259–273.
31. Chahrour MH, Yu TW, Lim ET, Ataman B, Coulter ME, Hill RS, Stevens CR, Schubert CR, Greenberg ME, Gabriel SB *et al.*: Whole-exome sequencing and homozygosity analysis implicate depolarization-regulated neuronal genes in autism. *PLoS Genet* 2012, **8**(4):e1002635.
32. Lim ET, Raychaudhuri S, Sanders SJ, Stevens C, Sabo A, Macarthur DG, Neale BM, Kirby A, Ruderfer DM, Fromer M *et al.*: Rare complete knockouts in humans: population distribution and significant role in autism spectrum disorders. *Neuron* 2013, **77**(2):235–242.
33. Liu L, Sabo A, Neale BM, Nagaswamy U, Stevens C, Lim E, Bodea CA, Muzny D, Reid JG, Banks E *et al.*: Analysis of rare, exonic variation amongst subjects with autism spectrum disorders and population controls. *PLoS Genet* 2013, **9**(4):e1003443.
34. Talkowski ME, Rosenfeld JA, Blumenthal I, Pillalamarri V, Chiang C, Heilbut A, Ernst C, Hanscom C, Rossin E, Lindgren AM *et al.*: Sequencing chromosomal abnormalities reveals neurodevelopmental loci that confer risk across diagnostic boundaries. *Cell* 2012, **149**(3):525–537.
35. Bi C, Wu J, Jiang T, Liu Q, Cai W, Yu P, Cai T, Zhao M, Jiang YH, Sun ZS: Mutations of ANK3 identified by exome sequencing are associated with autism susceptibility. *Hum Mutat* 2012, **33**(12):1635–1638.

36. Shi L, Zhang X, Golhar R, Otieno FG, He M, Hou C, Kim C, Keating B, Lyon GJ, Wang K *et al.*: Whole-genome sequencing in an autism multiplex family. *Mol Autism* 2013, **4**(1):8.
37. Walker S, Scherer SW: Identification of candidate intergenic risk loci in autism spectrum disorder. *BMC Genomics* 2013, **14**(1):499.
38. Wisniowiecka-Kowalnik B, Nesteruk M, Peters SU, Xia Z, Cooper ML, Savage S, Amato RS, Bader P, Browning MF, Haun CL *et al.*: Intragenic rearrangements in NRXN1 in three families with autism spectrum disorder, developmental delay, and speech delay. *Am J Med Genet B Neuropsychiatr Genet* 2010, **153B**(5):983–993.
39. Noor A, Whibley A, Marshall CR, Gianakopoulos PJ, Piton A, Carson AR, Orlic-Milacic M, Lionel AC, Sato D, Pinto D *et al.*: Disruption at the PTCHD1 Locus on Xp22.11 in Autism spectrum disorder and intellectual disability. *Sci Transl Med* 2010, **2**(49):49ra68.
40. Pang AW, Macdonald JR, Yuen RK, Hayes VM, Scherer SW: Performance of High-Throughput Sequencing for the Discovery of Genetic Variation Across the Complete Size Spectrum. *G3 (Bethesda)* 2013, **4**(1):63–65.
41. Pang AW, MacDonald JR, Pinto D, Wei J, Rafiq MA, Conrad DF, Park H, Hurles ME, Lee C, Venter JC *et al.*: Towards a comprehensive structural variation map of an individual human genome. *Genome Biol* 2010, **11**(5):R52.
42. Michaelson JJ, Shi Y, Gujral M, Zheng H, Malhotra D, Jin X, Jian M, Liu G, Greer D, Bhandari A *et al.*: Whole-genome sequencing in autism identifies hot spots for de novo germline mutation. *Cell* 2012, **151**(7):1431–1442.
43. Londin ER, Keller MA, D'Andrea MR, Delgrosso K, Ertel A, Surrey S, Fortina P: Whole-exome sequencing of DNA from peripheral blood mononuclear cells (PBMC) and EBV-transformed lymphocytes from the same donor. *BMC Genomics* 2011, **12**:464.
44. Schafer CM, Campbell NG, Cai G, Yu F, Makarov V, Yoon S, Daly MJ, Gibbs RA, Schellenberg GD, Devlin B *et al.*: Whole exome sequencing reveals minimal differences between cell line and whole blood derived DNA. *Genomics* 2013, **102**(4):270–277.
45. Ali-Khan SE, Daar AS, Shuman C, Ray PN, Scherer SW: Whole genome scanning: resolving clinical diagnosis and management amidst complex data. *Pediatr Res* 2009, **66**(4):357–363.
46. Li H, Homer N: A survey of sequence alignment algorithms for next-generation sequencing. *Brief Bioinform* 2010, **11**(5):473–483.
47. Li B, Chen W, Zhan X, Busonero F, Sanna S, Sidore C, Cucca F, Kang HM, Abecasis GR: A likelihood-based framework for variant calling and de novo mutation detection in families. *PLoS Genet* 2012, **8**(10):e1002944.
48. Miller DT, Adam MP, Aradhya S, Biesecker LG, Brothman AR, Carter NP, Church DM, Crolla JA, Eichler EE, Epstein CJ *et al.*: Consensus statement: chromosomal microarray is a first-tier clinical diagnostic test for individuals

with developmental disabilities or congenital anomalies. *Am J Hum Genet* 2010, **86**(5):749–764.

49. Pang AW, Migita O, Macdonald JR, Feuk L, Scherer SW: Mechanisms of formation of structural variation in a fully sequenced human genome. *Hum Mutat* 2013, **34**(2):345–354.
50. Teo SM, Pawitan Y, Ku CS, Chia KS, Salim A: Statistical challenges associated with detecting copy number variations with next-generation sequencing. *Bioinformatics* 2012, **28**(21):2711–2718.
51. Nelson MR, Wegmann D, Ehm MG, Kessner D, St Jean P, Verzilli C, Shen J, Tang Z, Bacanu SA, Fraser D *et al.*: An abundance of rare functional variants in 202 drug target genes sequenced in 14,002 people. *Science* 2012, **337**(6090):100–104.
52. Keinan A, Clark AG: Recent explosive human population growth has resulted in an excess of rare genetic variants. *Science* 2012, **336**(6082):740–743.
53. Tennessen JA, Bigham AW, O'Connor TD, Fu W, Kenny EE, Gravel S, McGee S, Do R, Liu X, Jun G *et al.*: Evolution and functional impact of rare coding variation from deep sequencing of human exomes. *Science* 2012, **337**(6090):64–69.
54. Piton A, Gauthier J, Hamdan FF, Lafreniere RG, Yang Y, Henrion E, Laurent S, Noreau A, Thibodeau P, Karemera L *et al.*: Systematic resequencing of X-chromosome synaptic genes in autism spectrum disorder and schizophrenia. *Mol Psychiatry* 2011, **16**(8):867–880.
55. Nava C, Lamari F, Heron D, Mignot C, Rastetter A, Keren B, Cohen D, Faudet A, Bouteiller D, Gilleron M *et al.*: Analysis of the chromosome X exome in patients with autism spectrum disorders identified novel candidate genes, including TMLHE. *Transl Psychiatry* 2012, **2**:e179.
56. Dorfman R, Nalpathamkalam T, Taylor C, Gonska T, Keenan K, Yuan XW, Corey M, Tsui LC, Zielenski J, Durie P: Do common in silico tools predict the clinical consequences of amino-acid substitutions in the CFTR gene? *Clin Genet* 2010, **77**(5):464–473.
57. Carter MT, Scherer SW: Autism spectrum disorder in the genetics clinic: a review. *Clin Genet* 2013, **83**(5):399–407.
58. Wu JY, Kuban KC, Allred E, Shapiro F, Darras BT: Association of Duchenne muscular dystrophy with autism spectrum disorder. *J Child Neurol* 2005, **20**(10):790–795.
59. Erturk O, Bilguvar K, Korkmaz B, Bayri Y, Bayrakli F, Arlier Z, Ozturk AK, Yalcinkaya C, Tuysuz B, State MW *et al.*: A patient with Duchenne muscular dystrophy and autism demonstrates a hemizygous deletion affecting Dystrophin. *Am J Med Genet A* 2010, **152A**(4):1039–1042.
60. Pallais JC, Au M, Pitteloud N, Seminara S, Crowley WF: Kallmann Syndrome. In: *GeneReviews*. Edited by Pagon RA, Bird TD, Dolan CR, Stephens K, Adam MP. Seattle (WA); 1993.

61. Trider CL, Corsten G, Morrison D, Hefner M, Davenport S, Blake K: Understanding obstructive sleep apnea in children with CHARGE syndrome. *Int J Pediatr Otorhinolaryngol* 2012, **76**(7):947–953.
62. Edwards BM, Van Riper LA, Kileny PR: Clinical manifestations of CHARGE Association. *Int J Pediatr Otorhinolaryngol* 1995, **33**(1):23–42.
63. Bergman JE, Janssen N, Hoefsloot LH, Jongmans MC, Hofstra RM, van Ravenswaaij-Arts CM: CHD7 mutations and CHARGE syndrome: the clinical implications of an expanding phenotype. *J Med Genet* 2011, **48**(5):334–342.
64. Silverman JL, Smith DG, Rizzo SJ, Karras MN, Turner SM, Tolu SS, Bryce DK, Smith DL, Fonseca K, Ring RH *et al.*: Negative allosteric modulation of the mGluR5 receptor reduces repetitive behaviors and rescues social deficits in mouse models of autism. *Sci Transl Med* 2012, **4**(131):131ra151.
65. Han S, Tai C, Westenbroek RE, Yu FH, Cheah CS, Potter GB, Rubenstein JL, Scheuer T, de la Iglesia HO, Catterall WA: Autistic-like behaviour in Scn1a+/- mice and rescue by enhanced GABA-mediated neurotransmission. *Nature* 2012, **489**(7416):385–390.
66. Henderson C, Wijetunge L, Kinoshita MN, Shumway M, Hammond RS, Postma FR, Brynczka C, Rush R, Thomas A, Paylor R *et al.*: Reversal of disease-related pathologies in the fragile X mouse model by selective activation of GABA(B) receptors with arbaclofen. *Sci Transl Med* 2012, **4**(152):152ra128.
67. Rundfeldt C, Netzer R: The novel anticonvulsant retigabine activates M-currents in Chinese hamster ovary-cells tranfected with human KCNQ2/3 subunits. *Neurosci Lett* 2000, **282**(1–2):73–76.
68. Chu A, Caldwell JS, Chen YA: Identification and characterization of a small molecule antagonist of human VPAC(2) receptor. *Mol Pharmacol* 2010, **77**(1):95–101.
69. O'Rawe J, Guangqing S, Wang W, Hu J, Bodily P, Tian L, Hakonarson H, Johnson E, Wei Z, Jiang T *et al.*: Low concordance of multiple variant-calling pipelines: practical implications for exome and genome sequencing. *Genome Med* 2013, **5**(3):28.
70. Cook EH, Jr., Scherer SW: Copy-number variations associated with neuropsychiatric conditions. *Nature* 2008, **455**(7215):919–923.
71. Xue Y, Chen Y, Ayub Q, Huang N, Ball EV, Mort M, Phillips AD, Shaw K, Stenson PD, Cooper DN *et al.*: Deleterious- and Disease-Allele Prevalence in Healthy Individuals: Insights from Current Predictions, Mutation Databases, and Population-Scale Resequencing. *Am J Hum Genet* 2012, **91**(6):1022–1032.
72. Macdonald JR, Ziman R, Yuen RK, Feuk L, Scherer SW: The database of genomic variants: a curated collection of structural variation in the human genome. *Nucleic Acids Res* 2013, **42**(Database issue):D986–992.
73. Dunham I, Kundaje A, Aldred SF, Collins PJ, Davis CA, Doyle F, Epstein CB, Frietze S, Harrow J, Kaul R *et al.*: An integrated encyclopedia of DNA elements in the human genome. *Nature* 2012, **489**(7414):57–74.

74. Kerin T, Ramanathan A, Rivas K, Grepo N, Coetzee GA, Campbell DB: A noncoding RNA antisense to moesin at 5p14.1 in autism. *Sci Transl Med* 2012, **4**(128):128ra140.
75. Thomas-Chollier M, Defrance M, Medina-Rivera A, Sand O, Herrmann C, Thieffry D, van Helden J: RSAT 2011: regulatory sequence analysis tools. *Nucleic Acids Res* 2011, **39**(Web Server issue):W86–91.

Chapter 5

The Impact of Integrative Unconventional Data Analysis Approaches on Advancing Autism Genetics Research

Zohreh Talebizadeh* and Ayten Shah

Abstract

The progress made in genomics technology provides unprecedented opportunity for the research community to accelerate discovering the genetic etiology of autism. A clear trend in the literature from gene-centric studies to whole-genome approaches is the best illustration of the incorporation of the state-of-the art techniques in autism research, which has progressed from the era of linkage scans, single gene and genome-wide association studies, and copy number variations to next-generation sequencing of the entire human genome. However, with the exception of studies that identified a few causative genes and a limited number of replicated findings, the rest have mainly contributed to the ever growing list of potential autism candidate genes, awaiting replication. This unexpected outcome has now become an additional challenge in moving forward from gene discoveries to diagnostic tests and treatment options in autism. In this chapter, we highlight findings derived from a few integrative approaches using unconventional models, which have resulted in promising discoveries and provided practical examples of how uncovering genetic causes of autism may be accelerated.

*Correspondence to: Children's Mercy Hospital, Kansas City, MO, USA; University of Missouri-Kansas City; School of Medicine; Kansas City; MO; USA. Email: ztalebi@cmh.edu

Keywords: Autism; Integrative approaches; Genetics; Epigenetics; Gene regulatory processes; Gender disparity; Phenotypic stratification; Systems biology; Interactomes.

Introduction

Various susceptibility genes and chromosomal abnormalities have been associated with autism spectrum disorder (ASD), but most discoveries either failed replication or accounted for a small effect. Although, it is widely accepted that multiple genes (and environmental factors) may contribute to the etiology of autism, the question of how many genetic susceptibility factors are involved and how they may be related cumulatively to the respective phenotypic subsets of ASD still remains unsolved. One of the main contributing factors to the low reproducibility rate in autism genetic studies is the lack of sufficient connectivity and integration among research findings. One way to address this gap is by employing integrated system biology-oriented research protocols. To demonstrate the importance of such approaches, a number of successfully applied methods are discussed here. The selected examples describe integrating different layers of genetic and epigenetic factors, gender disparity, and phenotypic subject stratification in autism studies as well as in bioinformatics pipelines thereby connecting the dots, which is expected to lead to the construction of autism-specific system biology networks (i.e., autism interactomes).

Whole Genome Studies

Numerous candidate gene and genome-wide association (GWAS) studies have been reported for autism with the goal of detecting common genetic risk variants, but with inconclusive findings. For instance, GWAS on two independent data sets, including family-based (n = 943 families) and case-control (n = 1,204 and 6,491 subjects, respectively) autism cohorts, did not detect genome-wide significant association initially. However, application of a meta-analysis approach recommended by de Bakker *et al.*[1] on both of these independent data sets revealed genome-wide significance with the rs4307059 SNP on 5p14.1.[2] Evidence for association of autism with 5p14.1 SNPs was also reported by Ma *et al.*[3]

While replication of 5p14.1 association was intriguing, the remaining challenge was to uncover a potential role of this chromosomal region in the etiology of autism. The associated SNPs on 5p14.1 do not reside within known genes; therefore, assessment of the potential candidate genes (i.e., cadherin genes) did not provide conclusive results.[2] While conventional candidate gene screening was not fruitful, using a combination of bioinformatics analysis and expression studies Kerin *et al.*[4] discovered that a large noncoding RNA is transcribed from this autism associated region. Surprisingly, this product was encoded by a pseudogene (*MSNP1AS*), which was highly identical and antisense to the protein coding gene (*MSN*) located on the X chromosome. Individuals who carry the autism associated rs4307059 allele showed an elevation in the expression of this pseudogene. *MSNP1AS* can bind to the complementary transcript on the X chromosome resulting in down regulation of *MSN*. This innovative approach shed light on the interconnected nature of genetic factors and the pivotal role that understudied elements, such as noncoding RNAs and pseudogenes, may play in filling the gaps and unveiling the underlying pathogenic mechanisms of this complex condition.

Weiss *et al.*[5] conducted another large GWAS study using a combination of the transmission disequilibrium test (TDT) and case-control approaches, but found no support for association with the reported SNP at the 5p14.1 locus; instead, their study provided an evidence for association with an intergenic SNP near *SEMA5A* on chromosome 5p15. Initial gene expression studies provided evidence for reduced expression of *SEMA5A* in lymphoblastoid cell line (LCL), blood and brain samples in subjects with autism compared with controls.[5,6] However, a low minor allele frequency (MAF < 0.05) of the associated 5p15 SNP suggested that factors besides reduced expression of this single flanking gene might be involved in susceptibility to autism.

Later, the Autism Genome Project (AGP) consortium, representing more than 50 centers in North America and Europe, provided the largest family-based GWAS study to date, which included 1558 family trios.[7] No evidence was found for association with either 5p14.1 or 5p15 loci, but one SNP located at 20p12.1 (rs4141463 in a *MACROD2* intron) reached genome-wide association significance. However, a later attempt to replicate this finding in an independent case-control association study failed.[8] This lack of replication reflects the fundamental obstacle introduced by the extensive heterogeneity of ASD, which may even increase when subjects are recruited/compiled from diverse genetic backgrounds (e.g., ethnicities).

A follow-up *in silico* eQTL approach using a combination of expression and genotyping data sets revealed that rare variants in the *SEMA5A* regulatory network, including previous autism candidate genes and regions, such as *MACROD2*, may impact autism risk. This integrated approach provided evidence that the same pathway may contain both rare and common autism susceptibility genetic variants.[9] Therefore, collective assessments of different types of susceptibility factors enabled the detection of a potential connectivity between previously scattered findings.

Recent studies provided new insight into interpreting GWAS data and understanding genetic architecture of complex conditions, such as *cis* regulatory gene expression SNPs, by enrichment for expression quantitative trait loci (eQTL) and methylation QTL (mQTL) in relevant tissues.[10,11] Davis and colleagues[12] used existing GWAS data generated by the AGP,[7] genome-wide expression data sets in brain from Gamazon *et al.*,[11] and the SNP and CNV annotation databases.[13] This group adopted the approach previously used on bipolar disorder[11] to test for enrichment of identified eQTLs (in brain and LCLs) among top susceptibility SNPs from autism GWAS reports. This approach showed a global enrichment of brain eQTL (but not LCL eQTL) among top autism associated SNPs previously identified by GWAS.[7] Application of this annotation-based approach[12] also identified three genes as being strongly implicated in autism: *PANX1*, *PANX2*, and *SLC25A12*, two of which (*PANX2*, and *SLC25A12*) were previously reported to be differentially expressed in autism.[14] This type of research approach shows the direct and positive impact that integrated approaches provide for increasing the power of GWAS studies on neurodevelopmental disorders.

In addition to the necessity of providing functional perspective for susceptibility factors as discussed above, another obstacle in detecting genome-wide associated genes/alleles is the additive effect of other risk factors (i.e., genes, gender, and environmental factors), requiring assessment of interactions among multiple elements.[15] Because of the sex bias in autism, gender is one of the secondary factors that need to be assessed in GWAS studies. To improve detection power in autism GWAS studies, Lu and Cantor[15] conducted joint association tests[16] in a cohort of 990 multiplex autism families from the Autism Genetic Resource Exchange (AGRE) by incorporating gender as a risk factor in data analysis. The rationale behind this approach was to investigate the potential gender-specific differential penetrance of some ASD risk alleles. Stratification of the families into two groups based on the sex of the affected member [i.e., male-only (MO) and female-containing (FC)] revealed two associations exceeding genome-wide significance. The two susceptibility intronic alleles, rs6683048 (within *RyR2*) and rs1740138 (within *UPP2*), exhibited an over-transmission in both stratified groups (74% and 75% in the MO, 66% and 77% in the FC families, respectively). Although, the association was observed in both groups, it would not have been detected in a standard family based association test (i.e., without considering the sex factor) using the genome-wide level of significance.

More recent advancements in genomics technology provided a high throughput method for sequencing the entire protein coding regions of the genome using whole exome sequencing (WES). Detection of rare *de novo* loss of function (LoF) mutations has been the central point of most WES reports. However, considering the disease complexity and involvement of multiple genes, other types of *de novo* mutations as well as inherited events need to be assessed, in addition to LoF mutations, but they have been usually excluded in conventional autism WES studies.[17] To address this critical gap, He *et al.*[17] developed a novel integrated statistical method, called TADA for "transmission and *de novo* association". This method yields gene-specific p-values and utilizes LoF *de novo* mutations in conjunction with inherited variations to identify autism risk genes. Such a method also allowed for incorporation of data from family-based and case-control studies, not possible in conventional analytical methods. Application of the TADA method to previously published WES data from about 1000 ASD trios[18–21] and 2000 subjects from a case-control study[22] demonstrated a dramatically higher detection power of this method compared with the commonly used methods (i.e., assessment of *de novo* LoF). Furthermore, some of the detected genes met genome-wide significance in this integrated model, while such a finding would be impossible to obtain without combining the *de novo*, transmitted and case-control data. This improvement in analyzing sequencing data illustrates that, unless used in conjunction with integrated methods, even the most advanced genomic methods (e.g., WES) would not be able to uncover the full complex picture of autism pathogenesis.

Gender Bias in Autism

Autism has a male preponderance of at least 4:1.[23] However, the biologic underpinnings of this gender discrepancy have not been fully elucidated. Among the most prominent proposed theories is the role of X-linked genes and epigenetic factors such as imprinting and X chromosome inactivation.[24] One of the first proposed theories to explain the increased ASD risk in males was a protective role of the X chromosome.[25] A premature speculation was that ASD might be an X-linked disorder, in which case females would have a protective mechanism provided by their second X chromosome.[26] Despite the emerging evidence in favor of a multi-factorial inheritance of ASD (ruling out an X-linked mode of transmission), prenatal sex hormone effects on autistic traits,[27–29] some compensatory effect mechanisms through sex chromosomal gene dosage or sex hormone levels[30] may still exist. Additional area for consideration in relation to the X chromosome theory is the role of the X-linked microRNAs. The excessively high number (~10%) of all human microRNAs located on X versus autosomal chromosomes and their role in immune functions has provided support for another emerging hypothesis (i.e., an immunological advantage of females over males with respect to disease etiology).[31] Therefore, the degree to which X-linked microRNAs may account for the sex differences in autism should also be assessed.

Another proposed explanation for the gender bias in the prevalence of autism is a theory that males on average require fewer genetic risk factors necessary to reach an equivalent impairment threshold.[19,32] This hypothesis was tested from different perspectives; for instance, one study[33] demonstrated that in simplex autistic families, females were more severely affected (based on nonverbal IQ) than boys. This finding was not seen in multiplex families, which suggests the existence of underlying differences with respect to gender and impairment severity in simplex and multiplex families. The study conducted by Szatmari *et al.*[34] tested whether there are gender differences in severity of autism traits in multiplex families (divided into MO and FC) by focusing on repetitive behavior. This study demonstrated that in FC families autistic male relatives are more severely affected than in MO families, a finding also supported by the largest (n = 192 pairs) population based autism twin study.[35] These observations were replicated in a recent study,[36] further supporting the above discussed hypothesis of a female protective effect against autism.

Even though contributing factors to the observed gender bias still need to be elucidated, the emerging evidence has shown that implementing gender-base stratifications in autism genetic studies will increase detection power. This increase in power has been shown in SNP and candidate gene association studies, as well as in the detection of chromosomal abnormalities. For example, copy number variations at the chromosome 16p13 region have been frequently associated with neurodevelopmental disorders, including autism. Assessment of 46 cases with chromosome 16p13 CNVs (both deletions and duplications) showed a sex-biased effect, with a significant enrichment of CNVs only in males, but not in females.[37] This finding demonstrates that taking into account the gender bias is crucial in investigating the potential pathological role of the autism associated CNVs.

By applying a network-based analysis on genome-wide *de novo* CNV data[32] generated from a large cohort of simplex families Gilman *et al.*[38] tested a hypothesis that underlying disease-causing genetic disturbances in females were much stronger than in males. This analysis provided support for two contributing mechanisms: (1) significantly larger CNVs were observed in females than males (harboring about 10 and 3 genes per CNV, respectively) and (2) genes affected by CNVs in females had higher impact on the overall network cluster score than those in males.

In addition to CNVs, the impact of gender bias has also been shown in other autism genetic studies, including linkage and GWAS. In one of the earlier whole genome scans addressing sex bias, a linkage study was conducted on the basis of the sex of affected individuals (i.e., MO and FC families), which identified a male-specific linkage peak.[39] The importance of the applied gender stratification in this linkage study is further highlighted by the fact that the original non-stratified genome scan[40] failed to differentiate the autistic cohort from the general population. Applying sex as an additional risk factor in GWAS of multiplex autism families enhanced the study power and resulted in the detection of sex-specific genetic variants of genome-wide significance.[15]

Not only did inclusion of gender effect increase the detection power in finding autism susceptibility factors, it also improved autism risk assessment models. Carayol *et al.*[41] improved their previously proposed risk assessment genetic scoring system[42] by developing sex-specific score models to identify siblings with significantly higher risk of autism. This approach identified specific sets of SNPs with a high reproducibility index in males and females. Comparing the previous model[42] with this sex-specific scoring system showed a higher detection power in the latter.

Phenotypic Stratification

Autism is often accompanied by other symptom or comorbidity; such co-occurring symptoms may serve as biomarkers to help identify subsets with a greater chance of having an underlying mechanism that plays a role in both ASD and the concurrent non-ASD phenotypes. These co-occurring symptoms have been used in subject stratification for genetic linkage, association, and mutation screening studies. The majority of the findings for these stratified ASD subsets have been successfully replicated, further demonstrating the value of implementing stratification methods for better understanding of the genetic etiology of this heterogeneous condition. Two highly replicated discoveries of causative genes in autism (*PTEN* and *MET*) were made by focusing on co-occurring symptoms (i.e., macrocephaly and gastrointestinal dysfunction, respectively) in subsets of autistic subjects.[43–46]

In addition to co-occuring symptoms, behavioral profiles may be useful in narrowing down the existing phenotypic heterogeneity in autism. Since individuals with ASD manifest deficiencies across a broad range of behaviors, addressing phenotypic heterogeneity requires more inclusive methods than subject stratification based on severity along a single domain, such as language impairment or nonverbal communication. In an attempt to address this gap, Hu and Steinberg[47] applied multiple clustering methods on more than 100 ADI-R scores across a broad range of symptoms, which identified four subgroups of autistic individuals representing distinct phenotypes. Heterogeneity in ASD is also reflected at the family level and, thus, in multiplex families autistic symptoms may vary among affected siblings. Therefore, to address inter- and intra-family heterogeneity we developed a multi-step subject stratification approach[48] by incorporating the proposed ADI-R classification[47] on multiplex families.

To test the detection power of this model in linkage scans, we re-analyzed existing SNP genotyping data for 392 multiplex AGRE families from published studies.[5,49] The affected subject's ADI-R sub-phenotype[47] was used to create group-specific SNP data sets using a multi-step stratification process. The applied deep stratification method provided a pipeline to further filter the original heterogeneous ASD pedigree data file into more homogeneous data sets by first using ADI-R cluster analysis, followed by removal of sub-phenotypically discordant siblings, and finally by separation of MO and FC pedigrees.[48]

No locus reached significance for the combined non-stratified cohort (n = 392 families). However, study-wide significant linkage scores (i.e., LOD score greater than 4) were reached for chromosomes 22q11 and 13q21 for two ASD subsets (n = 232 and 63 families, respectively) representing the most severely language-impaired individuals. Notably, 13q21 has been previously linked to autism with language impairment, and 22q11 has been separately associated with either autism or language disorders. Linkage analysis on chromosome 5p15 for a combination of two stratified FC subgroups (n = 23 families) demonstrated suggestive linkage (LOD = 3.5), which replicates previous linkage results for FC pedigrees.[49] A trend was also found for the association of previously reported autism associated 5p14-p15 SNPs[2,5] in the same combined FC cohort.

Interestingly, conducting a similar analysis on all FC families (n = 166), without applying the ADI-R subtyping, did not detect associations to this chromosomal region. Our findings in a given ADI-R stratified FC cohort not only confirmed previous linkage[49] and association[2] reports, but also further refined the proposed link between the 5p region and autism. Furthermore, examination of the associated noncoding RNA (*MSNP1AS*) in the identified homogeneous FC cohort in our study (that showed positive associations to the susceptibility SNPs at 5p) may enhance the revelation of other autism risk factors in the intriguing model proposed by Kerin *et al.*[4]

This integrated approach illustrated the added likelihood of detecting significant linkage when the heterogeneity of the ASD population is reduced by proper sample stratification. It also provided first lines of evidence at the gene scan level for both inter- and intra-family heterogeneity, reflecting both shared and distinct genetic makeup in the autism population. Inter- and intra-family heterogeneity is another important factor that has been overlooked in whole-genome screening of multiplex families.

Network Analysis

It has been more than a decade since researchers became interested in developing autism network models to predict gene/environment or gene/gene interactions by evaluating experimental findings using computational tools and data-mining. In one of the earliest papers on this topic[50] multiple bioinformatics tools, including the Pathwayassist software (Ariadne Genomics, Rockville, MD) and the ResNet database, were applied to positional autism candidate genes identified by an earlier genome-wide linkage study.[40] The objectives of the data-mining study were to prioritize candidate genes, predict pathways of interacting genes, and identify biologically meaningful autism candidates. Since then, a wider spectrum of analytical tools have been developed and applied to extensive autism genetic data, and some of them will be highlighted here.

Lee *et al.*[51] performed an integrative autism gene network analysis based on seven phenotypic features commonly reported in autism, in conjunction with multiple sets of autism genetic data (i.e., GWAS, CNV, and gene expression). Genes related to these seven restricted phenotypes were retrieved from the curated databases (Ingenuity, GeneGo, and HuGe navigator). A novel CNV-centric node network analysis method, as well as GWAS and expression data were then used to identify potential regulatory networks in autism.

Later, literature-based curated data for autism-specific rare and syndromic candidate genes were used to build a composite reference profile based on functional and expression analyses using an integrated bioinformatics approach.[52] A brain region-specific predictive gene map for autism was built by creating a functional profile of the reference set of ASD candidate genes using GO enrichment analysis, along with an expression profile utilizing DAVID analysis. The functional profile identified enrichment of three biological processes critical for synaptic transmission in ASD genes. Such a predictive gene map may facilitate prioritizing autism-related genomic data.

Poelmans *et al.*[29] constructed protein networks by conducting bioinformatics analysis and integrating SNP data from six GWAS studies.[2,3,5,7,53,54] This finding was further corroborated by obtaining evidence from exome sequencing, CNV, expression studies (gene and microRNA), as well as genetic animal studies, which resulted in the identification of three ASD-related signaling networks. Furthermore, this bioinformatics and literature analyses data integration study showed overlap between rare and common ASD implicated genetic variants.

More recently, to examine the multi-factorial nature of autism, Zeidan-Chulia and colleagues[55] developed a computational method using a combination of thorough literature search and a system biology approach. One of their objectives was to determine the main biological processes and the dominant contributing factors (genetic and/or environmental) integral to the disorder. Such *in silico* models may offer a way to identify potentially relevant genetic and environmental factors as well as their interactions in autism.

Biological relevance of disease associated genes needs to be confirmed by functional experiments. One of the challenges in pursuing productive functional studies for autism genes is to know the relevant spatial and temporal gene expression pattern in brain. To address this gap, Willsey *et al.*[56] constructed co-expressed networks based on a list of high confidence ASD genes identified by exome and genome-wide sequencing studies and using a publicly available spatio-temporal transcriptome of the human brain data set[57] as the substrate. Such an integrated analytic approach provided a potential basis for identification of pathophysiological mechanisms of ASD-related mutations by guiding when, where, and in what cell type should a candidate gene be studied.[56]

Interactome Networks

The importance of employing integrated approaches in understanding the pathophysiology of human diseases has been further highlighted in the genetic field by introducing a new concept of interactome networks. This concept refers to the interactions between biological systems and complex cellular networks, and has been proposed as a next step toward predicting genotype-phenotype related outcomes.[58] There are different types of interactome networks in cellular systems, including metabolic, protein-protein interaction, gene regulatory systems, and networks of functional links (transcriptional profiling, phenotypic and genetic interaction networks). To avoid "simplifying" compound systems and losing the functional influence of each component, assessment of a combination of networks is essential. Because of the overwhelming intricacy of biological network perturbations within cells, the interactome network mapping has not yet been fully completed; however, mapping has been compiled for several model organisms and more advancements are underway for human as well.[58]

As anticipated, the concept of interactomes has begun to receive attention in the autism research community as shown by the development of an autism-specific protein interactome.[59] Such efforts can be further expanded by incorporating other networks, such as microRNA inter-regulatory human interactomes.[60] Integrated approaches in autism genetic studies such as those discussed in this chapter will provide additional knowledge-base for advancing autism interactomes re-construction.

Future Directions

Despite a growing recognition of the multi-factorial architecture of autism and the need to consider gender disparity and phenotypic heterogeneity, these essential elements are still often overlooked in study designs and research protocols. A substantial portion of the research funding has already been devoted to generating the rapidly growing amount of different type of genomic data from autistic subjects. Autism initiatives (i.e., the Autism Genetic Resource Exchange (AGRE), Simons Simplex Collection (SSC), and the National Database for Autism Research (NDAR)[61]) and consortiums (i.e., Autism Genome Project (AGP) and Autism Sequencing) have provided invaluable resources to the research community by collecting phenotypic data and supporting the ongoing generation of genomic data (e.g., genotyping, CNV, sequencing, and expression) from subjects with autism.

A long-needed shift from generating new data to re-assessment of already existing phenotypic and genomic data along with developing novel integrated research models would accelerate the uncovering of previously undetected autism risk factors. If incorporated in genomic data re-assessments, established concepts such as sex bias, comorbidity, and phenotypic heterogeneity have a great potential to accelerate replication of findings, subtype detection, and biomarker identification, as well as to facilitate the development of evidence-based and more effective treatment strategies for autism.

Executive Summary

- Multiple genes (and environmental factors) may contribute to the etiology of non-oligogenic forms of autism, which necessitates cumulative assessment of potential genetic and epigenetic contributing factors.
- Integrated approaches (i.e., utilizing two or more phenotypic and genetic factors as well as bioinformatics and data mining pipelines) would be beneficial in detecting biologically relevant and replicated findings that move the autism genetic field forward.
- The examples discussed in this chapter illustrate how connectivity among multiple lines of scattered autism research findings can be established by employing integrated system biology-oriented research protocols.
- Abnormalities in gene regulatory processes such as noncoding RNAs, alternative splicing, and DNA methylation may play a critical role in pathogenesis of complex conditions. The existing and ongoing large-scale autism genetic sequencing and expression data need to be also analyzed with respect to assessing the gene regulatory factors.

- Considering the extensive crosstalk between genetic and epigenetic mechanisms assessment of cumulative and interactive effects of multiple factors in autism requires designing comprehensive research plans and data analysis strategies to unveil pathogenesis of multi-factorial forms of ASD.

References

1. de Bakker PI, Ferreira MA, Jia X, Neale BM, Raychaudhuri S, Voight BF: Practical aspects of imputation-driven meta-analysis of genome-wide association studies. *Hum Mol Genet* 2008, **17**(R2):R122–128.
2. Wang K, Zhang H, Ma D, Bucan M, Glessner JT, Abrahams BS, Salyakina D, Imielinski M, Bradfield JP, Sleiman PM *et al.*: Common genetic variants on 5p14.1 associate with autism spectrum disorders. *Nature* 2009, **459**(7246):528–533.
3. Ma D, Salyakina D, Jaworski JM, Konidari I, Whitehead PL, Andersen AN, Hoffman JD, Slifer SH, Hedges DJ, Cukier HN *et al.*: A genome-wide association study of autism reveals a common novel risk locus at 5p14.1. *Ann Hum Genet* 2009, **73**(Pt 3):263–273.
4. Kerin T, Ramanathan A, Rivas K, Grepo N, Coetzee GA, Campbell DB: A noncoding RNA antisense to moesin at 5p14.1 in autism. *Sci Transl Med* 2012, **4**(128):128ra140.
5. Weiss LA, Arking DE, Daly MJ, Chakravarti A: A genome-wide linkage and association scan reveals novel loci for autism. *Nature* 2009, **461**(7265): 802–808.
6. Melin M, Carlsson B, Anckarsater H, Rastam M, Betancur C, Isaksson A, Gillberg C, Dahl N: Constitutional downregulation of SEMA5A expression in autism. *Neuropsychobiology* 2006, **54**(1):64–69.
7. Anney R, Klei L, Pinto D, Regan R, Conroy J, Magalhaes TR, Correia C, Abrahams BS, Sykes N, Pagnamenta AT *et al.*: A genome-wide scan for common alleles affecting risk for autism. *Hum Mol Genet* 2010, **19**(20):4072–4082.
8. Curran S, Bolton P, Rozsnyai K, Chiocchetti A, Klauck SM, Duketis E, Poustka F, Schlitt S, Freitag CM, Lee I *et al.*: No association between a common single nucleotide polymorphism, rs4141463, in the MACROD2 gene and autism spectrum disorder. *Am J Med Genet B Neuropsychiatr Genet* 2011, **156B**(6):633–639.
9. Cheng Y, Quinn JF, Weiss LA: An eQTL mapping approach reveals that rare variants in the SEMA5A regulatory network impact autism risk. *Hum Mol Genet* 2013, **22**(14):2960–2972.
10. Nicolae DL, Gamazon E, Zhang W, Duan S, Dolan ME, Cox NJ: Trait-associated SNPs are more likely to be eQTLs: annotation to enhance discovery from GWAS. *PLoS Genet* 2010, **6**(4):e1000888.

11. Gamazon ER, Badner JA, Cheng L, Zhang C, Zhang D, Cox NJ, Gershon ES, Kelsoe JR, Greenwood TA, Nievergelt CM *et al.*: Enrichment of cis-regulatory gene expression SNPs and methylation quantitative trait loci among bipolar disorder susceptibility variants. *Mol Psychiatry* 2013, **18**(3):340–346.
12. Davis LK, Gamazon ER, Kistner-Griffin E, Badner JA, Liu C, Cook EH, Sutcliffe JS, Cox NJ: Loci nominally associated with autism from genome-wide analysis show enrichment of brain expression quantitative trait loci but not lymphoblastoid cell line expression quantitative trait loci. *Mol Autism* 2012, **3**(1):3.
13. Gamazon ER, Zhang W, Konkashbaev A, Duan S, Kistner EO, Nicolae DL, Dolan ME, Cox NJ: SCAN: SNP and copy number annotation. *Bioinformatics* 2010, **26**(2):259–262.
14. Voineagu I, Wang X, Johnston P, Lowe JK, Tian Y, Horvath S, Mill J, Cantor RM, Blencowe BJ, Geschwind DH: Transcriptomic analysis of autistic brain reveals convergent molecular pathology. *Nature* 2011, **474**(7351):380–384.
15. Lu AT, Cantor RM: Allowing for sex differences increases power in a GWAS of multiplex Autism families. *Mol Psychiatry* 2012, **17**(2):215–222.
16. Cordell HJ, Barratt BJ, Clayton DG: Case/pseudocontrol analysis in genetic association studies: a unified framework for detection of genotype and haplotype associations, gene-gene and gene-environment interactions, and parent-of-origin effects. *Genet Epidemiol* 2004, **26**(3):167–185.
17. He X, Sanders SJ, Liu L, De Rubeis S, Lim ET, Sutcliffe JS, Schellenberg GD, Gibbs RA, Daly MJ, Buxbaum JD *et al.*: Integrated model of de novo and inherited genetic variants yields greater power to identify risk genes. *PLoS Genet* 2013, **9**(8):e1003671.
18. Sanders SJ, Murtha MT, Gupta AR, Murdoch JD, Raubeson MJ, Willsey AJ, Ercan-Sencicek AG, DiLullo NM, Parikshak NN, Stein JL *et al.*: De novo mutations revealed by whole-exome sequencing are strongly associated with autism. *Nature* 2012, **485**(7397):237–241.
19. Neale BM, Kou Y, Liu L, Ma'ayan A, Samocha KE, Sabo A, Lin CF, Stevens C, Wang LS, Makarov V *et al.*: Patterns and rates of exonic de novo mutations in autism spectrum disorders. *Nature* 2012, **485**(7397):242–245.
20. O'Roak BJ, Vives L, Girirajan S, Karakoc E, Krumm N, Coe BP, Levy R, Ko A, Lee C, Smith JD *et al.*: Sporadic autism exomes reveal a highly interconnected protein network of de novo mutations. *Nature* 2012, **485**(7397):246–250.
21. Iossifov I, Ronemus M, Levy D, Wang Z, Hakker I, Rosenbaum J, Yamrom B, Lee YH, Narzisi G, Leotta A *et al.*: De novo gene disruptions in children on the autistic spectrum. *Neuron* 2012, **74**(2):285–299.
22. Liu L, Sabo A, Neale BM, Nagaswamy U, Stevens C, Lim E, Bodea CA, Muzny D, Reid JG, Banks E *et al.*: Analysis of rare, exonic variation amongst subjects with autism spectrum disorders and population controls. *PLoS Genet* 2013, **9**(4):e1003443.

23. Volkmar FR, Szatmari P, Sparrow SS: Sex differences in pervasive developmental disorders. *J Autism Dev Disord* 1993, **23**(4):579–591.
24. Marco EJ, Skuse DH: Autism-lessons from the X chromosome. *Soc Cogn Affect Neurosci* 2006, **1**(3):183–193.
25. Skuse DH: Imprinting, the X-chromosome, and the male brain: explaining sex differences in the liability to autism. *Pediatr Res* 2000, **47**(1):9–16.
26. Baron-Cohen S, Lombardo MV, Auyeung B, Ashwin E, Chakrabarti B, Knickmeyer R: Why are autism spectrum conditions more prevalent in males? *PLoS Biol* 2011, **9**(6):e1001081.
27. Auyeung B, Baron-Cohen S, Ashwin E, Knickmeyer R, Taylor K, Hackett G: Fetal testosterone and autistic traits. *Br J Psychol* 2009, **100**(Pt 1):1–22.
28. Auyeung B, Taylor K, Hackett G, Baron-Cohen S: Foetal testosterone and autistic traits in 18 to 24-month-old children. *Mol Autism* 2010, **1**(1):11.
29. Poelmans G, Franke B, Pauls DL, Glennon JC, Buitelaar JK: AKAPs integrate genetic findings for autism spectrum disorders. *Transl Psychiatry* 2013, **3**:e270.
30. Sarachana T, Xu M, Wu RC, Hu VW: Sex hormones in autism: androgens and estrogens differentially and reciprocally regulate RORA, a novel candidate gene for autism. *PLoS One* 2011, **6**(2):e17116.
31. Pinheiro I, Dejager L, Libert C: X-chromosome-located microRNAs in immunity: might they explain male/female differences? The X chromosome-genomic context may affect X-located miRNAs and downstream signaling, thereby contributing to the enhanced immune response of females. *Bioessays* 2011, **33**(11):791–802.
32. Levy D, Ronemus M, Yamrom B, Lee YH, Leotta A, Kendall J, Marks S, Lakshmi B, Pai D, Ye K *et al.*: Rare de novo and transmitted copy-number variation in autistic spectrum disorders. *Neuron* 2011, **70**(5):886–897.
33. Banach R, Thompson A, Szatmari P, Goldberg J, Tuff L, Zwaigenbaum L, Mahoney W: Brief Report: relationship between non-verbal IQ and gender in autism. *J Autism Dev Disord* 2009, **39**(1):188–193.
34. Szatmari P, Liu XQ, Goldberg J, Zwaigenbaum L, Paterson AD, Woodbury-Smith M, Georgiades S, Duku E, Thompson A: Sex differences in repetitive stereotyped behaviors in autism: implications for genetic liability. *Am J Med Genet B Neuropsychiatr Genet* 2012, **159B**(1):5–12.
35. Hallmayer J, Cleveland S, Torres A, Phillips J, Cohen B, Torigoe T, Miller J, Fedele A, Collins J, Smith K *et al.*: Genetic heritability and shared environmental factors among twin pairs with autism. *Arch Gen Psychiatry* 2011, **68**(11):1095–1102.
36. Robinson EB, Lichtenstein P, Anckarsater H, Happe F, Ronald A: Examining and interpreting the female protective effect against autistic behavior. *Proc Natl Acad Sci U S A* 2013, **110**(13):5258–5262.
37. Tropeano M, Ahn JW, Dobson RJ, Breen G, Rucker J, Dixit A, Pal DK, McGuffin P, Farmer A, White PS *et al.*: Male-biased autosomal effect of

16p13.11 copy number variation in neurodevelopmental disorders. *PLoS One* 2013, **8**(4):e61365.
38. Gilman SR, Iossifov I, Levy D, Ronemus M, Wigler M, Vitkup D: Rare de novo variants associated with autism implicate a large functional network of genes involved in formation and function of synapses. *Neuron* 2011, **70**(5):898–907.
39. Stone JL, Merriman B, Cantor RM, Yonan AL, Gilliam TC, Geschwind DH, Nelson SF: Evidence for sex-specific risk alleles in autism spectrum disorder. *Am J Hum Genet* 2004, **75**(6):1117–1123.
40. Yonan AL, Alarcon M, Cheng R, Magnusson PK, Spence SJ, Palmer AA, Grunn A, Juo SH, Terwilliger JD, Liu J *et al.*: A genomewide screen of 345 families for autism-susceptibility loci. *Am J Hum Genet* 2003, **73**(4):886–897.
41. Carayol J, Schellenberg GD, Dombroski B, Genin E, Rousseau F, Dawson G: Autism risk assessment in siblings of affected children using sex-specific genetic scores. *Mol Autism* 2011, **2**(1):17.
42. Carayol J, Schellenberg GD, Tores F, Hager J, Ziegler A, Dawson G: Assessing the impact of a combined analysis of four common low-risk genetic variants on autism risk. *Mol Autism* 2010, **1**(1):4.
43. Butler MG, Dasouki MJ, Zhou XP, Talebizadeh Z, Brown M, Takahashi TN, Miles JH, Wang CH, Stratton R, Pilarski R *et al.*: Subset of individuals with autism spectrum disorders and extreme macrocephaly associated with germline PTEN tumour suppressor gene mutations. *J Med Genet* 2005, **42**(4):318–321.
44. McBride KL, Varga EA, Pastore MT, Prior TW, Manickam K, Atkin JF, Herman GE: Confirmation study of PTEN mutations among individuals with autism or developmental delays/mental retardation and macrocephaly. *Autism Res* 2010, **3**(3):137–141.
45. Herman GE, Butter E, Enrile B, Pastore M, Prior TW, Sommer A: Increasing knowledge of PTEN germline mutations: two additional patients with autism and macrocephaly. *Am J Med Genet A* 2007, **143**(6):589–593.
46. Campbell DB, Buie TM, Winter H, Bauman M, Sutcliffe JS, Perrin JM, Levitt P: Distinct genetic risk based on association of MET in families with co-occurring autism and gastrointestinal conditions. *Pediatrics* 2009, **123**(3):1018–1024.
47. Hu VW, Steinberg ME: Novel clustering of items from the Autism Diagnostic Interview-Revised to define phenotypes within autism spectrum disorders. *Autism Res* 2009, **2**(2):67–77.
48. Talebizadeh Z, Arking DE, Hu VW: A novel stratification method in linkage studies to address inter- and intra-family heterogeneity in autism. *PLoS One* 2013, **8**(6):e67569.
49. Szatmari P, Paterson AD, Zwaigenbaum L, Roberts W, Brian J, Liu XQ, Vincent JB, Skaug JL, Thompson AP, Senman L *et al.*: Mapping autism risk loci using genetic linkage and chromosomal rearrangements. *Nat Genet* 2007, **39**(3):319–328.

50. Yonan AL, Palmer AA, Smith KC, Feldman I, Lee HK, Yonan JM, Fischer SG, Pavlidis P, Gilliam TC: Bioinformatic analysis of autism positional candidate genes using biological databases and computational gene network prediction. *Genes Brain Behav* 2003, **2**(5):303–320.
51. Lee TL, Raygada MJ, Rennert OM: Integrative gene network analysis provides novel regulatory relationships, genetic contributions and susceptible targets in autism spectrum disorders. *Gene* 2012, **496**(2):88–96.
52. Kumar A, Swanwick CC, Johnson N, Menashe I, Basu SN, Bales ME, Banerjee-Basu S: A brain region-specific predictive gene map for autism derived by profiling a reference gene set. *PLoS One* 2011, **6**(12):e28431.
53. Salyakina D, Ma DQ, Jaworski JM, Konidari I, Whitehead PL, Henson R, Martinez D, Robinson JL, Sacharow S, Wright HH *et al.*: Variants in several genomic regions associated with asperger disorder. *Autism Res* 2010, **3**(6):303–310.
54. Hussman JP, Chung RH, Griswold AJ, Jaworski JM, Salyakina D, Ma D, Konidari I, Whitehead PL, Vance JM, Martin ER *et al.*: A noise-reduction GWAS analysis implicates altered regulation of neurite outgrowth and guidance in autism. *Mol Autism* 2011, **2**(1):1.
55. Zeidan-Chulia F, Rybarczyk-Filho JL, Salmina AB, de Oliveira BH, Noda M, Moreira JC: Exploring the multifactorial nature of autism through computational systems biology: calcium and the Rho GTPase RAC1 under the spotlight. *Neuromolecular Med* 2013, **15**(2):364–383.
56. Willsey AJ, Sanders SJ, Li M, Dong S, Tebbenkamp AT, Muhle RA, Reilly SK, Lin L, Fertuzinhos S, Miller JA *et al.*: Coexpression networks implicate human midfetal deep cortical projection neurons in the pathogenesis of autism. *Cell* 2013, **155**(5):997–1007.
57. Kang HJ, Kawasawa YI, Cheng F, Zhu Y, Xu X, Li M, Sousa AM, Pletikos M, Meyer KA, Sedmak G *et al.*: Spatio-temporal transcriptome of the human brain. *Nature* 2011, **478**(7370):483–489.
58. Vidal M, Cusick ME, Barabasi AL: Interactome networks and human disease. *Cell* 2011, **144**(6):986–998.
59. Sakai Y, Shaw CA, Dawson BC, Dugas DV, Al-Mohtaseb Z, Hill DE, Zoghbi HY: Protein interactome reveals converging molecular pathways among autism disorders. *Sci Transl Med* 2011, **3**(86):86ra49.
60. Sengupta D, Bandyopadhyay S: Participation of microRNAs in human interactome: extraction of microRNA-microRNA regulations. *Mol Biosyst* 2011, **7**(6):1966–1973.
61. Hall D, Huerta MF, McAuliffe MJ, Farber GK: Sharing heterogeneous data: the national database for autism research. *Neuroinformatics* 2012, **10**(4):331–339.

Chapter 6

Construction of a Genetic Classifier for ASD Using Gene Pathway Analysis

Gursharan Chana,* Renee Testa, Piers Gillett, Daniel Williams, Chad A. Bousman, Daniela Zantomio, Ian P. Everall, Christos Pantelis and Efstratios Skafidas

Abstract

A number of genetic investigations in autism spectrum disorders (ASD) have identified single nucleotide polymorphisms (SNPs) and copy number variants (CNVs) that are associated with this neurodevelopmental disorder. However, as research progresses the evidence points to these genetic alterations likely being confined to specific families or populations with their penetrance in larger populations being somewhat lower. This has led researchers to postulate that the genetics of ASD may consist of an additive collection of common genetic variations that when considered in combination with each other are able to explain more of the heritability of the disorder. In addition to taking a more global view of ASD genetics the contribution of these alterations to the clinical landscape of ASD must also be considered. This chapter begins by summarising non-genetic and genetic findings relating to diagnosing or classifying individuals falling within the ASD continuum. We then discuss the utility of using gene pathway analysis to build a genetic classifier for ASD based

*Correspondence to: Centre for Neural Engineering, The University of Melbourne, Parkville Victoria, 3010. Email: gchana@unimelb.edu.au

on our recent published work including strategies to control for population stratification and inclusion of both risk and protective contributors of ASD status. Finally, we will discuss how this approach may be applied to other neuropsychiatric disorders and ways to improve on classification accuracy.

Keywords: Genetic; Autism; Pathway; Diagnosis; Classification.

Introduction

Autism Spectrum Disorders (ASD) are a heterogenous group of lifelong neurodevelopmental disabilities with a prevalence of 1–3%.[1] ASD is a broad diagnostic classification that includes autistic disorders, Asperger Disorder (AD) and Pervasive Developmental Disorder Not Otherwise Specified (PDD-NOS). In both the recent diagnostic and statistical manual-5 (DSM-5) and the draft of the international classification of diseases-11 (ICD-11), the core features of ASD have been grouped into two behavioral dimensions: (1) social communication and (2) restricted, repetitive behaviors and interests.[2] To date, the aetiology of ASD remains to be resolved with diagnosis relying solely on clinical interview. Nevertheless, ASD have been shown to be highly heritable with concordance rates amongst monozygotic twins reaching at least 60% and 30% for dizygotic twins.[3] In addition, genetic investigations have revealed significant evidence for mutations in several candidate genes for ASD including genes encoding the SH3 and multiple ankyrin repeat domains (SHANK) family of proteins[4–7] with dysregulation in methylation status of *SHANK3* also demonstrated in post-mortem brains of individuals with ASD versus controls.[8] These findings for the SHANK family of proteins are of interest as they are downstream of metabotropic glutamate receptor-5 (GRM5) that has also been heavily implicated in ASD.[9, 10] Evidence for the involvement of other loci in ASD have come from studies demonstrating copy number variations (CNVs) as well as *de novo* events in 5–15% of individuals with ASD.[11] Taken individually, alterations in these genes and loci have provided strong signals for association; however, on their own they fail to achieve accurate diagnostic classification for ASD in meta-analyses or larger cohorts. This result has led researchers to the conclusion that ASD comprise a group of polygenic and heterogeneous disorders with a multi-factorial aetiology. Indeed, a recent investigation estimates that there may be between 400–1000 gene loci involved in ASD.[12] Ontological grouping of these genes according to their cellular and molecular functioning has provided a way to cohesively look at ASD risk genes and to highlight key pathways that may be involved with the disorder. Taking this approach, it is clear that a number of different pathways stand out as being associated with ASD. These pathways include synaptic and neuronal activity, glutamatergic functioning, cell adhesion as well as genes involved in gamma-amino-butyric acid (GABA) functioning and genes associated with other neurodevelopment disorders. This chapter will initially provide an overview of non-genetic approaches to classifying ASD, including analysis of behavior and cognition, followed by a discussion of genetic findings for ASD. We will then cover strategies for developing genetic diagnostic classifiers for ASD and allude to future directions to improve these classifiers while accounting for varying degrees of ethnic divergence across and within populations under study. Finally, the importance of validating and extending on genetic findings for ASD by investigation of the mRNA and proteins they encode will be discussed.

Clinical Diagnosis of ASD

There are many challenges within the diagnostic field of autism spectrum disorders. At present there is no instrument or test (psychological, behavioral, or biological) that allows for a definitive diagnosis of an autism spectrum disorder (ASD) in the absence of clinical expertise.[13] Nevertheless, the DSM-IV-TR[14] and more recently the DSM-5[15] provide descriptions of key domains of dysfunction that are present in those with an ASD including social impairment and repetitive and restrictive behaviors and interests, in addition to the more recently recognized sensory processing disturbances.[16,17] A range of assessment measurements aimed at identifying this set of clinical and behavioral characteristics has arisen. These measurements are generally designed to allow the clinician (i.e. paediatrician, psychologist, psychiatrist, general medical practitioner) to calculate an 'Autism' score in order to determine whether the child would meet criteria for an ASD.

Measurements available have ranged from short questionnaire measures such as the Modified Checklist for Autism in Toddlers (M-CHAT),[18–20] Social Communication Questionnaire (SCQ),[21–23] the Autism Behavior Checklist (ABC),[24,25] and the Social Responsiveness Scale (SRS), which are directed at specific behaviors thought to be lacking or characteristic of ASD. These include protodeclarative pointing, following a point, bringing objects to show parents, social interest, imitation, and responding to one's name; these behaviors have been found to discriminate between toddlers later diagnosed with or without autism or pervasive developmental disorder.[26–28] Such short behavioral checklists have at times been found to be inadequate and have not provided an in-depth assessment of "more difficult to diagnose" children. In response to this, the Autism Diagnostic Interview (now revised; ADI-R)[29] was developed. This is a structured, standardized parent interview devised to assess the presence and severity of symptoms of autism in early childhood across three main symptom areas of social relatedness, communication, and repetitive, restrictive behaviors. Whilst this was considered the gold standard of autism assessment, the long administration time of two to three hours, in addition to the reliance on parental retrospective reporting meant that additional revisions of ASD assessments were deemed necessary. The lack of clinician interaction with the child being assessed was most prominent, as previous measurements had solely relied on the parents or caregiver to provide the necessary information to make a diagnosis. Further, reliance on developmental history and lack of direct face-to-face clinical assessment of the child meant that poor developmental historians (i.e., parents or caregivers), or children that did not exhibit the 'classic' characteristic ASD behaviors listed on these checklists, may have led to diagnostic uncertainty. In addition, those presenting with confounding variables, such as intellectual delay or developmental language disorder, made assessment extremely difficult or ambiguous at best. This had implications for early intervention and access to government resources that require a diagnosis. In view of these limitations, the most recent advancement in the clinical diagnosis of ASD, the Autism Diagnostic Observation Schedule (ADOS),[30–32] was developed. The ADOS is a semi-structured standardized interview using developmentally appropriate social and toy-based interactions in a 30- to 45-min interview to elicit symptoms of autism in four areas: social interaction, communication, play, and repetitive behaviors. It places the child in several situations that stress social and communicative interactions and allows for direct observation of ASD symptoms. Four modules, each directed at a particular level of language ability are available.

Regardless of the behavioral measurements that are utilised, what is evident is that a key series of assessment steps are required before any diagnosis of a child is made. This allows children of all ages and levels of cognitive and language functioning to be assessed, as this process can be adapted according to the level of need.[7,33,34] Firstly, a detailed account of the child's early developmental progression and symptom presentation, in addition to direct observation and interaction with the child, is required. Given that parental report can be biased or not provide enough information, gathering data from multiple settings and sources is imperative (such as school, teacher, tutor, home environment, two parents). Children with autism are very sensitive to their environment as well as the structure or support they are provided with during assessment. Therefore, they can behave and exhibit vastly different behaviors across different settings. In addition to this, the assessment of cognitive, language, and adaptive skills for the presence of overall developmental delay or language impairments is necessary. This may be carried out by a team of clinicians, but is required to ensure that the presenting behaviors are not consistent with another neurodevelopmental condition. Many ASD behaviors are not specific to only autism and therefore a comprehensive and wide-ranging assessment needs to occur to rule-out alternative diagnoses.[35,36]

A significant obstacle to the diagnosis of ASD is the wide-ranging spectrum of symptoms that are present over the period from birth to adulthood.[26] It has been reported that any individual can present with specific ASD-like behaviors over a lifetime, which may not be consistently evident. For example, repetitive behaviors of play observed in the early toddler years may not be seen from early childhood and may manifest as a circumscribed interest instead. Alternatively, some children may demonstrate no anxiety in relation to new events or social situations and exhibit an interest in peer friendships, but later demonstrate very specific routines of behavior in their home environment that must be adhered to otherwise significant anxiety-based behaviors are observed.[2,37] The very specific nature of these ASD traits, which are dependent on the profile of the individual and their stage of development and age, mean that evaluation of the neurodevelopmental trajectories of ASD behaviors over time would be extremely informative for the assessment and diagnosis of ASD.[38] If we were able to develop a profile of possible ASD behaviors exhibited in different sub-groups of ASD individuals from birth to adulthood, we would be able to enquire more specifically about behaviors, characteristics or traits that we could expect from each individual that may easily be overlooked. Indeed, in a prospective longitudinal study, Ozonoff and colleagues (2010) reported that children at high-risk of developing ASD showed overt behavioral signs of ASD between 6–12 months of age, including a decrease in eye contact, social smiling, and social responsiveness, with significant differences from typically developing children apparent by 12–18 months.[28] The early identification and recognition of emerging autism-specific behaviors has obvious implications for the early detection of children at risk. Greater understanding of the developmental trajectory of ASD would assist the diagnostic process by informing clinicians of expected symptom presentation at different ages and would also help families by offering a clearer view of an affected individual's expected development over time.[39,40]

Whilst mapping the neurodevelopmental trajectories of ASD behaviors would be invaluable, the analysis and identification of behaviors as the principal source for diagnosis of ASD still has ongoing challenges. For example, the behavior in question may not be evident at the specific time of assessment, or the behavior could be misinterpreted, mislabelled or not identified. Given this ambiguity, other means to support a clinical behavioral diagnosis, such as biological biomarkers, would help facilitate the assessment process and increase the reliability of the available diagnostic tools.

Genetic Studies in ASD

Monogenic disorders with increased incidence of ASD

Strong evidence for a genetic basis to ASD has come from monogenic disorders that show significantly higher rates of ASD incidence than the general population. This includes Fragile-X syndrome (FXS), Rett syndrome and Tuberous Sclerosis. Together these disorders are associated with ~5% of all ASD cases.[41] Of these disorders, FXS is the most common cause of inherited intellectual disability with 40–60% of male patients and 20% of female patients meeting diagnostic criteria for ASD.[42,43] FXS results from a mutation in the fragile X mental retardation 1 (*FMR1*) gene on the X-chromosome which encodes for a protein, FMRP, that has a role in RNA-binding and blocking off translation of particular mRNAs at synapses.[44] This is of particular relevance to ASD with synaptic defects potentially being a hallmark of pathogenesis and a likely treatment target.[45] Dysregulation of mRNA translation is not unique to FXS though as in tuberous sclerosis mutations in tuberous sclerosis-1 (*TSC-1*) and *TSC-2* genes also lead to dysregulation of mRNA translation that are responsible for the disorder. Tuberous sclerosis likewise has a high incidence of ASD among sufferers which is estimated at between 20–60%.[46] The incorrectly functioning TSC complex as a result of the mutations leads to abnormal protein synthesis in the neurons of afflicted individuals and thus negatively affects synaptic development and plasticity.[44]

Another neurodevelopmental disorder with a higher incidence of ASD is Rett syndrome. The disorder is found almost exclusively in females and ASD symptoms are typically part of the overall symptomatology.[47] The condition arises from a mutation in methyl CpG binding protein 2 (*MECP2)* gene, which encodes a protein product that is thought to be involved in chromatin structure and gene regulation and with mutations present in over 80% of all females with ASD.[44] Disruption of *MECP2*, a transcriptional regulator, causes Rett syndrome and is associated with reduced expression of the mammalian target of rapamycin (mTOR) and reduced phosphorylated S6 kinase (RPS6KB1), important for translation of synaptic proteins.[48] MeCP2 regulates brain-derived neurotrophic factor (BDNF), cyclic-AMP responsive element binding protein-1 (CREB-1) and insulin growth factor-1 (IGF-1), which mediate the PI3K/Akt/mTOR/MAPK pathway.[49] Inability to phosphorylate MeCP2 protein in response to neuronal activity in cultured hippocampal neurons from the $Mecp2^{S421A;S424A/y}$ mouse resulted in inhibited bicuculline-induced reduced synaptic scaling and absence of MeCP2 results in reduced GRM5 signalling.[50]

Copy number variants (CNVs) in ASD

Over the last decade large, rare and *de novo* CNVs have been identified in individuals with ASD at a number of different sites on various chromosomes including 15q11–13, 16p11.2, 22q11.2, 7q11.23 and 1q21.1.[51–57] Although initial studies pointed to these CNVs being increased in frequency across the genome in individuals with ASD in comparison to controls,[58,59] follow-up studies have not confirmed these global alterations.[60–62] Therefore, global genomic instability may not be present across all individuals with ASD and instead may be confined to certain families. While most CNVs are private, recurrent microdeletions have also been identified in individuals with ASD or autistic traits.[63] As these CNVs are associated with a number of different autism phenotypes, it is difficult to estimate the individual penetrance of CNVs to ASD as a whole but nevertheless the presence or absence of these CNVs in ASD represent a useful genetic tool to enrich our understanding of the overall landscape of the genetics of ASD.

Single nucleotide polymorphisms (SNPs) in ASD

Given that the heritability of ASD does not follow a classical Mendelian inheritance, it has long been postulated that ASD are polygenic disorders. Unfortunately, SNP-based approaches such as Genome-Wide Association Studies (GWAS) that focus on associating individual SNPs in genes to the risk of developing ASD have not provided targets that hold up to the scrutiny of meta-analyses or investigation of larger populations. The premise behind these studies was that while many genes may be associated with ASD, there may be key SNPs in genes with much greater contribution and, hence, ability to explain more of the variance in heritability. Given a lack of consistency in candidate SNPs identified across studies, researchers have begun to think about risk for ASD in terms of the presence or absence of a number of common genetic variants that likely act additively to produce a disease phenotype,[64] an approach that has also been adopted for other neuropsychiatric disorders such as schizophrenia. However, while risk for developing ASD may be associated with a number of common variants in many different genes, it is likely that some biological pathways and processes may be over-represented by these variants. Therefore, taking an additive pathway analysis approach may provide additional dividends. Another important aspect of assessing contribution or association of SNPs to a disease is the presence or absence of protective rather than risk SNPs that can also decrease or increase the risk for disease.

Development of a Genetic Classifier for ASD using Gene Pathway Analysis

In our recent publication, we developed and validated a predictive diagnostic genetic classifier for ASD in three independent cohorts totalling 3,346 individuals with ASD and 3,352 controls, with an overall predictive accuracy for ASD of between 70–80%.[65] This classifier was generated using SNP data obtained from GWAS investigations and utilising ASD and control cohorts from the Autism

Genetic Research Exchange (AGRE), Simon Foundation for Autism Research (SFARI) and the Wellcome Trust Birth Cohort (WTBC). The structure of the classifier was composed of 237 SNPs in 146 genes and contained both risk SNPs as well as resilience SNPs that conferred protection. Our methods incorporated the following strategy to achieve our results.

1. **Use of a case-control study design** — While there is clear knowledge to be gained by using family-based designs for investigating genes for association to neuropsychiatric disorders such as ASD, the need for case-control studies is paramount as parents and unaffected siblings can exhibit autistic features but fall short of diagnosis.[65] This finding is mirrored by our previous analysis demonstrating that first degree relatives fall between normal controls and probands for ASD.[65] A disadvantage of case-control studies is that they have the potential to be confounded by population differences, such as those due to different ancestries or geographical locations. While some investigators have postulated that our classifier was confounded by such population effects,[67,68] we have shown this not to be the case,[69] as we will discuss below.
2. **Enrichment of sample population** — Given that individuals from varying ancestral backgrounds will have differences in their genetic architecture, it is likely that a classifier built and trained on a homogeneous ethnic population will not be as accurate when applied to a population with different ancestry. This has been shown by numerous genetic association studies on different disorders demonstrating that, while certain genes may have strong association within defined populations, these genes do not necessarily have the same contribution when other populations of divergent ancestry are investigated. Based on these observations, we limited our analysis to individuals whose ancestors came from Central Europe (CEU) using known ancestral markers.
3. **Gene set enrichment and pathway analysis** — A common strategy to identify strong SNPs of interest is to plot the

logarithm of the odds (LOD) scores for individual candidate SNPs using a manhattan plot and enforce a stringent statistical cut-off to guard against Type I errors (false-positives). While this strategy may identify some strong candidate SNPs in genes for a particular disorder, it inflates the probability of Type II errors (false-negatives) and, as such, disregards many potentially informative classifier SNPs. Therefore, a strategy for identifying SNPs of interest is still required. Utilising the raw SNP data we first identified which biological pathways were over-represented by the data, and then performed a permutation analysis to generate a test statistic. In our model, weighting for SNPs were assigned as 0, 1, 3 and were calculated from whether a SNP had two copies of a dominant allele, whether they were heterozygous or had two copies of the minor allele. Assignment of these weightings provided greater accuracy over using 0, 1, 2. Such superadditive modelling have been utilised previously to understand genetic interactions.[70]

4. **Training and validation** — In order to train our classifier, we initially split the AGRE cohort into training and validation sets whereby the classifier was trained on 732 individuals with ASD and then validated on the remaining 243 individuals. While our classifier achieved a high accuracy of 85.6% in our AGRE validation set, it also achieved 84.3% in a related Tuscan population, but more importantly achieved 71.7% accuracy in an independent large data set comprising 737 ASD cases from the SFARI collection of families with simplex autism and 2,930 controls from the WTBC.

Since publication, our methods for generating a diagnostic classifier have come under scrutiny from experts within the field who postulated that our CEU populations used to derive individuals with ASD may be different, i.e. come from different geographical locations relative to those used for control individuals. Therefore, they argued that our results may have been biased by population stratification.[67,68] In order to address the issue of population stratification we have now responded to the initial letter by Belgard and colleagues[67] and demonstrated that SNPs utilised in our classifier are not merely an outcome of population stratification.[69] Importantly, in their analyses, the authors of both these papers utilised only 30 of the 237 SNPs contained in our original classifier, and their studies were thus substantially underpowered to replicate classification accuracy in their independent populations. An additional confounder of these analyses was the use of "related controls"; this is problematic as it skews and underestimates genetic diversity within ethnic populations when attempting to look for candidate risk and protective genes for disease. With regard to the issue of population stratification, it is an important and worthy consideration when building a genetic classifier or in assessing the contribution for any genetic association investigation. Therefore, in our published response,[68] we validated our classifier performance by correctly identifying control individuals from CEU, Finnish (FIN), Great British (GBR) and Iberian Spanish (IBS) populations. We found that greater differences in classifier score between these populations occurred when only part of the classifier was used (a difference as high as 25% was observed between the FIN and GBR groups). However, using the full classifier, the effects of ethnic population contributed to less than 6% of the total difference in classifier score. Using our classifier, we then examined its predictive accuracy in classifying control individuals from the FIN and GBR (non ASD) populations, as well as SFARI ASD probands (the independent validation sample in our paper). We plotted the percentage of individuals classified as ASD against the number of SNPs used in the classifier, with SNPs ordered by absolute magnitude of their weightings. As can be seen in Figure 1, while population stratification may have an influence on classifier accuracy at lower SNP numbers with regard to differences in classifier accuracy between populations, such an effect is diminished as a greater number of SNPs are included. The separation in percentage classified as ASD between the SFARI/ASD and the FIN/GBR groups occurred with increasing gradient between 50–100 SNPs, while at >150 SNPs the separation between these groups plateaus.[69]

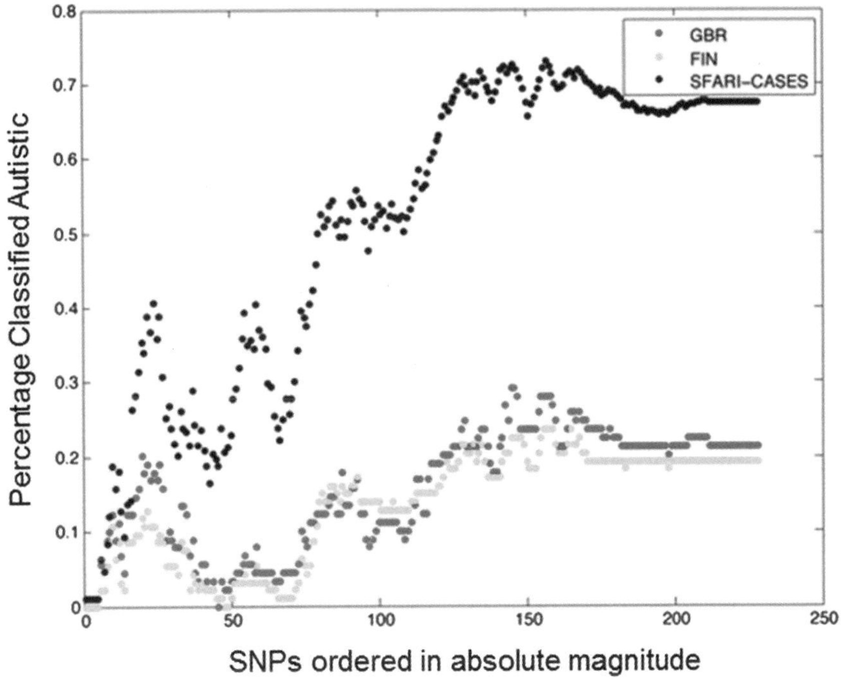

Figure 1. Percentage of individuals classified as ASD as a function of the number of SNPs ordered in decreasing absolute magnitude. Significant variance was observed at smaller number of SNPs (not plotted). Note the gradient differential between SFARI cases versus FIN and GBR between SNPs 80 and 150. Key: ASD, autism spectrum disorder; SNPs — single nucleotide polymorphisms; SFARI-CASES — Simons Foundation Autism Research Initiative ASD probands; population samples from the 1000 genome cohort: GBR — Great British; FIN — Finnish.

Our further analyses therefore illustrate that when constructing a genetic classifier, there is a threshold at which adding more information with regard to genetic alterations will not necessarily improve the performance of the classifier as illustrated above. We further show that confounding effects due to population stratification can be controlled with careful population enrichment. However, we argue that ethnic differences necessitate the development of separate and ethnically distinct genetic classifiers. Importantly, our analysis identified overlapping cellular and molecular pathways across ethnically distinct populations. This suggests that, while SNPs coding for the genes relevant to these pathways may differ between ethnic groups, the pathways relevant to the underlying pathophysiology of ASD may be common.

SNP quality control

An important consideration in building a single nucleotide polymorphism (SNP)-based classifier is to ensure that any allelic SNP differences between the cases and controls is representative of real signal related to the phenotype and not to genotyping, experimental, population or other artifacts. Although rare variants play an important role in the pathogenesis of any disorder, here we consider only the interaction of common genetic variants. Usually, not all data is available in one set and this requires merging of multiple data sets. It is of paramount importance to ensure that only SNPs that have been genotyped across all platforms are used in the analysis. Imputation[71] of missing genotypic markers may be considered but should only be used with extreme caution. During the imputation process, the missing SNPs are estimated based on linkage disequilibrium statistics from other ethnically similar genotype data sets, such as the HapMap data sets.[72] Care must be exercised in imputing SNPs based on such data as many disorders, especially neuropsychiatric disorders, are hypothesized to involve the complex interaction of many SNPs.

When merging data sets, it is important to ensure that the same strand is used for each SNP across data sets. Genetic association tools, such as Plink, can be used to test consistency of DNA strand across data sets. If a strand inconsistency is detected, the strand can be flipped. Other important potential confounds include differences in missed data and potential call rate differences between data sets genotyped on different platforms. We recommend that data be discarded for genotype markers that are differentially missed between cases and controls to an extent greater than 1 percent, markers with minimum allele frequency less than 5 percent, and

Hardy Weinberg Equilibrium in controls $p > 0.0001$. Anderson *et al.* describes a similar protocol.[71] It is also critical to test for potential call rate differences between platforms. A simple way in which this can be achieved is to undertake a two locus test between the platforms and discard any SNPs that indicate a statistically significant difference between platforms at the predetermined significance level ($p = 0.001$).

Determining Ethnicity/Ancestry and Controlling for Population Stratification

It is well-known that SNP allele frequencies show a strong ethnic and geographic correlation. After SNP quality control procedures have been performed, an important next step is to ensure the population is as ethnically homogeneous as possible. Two main mechanisms to do this involve using ancestral informative markers or tools such as Spatial Ancestry Analysis (SPA)[73] that are able to estimate an individual's geographical ancestry within a reported accuracy of five hundred kilometers. In our analysis, we used SPA with the provided European model to estimate the ancestry of individuals in our cohort. In our data set, there were a total of 63,840 SNPs in common with the SPA European model and the combined Autism Genetic Research Exchange (AGRE) and Simons Foundation (SFARI) white non-Hispanic (WNH) data set. We retained 3,452 individuals whose ancestry was determined to originate geographically from 48N to 53N in latitude and 5W to 8E in longitude. These geographic coordinates correspond to the northern European region depicted in Figure 2, which includes parts of Great Britain, France, Belgium, the Netherlands, and Germany.

Although the SFARI cohort is an ethnically homogenous collection of trios, parents and their affected offspring, we also performed principal components analysis to ascertain that there was no obvious population stratification. The principal component analysis[74] was performed in Matlab using the probabilistic principal components analysis (PPCA) function on all the 439,264 genotyped markers. We plotted the value of each individual's genotype

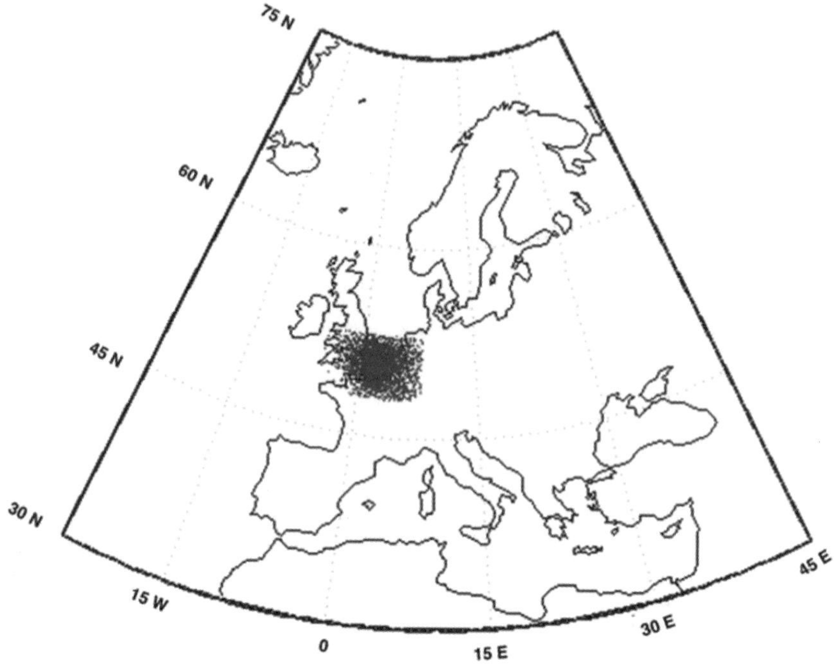

Figure 2. Using the Spatial Ancestry Analysis (SPA) method, the geographic ancestral origin of 4,970 individuals described as of 'White Non-Hispanic' origin (in SFARI and AGRE databases) was determined. The figure shows the results plotted on a map of Europe, demonstrating that most of these individuals have an ancestry originating from a defined region bounded by the square with coordinates 5W to 8E in longitude and 48 to 53N in latitude.

projected onto the two major principal components. We also tested the distributions of cases versus controls along each principal component dimension using a Mann-U test. We found no statistically significant differences.

Development of Genetic Classifiers for Other Psychiatric Disorders

Similar to ASD, the causes and mechanisms underlying other major psychiatric disorders, such as schizophrenia (SZ), bipolar disorder (BPD), and major depressive disorder (MDD), are not well understood, making diagnosis difficult. Thus, the development of reliable genetic classifiers for these disorders would have important ramifications for diagnosis and prevention.

Twin studies have suggested heritability of SZ, BPD, and MDD as 80%,[75] 71%,[76] and 37%,[77] respectively. However, identification of the actual heritable units (e.g. SNPs, genes, CNVs, etc.) that account for this estimated heritability has been elusive. For example, over 1,000 GWA studies related to schizophrenia have been published to date but the results have been largely inconsistent.[78] Furthermore, the SNP-based heritability for schizophrenia as well as other common psychiatric disorders is estimated to be <30%.[79] This heritability discordance between twin studies and GWAS, often referred to as the "missing heritability", is common across most psychiatric disorders and supports the notion that major psychiatric disorders have high levels of polygenicity and heterogeneity. As such, the traditional GWAS that uses single-variant association testing procedures and stringent statistical thresholds (e.g. $p\text{-value} < 1.0 \times 10^{-5}$) is sub-optimal and unable to capture the complex interactions that likely underlie these disorders. A promising alternative approach, such as the one employed in the development of our ASD classifier, exploits gene networks or pathways and, in so doing, is more suitable for capturing biological complexities. Thus, this approach could be adapted and used to develop classifiers for a range of psychiatric disorders with heritability and phenotypic heterogeneity levels comparable to ASD.

Future Directions

Future genetic studies of ASD will be greatly aided through more detailed mapping of the human genome via advancements in technology such as next-generation sequencing (as described in Chapter 4). This more detailed and closer inspection of the genome may lead to novel candidate loci, genes, SNPs or CNVs and, more importantly, will provide the ability to further characterise genetic hotspots already implicated. These data will have a great potential to provide an additive effect in the development of future genetic diagnostic classifiers. Such increased knowledge will provide further detail on differences in genetic architecture between ethnicities, thereby providing researchers with a greater ability to control for effects such as population stratification. It will also aid in identifying common and unique genetic signatures for disease that may exist both across and within ethnicities. While genetic studies provide great utility in identifying candidate markers for disease, it is important to pair these studies with more functional gene and protein expression investigations that utilise both blood from patients with ASD and post-mortem brain samples. In doing so, we can begin to better understand the aetiological processes involved in ASD and how better to target them with treatments versus continuing solely on a statistically genetic perspective without consideration of the biology.[80]

Executive Summary

- A number of genetic investigations in autism spectrum disorders (ASD) have identified single nucleotide polymorphisms (SNPs) and copy number variants (CNVs) that are associated with this neurodevelopmental disorder. However, as research progresses the evidence points to these genetic alterations likely being confined to specific families or populations with their penetrance in larger populations being somewhat lower.
- Taking a more expansive view, some researchers postulate that the genetics of ASD may consist of an additive collection of common genetic variations that, when considered in combination with each other, are able to explain more of the heritability of the disorder. In addition to taking a more global view of ASD genetics, the contribution of these alterations to the clinical landscape of ASD must also be considered.
- In order to identify common variants that contribute to ASD genetics, a logical approach is to utilise gene pathway analysis in order to assess which pathways and variants may be used to build a genetic classifier for ASD. When employing this strategy, it is important for investigations to not only

focus on variants that may confer risk but also those that may be protective while also controlling for other potential confounds such as population stratification. While this approach has already proven useful in classifying ASD individuals from controls, there is scope to adopt similar approaches to other complex psychiatric disorders such as schizophrenia, where large genetic databases also exist.

Acknowledgements

AGRE: We gratefully acknowledge the resources provided by the Autism Genetic Resource Exchange (AGRE) Consortium and the participating AGRE families. The Autism Genetic Resource Exchange is a program of Autism Speaks and is supported, in part, by grant 1U24MH081810 from the National Institute of Mental Health to Clara M Lajonchere (PI).

SFARI: We are grateful to all of the families at the participating Simons Simplex Collection (SSC) sites, as well as the principal investigators (A Beaudet, R Bernier, J Constantino, E Cook, E Fombonne, D Geschwind, R Goin-Kochel, E Hanson, D Grice, A Klin, D Ledbetter, C Lord, C Martin, D Martin, R Maxim, J Miles, O Ousley, K Pelphrey, B Peterson, J Piggot, C Saulnier, M State, W Stone, J Sutcliffe, C Walsh, Z Warren, E Wijsman).

WTBC: We acknowledge use of the British 1958 Birth Cohort DNA collection, funded by the Medical Research Council grant G1234567 and the Wellcome Trust grant 012345.

Christos Pantelis was supported by a NHMRC Senior Principal Research Fellowship (ID 628386).

References

1. Baron-Cohen S, Lombardo MV, Auyeung B, Ashwin E, Chakrabarti B, Knickmeyer R: Why are autism spectrum conditions more prevalent in males? *PLoS Biol* 2011, **9**(6):e1001081.
2. Lord C, Jones RM: Annual research review: re-thinking the classification of autism spectrum disorders. *J Child Psychol Psychiatry* 2012, **53**(5):490–509.

3. Hallmayer J, Cleveland S, Torres A, Phillips J, Cohen B, Torigoe T, Miller J, Fedele A, Collins J, Smith K *et al.*: Genetic heritability and shared environmental factors among twin pairs with autism. *Arch Gen Psychiatry* 2011, **68**(11):1095–1102.
4. Leblond CS, Heinrich J, Delorme R, Proepper C, Betancur C, Huguet G, Konyukh M, Chaste P, Ey E, Rastam M *et al.*: Genetic and functional analyses of SHANK2 mutations suggest a multiple hit model of autism spectrum disorders. *PLoS Genet* 2012, **8**(2):e1002521.
5. Sykes NH, Toma C, Wilson N, Volpi EV, Sousa I, Pagnamenta AT, Tancredi R, Battaglia A, Maestrini E, Bailey AJ *et al.*: Copy number variation and association analysis of SHANK3 as a candidate gene for autism in the IMGSAC collection. *Eur J Hum Genet* 2009, **17**(10):1347–1353.
6. Gauthier J, Spiegelman D, Piton A, Lafreniere RG, Laurent S, St-Onge J, Lapointe L, Hamdan FF, Cossette P, Mottron L *et al.*: Novel de novo SHANK3 mutation in autistic patients. *Am J Med Genet B Neuropsychiatric Genet* 2009, **150B**(3):421–424.
7. Durand CM, Betancur C, Boeckers TM, Bockmann J, Chaste P, Fauchereau F, Nygren G, Rastam M, Gillberg IC, Anckarsater H *et al.*: Mutations in the gene encoding the synaptic scaffolding protein SHANK3 are associated with autism spectrum disorders. *Nat Genet* 2007, **39**(1):25–27.
8. Zhu L, Wang X, Li XL, Towers A, Cao X, Wang P, Bowman R, Yang H, Goldstein J, Li YJ *et al.*: Epigenetic Dysregulation of SHANK3 in Brain Tissues from Individuals with Autism Spectrum Disorders. *Hum Mol Genet* 2013.
9. Anagnostou E: Translational medicine: Mice and men show the way. *Nature* 2012, **491**(7423):196–197.
10. Auerbach BD, Osterweil EK, Bear MF: Mutations causing syndromic autism define an axis of synaptic pathophysiology. *Nature* 2011, **480**(7375):63–68.
11. Huguet G, Ey E, Bourgeron T: The genetic landscapes of autism spectrum disorders. *Ann Rev Genomics Hum Genet* 2013, **14**:191–213.
12. O'Roak BJ, Deriziotis P, Lee C, Vives L, Schwartz JJ, Girirajan S, Karakoc E, Mackenzie AP, Ng SB, Baker C *et al.*: Exome sequencing in sporadic autism spectrum disorders identifies severe de novo mutations. *Nat Genet* 2011, **43**(6):585–589.
13. Volkmar FR, State M, Klin A: Autism and autism spectrum disorders: diagnostic issues for the coming decade. *J Child Psychol Psychiatry* 2009, **50**(1–2):108–115.
14. American Psychiatric Association., American Psychiatric Association. Task Force on DSM-IV.: *Diagnostic and Statistical Manual of Mental Disorders : DSM-IV-TR*, 4th edn. Washington, DC: American Psychiatric Association; 2000.
15. American Psychiatric Association., American Psychiatric Association. DSM-5 Task Force.: *Diagnostic and Statistical Manual of Mental Disorders : DSM-5*, 5th edn. Washington, D.C.: American Psychiatric Association; 2013.

16. McPartland JC, Reichow B, Volkmar FR: Sensitivity and specificity of proposed DSM-5 diagnostic criteria for autism spectrum disorder. *J Am Acad Child Adolesc Psychiatry* 2012, **51**(4):368–383.
17. Huerta M, Bishop SL, Duncan A, Hus V, Lord C: Application of DSM-5 criteria for autism spectrum disorder to three samples of children with DSM-IV diagnoses of pervasive developmental disorders. *Am J Psychiatry* 2012, **169**(10):1056–1064.
18. Baird G, Charman T, Baron-Cohen S, Cox A, Swettenham J, Wheelwright S, Drew A: A screening instrument for autism at 18 months of age: a 6-year follow-up study. *J Am Acad Child Adolesc Psychiatry* 2000, **39**(6):694–702.
19. Sunita, Bilszta JL: Early identification of autism: a comparison of the Checklist for Autism in Toddlers and the Modified Checklist for Autism in Toddlers. *J Paediatrics Child Health* 2013, **49**(6):438–444.
20. Robins DL, Fein D, Barton ML, Green JA: The Modified Checklist for Autism in Toddlers: an initial study investigating the early detection of autism and pervasive developmental disorders. *J Autism Developmental Disord* 2001, **31**(2):131–144.
21. Chandler S, Charman T, Baird G, Simonoff E, Loucas T, Meldrum D, Scott M, Pickles A: Validation of the social communication questionnaire in a population cohort of children with autism spectrum disorders. *J Am Acad Child Adolesc Psychiatry* 2007, **46**(10):1324–1332.
22. Eaves LC, Wingert HD, Ho HH, Mickelson EC: Screening for autism spectrum disorders with the social communication questionnaire. *J Developmental Behavioral Pediatrics* 2006, **27**(2 Suppl):S95–S103.
23. Allen CW, Silove N, Williams K, Hutchins P: Validity of the social communication questionnaire in assessing risk of autism in preschool children with developmental problems. *J Autism Developmental Disord* 2007, **37**(7):1272–1278.
24. Volkmar FR, Cicchetti DV, Dykens E, Sparrow SS, Leckman JF, Cohen DJ: An evaluation of the Autism Behavior Checklist. *J Autism Developmental Disord* 1988, **18**(1):81–97.
25. Rellini E, Tortolani D, Trillo S, Carbone S, Montecchi F: Childhood Autism Rating Scale (CARS) and Autism Behavior Checklist (ABC) correspondence and conflicts with DSM-IV criteria in diagnosis of autism. *J Autism Developmental Disord* 2004, **34**(6):703–708.
26. Bolton PF, Golding J, Emond A, Steer CD: Autism spectrum disorder and autistic traits in the Avon Longitudinal Study of Parents and Children: precursors and early signs. *J Am Acad Child Adolesc Psychiatry* 2012, **51**(3):249–260 e225.
27. Landa RJ, Holman KC, Garrett-Mayer E: Social and communication development in toddlers with early and later diagnosis of autism spectrum disorders. *Arch Gen Psychiatry* 2007, **64**(7):853–864.
28. Ozonoff S, Iosif AM, Baguio F, Cook IC, Hill MM, Hutman T, Rogers SJ, Rozga A, Sangha S, Sigman M *et al.*: A prospective study of the emergence of

early behavioral signs of autism. *J Am Acad Child Adolesc Psychiatry* 2010, **49**(3):256–266 e251–252.

29. Lord C, Rutter M, Le Couteur A: Autism Diagnostic Interview-Revised: a revised version of a diagnostic interview for caregivers of individuals with possible pervasive developmental disorders. *J Autism Developmental Disorders* 1994, **24**(5):659–685.
30. Lord C, Risi S, Lambrecht L, Cook EH, Jr., Leventhal BL, DiLavore PC, Pickles A, Rutter M: The autism diagnostic observation schedule-generic: a standard measure of social and communication deficits associated with the spectrum of autism. *J Autism Developmental Disord* 2000, **30**(3):205–223.
31. Oosterling I, Rommelse N, de Jonge M, van der Gaag RJ, Swinkels S, Roos S, Visser J, Buitelaar J: How useful is the Social Communication Questionnaire in toddlers at risk of autism spectrum disorder? *J Child Psychol Psychiatry* 2010, **51**(11):1260–1268.
32. Corsello CM, Akshoomoff N, Stahmer AC: Diagnosis of autism spectrum disorders in 2-year-olds: a study of community practice. *J Child Psychol Psychiatry* 2013, **54**(2):178–185.
33. Falkmer T, Anderson K, Falkmer M, Horlin C: Diagnostic procedures in autism spectrum disorders: a systematic literature review. *Eur Child Adolesc Psychiatry* 2013, **22**(6):329–340.
34. Charman T, Gotham K: Measurement Issues: Screening and diagnostic instruments for autism spectrum disorders - lessons from research and practice. *Child Adolesc Mental Health* 2013, **18**(1):52–63.
35. Mahjouri S, Lord CE: What the DSM-5 portends for research, diagnosis, and treatment of autism spectrum disorders. *Curr Psychiatry Rep* 2012, **14**(6):739–747.
36. Bitsika V, Sharpley CF, Orapeleng S: An exploratory analysis of the use of cognitive, adaptive and behavioural indices for cluster analysis of ASD subgroups. *J Intellectual Disability Res* 2008, **52**(11):973–985.
37. Turner LM, Stone WL: Variability in outcome for children with an ASD diagnosis at age 2. *J Child Psychol Psychiatry, and allied disciplines* 2007, **48**(8):793–802.
38. Siegel B: Toward DSM-IV: a developmental approach to autistic disorder. *Psychiatric Clinics N Am* 1991, **14**(1):53–68.
39. Lord C, Petkova E, Hus V, Gan W, Lu F, Martin DM, Ousley O, Guy L, Bernier R, Gerdts J *et al.*: A multisite study of the clinical diagnosis of different autism spectrum disorders. *Arch Gen Psychiatry* 2012, **69**(3):306–313.
40. Ekinci O, Arman AR, Melek I, Bez Y, Berkem M: The phenomenology of autistic regression: subtypes and associated factors. *Eur Child Adolesc Psychiatry* 2012, **21**(1):23–29.
41. Carter MT, Scherer SW: Autism spectrum disorder in the genetics clinic: a review. *Clin Genet* 2013, **83**(5):399–407.
42. Clifford S, Dissanayake C, Bui QM, Huggins R, Taylor AK, Loesch DZ: Autism spectrum phenotype in males and females with fragile X full mutation and premutation. *J Autism Developmental Disord* 2007, **37**(4):738–747.

43. Chaste P, Betancur C, Gerard-Blanluet M, Bargiacchi A, Kuzbari S, Drunat S, Leboyer M, Bourgeron T, Delorme R: High-functioning autism spectrum disorder and fragile X syndrome: report of two affected sisters. *Mol Autism* 2012, **3**(1):5.
44. Ebert DH, Greenberg ME: Activity-dependent neuronal signalling and autism spectrum disorder. *Nature* 2013, **493**(7432):327–337.
45. Delorme R, Ey E, Toro R, Leboyer M, Gillberg C, Bourgeron T: Progress toward treatments for synaptic defects in autism. *Nat Med* 2013, **19**(6):685–694.
46. Ehninger D, Silva AJ: Rapamycin for treating Tuberous sclerosis and Autism spectrum disorders. *Trends Mol Med* 2011, **17**(2):78–87.
47. Lintas C, Sacco R, Persico AM: Genome-wide expression studies in autism spectrum disorder, Rett syndrome, and Down syndrome. *Neurobiol Dis* 2012, **45**(1):57–68.
48. Ricciardi S, Boggio EM, Grosso S, Lonetti G, Forlani G, Stefanelli G, Calcagno E, Morello N, Landsberger N, Biffo S *et al.*: Reduced AKT/mTOR signaling and protein synthesis dysregulation in a Rett syndrome animal model. *Hum Mol Genet* 2011, **20**(6):1182–1196.
49. Castro J, Mellios N, Sur M: Mechanisms and therapeutic challenges in autism spectrum disorders: insights from Rett syndrome. *Curr Opin Neurol* 2013, **26**(2):154–159.
50. Zhong X, Li H, Chang Q: MeCP2 phosphorylation is required for modulating synaptic scaling through mGluR5. *J Neurosci* 2012, **32**(37):12841–12847.
51. Cook EH, Jr., Lindgren V, Leventhal BL, Courchesne R, Lincoln A, Shulman C, Lord C, Courchesne E: Autism or atypical autism in maternally but not paternally derived proximal 15q duplication. *Am J Hum Genet* 1997, **60**(4): 928–934.
52. Bolton PF, Dennis NR, Browne CE, Thomas NS, Veltman MW, Thompson RJ, Jacobs P: The phenotypic manifestations of interstitial duplications of proximal 15q with special reference to the autistic spectrum disorders. *Am J Med Genet* 2001, **105**(8):675–685.
53. Shinawi M, Liu P, Kang SH, Shen J, Belmont JW, Scott DA, Probst FJ, Craigen WJ, Graham BH, Pursley A *et al.*: Recurrent reciprocal 16p11.2 rearrangements associated with global developmental delay, behavioural problems, dysmorphism, epilepsy, and abnormal head size. *J Med Genet* 2010, **47**(5): 332–341.
54. Jacquemont S, Reymond A, Zufferey F, Harewood L, Walters RG, Kutalik Z, Martinet D, Shen Y, Valsesia A, Beckmann ND *et al.*: Mirror extreme BMI phenotypes associated with gene dosage at the chromosome 16p11.2 locus. *Nature* 2011, **478**(7367):97–102.
55. Courtens W, Schramme I, Laridon A: Microduplication 22q11.2: a benign polymorphism or a syndrome with a very large clinical variability and reduced penetrance? — Report of two families. *Am J Med Genet A* 2008, **146A**(6):758–763.

56. Van der Aa N, Rooms L, Vandeweyer G, van den Ende J, Reyniers E, Fichera M, Romano C, Delle Chiaie B, Mortier G, Menten B *et al.*: Fourteen new cases contribute to the characterization of the 7q11.23 microduplication syndrome. *Eur J Med Genet* 2009, **52**(2–3):94–100.
57. Mefford HC, Sharp AJ, Baker C, Itsara A, Jiang Z, Buysse K, Huang S, Maloney VK, Crolla JA, Baralle D *et al.*: Recurrent rearrangements of chromosome 1q21.1 and variable pediatric phenotypes. *New Eng J Med* 2008, **359**(16):1685–1699.
58. Marshall CR, Noor A, Vincent JB, Lionel AC, Feuk L, Skaug J, Shago M, Moessner R, Pinto D, Ren Y *et al.*: Structural variation of chromosomes in autism spectrum disorder. *Am J Hum Genet* 2008, **82**(2):477–488.
59. Sebat J, Lakshmi B, Malhotra D, Troge J, Lese-Martin C, Walsh T, Yamrom B, Yoon S, Krasnitz A, Kendall J *et al.*: Strong association of de novo copy number mutations with autism. *Science* 2007, **316**(5823):445–449.
60. Bucan M, Abrahams BS, Wang K, Glessner JT, Herman EI, Sonnenblick LI, Alvarez Retuerto AI, Imielinski M, Hadley D, Bradfield JP *et al.*: Genome-wide analyses of exonic copy number variants in a family-based study point to novel autism susceptibility genes. *PLoS Genet* 2009, **5**(6):e1000536.
61. Pinto D, Pagnamenta AT, Klei L, Anney R, Merico D, Regan R, Conroy J, Magalhaes TR, Correia C, Abrahams BS *et al.*: Functional impact of global rare copy number variation in autism spectrum disorders. *Nature* 2010, **466**(7304):368–372.
62. Glessner JT, Wang K, Cai G, Korvatska O, Kim CE, Wood S, Zhang H, Estes A, Brune CW, Bradfield JP *et al.*: Autism genome-wide copy number variation reveals ubiquitin and neuronal genes. *Nature* 2009, **459**(7246):569–573.
63. Persico AM, Napolioni V: Autism genetics. *Behavioural Brain Res* 2013, **251**: 95–112.
64. Klei L, Sanders SJ, Murtha MT, Hus V, Lowe JK, Willsey AJ, Moreno-De-Luca D, Yu TW, Fombonne E, Geschwind D *et al.*: Common genetic variants, acting additively, are a major source of risk for autism. *Mol Autism* 2012, **3**(1):9.
65. Skafidas E, Testa R, Zantomio D, Chana G, Everall IP, Pantelis C: Predicting the diagnosis of autism spectrum disorder using gene pathway analysis. *Mol Psychiatry* 2012.
66. Losh M, Adolphs R, Poe MD, Couture S, Penn D, Baranek GT, Piven J: Neuropsychological profile of autism and the broad autism phenotype. *Arch Gen Psychiatry* 2009, **66**(5):518–526.
67. Belgard TG, Jankovic I, Lowe JK, Geschwind DH: Population structure confounds autism genetic classifier. *Mol Psychiatry* 2013.
68. Robinson EB, Howrigan D, Yang J, Ripke S, Anttila V, Duncan LE, Jostins L, Barrett JC, Medland SE, Macarthur DG *et al.*: Response to 'Predicting the diagnosis of autism spectrum disorder using gene pathway analysis'. *Mol Psychiatry* 2013.

69. Skafidas E, Testa R, Zantomio D, Chana G, Everall IP, Pantelis C: Response to Belgard *et al. Mol Psychiatry* 2014.
70. Perez-Perez JM, Candela H, Micol JL: Understanding synergy in genetic interactions. *Trends Genet* 2009, **25**(8):368–376.
71. Marchini J, Howie B: Genotype imputation for genome-wide association studies. *Nature Rev Genet* 2010, **11**(7):499–511.
72. International HapMap C, Altshuler DM, Gibbs RA, Peltonen L, Altshuler DM, Gibbs RA, Peltonen L, Dermitzakis E, Schaffner SF, Yu F *et al.*: Integrating common and rare genetic variation in diverse human populations. *Nature* 2010, **467**(7311):52–58.
73. Yang WY, Novembre J, Eskin E, Halperin E: A model-based approach for analysis of spatial structure in genetic data. *Nat Genet* 2012, **44**(6):725–731.
74. Price AL, Patterson NJ, Plenge RM, Weinblatt ME, Shadick NA, Reich D: Principal components analysis corrects for stratification in genome-wide association studies. *Nat Genet* 2006, **38**(8):904–909.
75. Heckers S: Bleuler and the neurobiology of schizophrenia. *Schizophr Bull* 2011, **37**(6):1131–1135.
76. Edvardsen J, Torgersen S, Roysamb E, Lygren S, Skre I, Onstad S, Oien PA: Heritability of bipolar spectrum disorders. Unity or heterogeneity? *J Affect Disord* 2008, **106**(3):229–240.
77. Sullivan PF, Neale MC, Kendler KS: Genetic epidemiology of major depression: review and meta-analysis. *Am J Psychiatry* 2000, **157**(10):1552–1562.
78. Allen NC, Bagade S, McQueen MB, Ioannidis JP, Kavvoura FK, Khoury MJ, Tanzi RE, Bertram L: Systematic meta-analyses and field synopsis of genetic association studies in schizophrenia: the SzGene database. *Nat Genet* 2008, **40**(7):827–834.
79. Lee SH, Ripke S, Neale BM, Faraone SV, Purcell SM, Perlis RH, Mowry BJ, Thapar A, Goddard ME, Witte JS *et al.*: Genetic relationship between five psychiatric disorders estimated from genome-wide SNPs. *Nat Genet* 2013, **45**(9):984–994.
80. Niculescu AB, 3rd: DISCovery in psychiatric genetics. *Mol Psychiatry* 2014, **19**(2):145.

Gene Dysregulation in ASD: From Transcription to Translation

Chapter 7

Genome-Wide Expression Studies of Blood and Lymphoblastoid Cell Lines in Autism Spectrum Disorders

Boryana Stamova* and Frank R. Sharp

Abstract

mRNA and miRNA expression studies of blood and lymphoblastoid cell lines derived from individuals with autism spectrum disorder (ASD) have shown few genes that are shared between studies. Some that have been replicated in at least three studies include: BTG1, HOOK3, FCGBP, REV3L, CCDC47 and MAP4, all of which are implicated in cell proliferation. Pathways that have been replicated in at least two studies include: mTOR, neurotrophin signaling, actin cytoskeleton/ILK signaling, VEGF and Natural Killer cell signaling. Abnormalities of mTOR signaling are implicated in idiopathic ASD as well as in genetic syndromes associated with ASD, including Fragile X, TSC1, TSC2, NF-1 and PTEN. This chapter reviews the results of gene expression studies on blood, lymphoblastoid cell lines, and brain tissues that, despite the lack of shared genes, reveal pathways common to subgroups of idiopathic and genetic cases of

*Correspondence to: MIND Institute Wet Labs, University of California at Davis Medical Center, 2805 50th Street, Sacramento, CA 95817. Email: boryana.stamova@ucdmc.ucdavis.edu

ASD, and suggests that a pathway-oriented approach may identify better treatment targets.

Keywords: Autism; Gene expression; RNA expression; microRNA; Microarrays; Blood; Leukocytes; Lymphoblastoid cell lines; Alternative splicing; Pathways.

Introduction

Though a genetic cause of autism spectrum disorder (ASD) can now be ascribed to nearly 25% of subjects, the "cause" of the ASD in the remaining subjects remains unclear.[1] ASD is highly heritable though the nature of this heritability is still controversial.[1] One possibility is that hundreds of genetic causes remain to be identified, possibly involving non-coding portions of the genome. Another possibility is that ASD is a complex genetic disorder where a group of 10–50 genes interact with an environmental factor(s) which results in ASD.

Whichever model one espouses — hundreds of individual rare mutations vs. complex genetic disease — they all cause a similar clinical syndrome associated with impaired social interactions, impaired language function and stereotyped behaviors. It is generally assumed that these deficits are associated with abnormalities of specific brain regions that mediate these behaviors which, in turn, are associated with specific sets of molecules and molecular pathways in those brain regions. Identifying common sets of molecules and molecular pathways would be extremely useful for: (1) developing diagnostic biomarkers;[2] (2) developing prognostic biomarkers; (3) potentially identifying environmental factors relevant for ASD[3,4] (4) identifying therapeutic targets that might be common to several ASD causes; and (5) identifying markers that respond to treatment.[5]

One of the great problems in the field has been the paucity of human brain material to study these pathways. The brain studies that have been performed have been limited for a variety of reasons, and approaches are needed to broach this problem. A small number of investigators have begun to determine whether peripheral blood and/or lymphoblastoid cells lines (LCL) derived from peripheral blood might be able to provide information about ASD phenotypes and perhaps the brain abnormalities, genetic abnormalities and/or environmental contributor(s) that underlie ASD.

Though some question the value of studying blood and LCL in ASD, there are a number of reasons why the study of peripheral tissues might contribute to understanding the pathogenesis of ASD. These include: (1) a number of genes expressed in blood are also expressed in brain; (2) peripheral immune abnormalities have been observed in a subpopulation of ASD subjects suggesting alterations of gene expression in blood lymphocytes;[6] (3) monocytes in blood and microglia in brain have very similar transcriptiomes, and abnormalities of microglia have been observed in ASD brain;[7,8] (4) RNA expression of a group of immune molecules, known to be expressed in peripheral blood, has been shown to be altered in ASD brain;[9] (5) genetic and environmental factors implicated in ASD might both affect gene expression in peripheral blood and these might change over time; (6) environmental toxicants elevating risk for ASD might produce characteristic changes of gene expression in peripheral blood; (7) subgroups of individuals with similar pathogenesis might have similar gene expression profiles in blood and brain; and (8) for diagnostic purposes, blood is readily accessible whereas brain tissue is not.

LCL have been used by a number of groups to study gene expression in ASD. These cell lines have the advantage that they do not require blood draws, they presumably should reflect many of the differences of ASD and controls based upon genetic differences between the groups, they are reasonably standardized across laboratories and can be readily cultured for a given experiment. Furthermore, they allow for experimental manipulation, such as treatment with toxicants, as well as allow for verification of findings between different laboratories. Another advantage of LCL is that they are often obtained from cell repositories that have not only validated the diagnosis of ASD, but have also obtained comprehensive phenotypic data on the patients.

This review will summarize the findings from whole genome studies of RNA expression from whole blood and LCL from children with ASD. This will be followed by a discussion of the pathways associated with ASD subjects based upon current studies.

mRNA Expression Studies of LCL in ASD

LCL are produced by incubating EBV (Epstein Barr Virus) with blood samples which results in immortalization of primarily B cells which can be cultured, cryogenically preserved, and studied over extended periods of time. They offer convenience in terms of obtaining samples and the potential for manipulating the cells with various treatments.

A number of studies of ASD have now been performed using LCL and are summarized in Table 1. It is important to note that many different experimental protocols have been used, including comparisons of individuals with ASD to unrelated controls or to sibling controls, and comparisons of discordant twins. Sample sizes are generally small and many different expression platforms have been used. These factors, in addition to the clinical or phenotypic heterogeneity of the different sample populations studies, likely account for the finding that very few genes have been shown to be validated/confirmed between any of these studies. In spite of the failure to confirm individual genes, almost all of the studies have confirmed similar pathways including neurodevelopment, alternative splicing/transcriptional regulation, and immune-related genes. CD38, for example, has been confirmed in two separate studies.[10,11] CD38 is a multifunctional ectoenzyme found on the surface of B cells, Natural Killer (NK) cells, and other immune cells that catalyzes the synthesis and hydrolysis of cyclic ADP-ribose (cADPR) from NAD^+ to ADP-ribose. Decreased CD38 is associated with impaired immune function in humans and could relate to impaired immune function reported in ASD.[6] A number of the genes reported to be differentially expressed in ASD versus control LCL in Table 1 have been directly implicated in ASD or at least in neurodevelopment, including SEMA5A, NLGN3, SHANK3 and

Table 1. Whole genome studies of lymphoblastoid cell lines in ASD subjects

Study	Samples	Arrays Used	Comments
[10]	5 pairs ASD twins discordant for ASD severity	TIGR 40 K Human Set	43 genes including ROBO1, CYP1B1, mitochondrial aconitase, ECAT11, IL6ST, argininosuccinate synthetase, NAGLU, P2X5a, CD38
[84]	6 ASD, 6 controls	Invitrogen cDNA arrays	9 down regulated genes including SEMA5A
[85]	3 ASD, 3 siblings	Affymetrix U95	48 genes, including CD44, ITGB7, FLOT2, and TPM4
[86]	15 ASD, 27 controls	Agilent	68 genes common to FMR1-FM and dup (15q). JAKMIP1, GGPR155
[87]	20 sib pairs — ASD and controls	TIGR 40 K Human Set arrays, PCR	Neuronal development and steroid biosynthesis pathways
[31]	116 subjects — ASD and controls	TIGR 40 K Human Set arrays, PCR	ITGAM, NFKB1, MBD2, RHOA, SLIT2 decreased in ASD by PCR Circadian rhythm genes
[11]	40 ASD, 40 unaffected parents	PCR	CD38 reduced
[88]	20 ASD, 22 control siblings	Illumina microarrays	HEY1, SOX9, IFITM3, TPM2, ARHGAP24, LMO2, HLA-DRB1, IL32, UCHL1, MFGE8, LEF1, TGFBR2
[89]	82 ASD, 64 controls	Affy U133 Microarrays	Alternative splicing, zinc, ubiquitin, transcription pathways
[90]	35 ASD, 35 controls	PCR	NLGN3, SHANK3
[25]	244 families with discordant children	Illumina arrays	Structural variants (CNVs) have effects on RNA expression in ASD
[91]	16 ASD and 16 controls (male)	Affy U133 Microarrays	202 genes, 89 upregulated FOXP1 confirmed by PCR
[27]	Public Data Bases	Affymetrix Human Exon 1.0 ST Array	SNP near SEMA5A associated with reduced expression same gene and with an autism regulatory network

FOXP1. Possible reasons for the relative failure to replicate individual genes are discussed at the end of this chapter.

Of interest is the finding that a number of genes reported to be differentially expressed in LCL derived from individuals with ASD have also been reported to be differentially expressed in studies on ASD brain.[12,13] These are summarized by the Voineagu review[12] and include: ARMC8, ATP2B2, DGCR9, IFITM3, MAP4, NFKBIZ, PDE4DIP, PFTK1, PRKCE, RBMS1, SOX9, TNPO1, TOX, TSPO, UCHL1, and YAP1 (also summarized in Table 4). Of these genes, several are relevant for ASD. IFITM3 plays a role in the defense against influenza; NFKBIZ is related to I kappa B family of proteins that regulate responses to bacteria; PRKCE modulates behavior and immune responses; SOX9 modulates male differentiation; Translocator protein (TSPO) is found on the outer mitochondrial membrane where it interacts with StAR (steroidogenic acute regulatory protein) to transport cholesterol into mitochondria; and UCHL1, found in all neurons, hydrolyzes small C-terminal adducts of ubiquitin to generate the ubiquitin monomer during process of protein degradation. Infection, immune responses, mitochondrial abnormalities, male gender, and abnormalities of protein translation and degradation have all been implicated in ASD.

microRNA (miRNA) Expression Studies of LCL in ASD

Similarly, there have been several studies that have examined microRNA expression differences in individuals with ASD compared to that in various types of controls. These are summarized in Table 2. Though few shared miRNAs have been reported, three studies found dysregulation of miR-23a.[14–16] miR-23a plays a role in immune function, which decreases in the blood of multiple sclerosis patients[17] and plays a role in B cell development.[18] It targets interferon regulatory factor 1 to regulate cellular proliferation,[19] and also regulates DNA damage repair.[20] The NF-κB member p65 controls glutamine metabolism through miR-23a, and could relate

Table 2. Whole genome studies of miRNA in LCL and brain tissues of individuals with ASD

Study	Samples	Tissue Source	Arrays Used	Comments
[88]	ASD and controls	Lymphoblastoid Cell Lines	Illumina Universal-12 BeadChips	miR-486, miR-181b
[14]	8 ASD twins, 6 unaffected controls	Lymphoblastoid Cell Lines	Custom Plated Microarrays	miR-23a, miR-106b, miR-29b, miR-2195–p
[15]	6 ASD, 6 controls	Lymphoblastoid Cell Lines	μParaflo™ miRNA microarray	9 miRNA (23a,b ; 132; 146a,b; 663; 92, 320)
[16]	13 ASD, 13 control	Cerebellar Cortex	Multiplex PCR	28 miRNA — with miR-23a

to the glutamine-glutamate shuttle in brain.[21] miR-23a regulation of the X-linked inhibitor of apoptosis (XIAP) contributes to sex differences in the response to cerebral ischemia.[22] Finally, miR-23a is differentially expressed in cerebellum of ASD compared to control subjects.[16]

miR-106b is another microRNA dysregulated in LCL and in the cerebellum of individuals with ASD.[14,16] MiR-106b impairs cholesterol efflux and increases Aβ levels by repressing ABCA1 expression.[23] The microRNA cluster miR-106b~25 also regulates adult neural stem/progenitor cell proliferation and neuronal differentiation.[24] Thus, these studies on LCL have identified some differentially expressed mRNAs and miRNAs that are either reflected in the brain of individuals with ASD, or may be relevant to ASD pathogenesis.

RNA Expression Studies in Relation to Genotype

Only recently have expression studies begun to consider possible underlying genotypes. For example, Luo *et al.* examined the effect of rare *de novo* and recurrent CNVs on gene expression in ASD using LCL.[25] They found that outlier genes cluster with the most pathogenic CNVs (rare *de novo* CNVs). Several nonrecurrent CNVs with significant gene expression alterations were identified.[25] Moreover, specific gene expression patterns were identified with specific microdeletions and microduplications, and some of these correlate with the clinical phenotype of ASD.[25] Similarly, reduced transcript expression of genes is affected by inherited and *de novo* CNVs in autism.[26] An eQTL mapping approach revealed that rare variants in the SEMA5A regulatory network impact autism risk,[27] an eQTL being a gene where specific SNPs within a gene correlate with expression of the gene.

RNA Expression Studies of Blood in ASD

The advantage of studying blood over LCL or other cultured cells is that such studies might capture the "physiological status" and/or "environmental effects" being experienced by subjects with ASD at the time of blood collection. In addition, using blood collection tubes that stabilize RNA immediately at the time the blood is drawn also helps to ensure the proper capture of such physiological or environmental effects. Thus, an increasing number of studies have now been done using primary blood cells. These studies are summarized in Table 3.

The blood studies to date show that there are relatively few genes (discussed in the following paragraphs) shared between the studies. Some of the reasons for the lack of agreement are similar to those for the LCL: use of different microarray platforms; different blood collection methods — EDTA, PAXgene tubes; peripheral blood monocytes versus whole blood; different ages of the subjects; different demographics, and different clinical phenotypes of the subjects (Table 3). In addition, most of the sample sizes are small.

In spite of the modest advances from the above studies of blood, there are shared genes and pathways when one compares the expression studies on LCL, blood, and brain. To demonstrate this, we have compared our most recent study of 2–4 year old boys with ASD (n=30) and age-matched, typically developing control boys (n=20).[28] Blood was collected into PAXgene tubes to

Table 3. Whole genome studies of blood from individuals with ASD

Study	Samples	Arrays Used	Comments
[30]	36 ASD, 12 controls	Affymetrix U133, PCR	PAM, SPON2, IL2RB, PRF1, GZM, CS3CR1, EAT2, EDG8, IGFBP7, ZNF145, KIR3DL2
[61]	52 ASD, 27 controls- PAXgene tubes	Affymetrix U133	Natural Killer Cell genes, functional dysfunction of NK cells
[92]	21 ASD adults and controls; 21 mothers of ASD children and control mothers- PAXgene tubes	Agilent	19 genes different for ASD vs controls; ASD mothers had some expression differences compared to control mothers.
[93]	8 ASD/SPAD, 39 ASD, 37 controls, 12 SPAD	Agilent	Peripheral Blood Monocytes — TGFBR, Notch, EGFR1
[94]	60 ASD, 68 controls; EDTA tubes	Illumina	48 probe classifier with Area Under Curve of 0.91. Immune response genes, hemoglobin complex, and guanine or guananylate binding
[29]	20 ASD, 20 sibling, 18 controls — PAXgene tubes	Affymetrix U133	Up regulation of Ribosomal, splicesomal and mitochondrial pathways in probands vs siblings and down regulation of neuroreceptor-ligand, immune and calcium signals
[2]	66 male ASD and 33 controls in training set; 104 ASD and 82 controls in test set — PAXgene tubes	Affymetrix U133, and Affy GeneST	55 gene signature predicted ASD vs controls with Area Under Curve of 0.7. Pathways: neutrotrophins; FC gamma and epsilon; chemokine; mTOR; B, T and Natural Killer signaling; Erb B, VEGF, Notch
Letovsky *et al.* 2013 — in review	ASD and Controls; training and test cohorts-PAXgene tubes	RNAseq	Area Under Curve for prediction of ASD ~ 0.60.
[28]	30 ASD, 20 controls — ages 2–4; PAXgene tubes	Affymetrix Exon Expression arrays	53 genes distinguished ASD; Pathways: ROS, Nrf2, ILK, NK cell, FCgamma, HMGB1, TREM1, Jak/Stat, VEGF, PI3K/AKT/mTOR, NFkappaB, PAK, NGF, GF-CSF, Ceramide, Actin, CTLA4

immediately stabilize RNA, and RNA expression and alternative splicing were assessed on Affymetrix Exon expression arrays.[28] A total of 53 genes were predicted to have differential alternative splicing for ASD versus controls after correction for multiple comparisons. However, we used a larger list of genes (n=477) to assess pathways (Figure 1) and to compare with other gene expression studies (Table 4). These 477 genes demonstrated significant

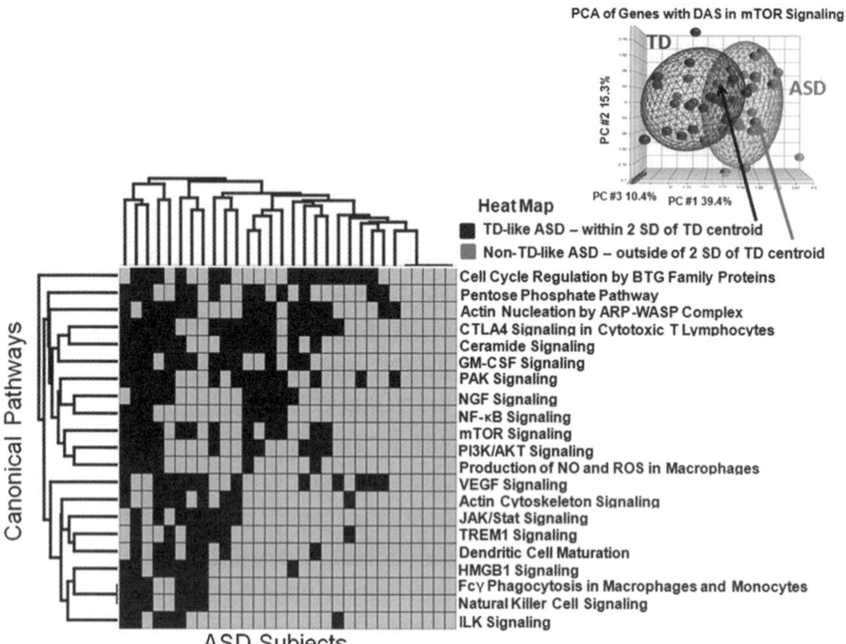

Figure 1. Pathways in ASD versus Typically Developing Controls (TD). This figure was published in the journal *Molecular Autism* in 2013 (Stamova *et al.* 2013). Individual-subject level analysis. Hierarchical clustering of ASD subjects and significant pathways ($p<0.05$). Each ASD subject in each of the significant pathways was scored as either TD-like, if its PCA position on the 3-D PCA plot for the genes with DAS/DEU in the particular pathway was within 2 standard deviations (SD) of the TD centroid (within the TD ellipsoid); or as non-TD-like, if its PCA position on the 3-D PCA components space was outside 2 SD of the TD centroid (outside of the TD ellipsoid). An example of the scoring scheme — in the upper right-hand corner — is presented for the significant genes in the mTOR signaling pathway. Reproduced from Stamova *et al.*[28]

Table 4. Genes dysregulated in ASD and that are shared between at least two studies or shared between three studies (bold)

Study	Tissue and Platform	Genes Shared with Stamova *et al.*[28] Study Whole Blood — PAXgene Tubes — Exon Arrays
[31]	Lymphoblastoid Cell Lines — TIGR arrays	**BTG1**, DSE, **HOOK3**, KRCC1, LRP11. MCRS1, MCTP2, ANKRD17, CD83, DHX15, DSE, EHD4, FAF1, **FCGBP**, FNBP1, LRRC41, NF1, PPP2R2A, **REV3L**, RNF32, SFPQ, SND1, STAT5B, TSSK4, COX4NB, AKT3, **CCDC47**, CTSO, GFER, **MAP4**
[30]	Whole Blood — PAXgene tubes U133 arrays	DNMBP. PALLD, STX5
[29]	Whole Blood — PAXgene tubes U133 arrays	ARL4C, **CCDC47**, CEBPG, EEA1, EIF4A2, FXR1, ROCK2, USP9Y, CAB39, CMYA5, DNMBP, LSM14B, MAML3, PCF11, SPOPL, TMCO3, TOR1AIP1, TTC14, WAPAL, YTHDC1, ZBED4
[32]	Lymphoblastoid Cell Lines — Exon Arrays	EZR, **HOOK3**, MACF1. MRPS27, **REV3L**, RPN1, TARS
[9]	Brain Tissue — Illumina arrays	ALDH18A1, ANKRA2, BRPF1, **BTG1**, C1orf131, CCDC90B, CHD5, COL7A1, EPHX1, FAIM2, **FCGBP**, GPNMB, INA, INTU, ITGB1BP1, LRP1B, LYN, MAP3K12, **MAP4**, MIB2, PALLD, PFKP, PTK2B, SLC4A8, STAT4, UCHL1, ULK1, ZBTB40, BAI2, BIN1, CHPT1, DYNC1I2, FEZ2, GPATCH4, INTS10, LSM14A, LSM14B, MADD, MFSD6, PLEKHA5, SCRIB, SYNE1

alternative splicing ($p<0.05$) and significant expression differences of at least one exon in the genes ($p<0.005$ and |fold-change| >1.2).[28] Since the exon array used does not contain probe sets covering exon-exon junctions, deducing differential alternative splicing (DAS) is indirect given that the unit measured is exon expression which provides a measure of exon usage. Thus, we used the term predicted 'differential alternative splicing/differential exon usage (DAS/DEU)' to emphasize the fact that differential exon expression is used to predict DAS in these studies.

We compared the 477 genes with those identified by selected gene expression studies of blood, LCL, and brain tissue in the literature (Table 4), and arrived at several conclusions from the overlap comparisons. There are relatively few shared genes between the Stamova study and any of the other studies. This lack of overlap is even more apparent when one realizes that, despite the nearly identical technical methods of the Kong and Gregg studies,[29,30] there were no shared genes. However, there are some shared genes between the Stamova *et al.* study[28] and that of Hu *et al.*[31] with LCL, and even more shared genes between the Stamova *et al.* study and the study of brain tissue by Voineagu *et al.*[9] The genes in bold (Table 4) are those that are shared between at least three studies (n=6 genes). These results indicate tremendous heterogeneity in ASD causes/pathophysiology since there are so few shared differentially expressed genes.

With that said, there are also some common genes that deserve to be mentioned. HOOK3 and REV3L (Table 4) are both predicted to be alternatively spliced in ASD compared to controls in the Stamova and Talebizadeh studies,[28,32] both based on Affymetrix exon array data. These results need to be confirmed using direct methods on independent samples. Hook proteins have N-terminal domains which attach to microtubules, and more divergent C-terminal domains that mediate binding to organelles.[33] Hook3 interacts with PCM1 to regulate pericentriolar material assembly and the timing of neurogenesis,[34] and interacts with scavenger receptor A to modulate innate immunity.[35] The Rev3L subunit interacts with Rev7 to form Pol ζ, a B family polymerase. Pol ζ lacks 3′ to 5′ exonuclease activity and is a moderate fidelity polymerase. It cannot add nucleotides across from DNA lesions, yet it can extend from primers with terminal mismatches. DNA polymerase zeta is required for proliferation of normal mammalian cells[36] and disruption of the polymerase is embryonically lethal.[37] Thus, both genes predicted to be differentially alternatively spliced in ASD, Hook3 and Rev3L, regulate cell proliferation and could be related to reports of altered neuronal numbers in some regions of ASD brain.

BTG1, FCGBP, and MAP4 were dysregulated in three different ASD studies, including a study on brain tissues, and are predicted to be alternatively spliced in ASD blood.[9,28,31] BTG1 negatively regulates cell proliferation and plays a role in regulating neurogenesis.[38] Btg1 (B cell translocation gene 1) is required to maintain the pool of stem and progenitor cells of the brain Dentate Gyrus and Subventricular Zone.[39] Polymorphisms in the BTG1 gene are associated with multiple sclerosis, implying a role for the gene in immune regulation.[40] BTG1 also sensitizes brain microglia to apoptosis.[41] A BTG1 analogue controls sexual fate in Caenorhabditis elegans germ cells.[42] Thus, BTG1 could be involved in ASD pathogenesis at multiple levels — neurogenesis, immune function and possibly even sex determination. FCGBP (FC gamma fragment of IgG binding protein), an IgG-Fc receptor, can be associated with progenitor cells.[43] Elevated plasma levels of FCGBP antigen has been reported in the blood of patients with autoimmune diseases,[44] which may relate to the immune abnormalities observed in some children with ASD.[45,46] MAP4 binds septins and controls the accessibility of septins to microtubules and serves to stabilize microtubules,[47] and MAP4 also binds to actin filaments and modulates their properties.[48] Over-expression of MAP4 on microtubules disrupts messenger ribonucleoprotein particles (mRNPs) to translation sites.[49] MAP4 helps stabilize mitochondrial permeability transition in hypoxia by microtubule enhancement and DYNLT1 interaction with VDAC1.[50] Phosphorylation of MAP4 affects microtubule properties and cell cycle progression.[51] MAP4 and CLASP1 operate as a safety mechanism to maintain a stable spindle position in mitosis.[52] MAP4 is highly enriched in cerebellar Purkinje cells and hippocampal pyramidal neurons.[53] Antidepressants upregulate MAP4 mRNA in brain hippocampus,[54] MAP4 is dysregulated in Alzheimer's disease brain[55] and MAP4 inhibits G protein receptor trafficking in neuroblastoma cells.[56] In addition, MAP4 is regulated by CTNNB1 (Catenin (Cadherin-Associated Protein), Beta 1). *De novo* variants in CTNNB1 has been identified in two separate reports using ASD probands from the Simons Simplex Collection.[57,58] In addition, another member of the microtubule-associated protein family, MAP2, has been implicated in ASD.[59,60]

Pathways Associated with ASD

Though confirmation of individual genes between whole genome expression studies has been relatively poor (Table 4), a bit more success has been realized in replicating various pathways between studies. We were the first to suggest that there was over-representation of genes associated with NK cells.[30,61] This finding has been confirmed in part by a recent study of whole blood with RNA processed on U133 arrays[29] and in our recent study of exon expression and predicted alternative splicing using whole blood RNA processed on Affymetrix exon arrays.[28]

Indeed, if one compares the Kong and Stamova studies,[28,29] there are several similar dysregulated pathways in blood from individuals with ASD compared to that of controls. Kong reports the following pathways: Neurotrophin, Fc gamma R-mediated phagocytosis, renal cell cancer, chemokine, regulation of actin cytoskeleton, mTOR, chromic myeloid leukemia, Fc episolon RI, B cell, T cell, focal adhesion, ErbB, NK cell, VEGF, Progesterone, Long-term potentiation/depression, notch, lysosomes, leukocyte transendothelial migration, endocytosis and MAPK signaling.[29] Stamova reports the following pathways: ILK (actin cytoskeleton), Natural Killer, Fc gamma, HMGB1, Dendritic cell, TREM1, JAK/Stat, actin cytoskeleton, VEGF, NO and ROS in macrophages, PI3K/AKT/mTOR, NFkappaB, NGF, PAK, GM-CSF, ceramide,

T cells, actin nucleation, pentose phosphate, and cell cycle regulation by BTG family proteins.[28] Of the 21 pathways identified by the Stamova study, eight were also identified in the Kong study. Thus, identification of pathways associated with ASD has been somewhat more consistent than identification of differentially expressed genes.

Examining Dysregulated Pathways in Individual ASD Subjects

We recently developed an approach for determining which pathways are dysregulated in a given individual with ASD.[28] As mentioned above, using exon expression arrays we identified 477 genes that were predicted to have differential alternative splicing (DAS) and to have significant differences in exon expression (differential exon usage, DEU) between individuals with ASD and typically developing (TD) individuals ($p<0.005$). A pathway analysis of these 477 genes revealed 21 pathways that were significantly different for ASD vs. TD ($p<0.05$).[28]

We then determined which of the 21 pathways were dysregulated in each individual with ASD. To do this, principal components analysis (PCA) mapping was performed based on exon-level expression of the genes predicted to have DAS/DEU in each over-represented pathway. For example, the PCA plot of the mTOR signaling pathway (Figure 1, upper right inset) was based on the expression level of eight exons in mTOR genes predicted to have DAS/DEU. Based on the position of each ASD subject within the PCA plot, the mTOR signaling pathway was scored as TD-like (if within 2 standard deviations of the TD centroid), or non-TD-like (if outside the 2 standard deviations of the TD centroid) (Figure 1, upper right inset). The same procedure was then performed for each of the 21 significantly over-represented canonical pathways (Figure 1). Hierarchical clustering of the ASD subjects and the 21 Canonical Pathways (Figure 1) showed that different combinations of pathways were affected (non-TD-like ASD, gray rectangles) in different ASD subjects (Figure 1). No pathway was affected in every subject. Five ASD subjects had all 21 pathways affected (Figure 1, right side of X-axis cluster), but there were some ASD subjects with only several pathways affected (Figure 1, left side of X-axis cluster). The pathways affected in the majority of the ASD subjects included ILK Signaling, NK Cell Signaling and FCγ Receptor-mediated phagocytosis in macrophages and monocytes (Figure 1, lower portion of cluster). The mTOR pathway was altered in 18 out of 30 ASD subjects (Figure 1).

These data are important for several reasons. They suggest that no single pathway will be shared by all ASD subjects. In addition, even amongst a number of common pathways, each ASD individual seems to have a different assortment of those affected pathways. ASD is thus associated with a composite of pathways that appear to differ from individual to individual.

Importance of Pathway Studies for ASD

Figure 2 shows the relationship between some of the pathways identified in Figure 1. The convergence of several pathways that are consistently observed in some ASD subjects provides additional support for the involvement of these pathways, which may indicate a consistent biology underlying ASD. Though the data from Figure 1 appears discouraging with respect to the development of biomarkers for diagnosis or prognosis, they do suggest that there may be subgroups of ASD subjects that have dysregulation of the SAME pathway — even if there might be different causes for this dysregulation.

An example is the mTOR signaling pathway. The data from Figure 1 suggests that ~18 of the 30 ASD subjects have dysregulation of mTOR pathways. It is notable that the study of Kong *et al.* also found that the mTOR signaling pathway was associated with ASD ($p=0.014$).[29] mTOR is particularly interesting because there are both human studies and animal studies to support a role for mTOR signaling in at least some ASD subjects.

Aberrant mTOR signaling has been implicated in single gene disorders which can be associated with ASD in some subjects. The

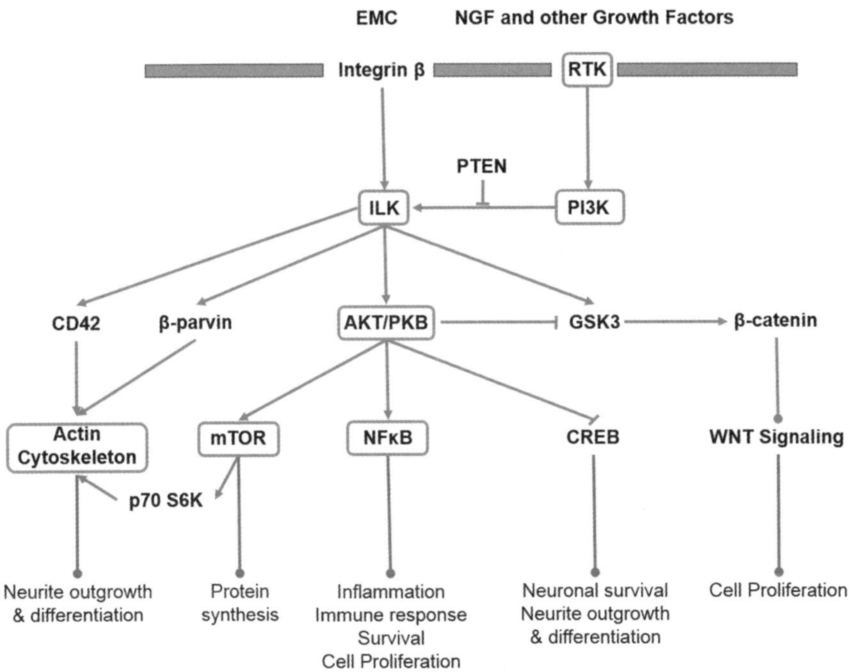

Figure 2. Relationship between some of the pathways identified in Figure 1 (boxed). ILK (Integrin linked kinase) couples actions on integrin receptors to regulation of the actin cytoskeleton and to AKT actions on mTOR and NFkappaB pathways.

prototypical example is Tuberous Sclerosis (TSC) where some children with TSC also have ASD, and both genetic forms of TSC are associated with up-regulation of mTOR pathways.[62] Inhibition of mTOR pathways in TSC mice with rapamycin reverses impaired social interactions,[63] and rapamycin, which is used to treat tumors associated with TSC, has also been suggested as possible treatment for ASD-like features in TSC patients.[64,65]

Another example is fragile X syndrome, which is associated with ASD in about a third of cases. Increased mTOR signaling has been demonstrated in humans with fragile X as well as animal models of fragile X.[66,67] Genetic removal of p70 S6 kinase 1, which corrects abnormalities of mTOR signaling, corrects molecular, synaptic, and behavioral phenotypes in fragile X syndrome mice.[68] Neurofibromatosis (NF-1), which is associated with ASD in some subjects, also demonstrates up-regulation of mTOR pathways.[69,70] Consequently, rapamycin is being considered for NF treatment trials.[71] PTEN mutations that cause dysregulation of mTOR[72–74] can also be associated with ASD and macrocephaly, a well-reported anatomical feature of some cases of ASD.

Thus, a number of syndromic disorders associated with dysregulation of mTOR, including TSC1, TSC2, Fragile X, NF-1 and PTEN, are also associated with ASD.[74] These clinical findings have prompted various studies showing that perturbing protein synthesis or protein synthesis regulators, such as elongation factors, can be associated with clinical features of ASD in humans and in animal models.[1,62,64,65,68,75–79]

The implications of these studies of single gene causes of ASD associated with mTOR abnormalities coupled with recent gene profiling of cases of idiopathic ASD[2,28] is that a subgroup of idiopathic ASD cases may have abnormalities of mTOR signaling (Figures 1 and 2). If so, these ASD subjects could be potentially studied for effects of mTOR inhibitors, like rapamycin, on behavioral phenotypes in future trials. There are examples of other pathways from the existing studies on ASD that suggest novel targeted therapies, including NGF/neurotrophin signaling, NK Cell signaling, FCgamma phagocytosis in macrophages and monocytes, and RORA signaling that have been observed in at least several independent studies.[2,28–30,61,80,81] If it can be confirmed that there are subgroups of ASD subjects with abnormal signaling of these pathways, these may be potential treatment targets in these individuals.

Future Directions

Since common pathways have been identified as dysregulated in ASD studies despite few shared genes, a more promising avenue for the near term is to focus on pathways that have already been implicated in ASD. Designing focused expression studies targeting 10–30 potential pathways might be more fruitful. Care in designing these studies will be needed, however, and most genes within a pathway probably should be included since individual genes do not seem to replicate very well. Confirmation of regulated pathways in ASD offers the advantage of then being able to use animal models to see if perturbation of those pathways results in behavioral abnormalities relevant for ASD — similar to what has been done already for AKT/PI3K/mTOR studies of translation. Moreover, careful attention to clinical features of ASD might also make the pathway studies more fruitful since they may help to enrich subjects for certain pathways. An example is macrocephaly where differences of alternative splicing were more apparent in ASD children with normal cerebral volumes compared to those with large cerebral volumes.[28]

Recent studies of dysregulation of alternative splicing in ASD compared to controls in brain,[9] blood[28] and lymphoblastoid cell lines[32] point to a number of shared genes in these studies. It is likely that alternative splicing is different in blood and brain, but it is also possible that there are core genes or pathways where dysregulation of alternative splicing is particularly relevant for ASD. Direct confirmation of differential alternative splicing in ASD would be useful, particularly if brain and blood were compared. This might point to some abnormality of alternative splicing machinery or abnormality of specific pathways shared by the blood and brain. Studies of other epigenetic mechanisms might be similarly fruitful.[82,83]

Executive Summary

- Very few genes are dysregulated in common across published ASD studies.
- There are some commonalities of dysfunctional pathways across published ASD studies.
- Subgroups of ASD subjects may have similar dysregulaton of a given pathway.
- The mTOR pathway appears to be dysregulated in some ASD subjects.
- Alternative splicing is involved in the pathophysiology of ASD.

References

1. Huguet G, Ey E, Bourgeron T: The genetic landscapes of autism spectrum disorders. *Annu Rev Genomics Hum Genet* 2013, **14**:191–213.
2. Kong SW, Shimizu-Motohashi Y, Campbell MG, Lee IH, Collins CD, Brewster SJ, Holm IA, Rappaport L, Kohane IS, Kunkel LM: Peripheral blood gene expression signature differentiates children with autism from unaffected siblings. *Neurogenetics* 2013, **14**(2):143–152.
3. Stamova B, Green PG, Tian Y, Hertz-Picciotto I, Pessah IN, Hansen R, Yang X, Teng J, Gregg JP, Ashwood P *et al.*: Correlations between gene expression and mercury levels in blood of boys with and without autism. *Neurotox Res* 2011, **19**(1):31–48.
4. Tian Y, Green PG, Stamova B, Hertz-Picciotto I, Pessah IN, Hansen R, Yang X, Gregg JP, Ashwood P, Jickling G *et al.*: Correlations of gene expression with blood lead levels in children with autism compared to typically developing controls. *Neurotox Res* 2011, **19**(1):1–13.
5. Lit L, Sharp FR, Bertoglio K, Stamova B, Ander BP, Sossong AD, Hendren RL: Gene expression in blood is associated with risperidone response in children with autism spectrum disorders. *Pharmacogenomics J* 2012, **12**(5):368–371.
6. Careaga M, Van de Water J, Ashwood P: Immune dysfunction in autism: a pathway to treatment. *Neurotherapeutics* 2010, **7**(3):283–292.
7. Vargas DL, Nascimbene C, Krishnan C, Zimmerman AW, Pardo CA: Neuroglial activation and neuroinflammation in the brain of patients with autism. *Ann Neurol* 2005, **57**(1):67–81.
8. Morgan JT, Chana G, Pardo CA, Achim C, Semendeferi K, Buckwalter J, Courchesne E, Everall IP: Microglial activation and increased microglial density observed in the dorsolateral prefrontal cortex in autism. *Biol Psychiatry* 2010, **68**(4):368–376.
9. Voineagu I, Wang X, Johnston P, Lowe JK, Tian Y, Horvath S, Mill J, Cantor RM, Blencowe BJ, Geschwind DH: Transcriptomic analysis of autistic brain reveals convergent molecular pathology. *Nature* 2011, **474**(7351):380–384.
10. Hu VW, Frank BC, Heine S, Lee NH, Quackenbush J: Gene expression profiling of lymphoblastoid cell lines from monozygotic twins discordant in severity of autism reveals differential regulation of neurologically relevant genes. *BMC Genomics* 2006, **7**:118.
11. Lerer E, Levi S, Israel S, Yaari M, Nemanov L, Mankuta D, Nurit Y, Ebstein RP: Low CD38 expression in lymphoblastoid cells and haplotypes are both associated with autism in a family-based study. *Autism Res* 2010, **3**(6):293–302.
12. Voineagu I: Gene expression studies in autism: moving from the genome to the transcriptome and beyond. *Neurobiol Dis* 2012, **45**(1):69–75.
13. Garbett K, Ebert PJ, Mitchell A, Lintas C, Manzi B, Mirnics K, Persico AM: Immune transcriptome alterations in the temporal cortex of subjects with autism. *Neurobiol Dis* 2008, **30**(3):303–311.

14. Sarachana T, Zhou R, Chen G, Manji HK, Hu VW: Investigation of post-transcriptional gene regulatory networks associated with autism spectrum disorders by microRNA expression profiling of lymphoblastoid cell lines. *Genome Med* 2010, **2**(4):23.
15. Talebizadeh Z, Butler MG, Theodoro MF: Feasibility and relevance of examining lymphoblastoid cell lines to study role of microRNAs in autism. *Autism Res* 2008, **1**(4):240–250.
16. Abu-Elneel K, Liu T, Gazzaniga FS, Nishimura Y, Wall DP, Geschwind DH, Lao K, Kosik KS: Heterogeneous dysregulation of microRNAs across the autism spectrum. *Neurogenetics* 2008, **9**(3):153–161.
17. Ridolfi E, Fenoglio C, Cantoni C, Calvi A, De Riz M, Pietroboni A, Villa C, Serpente M, Bonsi R, Vercellino M *et al.*: Expression and genetic analysis of microRNAs involved in multiple sclerosis. *Int J Mol Sci* 2013, **14**(3):4375–4384.
18. Kong KY, Owens KS, Rogers JH, Mullenix J, Velu CS, Grimes HL, Dahl R: MIR-23A microRNA cluster inhibits B-cell development. *Exp Hematol* 2010, **38**(8):629–640 e621.
19. Liu X, Ru J, Zhang J, Zhu LH, Liu M, Li X, Tang H: miR-23a targets interferon regulatory factor 1 and modulates cellular proliferation and paclitaxel-induced apoptosis in gastric adenocarcinoma cells. *PLoS One* 2013, **8**(6):e64707.
20. Guo Z, Zhou B, Liu W, Xu Y, Wu D, Yin Z, Permatasari F, Luo D: MiR-23a regulates DNA damage repair and apoptosis in UVB-irradiated HaCaT cells. *J Dermatol Sci* 2013, **69**(1):68–76.
21. Rathore MG, Saumet A, Rossi JF, de Bettignies C, Tempe D, Lecellier CH, Villalba M: The NF-kappaB member p65 controls glutamine metabolism through miR-23a. *Int J Biochem Cell Biol* 2012, **44**(9):1448–1456.
22. Siegel C, Li J, Liu F, Benashski SE, McCullough LD: miR-23a regulation of X-linked inhibitor of apoptosis (XIAP) contributes to sex differences in the response to cerebral ischemia. *Proc Natl Acad Sci USA* 2011, **108**(28): 11662–11667.
23. Kim J, Yoon H, Ramirez CM, Lee SM, Hoe HS, Fernandez-Hernando C: MiR-106b impairs cholesterol efflux and increases Abeta levels by repressing ABCA1 expression. *Exp Neurol* 2012, **235**(2):476–483.
24. Brett JO, Renault VM, Rafalski VA, Webb AE, Brunet A: The microRNA cluster miR-106b~25 regulates adult neural stem/progenitor cell proliferation and neuronal differentiation. *Aging (Albany NY)* 2011, **3**(2):108–124.
25. Luo R, Sanders SJ, Tian Y, Voineagu I, Huang N, Chu SH, Klei L, Cai C, Ou J, Lowe JK *et al.*: Genome-wide transcriptome profiling reveals the functional impact of rare de novo and recurrent CNVs in autism spectrum disorders. *Am J Hum Genet* 2012, **91**(1):38–55.
26. Nord AS, Roeb W, Dickel DE, Walsh T, Kusenda M, O'Connor KL, Malhotra D, McCarthy SE, Stray SM, Taylor SM *et al.*: Reduced transcript expression of genes affected by inherited and de novo CNVs in autism. *Eur J Hum Genet* 2011, **19**(6):727–731.

27. Cheng Y, Quinn JF, Weiss LA: An eQTL mapping approach reveals that rare variants in the SEMA5A regulatory network impact autism risk. *Hum Mol Genet* 2013, **22**(14):2960–2972.
28. Stamova B, Tian Y, Nordahl CW, Shen MD, Rogers S, Amaral DG, Sharp FR: Evidence for differential alternative splicing in blood of young boys with autism spectrum disorders. *Mol Autism* 2013, **4**(1):30.
29. Kong SW, Collins CD, Shimizu-Motohashi Y, Holm IA, Campbell MG, Lee IH, Brewster SJ, Hanson E, Harris HK, Lowe KR *et al.*: Characteristics and predictive value of blood transcriptome signature in males with autism spectrum disorders. *PLoS One* 2012, **7**(12):e49475.
30. Gregg JP, Lit L, Baron CA, Hertz-Picciotto I, Walker W, Davis RA, Croen LA, Ozonoff S, Hansen R, Pessah IN *et al.*: Gene expression changes in children with autism. *Genomics* 2008, **91**(1):22–29.
31. Hu VW, Sarachana T, Kim KS, Nguyen A, Kulkarni S, Steinberg ME, Luu T, Lai Y, Lee NH: Gene expression profiling differentiates autism case-controls and phenotypic variants of autism spectrum disorders: evidence for circadian rhythm dysfunction in severe autism. *Autism Res* 2009, **2**(2):78–97.
32. Talebizadeh Z, Aldenderfer R, Wen Chen X: A proof-of-concept study: exon-level expression profiling and alternative splicing in autism using lymphoblastoid cell lines. *Psychiatr Genet* 2013, Jul. 3, Epub ahead of print.
33. Walenta JH, Didier AJ, Liu X, Kramer H: The Golgi-associated hook3 protein is a member of a novel family of microtubule-binding proteins. *J Cell Biol* 2001, **152**(5):923–934.
34. Ge X, Frank CL, Calderon de Anda F, Tsai LH: Hook3 interacts with PCM1 to regulate pericentriolar material assembly and the timing of neurogenesis. *Neuron* 2010, **65**(2):191–203.
35. Sano H, Ishino M, Kramer H, Shimizu T, Mitsuzawa H, Nishitani C, Kuroki Y: The microtubule-binding protein Hook3 interacts with a cytoplasmic domain of scavenger receptor A. *J Biol Chem* 2007, **282**(11):7973–7981.
36. Lange SS, Wittschieben JP, Wood RD: DNA polymerase zeta is required for proliferation of normal mammalian cells. *Nucleic Acids Res* 2012, **40**(10): 4473–4482.
37. Wittschieben J, Shivji MK, Lalani E, Jacobs MA, Marini F, Gearhart PJ, Rosewell I, Stamp G, Wood RD: Disruption of the developmentally regulated Rev3l gene causes embryonic lethality. *Curr Biol* 2000, **10**(19):1217–1220.
38. Tirone F, Farioli-Vecchioli S, Micheli L, Ceccarelli M, Leonardi L: Genetic control of adult neurogenesis: interplay of differentiation, proliferation and survival modulates new neurons function, and memory circuits. *Front Cell Neurosci* 2013, **7**:59.
39. Farioli-Vecchioli S, Micheli L, Saraulli D, Ceccarelli M, Cannas S, Scardigli R, Leonardi L, Cina I, Costanzi M, Ciotti MT *et al.*: Btg1 is required to maintain the pool of stem and progenitor cells of the dentate gyrus and subventricular zone. *Front Neurosci* 2012, **6**:124.

40. Camina-Tato M, Morcillo-Suarez C, Navarro A, Fernandez M, Horga A, Montalban X, Comabella M: Genetic association between polymorphisms in the BTG1 gene and multiple sclerosis. *J Neuroimmunol* 2009, **213**(1–2):142–147.
41. Lee H, Cha S, Lee MS, Cho GJ, Choi WS, Suk K: Role of antiproliferative B cell translocation gene-1 as an apoptotic sensitizer in activation-induced cell death of brain microglia. *J Immunol* 2003, **171**(11):5802–5811.
42. Chen PJ, Singal A, Kimble J, Ellis RE: A novel member of the tob family of proteins controls sexual fate in Caenorhabditis elegans germ cells. *Dev Biol* 2000, **217**(1):77–90.
43. Stamp LA, Braxton DR, Wu J, Akopian V, Hasegawa K, Chandrasoma PT, Hawes SM, McLean C, Petrovic LM, Wang K *et al.*: The GCTM-5 epitope associated with the mucin-like glycoprotein FCGBP marks progenitor cells in tissues of endodermal origin. *Stem Cells* 2012, **30**(9):1999–2009.
44. Kobayashi K, Yagasaki M, Harada N, Chichibu K, Hibi T, Yoshida T, Brown WR, Morikawa M: Detection of Fcgamma binding protein antigen in human sera and its relation with autoimmune diseases. *Immunol Lett* 2001, **79**(3):229–235.
45. Onore C, Careaga M, Ashwood P: The role of immune dysfunction in the pathophysiology of autism. *Brain Behav Immun* 2012, **26**(3):383–392.
46. Ashwood P, Wills S, Van de Water J: The immune response in autism: a new frontier for autism research. *J Leukoc Biol* 2006, **80**(1):1–15.
47. Ghossoub R, Hu Q, Failler M, Rouyez MC, Spitzbarth B, Mostowy S, Wolfrum U, Saunier S, Cossart P, Jamesnelson W *et al.*: Septins 2, 7 and 9 and MAP4 colocalize along the axoneme in the primary cilium and control ciliary length. *J Cell Sci* 2013, **126** (Pt 12):2583–2594.
48. Matsushima K, Tokuraku K, Hasan MR, Kotani S: Microtubule-associated protein 4 binds to actin filaments and modulates their properties. *J Biochem* 2012, **151**(1):99–108.
49. Scholz D, Baicu CF, Tuxworth WJ, Xu L, Kasiganesan H, Menick DR, Cooper Gt: Microtubule-dependent distribution of mRNA in adult cardiocytes. *Am J Physiol Heart Circ Physiol* 2008, **294**(3):H1135–1144.
50. Fang YD, Xu X, Dang YM, Zhang YM, Zhang JP, Hu JY, Zhang Q, Dai X, Teng M, Zhang DX *et al.*: MAP4 mechanism that stabilizes mitochondrial permeability transition in hypoxia: microtubule enhancement and DYNLT1 interaction with VDAC1. *PLoS One* 2011, **6**(12):e28052.
51. Chang W, Gruber D, Chari S, Kitazawa H, Hamazumi Y, Hisanaga S, Bulinski JC: Phosphorylation of MAP4 affects microtubule properties and cell cycle progression. *J Cell Sci* 2001, **114**(Pt 15):28792–887.
52. Samora CP, Mogessie B, Conway L, Ross JL, Straube A, McAinsh AD: MAP4 and CLASP1 operate as a safety mechanism to maintain a stable spindle position in mitosis. *Nat Cell Biol* 2011, **13**(9):1040–1050.

53. Tokuraku K, Okuyama S, Matsushima K, Ikezu T, Kotani S: Distinct neuronal localization of microtubule-associated protein 4 in the mammalian brain. *Neurosci Lett* 2010, **484**(2):143–147.
54. Yang HJ, Li YF, Zhang HT, Zhang FQ, Zhao N, Gong ZH, Luo ZP: Up-regulation of microtubule-associated protein 4 and drebrin A mRNA expression by antidepressants in rat hippocampus following chronic stress. *Neurosci Lett* 2003, **351**(3):206–208.
55. Ray M, Ruan J, Zhang W: Variations in the transcriptome of Alzheimer's disease reveal molecular networks involved in cardiovascular diseases. *Genome Biol* 2008, **9**(10):R148.
56. Cheng G, Iijima Y, Ishibashi Y, Kuppuswamy D, Cooper Gt: Inhibition of G protein-coupled receptor trafficking in neuroblastoma cells by MAP 4 decoration of microtubules. *Am J Physiol Heart Circ Physiol* 2002, **283**(6):H2379–2388.
57. O'Roak BJ, Vives L, Girirajan S, Karakoc E, Krumm N, Coe BP, Levy R, Ko A, Lee C, Smith JD *et al.*: Sporadic autism exomes reveal a highly interconnected protein network of de novo mutations. *Nature* 2012, **485**(7397):246–250.
58. Sanders SJ, Murtha MT, Gupta AR, Murdoch JD, Raubeson MJ, Willsey AJ, Ercan-Sencicek AG, DiLullo NM, Parikshak NN, Stein JL *et al.*: De novo mutations revealed by whole-exome sequencing are strongly associated with autism. *Nature* 2012, **485**(7397):237–241.
59. Pescucci C, Meloni I, Bruttini M, Ariani F, Longo I, Mari F, Canitano R, Hayek G, Zappella M, Renieri A: Chromosome 2 deletion encompassing the MAP2 gene in a patient with autism and Rett-like features. *Clin Genet* 2003, **64**(6):497–501.
60. Mukaetova-Ladinska EB, Arnold H, Jaros E, Perry R, Perry E: Depletion of MAP2 expression and laminar cytoarchitectonic changes in dorsolateral prefrontal cortex in adult autistic individuals. *Neuropathol Appl Neurobiol* 2004, **30**(6):615–623.
61. Enstrom AM, Lit L, Onore CE, Gregg JP, Hansen RL, Pessah IN, Hertz-Picciotto I, Van de Water JA, Sharp FR, Ashwood P: Altered gene expression and function of peripheral blood natural killer cells in children with autism. *Brain Behav Immun* 2009, **23**(1):124–133.
62. Hoeffer CA, Klann E: mTOR signaling: at the crossroads of plasticity, memory and disease. *Trends Neurosci* 2010, **33**(2):67–75.
63. Sato A, Kasai S, Kobayashi T, Takamatsu Y, Hino O, Ikeda K, Mizuguchi M: Rapamycin reverses impaired social interaction in mouse models of tuberous sclerosis complex. *Nat Commun* 2012, **3**:1292.
64. Ehninger D: From genes to cognition in tuberous sclerosis: implications for mTOR inhibitor-based treatment approaches. *Neuropharmacology* 2013, **68**:97–105.
65. Sahin M: Targeted treatment trials for tuberous sclerosis and autism: no longer a dream. *Curr Opin Neurobiol* 2012, **22**(5):895–901.

66. Sharma A, Hoeffer CA, Takayasu Y, Miyawaki T, McBride SM, Klann E, Zukin RS: Dysregulation of mTOR signaling in fragile X syndrome. *J Neurosci* 2010, **30**(2):694–702.
67. Hoeffer CA, Sanchez E, Hagerman RJ, Mu Y, Nguyen DV, Wong H, Whelan AM, Zukin RS, Klann E, Tassone F: Altered mTOR signaling and enhanced CYFIP2 expression levels in subjects with fragile X syndrome. *Genes Brain Behav* 2012, **11**(3):332–341.
68. Bhattacharya A, Kaphzan H, Alvarez-Dieppa AC, Murphy JP, Pierre P, Klann E: Genetic removal of p70 S6 kinase 1 corrects molecular, synaptic, and behavioral phenotypes in fragile X syndrome mice. *Neuron* 2012, **76**(2):325–337.
69. Banerjee S, Crouse NR, Emnett RJ, Gianino SM, Gutmann DH: Neurofibromatosis-1 regulates mTOR-mediated astrocyte growth and glioma formation in a TSC/Rheb-independent manner. *Proc Natl Acad Sci USA* 2011, **108**(38):15996–16001.
70. Ma J, Li M, Hock J, Yu X: Hyperactivation of mTOR critically regulates abnormal osteoclastogenesis in neurofibromatosis Type 1. *J Orthop Res* 2012, **30**(1):144–152.
71. Gipson TT, Johnston MV: Plasticity and mTOR: towards restoration of impaired synaptic plasticity in mTOR-related neurogenetic disorders. *Neural Plast* 2012, **2012**:486402.
72. Mawrin C: Critical role of PTEN for development and progression of nerve sheath tumors in neurofibromatosis type 1. *Future Oncol* 2010, **6**(4):499–501.
73. Crino PB: mTOR: A pathogenic signaling pathway in developmental brain malformations. *Trends Mol Med* 2011, **17**(12):734–742.
74. Bourgeron T: A synaptic trek to autism. *Curr Opin Neurobiol* 2009, **19**(2):231–234.
75. Neves-Pereira M, Muller B, Massie D, Williams JH, O'Brien PC, Hughes A, Shen SB, Clair DS, Miedzybrodzka Z: Deregulation of EIF4E: a novel mechanism for autism. *J Med Genet* 2009, **46**(11):759–765.
76. Gkogkas CG, Khoutorsky A, Ran I, Rampakakis E, Nevarko T, Weatherill DB, Vasuta C, Yee S, Truitt M, Dallaire P *et al.*: Autism-related deficits via dysregulated eIF4E-dependent translational control. *Nature* 2013, **493**(7432):371–377.
77. Santini E, Huynh TN, MacAskill AF, Carter AG, Pierre P, Ruggero D, Kaphzan H, Klann E: Exaggerated translation causes synaptic and behavioural aberrations associated with autism. *Nature* 2013, **493**(7432):411–415.
78. Ginsberg MR, Rubin RA, Falcone T, Ting AH, Natowicz MR: Brain transcriptional and epigenetic associations with autism. *PLoS One* 2012, **7**(9):e44736.
79. Millan MJ: An epigenetic framework for neurodevelopmental disorders: from pathogenesis to potential therapy. *Neuropharmacology* 2013, **68**:2–82.
80. Hu VW: From genes to environment: using integrative genomics to build a "systems-level" understanding of autism spectrum disorders. *Child Dev* 2013, **84**(1):89–103.

81. Sarachana T, Hu VW: Genome-wide identification of transcriptional targets of RORA reveals direct regulation of multiple genes associated with autism spectrum disorder. *Mol Autism* 2013, **4**(1):14.
82. Hu VW: The expanding genomic landscape of autism: discovering the 'forest' beyond the 'trees'. *Future Neurol* 2013, **8**(1):29–42.
83. Nguyen A, Rauch TA, Pfeifer GP, Hu VW: Global methylation profiling of lymphoblastoid cell lines reveals epigenetic contributions to autism spectrum disorders and a novel autism candidate gene, RORA, whose protein product is reduced in autistic brain. *Faseb J* 2010, **24**(8):3036–3051.
84. Melin M, Carlsson B, Anckarsater H, Rastam M, Betancur C, Isaksson A, Gillberg C, Dahl N: Constitutional downregulation of SEMA5A expression in autism. *Neuropsychobiology* 2006, **54**(1):64–69.
85. Baron CA, Liu SY, Hicks C, Gregg JP: Utilization of lymphoblastoid cell lines as a system for the molecular modeling of autism. *J Autism Dev Disord* 2006, **36**(8):973–982.
86. Nishimura Y, Martin CL, Vazquez-Lopez A, Spence SJ, Alvarez-Retuerto AI, Sigman M, Steindler C, Pellegrini S, Schanen NC, Warren ST *et al.*: Genome-wide expression profiling of lymphoblastoid cell lines distinguishes different forms of autism and reveals shared pathways. *Hum Mol Genet* 2007, **16**(14):1682–1698.
87. Hu VW, Nguyen A, Kim KS, Steinberg ME, Sarachana T, Scully MA, Soldin SJ, Luu T, Lee NH: Gene expression profiling of lymphoblasts from autistic and nonaffected sib pairs: altered pathways in neuronal development and steroid biosynthesis. *PLoS One* 2009, **4**(6):e5775.
88. Ghahramani Seno MM, Hu P, Gwadry FG, Pinto D, Marshall CR, Casallo G, Scherer SW: Gene and miRNA expression profiles in autism spectrum disorders. *Brain Res* 2011, **1380**:85–97.
89. Alter MD, Kharkar R, Ramsey KE, Craig DW, Melmed RD, Grebe TA, Bay RC, Ober-Reynolds S, Kirwan J, Jones JJ *et al.*: Autism and increased paternal age related changes in global levels of gene expression regulation. *PLoS One* 2011, **6**(2):e16715.
90. Yasuda Y, Hashimoto R, Yamamori H, Ohi K, Fukumoto M, Umeda-Yano S, Mohri I, Ito A, Taniike M, Takeda M: Gene expression analysis in lymphoblasts derived from patients with autism spectrum disorder. *Mol Autism* 2011, **2**(1):9.
91. Chien WH, Gau SS, Chen CH, Tsai WC, Wu YY, Chen PH, Shang CY: Increased gene expression of FOXP1 in patients with autism spectrum disorders. *Mol Autism* 2013, **4**(1):23.
92. Kuwano Y, Kamio Y, Kawai T, Katsuura S, Inada N, Takaki A, Rokutan K: Autism-associated gene expression in peripheral leucocytes commonly observed between subjects with autism and healthy women having autistic children. *PLoS One* 2011, **6**(9):e24723.

93. Jyonouchi H, Geng L, Streck DL, Toruner GA: Immunological characterization and transcription profiling of peripheral blood (PB) monocytes in children with autism spectrum disorders (ASD) and specific polysaccharide antibody deficiency (SPAD): case study. *J Neuroinflammation* 2012, **9**:4.
94. Glatt SJ, Tsuang MT, Winn M, Chandler SD, Collins M, Lopez L, Weinfeld M, Carter C, Schork N, Pierce K *et al.*: Blood-based gene expression signatures of infants and toddlers with autism. *J Am Acad Child Adolesc Psychiatry* 2012, **51**(9):934–944 e932.

Chapter 8

Searching in the "Dark": Non-coding RNA as a New Avenue of Autism Research

Tewarit Sarachana* and Valerie W. Hu

Abstract

Autism spectrum disorder (ASD) describes a group of neurodevelopmental disorders characterized by deficits in social communication and presence of restricted interests and/or repetitive behaviors. Because of its high degree of heritability, genetic studies, focused mainly on the protein-coding regions of the genome, have attempted to identify gene mutations, both rare and common single nucleotide polymorphisms, and copy number variants that may cause or increase the susceptibility of this disorder. Nevertheless, no single gene, SNP, or DNA mutation can account for more than 1–2% of ASD cases. Recent studies have started to explore the genomic "dark matter" (DNA regions that do not directly code for proteins) which harbors many types of gene regulatory elements, including non-coding RNAs (ncRNA). However, ncRNA research in ASD is still in its infancy with only a few, albeit significant, ncRNA studies in the context of ASD. This article summarizes evidence that suggests future autism research needs to be expanded to cover the remaining 98% of the genome, especially ncRNAs, and emphasizes the need for subgrouping subjects to reduce the heterogeneity within study groups in order to draw comparisons between different independent cohorts.

Keywords: Autism; Noncoding RNA; miRNA; snoRNA; Biomarker.

---

*Correspondence to: Department of Clinical Chemistry, Faculty of Allied Health Sciences, Chulalongkorn University, Bangkok, 10330, Thailand. Emails: Tewarit Sarachana: tewarit.sa@chula.ac.th; Valerie Hu: valhu@gwu.edu

Introduction

Autism spectrum disorder (ASD) describes a group of neurodevelopmental disorders characterized by impaired reciprocal social interactions, deficits in verbal and non-verbal communication, and restricted interests and repetitive or stereotyped behaviors. With the 2013 publication of the DSM-5 diagnostic manual, ASD includes autistic disorder (sometimes referred to as autism or classic ASD), Rett syndrome, childhood disintegrative disorder (CDD), pervasive developmental disorder — not otherwise specified (PDD-NOS), and Asperger's syndrome under one diagnosis.[1] In many cases, behavioral abnormalities associated with ASD can be observed in the first two to three years of life, but the severity in the affected individuals can vary greatly. Individuals with autistic disorder, which describes the most severe form of ASD, are often completely nonverbal, exhibit severe stereotypies, social impairments, and intellectual deficits. In contrast, individuals with Asperger's syndrome are usually not diagnosed until preschool age or later because they have neither delays in early speech development nor intellectual deficits as assessed by routine tests. However, they still exhibit some difficulties in reciprocal social interactions, such as making little eye contact while talking or having trouble with understanding humor or teasing. Epidemiological surveys of ASD reveal that the prevalence of ASD has dramatically increased from 1 in 14,000 in the United States in 1970[2] to approximately 1 in 88 in recent years.[3] The prevalence of ASD is reported to be higher in males than females with various ratios depending on the population studied.[4–6] Although this higher rate of ASD may be a consequence of indirect factors including improved diagnosis and awareness, recent studies have shown that changes in diagnosis, inclusion of milder cases, and earlier age of diagnosis cannot fully explain this increase, suggesting the possibility that the actual incidence of ASD is indeed increasing.[7,8]

Despite little understanding of what causes ASD, both genetic and environmental factors are thought to play an important role in the etiology of, or the risk for, this disorder. Based on concordance data from twin and family studies, ASD is found to be among the most heritable of all psychiatric disorders with the heritability estimated to be as high as 90%.[9,10] Examples of ASD with a known genetic basis include fragile X syndrome and Rett syndrome. Fragile X syndrome is caused by mutations in the *FMR1* gene on the X chromosome which lead to the triplet repeat (CGG) expansion in the gene promoter.[11] This excess of CGG results in hypermethylation which eventually leads to transcriptional repression of *FMR1* which is required for neuronal development.[12,13] Similar to fragile X syndrome, Rett syndrome is caused by mutations in *MECP2*, a gene coding for methyl-CpG-binding protein 2 which binds methylated CpG sites of the target gene and recruits transcription factors and histone modification enzymes (e.g. histone deacetylase) to activate or repress the transcription of the gene.[14] However, only about 15% of ASD cases have been found to be etiologically related to known genetic abnormalities.[15]

In an attempt to gain a better understanding of idiopathic ASD which accounts for the majority (~80%) of ASD cases, a number of studies have searched for genetic risks of ASD using various approaches, from single nucleotide polymorphisms (SNPs) analyses focusing on specific gene(s) to genome-wide linkage and association analyses (see reviews in Refs. [16–19]). In addition, the next-generation sequencing technique was recently applied to ASD research in an attempt to identify rare and *de novo* genetic mutations as well as copy number variants associated with ASD in protein-coding regions of the human genome.[20–23] Together, these genetic studies have generated over one thousand potential ASD susceptibility loci on almost every chromosome. Yet, despite a few exceptions, there is little reproducibility in the genes/loci identified, suggesting the involvement of susceptibility factors beyond exomic sequence. Recognition of the role of other components to disease susceptibility has prompted research on gene regulatory epigenetic mechanisms, which include DNA methylation, histone modification, chromatin remodeling, and non-coding RNA (see review in Ref. [24]). In this chapter, we review a number of findings related to non-coding

RNAs (ncRNAs) that may play a role in the pathogenesis of ASD and that may be used as potential biomarkers for ASD.

The Genomic "Dark Matter"

The human genome contains 3 billion base pairs but it is known that only about 2% codes for proteins. The non-protein-coding region of the genome, which accounts for approximately 98% of the human DNA, was thought to be "junk" that was probably inactivated through evolutionary events. However, some studies have reported that greater than 90% of the human genome is transcribed,[25] although others have argued that the genome is not as actively transcribed as reported and that the majority of the detected transcription is due to technical artifacts and/or background biological noise.[26] This controversy about whether or not the non-coding portion of the genome (referred to as "dark matter" in analogy to the dark matter of the universe) is actively transcribed and functional is difficult to resolve due to current technological and analytical limitations combined with the massive expanse of the non-coding region. Using highly sensitive tiling arrays to target selected portions of the human transcriptome, Mercer and colleagues (2011) were able to identify and characterize unannotated transcripts whose rare or transient expression is below the detection limits of conventional sequencing approaches.[27] Their findings strongly support the existence of widespread and remarkably complex non-coding transcription in the non-coding region and help address the controversy about the genomic "dark matter" although its biological significance remains to be explored.

By studying nearly 150 human cell types, the ENCyclopedia Of DNA Elements (ENCODE) project involving 440 scientists from 32 laboratories around the world has started shedding light on the function of non-coding regions of the human genome. Findings from the studies in this enormous project reveal that approximately 80% of the genome is "functional", as defined by having specific biochemical activity (see Ref. [28] and other ENCODE companion papers). They estimated that there are over 70,000 promoter regions and nearly 400,000 enhancers in the non-coding regions.[28] In addition to the promoters and the enhancers, the ENCODE studies also found that thousands of RNA transcripts that do not encode a protein, referred to as non-coding RNAs (ncRNAs), are transcribed from the "dark matter" of the human genome.[28] Although there are still some arguments that the ENCODE's definition of "functional" may be too broad, it has become increasingly clear that this controversial portion of DNA is far more active and important than "junk" as previously thought.

Non-coding RNAs and ASD

Non-coding RNAs are broadly defined as a class of RNA transcripts that are not translated into proteins. Examples of ncRNAs with known functions that are found in humans include transfer RNA (tRNA), ribosomal RNA (rRNA), microRNA (miRNA), small nucleolar RNA (snoRNA), small nuclear RNA (snRNA), piwi-interacting RNA (piRNA), small Cajal body-specific RNA (scaRNA), and long noncoding RNA (lncRNA). Transfer RNA and ribosomal RNA are highly abundant in human cells and their functions in the RNA translation process have long been studied extensively and are widely known; thus, these will not be discussed here. So far, only miRNAs, snoRNAs, and lncRNAs have been associated with ASD and its comorbid disorders. The role of lncRNAs is also extensively discussed in another chapter of this book. Thus, in this review, we will focus only on miRNAs and snoRNAs.

miRNAs and ASD

miRNAs are endogenous single-stranded non-coding RNA molecules with 19–25 nucleotides in length that mediate transcriptional and post-transcriptional regulation of gene expression.[29] The biogenesis and gene regulatory mechanisms of miRNAs have been comprehensively described in many excellent review articles (see Refs. [30, 31] and references therein). Unlike other small RNAs, like tRNAs that are transcribed by RNA polymerase III, miRNA genes are transcribed by RNA polymerase II as primary miRNA transcripts (pri-miRNAs) that are usually several kilobases long and contain a hairpin structure. Pri-miRNAs are then trimmed into the hairpin intermediates of approximately 70 nucleotides in length (called pre-miRNAs) by the nuclear RNase III Drosha which interacts with the DiGeorge Syndrome Critical Region 8 (DGCR8) cofactor. Following Drosha-mediated trimming, pre-miRNAs are exported to the cytoplasm by the nuclear transport receptor exportin-5, then processed into ~22-nucleotide miRNA:miRNA* duplexes by the cytoplasmic RNase III Dicer. One of the miRNA strands in the duplexes is incorporated into the miRNA-containing RNA-induced silencing (miRISC) complex which binds and forms imperfect base-pairs with targeted mRNA transcripts, leading to mRNA degradation, inhibition of translational initiation, or inhibition of transcription.

miRNAs are known to play a critical role in nervous system development and functions as described in a number of excellent reviews[32,33] and references therein. Giraldez *et al.* (2005) were among the first to demonstrate that miRNAs are required for normal brain development. They found that zebrafish with a deficiency in mature miRNAs due to the lack of the key miRNA-generating enzyme Dicer exhibited specific developmental defects, including abnormal neuronal differentiation and brain morphogenesis.[34] Findings from many subsequent studies in mice also confirmed the role of miRNAs in nervous system development.[35,36] Moreover, miRNAs are also found to be essential for postmitotic neurons. Progressive loss of miRNAs caused by Dicer deficiency results in Purkinje cell death, followed by cerebellar degeneration and development of ataxia.[37] Furthermore, the localization of miRNA processing machinery at dendritic spines and post-synaptic densities further implicates miRNAs in neuroplasticity.[38] Cohen *et al.* (2011) demonstrated that miRNA, specifically miR-485, negatively regulated dendritic spine density, clustering of the post-synaptic dentity-95 (PSD-95), and neuronal surface expression of the glutamate receptor GluR2.[39] They also found that dysregulation of miR485 expression reduced spontaneous synaptic responses and transmitter release, confirming the role of miRNA in synaptic plasticity.

The roles of miRNAs in the etiology and the susceptibility of ASD remain unclear. At present, there are only six papers that are related to miRNAs in ASD and only four of them have directly focused on miRNA expression in tissues or cells derived from individuals with idiopathic ASD. The complete list of miRNAs reported in ASD is shown in Table 1. The first study was conducted by Abu-Elneel *et al.* (2008) using multiplex quantitative polymerase chain reaction (PCR) analysis of postmortem cerebellar cortex tissues from 13 individuals with ASD and a control set of 13 non-ASD cerebellar samples.[40] By investigating the expression of 466 human miRNAs, they found 28 miRNAs expressed at significantly different levels in at least one of the ASD samples compared to the non-ASD controls. Among those, nine of them were successfully validated when the analysis was reversed to detect differentially expressed miRNAs in the controls relative to the mean miRNA expression of combined ASD cases. However, Buyske (2009) reported that only five miRNAs were altered in only two individuals with ASD when more rigorous statistical analysis was applied to reevaluate the results from the study by Abu-Elneel *et al.* (2008).[41]

Talebizadeh *et al.* (2008) conducted expression profiling analysis of 470 mature human miRNAs using lymphoblastoid cell lines (LCLs) from six individuals with ASD and six sex- and age-matched controls.[42] Out of the 470 miRNAs interrogated, they observed differential expression of nine miRNAs in individuals with ASD relative to the controls. Interestingly, four out of nine miRNAs were found to overlap with those identified as differentially expressed miRNAs in the cerebellar cortex tissues from individuals with ASD by Abu-Elneel *et al.* (2008), suggesting that at least some of the molecular abnormalities detectable in peripheral tissues from individuals with ASD reflect those present in brain tissues. However, neither of these two studies experimentally validated gene targets of the differentially expressed miRNAs, making

Table 1. List of ncRNAs found to be differentially expressed or located in the CNV regions associated with ASD

ncRNAs	References	ncRNAs	References
hsa-mir-103	[43]	hsa-mir-153-1	[43]
hsa-mir-106a	[40]	hsa-mir-15a	[40]
hsa-mir-106b	[40], [43]	hsa-mir-15b	[40]
hsa-mir-107	[43]	hsa-mir-16-2	[43]
hsa-mir-10a	[45]	hsa-mir-17HG	[46]
hsa-mir-10a*	[45]	hsa-mir-181a	[45]
hsa-mir-1180	[47]	hsa-mir-181a*	[45]
hsa-mir-1182	[47]	hsa-mir-181b	[45]
hsa-mir-1236	[47]	hsa-mir-181c	[45]
hsa-mir-124-1	[47]	hsa-mir-181d	[40]
hsa-mir-124-3	[47]	hsa-mir-182-as	[43]
hsa-mir-1244	[47]	hsa-mir-185	[43], [47]
hsa-mir-1248*	[47]	hsa-mir-186	[43]
hsa-mir-125b	[45]	hsa-mir-188	[47]
hsa-mir-1268	[47]	hsa-mir-189	[43]
hsa-mir-128	[40]	hsa-mir-190	[43]
hsa-mir-1285-1	[47]	hsa-mir-191	[43]
hsa-mir-1286	[47]	hsa-mir-1910	[47]
hsa-mir-129	[40]	hsa-mir-1914	[47]
hsa-mir-1306	[47]	hsa-mir-193b	[40]
hsa-mir-132	[40], [42], [43]	hsa-mir-194	[43]
hsa-mir-1322	[47]	hsa-mir-195	[43], [47]
hsa-mir-133b	[43]	hsa-mir-196a	[45]
hsa-mir-136	[43]	hsa-mir-1972	[47]
hsa-mir-138-1	[47]	hsa-mir-199a-5p	[45]
hsa-mir-139	[43]	hsa-mir-199b	[43], [45]
hsa-mir-140	[40]	hsa-mir-200a	[47]
hsa-mir-142	[46]	hsa-mir-200b	[47]
hsa-mir-146a	[42]	hsa-mir-204	[47]
hsa-mir-146b	[40], [42]	hsa-mir-205	[43]
hsa-mir-148b	[40], [43]	hsa-mir-21	[40]
hsa-mir-149	[47]	hsa-mir-211	[43], [47]

(*Continued*)

Table 1. (*Continued*)

ncRNAs	References	ncRNAs	References
hsa-mir-212	[40]	hsa-mir-451	[43]
hsa-mir-216a	[47]	hsa-mir-455	[43]
hsa-mir-216b	[47]	hsa-mir-455-3p	[45]
hsa-mir-217	[47]	hsa-mir-484	[40], [47]
hsa-mir-219	[43]	hsa-mir-486-3p	[45]
hsa-mir-223	[46]	hsa-mir-486-5p	[45]
hsa-mir-23a	[40], [42], [43]	hsa-mir-489	[47]
hsa-mir-23b	[42], [43]	hsa-mir-495	[43]
hsa-mir-25	[43]	hsa-mir-497	[47]
hsa-mir-27a	[40]	hsa-mir-500	[47]
hsa-mir-28	[47]	hsa-mir-500*	[45]
hsa-mir-29b	[43]	hsa-mir-501	[47]
hsa-mir-29B1	[46]	hsa-mir-502	[47]
hsa-mir-30a*	[45]	hsa-mir-502-3p	[45]
hsa-mir-30c	[43]	hsa-mir-518a	[43]
hsa-mir-320	[42]	hsa-mir-519c	[43]
hsa-mir-320a	[40]	hsa-mir-520b	[43]
hsa-mir-324	[47]	hsa-mir-524	[43]
hsa-mir-326	[43]	hsa-mir-532	[47]
hsa-mir-338-3p	[45]	hsa-mir-539	[40]
hsa-mir-339	[47]	hsa-mir-548f-1	[47]
hsa-mir-33b	[47]	hsa-mir-548o	[45]
hsa-mir-342	[43]	hsa-mir-550	[40]
hsa-mir-346	[43]	hsa-mir-551a	[47]
hsa-mir-34a	[47]	hsa-mir-552	[47]
hsa-mir-362	[47]	hsa-mir-563	[47]
hsa-mir-363	[42]	hsa-mir-564	[47]
hsa-mir-367	[43]	hsa-mir-570	[47]
hsa-mir-376a-as	[43]	hsa-mir-571	[47]
hsa-mir-381	[40]	hsa-mir-577	[45]
hsa-mir-429	[47]	hsa-mir-589	[47]
hsa-mir-431	[40]	hsa-mir-590	[47]
hsa-mir-432	[40]	hsa-mir-591	[47]

(*Continued*)

Table 1. (*Continued*)

ncRNAs	References	ncRNAs	References
hsa-mir-597	[47]	SNORA30	[56]
hsa-mir-598	[40], [47]	SNORA41	[56]
hsa-mir-605	[47]	SNORA49	[46]
hsa-mir-629	[47]	SNORA59A	[46]
hsa-mir-630	[47]	SNORA67	[46]
hsa-mir-647	[47]	SNORA70F	[46]
hsa-mir-649	[47]	SNORD101	[46]
hsa-mir-650	[45]	SNORF116	[56]
hsa-mir-651	[47]	SNORD12C	[46]
hsa-mir-652	[40]	SNORD13	[46]
hsa-mir-653	[47]	SNORD14E	[46]
hsa-mir-660	[47]	SNORD15A	[46]
hsa-mir-663	[42]	SNORD1C	[46]
hsa-mir-671	[47]	SNORD28	[46]
hsa-mir-7	[40]	SNORD29	[56]
hsa-mir-7-3	[47]	SNORD31	[46]
hsa-mir-873	[47]	SNORD33	[46]
hsa-mir-876	[47]	SNORD35A	[46]
hsa-mir-92	[42]	SNORD3A	[46]
hsa-mir-920	[47]	SNORD41	[46]
hsa-mir-93	[40], [43]	SNORD44	[46]
hsa-mir-941-1	[47]	SNORD5	[56]
hsa-mir-941-2	[47]	SNORD50A	[46]
hsa-mir-941-3	[47]	SNORD54	[46]
hsa-mir-943	[47]	SNORD55	[46]
hsa-mir-944	[47]	SNORD57	[46]
hsa-mir-95	[40]	SNORD58A	[46]
hsa-mir-LET7F1	[46]	SNORD68	[46]
SNORA21	[46]	SNORD76	[46]
SNORA25	[56]	SNORD95	[46]

it very challenging to make a connection between the dysregulation of the miRNAs that they found and the biological functions that may be disrupted in ASD.

A subsequent study by our laboratory integrated miRNA expression profiling with mRNA expression profiling analysis using the same LCLs from five individuals with ASD and nine controls who are unaffected twins or siblings of the individuals in the ASD group.[43] Using LCLs with identical or similar genotypes allowed us to study epigenetic differences with minimal genetic heterogeneity; this study revealed 43 differentially expressed miRNAs in LCLs from individuals with ASD relative to their unaffected twins or siblings. Interestingly, at least 16 out of these miRNAs have been associated with the brain. In addition, the dysregulation of some miRNAs (e.g. the up-regulation of miR-23a and down-regulation of miR-106b) observed in this study is also consistent with previous findings reported in the study by Abu-Elneel *et al.* (2008) using post-mortem cerebellar tissues. These findings support the possibility that some of the molecular abnormalities observed in peripheral tissues, especially blood, of individuals with ASD can reflect the pathobiological conditions of this disorder in the brain. Moreover, lists of the potential target genes of each differentially expressed miRNA were generated using the miRBase database[44] and manually curated to select target genes whose mRNA expression levels were inversely correlated with the expression of the corresponding regulatory miRNA. Using this integrated-omics approach, we were able to identify as many as 1,053 genes that were potentially dysregulated by the differentially expressed miRNAs in ASD. To validate the miRNA microarray data, we selected four differentially expressed miRNAs (i.e. hsa-miR-219-5p, hsa-miR-139-5p, hsa-miR-29b, and hsa-miR-103) for quantitative reverse transcription-polymerase chain reaction (qRT-PCR) analysis. The qRT-PCR analysis confirmed the down-regulation of hsa-miR-219-5p and hsa-miR-139-5p and the up-regulation of hsa-miR-29b and hsa-miR-103 in LCLs from individuals with ASD relative to their unaffected twins or siblings. Moreover, putative gene targets (ID3 and PLK2) of two qRT-PCR-confirmed brain-specific miRNAs (hsa-miR-29b and hsa-miR-219-5p) were validated by miRNA overexpression or knockdown assays, respectively. Comparisons of these mRNA and miRNA expression levels between discordant twins and between case-control sib pairs show an inverse relationship, further suggesting that ID3 and PLK2 are *in vivo* targets of the respective miRNAs.

Similar to the study by Sarachana *et al.* (2010), Ghahramani Seno *et al.* (2011) conducted gene and miRNA expression profiling analyses of LCLs from 24 individuals with ASD and their unaffected sib-pairs.[45] They identified 12 miRNAs that were differentially expressed in at least 12 out of 24 sib-pair comparisons.[45] By reducing the heterogeneity of the samples based on their gene expression profiles, they were able to identify 13 miRNAs that were differentially expressed in a subgroup which exhibited a similar pattern of gene expression. Six of them (i.e. hsa-miR-486-5p, hsa-miR-486-3p, hsa-miR-548o, hsa-miR-10a, hsa-miR-10a*, and hsa-miR-199b-5p) overlapped with the previous analysis. Interestingly, the p-values of these six miRNAs were greatly improved, showing the advantage of reducing the heterogeneity of ASD samples, in this case, based on gene expression profiles.

The largest transcriptomic analysis of ASD blood samples to date was conducted by Kong *et al.* (2012) using 170 ASD cases and 115 age- and sex-matched controls.[46] The samples were divided into two cohorts. The first cohort of 66 male ASD cases and 33 age-matched male controls revealed 489 transcripts differentially expressed in individuals with ASD relative to controls. Although subsequent analyses in this study were focused mainly on differentially expressed protein-coding transcripts rather than ncRNAs, it is interesting to note that hsa-miR-17hg and hsa-miR-223 were among those identified to be differentially expressed in ASD blood samples in this. The validation cohort using 104 individuals with ASD and 82 controls reveals 610 transcripts differentially expressed between ASD cases and controls. Among those transcripts, hsa-miR-142, hsa-miR-29b, and hsa-miR-let7f1 were identified to be dysregulated in ASD blood samples relative to controls. Interestingly, hsa-miR-29b was also identified by Sarachana *et al.* (2010) to be significantly up-regulated in LCLs from individuals with ASD relative to controls,[43] suggesting that the dysregulation of at least some miRNAs observed in LCLs from ASD cases can reflect the pathobiological conditions in ASD blood and, therefore, support the feasibility of using LCLs to investigate the role of ncRNAs in ASD. In addition to miRNAs, Kong *et al.* also found a number of snoRNAs differentially expressed in the blood of individuals with ASD. Their findings and the role of snoRNAs in ASD will be discussed in detail in the next section.

As mentioned earlier, almost all genetic studies on ASD have focused mainly on protein-coding regions of the genome. However, it has become increasingly evident that non-coding regions are not negligible, as they are now known to play an important role in regulation of gene expression. Vaishnavi *et al.* (2013) therefore explored CNV loci that have been implicated in ASD and identified miRNAs located in those regions in an attempt to elucidate the biological and functional significance of miRNAs present in ASD-related CNV loci.[47] Using a series of bioinformatics tools and databases, including the Autism Database (AutDB)[48] and the UCSC Genome Browser (http://genome.ucsc.edu), they demonstrated that nearly 11% of the ASD-associated CNV loci contain miRNAs, some of which have been associated with ASD, such as hsa-miR-185, hsa-miR-195, and hsa-miR-211. Hsa-miR-185 and hsa-miR-195 are located in the duplicated CNV loci of the chromosome 22q11.2 and 17p13.1-p13.2 regions, respectively. Interestingly, the expression of these miRNAs has been reported to be up-regulated in LCLs from individuals of ASD by Sarachana *et al.*[43] On the other hand, the chromosome 15q13.1-q13.3 region which was previously reported to be a CNV locus deleted in ASD cases contains hsa-miR-211, a miRNA whose expression was found to be significantly reduced in LCLs from ASD cases.[43] Besides these three miRNAs, miRNAs that are located in ASD-associated CNV loci and have been found to be differentially expressed in ASD in independent cohorts by other groups include hsa-miR-7, hsa-miR-484, hsa-miR-500, hsa-miR-502, and hsa-miR-598.[40,45]

snoRNAs and ASD

snoRNAs are small non-coding RNA molecules that are 60–600 nucleotides in length (majority are in the range of 70–200 nucleotides) that primarily mediate post-transcriptional modifications of other RNAs, especially rRNA, tRNA or snRNAs (see reviews in Refs. [49, 50]). snoRNAs comprise two families, namely C/D and H/ACA, based on conserved sequence motifs in the RNA molecule. RNA in the box C/D family contains the PuUGAUGA (box C) and CUGA (box D) sequence motifs which are located near the 5′ and 3′ ends, respectively, whereas RNA in the box H/ACA group contains the sequence motif ANANNA (box H) in the middle and a highly conserved ACA (box ACA) near the 3′ end. Most snoRNA genes that have been identified and characterized in humans are located in intronic regions of genes that encode proteins involved in ribosome biogenesis. The discovery of snoRNAs in intergenic regions is more difficult compared with those in intronic regions due to the lack of sequence indicators such as the start and stop codons in these ncRNA genes. However, more and more snoRNAs are found to be derived from intergenic regions as polycistronic transcripts, each of which can yield as many as ten different mature ncRNAs after endonucleolytic cleavages. Some snoRNAs are also independently transcribed and undergo post-transcriptional modifications, specifically 5′ methylation and hypermethylation, like mRNAs and snRNAs. Thus, as sequencing technologies and bioinformatics tools are improved, it is speculated that there are more snoRNAs and other ncRNAs, both intronic and intergenic, yet to be found in the "dark matter."

Relatively little is known about the biological functions of snoRNAs compared to miRNAs. However, recent studies have begun to shed light on the role of snoRNAs in neurological and developmental functions. Several snoRNAs, including SNORA35 (HBI-36), SNORD64 (HBII-13), SNORD115 (HBII-52), and SNORD116 (HBII-85), have been reported to be expressed exclusively or in high abundance in the human central nervous system.[51] SNORD12, SNORD29, SNORD31, SNORD74, SNORD101, SNORD104, and SNORD115 are found to be highly enriched and differentially expressed during neural differentiation of embryonic stem cells,[52] suggesting that these snoRNAs are involved in neurological development. The mouse homolog of SNORD115 contains a highly conserved 18-nucleotide sequence complementary to serotonin 2C receptor mRNA transcript, suggesting that this snoRNA may play an important function in regulating the processing of serotonin 2C receptor mRNA.[51]

Disruptions of snoRNAs have been associated with neurodevelopmental disorders in humans. Examples are SNORD115 and SNORD116 which map to the Prader-Willi Syndrome locus on chromosome 15. Prader-Willi syndrome (PWS) is a genetic disorder caused by either the structural loss of material or the absence of gene expression from the paternally inherited copy of chromosome 15 in the q11-q13 region.[53] It is noteworthy that PWS is known to be one of the comorbid diseases of ASD with some behaviors in common, such as delays in language and motor development, and learning disabilities.[54] In addition to the deficits in those areas, many individuals with PWS also exhibit repetitive behaviors and social deficits reminiscent of ASD.[53] Microdeletion of SNORD116 has been associated with PWS and may lead to the pathological conditions shared between PWS and ASD, including neonatal hypotonia, feeding difficulties, sleep apnea, speech and global developmental delay.[55] It is interesting to note that the patient with SNORD116 microdeletion in the study by Sahoo and colleagues (2008) was also reported to meet the defined criteria for autism based on the Autism Diagnostic Observation Schedule (ADOS) and Autism Diagnostic Interview-Revised (ADI-R) assessment instruments.

There are only a few reports that directly associate dysregulation of snoRNAs to ASD, probably due to the lack of bioinformatics tools and comprehensive databases for human snoRNAs. Using custom ncRNA microarrays which contain oligonucleotide probes for human miRNAs and snoRNAs, our lab has observed dysregulation of many ncRNAs in LCLs from individuals with ASD.[43,56] In addition to the miRNAs mentioned in the previous section, we have found six snoRNAs significantly differentially expressed in individuals with ASD.[56] These snoRNAs are SNORA25 (ACA25), SNORA30 (ACA30), SNORA41 (ACA41), SNORD5 (mgh28S-2410), SNORD29 (U29), and SNORD116. It is noteworthy that SNORD116 was found to be significantly reduced in ASD relative to controls, suggesting the possibility that the loss-of-function disruption of SNORD116 may play an important role in the pathophysiology shared between PWS and ASD. The transcriptomics analysis of whole blood conducted by Kong and colleagues (2012) also revealed a number of snoRNAs differentially expressed in the blood of individuals with ASD (Table 1).[46] Interestingly, many of these snoRNAs, including SNORD12, SNORD31, and SNORD101, have been found to be differentially expressed in embryonic stem cells during neural differentiation,[52] suggesting that aberrant expression of snoRNAs observed in the blood may reflect similar changes in the brain, probably due to global or system-wide dysregulation of ncRNA expression.

Biological functions associated with ncRNAs reported in ASD

Interestingly, many brain-specific and brain-related miRNAs published by previous studies investigating miRNA expression profiles in the brain were found among the significantly differentially expressed ncRNAs in blood cells from individuals with ASD. Five miRNAs identified as brain-specific are hsa-miR-29b,[57] hsa-miR-153,[57] hsa-miR-199b,[58] hsa-miR-219,[57,58] and hsa-miR-326.[59] Some miRNAs have been found to be highly expressed in brain but can be found in other tissues with low expression as well, such as hsa-miR-132 and hsa-miR-139.[57] In addition, 10 miRNAs reported to be highly relevant to brain development and functions are hsa-miR-16,[57] hsa-miR-23a,[37] hsa-miR-23b,[60] hsa-miR-30c,[57] hsa-miR-103,[57] hsa-miR-106b,[40] hsa-miR-107,[61] hsa-miR-136,[37] hsa-miR-191,[59] and hsa-miR-211.[37] The dysregulation of the brain-specific and brain-related miRNAs in ASD strongly suggests that the ncRNAs that are disrupted in individuals with ASD may play an important role in neurological functions.

To gain insight into biological functions that may be disrupted in ASD as a consequence of altered ncRNA expression, the lists of ncRNAs identified by all of the ASD studies were combined and uploaded into the Ingenuity Pathway Analysis (IPA; www.ingenuity.com) which is a manually curated biological network prediction program. Significant biological functions and diseases highly represented in the networks were identified using Fisher's exact test ($P < 0.05$). Biological network prediction of the ncRNAs reported in ASD reveals a strong association with developmental disorders, hereditary disorders, inflammatory diseases, and neurological and psychological disorders (Table 2). Interestingly, developmental/hereditary disorders associated with the ncRNAs differentially expressed in ASD include Duchenne muscular dystrophy (e.g. hsa-miR-30, hsa-miR-133, hsa-miR-146, hsa-miR-148, and hsa-miR-199), X-linked hereditary disease (e.g. hsa-miR-30, hsa-miR-133, hsa-miR-146, hsa-miR-148, and hsa-miR-199), DiGeorge syndrome (e.g. hsa-miR-650, hsa-miR-1286, and hsa-miR-1306), schizophrenia (e.g. hsa-miR-29b, hsa-miR-30, hsa-miR-92, hsa-miR-103, and hsa-miR-146), Down's syndrome (e.g. hsa-miR-10 and hsa-miR-125),

Table 2. Diseases and functions predicted by the Ingenuity Pathway Analysis to be significantly associated with ncRNAs reported in ASD

	p-Values	#ncRNAs
Top Significant Diseases		
Developmental Disorders	4.89E-40–2.79E-02	36
Hereditary Disorders	4.89E-40–1.87E-02	42
Skeletal and Muscular Disorders	4.89E-40–2.79E-02	34
Cancer	4.33E-31–4.73E-02	64
Reproductive System Diseases	4.33E-31–4.73E-02	45
Other Significant Diseases		
Inflammatory Diseases	1.29E-22–9.32E-03	41
Neurological Diseases	1.06E-08–1.39E-02	21
Psychological Disorders	1.06E-08–1.03E-02	18
Nervous System Development and Functions	4.67E-08–2.73E-02	3

and epilepsy (e.g. hsa-miR-146), all of which are known to be comorbid disorders of ASD.[62–67] This finding suggests that aberrant expression of these ncRNAs may lead to the pathobiological conditions that these disorders share with ASD. Moreover, the IPA analysis also revealed as many as 353 genes that have been experimentally verified to be regulated by the ncRNAs differentially expressed in ASD. The IPA analysis of these target genes shows that these targets of the ncRNAs dysregulated in ASD are significantly associated with nervous system development and functions (p-values: 9.53E-16–3.94E-09), including development of cerebral cortex, neuronal differentiation, and neuritogenesis. Of particular relevance to behavioral deficits in ASD, as many as 62 genes related to learning and memory are also associated with these experimentally validated targets of the ncRNAs (p-values: 3.05E-14–6.20E-09). Moreover, the target genes of the ncRNAs are also involved in PTEN signaling (p-value = 2.23E-15), neuregulin signaling (p-value = 7.63E-12), PI3K/AKT/mTOR signaling (p-value = 1.28E-11), and neurotrophin/TRK signaling (p-value = 3.43E-11), all of which have been associated with ASD.[68–71] It is noteworthy that at least 21 among 353 genes are considered ASD candidate genes and have been deposited in the AutDB[48] database which is a manually curated database for ASD (Table 3). The enrichment of ASD candidate genes as well as genes related to biological functions and disorders associated with ASD in the list of ncRNA target genes suggests the possibility that the aberrant expression of the ncRNAs reported in ASD may lead to the dysregulation of ASD candidate genes which, in turn, can cause the pathobiological conditions reported in ASD.

Future Directions in ASD Biomarker Research

Although much of the research effort on ASD has been focused on searching for genes and genetic variation in the protein-coding portion of the genome, it becomes increasingly clear that the non-protein coding part also deserves increasing attention. As discussed in the preceding sections, the non-coding region of the genome contains a

Table 3. ASD candidate genes in the AutDB[48] database that have been experimentally verified by IPA to be targets of ncRNAs dysregulated in ASD

Symbol	Entrez Gene Name	Regulatory ncRNA
ALOX5AP	Arachidonate 5-lipoxygenase-activating protein	hsa-miR-199a-5p
ATRX	Alpha thalassemia/mental retardation syndrome X-linked	hsa-miR-30c
BCL2	B-cell CLL/lymphoma 2	hsa-miR-181a
CAMTA1	Calmodulin binding transcription activator 1	hsa-miR-129
DCX	Doublecortin	hsa-miR-128
ESR1	Estrogen receptor 1	hsa-miR-181a
FBXO33	F-box protein 33	hsa-miR-451
FMR1	Fragile Xmental retardation 1	hsa-miR-30a*
GPD2	Glycerol-3-phosphate dehydrogenase 2 (mitochondrial)	hsa-miR-30c
HDAC4	Histone deacetylase 4	hsa-miR-29b
ITGB3	Integrin, beta 3 (platelet glycoprotein llla, antigen CD61	hsa-miR-92
JARID2	Jumonji, AT rich interactive domain 2	hsa-miR-125b
MET	Met proto-oncogene (hepatocyte growth factor receptor)	hsa-miR-30c
NFIA	Nuclear factor I/A	hsa-miR-103
NTRK3	Neurotrophic tyrosine kinase, receptor, type 3	hsa-miR-128
PTEN	Phosphatase and tensin homolog	hsa-miR-92
RBM8A	RNA binding motif protein 8A	hsa-miR-125b
RELN	Reelin	hsa-miR-128
RERE	Arginine-glutamic acid dipeptide (RE) repeats	hsa-miR-29b
SLC4A10	Solute carrier family 4, sodium bicarbonate transporter, member 10	hsa-miR-30c
SYNE1	Spectrin repeat containing, nuclear envelope 1	hsa-miR-7

number of gene regulatory elements, especially a variety of ncRNAs, many of which have been found to be dysregulated in the blood and/or the brain of individuals with ASD. Changes in brain-related ncRNAs in blood cells of individuals with ASD suggests that alterations in ncRNA expression observable in the periphery may reflect some of the pathophysiological conditions occurring in the nervous system, thus supporting the use of peripheral cells, especially blood cells, as a surrogate tissue for investigating ncRNA and gene expression in autism research, which offers the additional advantage of identifying ncRNAs as biomarkers for this disorder.

Although searching for ASD biomarkers in the genomic dark matter may be a very promising avenue, one of the major challenges to ncRNA biomarker discovery is the lack of reproducibility in independent cohorts due to the high degree of heterogeneity of this disorder. As mentioned, the severity in affected individuals can vary greatly from person to person. When analyzing a combined case group exhibiting various clinical phenotypes, differing proportions of each "subphenotype" in different studies can lead to irreproducible or nonsignificant findings. Future research on ncRNA should therefore address this issue by subgrouping individuals with ASD according to their clinical phenotypes, specific traits, or even comorbidities to reduce the heterogeneity of the subjects. This strategy will not only enhance statistical power as demonstrated by many recent genome-wide linkage studies,[72–77] but also help identify ncRNA biomarkers specific to each ASD subtype. In fact, this subgrouping approach has been demonstrated to be very useful in identification of subphenotype-specific differentially expressed genes[78] and associated SNPs from genome-wide genetic analyses.[79] Thus, identification of ncRNA biomarkers for ASD following subgrouping of ASD subjects is likely to be extremely beneficial in future biomarker discovery research.

Another important challenge for ncRNA research in general is the lack of comprehensive databases for human ncRNAs. Until recently, most of ncRNA studies involving ASD have been focused only on miRNAs, partly because miRNA bioinformatics tools and databases have been extensively developed and are relatively more comprehensive than for other types of ncRNAs. However, an advance in computational technology has led to the improvement of bioinformatics tools for other types of ncRNAs in recent years. Examples are the snoRNABase which is a comprehensive database for human H/ACA and C/D box snoRNAs,[80] the lncRNA db which provides comprehensive annotations of eukaryotic long non-coding RNA,[81] and the sno/scaRNAbase which is a curated database for both snoRNAs and scaRNAs.[82] The availability of these new ncRNA databases will thus facilitate the interrogation and interpretation of changes in the expression of all types of ncRNAs in ASD. Inasmuch as ncRNAs, especially miRNAs, are known to be able to influence transcription and translation of a number of target genes which have variable expression in different tissues, it would be beneficial to correlate changes in ncRNA expression with concomitant changes in the expression of target genes in the same tissues. Integrating results from these studies would help researchers identify tissue-specific biological pathways that are negatively impacted in individuals with ASD as a consequence of ncRNA dysregulation, which in turn can lead to a better understanding of the pathobiological conditions associated with ASD.

Executive Summary

- ASD is one of the most heritable psychiatric disorders; yet to date, no single genetic mutation or SNP in the coding region of the genome can account for more than 1–2% of ASD cases.
- It is becoming increasingly clear that the non-coding region of the genome is important for gene regulation. However, very little ASD research has focused on the non-coding part of the genome.
- A few studies on ncRNAs have shown that ncRNAs are dysregulated in individuals with ASD. However, only a few ncRNAs are found to be reproducible in independent cohorts, likely due to the high degree of heterogeneity of ASD.
- Future research on ncRNA in ASD should consider reducing the heterogeneity by subgrouping individuals based on clinical phenotypes and/or other distinguishing criteria. This strategy would allow for identification of ncRNA biomarkers for different subgroup of ASD and also make different studies comparable.

- Improvement of bioinformatics tools and databases would aid in the investigation of many more types of ncRNAs besides miRNAs, such as snoRNAs, scaRNAs, and lncRNAs.
- Integrating ncRNA expression with biological phenotype of ASD, such as dysregulated target genes and biological functions, would be useful in identifying pathways and networks that are impacted as a consequence of ncRNA dysregulation.

Acknowledgements

The work that is described from Dr. Hu's laboratory was supported in part by Autism Speaks grant #2381 (VWH) and by NIMH grant R21 MH073393 (VWH). Dr. Sarachana was supported as a Master's and predoctoral student in Dr. Hu's laboratory by a Higher Educational Strategic Scholarship for Frontier Research from the Office of the Commission on Higher Education of the Royal Thai Government, Thailand.

References

1. American Psychiatric Association, American Psychiatric Association. DSM-5 Task Force: *Diagnostic and Statistical Manual of Mental Disorders: DSM-5:* 5th edn. Washington, D.C.: American Psychiatric Association; 2013.
2. Treffert DA: Epidemiology of infantile autism. *Arch Gen Psychiatry* 1970, **22**(5):431–438.
3. Baio J: Prevalence of Autism Spectrum Disorders — Autism and Developmental Disabilities Monitoring Network, 14 Sites, United States, 2008. *Centers for Disease Control and Prevention Morbidity and Mortality Weekly Report* 2012, **61**(3).
4. Fombonne E: Epidemiology of pervasive developmental disorders. *Pediatr Res* 2009, **65**(6):591–598.
5. Kim YS, Leventhal BL, Koh YJ, Fombonne E, Laska E, Lim EC, Cheon KA, Kim SJ, Kim YK, Lee H, Song DH, Grinker RR: Prevalence of autism spectrum disorders in a total population sample. *Am J Psychiatry* 2011, **168**(9):904–912.
6. Mattila ML, Kielinen M, Linna SL, Jussila K, Ebeling H, Bloigu R, Joseph RM, Moilanen I: Autism spectrum disorders according to DSM-IV-TR and comparison with DSM-5 draft criteria: an epidemiological study. *J Am Acad Child Adolesc Psychiatry* 2011, **50**(6):583–592.e11.

7. Hertz-Picciotto I, Delwiche L: The rise in autism and the role of age at diagnosis. *Epidemiology* 2009, **20**(1):84–90.
8. Grether JK, Rosen NJ, Smith KS, Croen LA: Investigation of shifts in autism reporting in the California Department of Developmental Services. *J Autism Dev Disord* 2009, **39**(10):1412–1419.
9. Bailey A, Le Couteur A, Gottesman I, Bolton P, Simonoff E, Yuzda E, Rutter M: Autism as a strongly genetic disorder: evidence from a British twin study. *Psychol Med* 1995, **25**(1):63–77.
10. Folstein S, Rutter M: Infantile autism: a genetic study of 21 twin pairs. *J Child Psychol Psychiatry* 1977, **18**(4):297–321.
11. Feng Y, Zhang F, Lokey LK, Chastain JL, Lakkis L, Eberhart D, Warren ST: Translational suppression by trinucleotide repeat expansion at FMR1. *Science* 1995, **268**(5211):731–734.
12. Sutcliffe JS, Nelson DL, Zhang F, Pieretti M, Caskey CT, Saxe D, Warren ST: DNA methylation represses FMR-1 transcription in fragile X syndrome. *Hum Mol Genet* 1992, **1**(6):397–400.
13. Hornstra IK, Nelson DL, Warren ST, Yang TP: High resolution methylation analysis of the FMR1 gene trinucleotide repeat region in fragile X syndrome. *Hum Mol Genet* 1993, **2**(10):1659–1665.
14. Chahrour M, Jung SY, Shaw C, Zhou X, Wong ST, Qin J, Zoghbi HY: MeCP2, a key contributor to neurological disease, activates and represses transcription. *Science* 2008, **320**(5880):1224–1229.
15. [http://www.genome.gov/25522099#al-5]
16. State MW, Levitt P: The conundrums of understanding genetic risks for autism spectrum disorders. *Nat Neurosci* 2011, **14**(12):1499–1506.
17. Geschwind DH: Autism: many genes, common pathways? *Cell* 2008, **135**(3): 391–395.
18. Freitag CM, Staal W, Klauck SM, Duketis E, Waltes R: Genetics of autistic disorders: review and clinical implications. *Eur Child Adolesc Psychiatry* 2010, **19**(3):169–178.
19. Grigorenko EL: Pathogenesis of autism: a patchwork of genetic causes. *Future Neurol* 2009, **4**(5):591–599.
20. Sanders SJ, Murtha MT, Gupta AR, Murdoch JD, Raubeson MJ, Willsey AJ, Ercan-Sencicek AG, DiLullo NM, Parikshak NN, Stein JL, Walker MF, Ober GT, Teran NA, Song Y, El-Fishawy P, Murtha RC, Choi M, Overton JD, Bjornson RD, Carriero NJ, Meyer KA, Bilguvar K, Mane SM, Sestan N, Lifton RP, Gunel M, Roeder K, Geschwind DH, Devlin B, State MW: De novo mutations revealed by whole-exome sequencing are strongly associated with autism. *Nature* 2012, **485**(7397):237–241.
21. Neale BM, Kou Y, Liu L, Ma'ayan A, Samocha KE, Sabo A, Lin CF, Stevens C, Wang LS, Makarov V, Polak P, Yoon S, Maguire J, Crawford EL, Campbell NG, Geller ET, Valladares O, Schafer C, Liu H, Zhao T, Cai G, Lihm J,

Dannenfelser R, Jabado O, Peralta Z, Nagaswamy U, Muzny D, Reid JG, Newsham I, Wu Y, Lewis L, Han Y, Voight BF, Lim E, Rossin E, Kirby A, Flannick J, Fromer M, Shakir K, Fennell T, Garimella K, Banks E, Poplin R, Gabriel S, DePristo M, Wimbish JR, Boone BE, Levy SE, Betancur C, Sunyaev S, Boerwinkle E, Buxbaum JD, Cook EH,Jr, Devlin B, Gibbs RA, Roeder K, Schellenberg GD, Sutcliffe JS, Daly MJ: Patterns and rates of exonic *de novo* mutations in autism spectrum disorders. *Nature* 2012, **485**(7397):242–245.

22. O'Roak BJ, Vives L, Girirajan S, Karakoc E, Krumm N, Coe BP, Levy R, Ko A, Lee C, Smith JD, Turner EH, Stanaway IB, Vernot B, Malig M, Baker C, Reilly B, Akey JM, Borenstein E, Rieder MJ, Nickerson DA, Bernier R, Shendure J, Eichler EE: Sporadic autism exomes reveal a highly interconnected protein network of de novo mutations. *Nature* 2012, **485**(7397):246–250.
23. Chahrour MH, Yu TW, Lim ET, Ataman B, Coulter ME, Hill RS, Stevens CR, Schubert CR, ARRA Autism Sequencing Collaboration, Greenberg ME, Gabriel SB, Walsh CA: Whole-exome sequencing and homozygosity analysis implicate depolarization-regulated neuronal genes in autism. *PLoS Genet* 2012, **8**(4):e1002635.
24. Hu VW: The expanding genomic landscape of autism: discovering the 'forest' beyond the 'trees'. *Future Neurol* 2013, **8**(1):29–42.
25. Lander ES: Initial impact of the sequencing of the human genome. *Nature* 2011, **470**(7333):187–197.
26. van Bakel H, Nislow C, Blencowe BJ, Hughes TR: Most "dark matter" transcripts are associated with known genes. *PLoS Biol* 2010, **8**(5):e1000371.
27. Mercer TR, Gerhardt DJ, Dinger ME, Crawford J, Trapnell C, Jeddeloh JA, Mattick JS, Rinn JL: Targeted RNA sequencing reveals the deep complexity of the human transcriptome. *Nat Biotechnol* 2011, **30**(1):99–104.
28. ENCODE Project Consortium, Bernstein BE, Birney E, Dunham I, Green ED, Gunter C, Snyder M: An integrated encyclopedia of DNA elements in the human genome. *Nature* 2012, **489**(7414):57–74.
29. Chen K, Rajewsky N: The evolution of gene regulation by transcription factors and microRNAs. *Nat Rev Genet* 2007, **8**(2):93–103.
30. Kim VN: MicroRNA biogenesis: coordinated cropping and dicing. *Nat Rev Mol Cell Biol* 2005, **6**(5):376–385.
31. Zamore PD, Haley B: Ribo-gnome: the big world of small RNAs. *Science* 2005, **309**(5740):1519–1524.
32. Cochella L, Hobert O: Diverse functions of microRNAs in nervous system development. *Curr Top Dev Biol* 2012, **99**:115–143.
33. Sun AX, Crabtree GR, Yoo AS: MicroRNAs: regulators of neuronal fate. *Curr Opin Cell Biol* 2013, **25**(2):215–221.
34. Giraldez AJ, Cinalli RM, Glasner ME, Enright AJ, Thomson JM, Baskerville S, Hammond SM, Bartel DP, Schier AF: MicroRNAs regulate brain morphogenesis in zebrafish. *Science* 2005, **308**(5723):833–838.

35. Davis TH, Cuellar TL, Koch SM, Barker AJ, Harfe BD, McManus MT, Ullian EM: Conditional loss of Dicer disrupts cellular and tissue morphogenesis in the cortex and hippocampus. *J Neurosci* 2008, **28**(17):4322–4330.
36. Huang T, Liu Y, Huang M, Zhao X, Cheng L: Wnt1-cre-mediated conditional loss of Dicer results in malformation of the midbrain and cerebellum and failure of neural crest and dopaminergic differentiation in mice. *J Mol Cell Biol* 2010, **2**(3):152–163.
37. Schaefer A, O'Carroll D, Tan CL, Hillman D, Sugimori M, Llinas R, Greengard P: Cerebellar neurodegeneration in the absence of microRNAs. *J Exp Med* 2007, **204**(7):1553–1558.
38. Ashraf SI, McLoon AL, Sclarsic SM, Kunes S: Synaptic protein synthesis associated with memory is regulated by the RISC pathway in Drosophila. *Cell* 2006, **124**(1):191–205.
39. Cohen JE, Lee PR, Chen S, Li W, Fields RD: MicroRNA regulation of homeostatic synaptic plasticity. *Proc Natl Acad Sci USA* 2011, **108**(28):11650–11655.
40. Abu-Elneel K, Liu T, Gazzaniga FS, Nishimura Y, Wall DP, Geschwind DH, Lao K, Kosik KS: Heterogeneous dysregulation of microRNAs across the autism spectrum. *Neurogenetics* 2008, **9**(3):153–161.
41. Buyske S: Comment on the article "Heterogeneous dysregulation of microRNAs across the autism spectrum" by Abu-Elneel *et al. Neurogenetics* 2009, **10**(2):167; author reply 169–170.
42. Talebizadeh Z, Butler MG, Theodoro MF: Feasibility and relevance of examining lymphoblastoid cell lines to study role of microRNAs in autism. *Autism Res* 2008, **1**(4):240–250.
43. Sarachana T, Zhou R, Chen G, Manji HK, Hu VW: Investigation of post-transcriptional gene regulatory networks associated with autism spectrum disorders by microRNA expression profiling of lymphoblastoid cell lines. *Genome Med* 2010, **2**(4):23.
44. Griffiths-Jones S, Grocock RJ, van Dongen S, Bateman A, Enright AJ: miRBase: microRNA sequences, targets and gene nomenclature. *Nucleic Acids Res* 2006, **34**(Database issue):D140–4.
45. Ghahramani Seno MM, Hu P, Gwadry FG, Pinto D, Marshall CR, Casallo G, Scherer SW: Gene and miRNA expression profiles in autism spectrum disorders. *Brain Res* 2011, **1380**:85–97.
46. Kong SW, Collins CD, Shimizu-Motohashi Y, Holm IA, Campbell MG, Lee IH, Brewster SJ, Hanson E, Harris HK, Lowe KR, Saada A, Mora A, Madison K, Hundley R, Egan J, McCarthy J, Eran A, Galdzicki M, Rappaport L, Kunkel LM, Kohane IS: Characteristics and predictive value of blood transcriptome signature in males with autism spectrum disorders. *PLoS One* 2012, **7**(12):e49475.
47. Vaishnavi V, Manikandan M, Tiwary BK, Munirajan AK: Insights on the functional impact of microRNAs present in autism-associated copy number variants. *PLoS One* 2013, **8**(2):e56781.

48. Basu SN, Kollu R, Banerjee-Basu S: AutDB: a gene reference resource for autism research. *Nucleic Acids Res* 2009, **37**(Database issue):D832–6.
49. Matera AG, Terns RM, Terns MP: Non-coding RNAs: lessons from the small nuclear and small nucleolar RNAs. *Nat Rev Mol Cell Biol* 2007, **8**(3): 209–220.
50. Bachellerie JP, Cavaille J, Huttenhofer A: The expanding snoRNA world. *Biochimie* 2002, **84**(8):775–790.
51. Cavaille J, Buiting K, Kiefmann M, Lalande M, Brannan CI, Horsthemke B, Bachellerie JP, Brosius J, Huttenhofer A: Identification of brain-specific and imprinted small nucleolar RNA genes exhibiting an unusual genomic organization. *Proc Natl Acad Sci* U S A 2000, **97**(26):14311–14316.
52. Skreka K, Schafferer S, Nat IR, Zywicki M, Salti A, Apostolova G, Griehl M, Rederstorff M, Dechant G, Huttenhofer A: Identification of differentially expressed non-coding RNAs in embryonic stem cell neural differentiation. *Nucleic Acids Res* 2012, **40**(13):6001–6015.
53. Dimitropoulos A, Schultz RT: Autistic-like symptomatology in Prader-Willi syndrome: a review of recent findings. *Curr Psychiatry Rep* 2007, **9**(2): 159–164.
54. Dykens EM, Lee E, Roof E: Prader-Willi syndrome and autism spectrum disorders: an evolving story. *J Neurodev Disord* 2011, **3**(3):225–237.
55. Sahoo T, del Gaudio D, German JR, Shinawi M, Peters SU, Person RE, Garnica A, Cheung SW, Beaud*et al*: Prader-Willi phenotype caused by paternal deficiency for the HBII-85 C/D box small nucleolar RNA cluster. *Nat Genet* 2008, **40**(6):719–721.
56. Sarachana T, Hu VW: Investigation of post-transcriptional gene regulatory networks associated with autism spectrum disorder (ASD) by miRNA expression profiling of lymphoblastoid cell lines. *Master's Thesis,* The George Washington University, 2009.
57. Sempere LF, Freemantle S, Pitha-Rowe I, Moss E, Dmitrovsky E, Ambros V: Expression profiling of mammalian microRNAs uncovers a subset of brain-expressed microRNAs with possible roles in murine and human neuronal differentiation. *Genome Biol* 2004, **5**(3):R13.
58. Liang Y, Ridzon D, Wong L, Chen C: Characterization of microRNA expression profiles in normal human tissues. *BMC Genomics* 2007, **8**:166.
59. Kim J, Krichevsky A, Grad Y, Hayes GD, Kosik KS, Church GM, Ruvkun G: Identification of many microRNAs that copurify with polyribosomes in mammalian neurons. *Proc Natl Acad Sci USA* 2004, **101**(1):360–365.
60. Miska EA, Alvarez-Saavedra E, Townsend M, Yoshii A, Sestan N, Rakic P, Constantine-Paton M, Horvitz HR: Microarray analysis of microRNA expression in the developing mammalian brain. *Genome Biol* 2004, **5**(9):R68.
61. Baskerville S, Bartel DP: Microarray profiling of microRNAs reveals frequent coexpression with neighboring miRNAs and host genes. *RNA* 2005, **11**(3): 241–247.

62. Wu JY, Kuban KC, Allred E, Shapiro F, Darras BT: Association of Duchenne muscular dystrophy with autism spectrum disorder. *J Child Neurol* 2005, **20**(10):790–795.
63. Hendriksen JG, Vles JS: Neuropsychiatric disorders in males with duchenne muscular dystrophy: frequency rate of attention-deficit hyperactivity disorder (ADHD), autism spectrum disorder, and obsessive--compulsive disorder. *J Child Neurol* 2008, **23**(5):477–481.
64. Vorstman JA, Morcus ME, Duijff SN, Klaassen PW, Heineman-de Boer JA, Beemer FA, Swaab H, Kahn RS, van Engeland H: The 22q11.2 deletion in children: high rate of autistic disorders and early onset of psychotic symptoms. *J Am Acad Child Adolesc Psychiatry* 2006, **45**(9):1104–1113.
65. Konstantareas MM, Hewitt T: Autistic disorder and schizophrenia: diagnostic overlaps. *J Autism Dev Disord* 2001, **31**(1):19–28.
66. Robinson SJ: Childhood epilepsy and autism spectrum disorders: psychiatric problems, phenotypic expression, and anticonvulsants. *Neuropsychol Rev* 2012, **22**(3):271–279.
67. Lesca G, Rudolf G, Labalme A, Hirsch E, Arzimanoglou A, Genton P, Motte J, de Saint Martin A, Valenti MP, Boulay C, De Bellescize J, Keo-Kosal P, Boutry-Kryza N, Edery P, Sanlaville D, Szepetowski P: Epileptic encephalopathies of the Landau-Kleffner and continuous spike and waves during slow-wave sleep types: genomic dissection makes the link with autism. *Epilepsia* 2012, **53**(9):1526–1538.
68. Zhou J, Parada LF: PTEN signaling in autism spectrum disorders. *Curr Opin Neurobiol* 2012, **22**(5):873–879.
69. Balciuniene J, Feng N, Iyadurai K, Hirsch B, Charnas L, Bill BR, Easterday MC, Staaf J, Oseth L, Czapansky-Beilman D, Avramopoulos D, Thomas GH, Borg A, Valle D, Schimmenti LA, Selleck SB: Recurrent 10q22-q23 deletions: a genomic disorder on 10q associated with cognitive and behavioral abnormalities. *Am J Hum Genet* 2007, **80**(5):938–947.
70. Bourgeron T: A synaptic trek to autism. *Curr Opin Neurobiol* 2009, **19**(2): 231–234.
71. Correia CT, Coutinho AM, Sequeira AF, Sousa IG, Lourenco Venda L, Almeida JP, Abreu RL, Lobo C, Miguel TS, Conroy J, Cochrane L, Gallagher L, Gill M, Ennis S, Oliveira GG, Vicente AM: Increased BDNF levels and NTRK2 gene association suggest a disruption of BDNF/TrkB signaling in autism. *Genes Brain Behav* 2010, **9**(7):841–848.
72. Liu XQ, Paterson AD, Szatmari P, Autism Genome Project Consortium: Genome-wide linkage analyses of quantitative and categorical autism subphenotypes. *Biol Psychiatry* 2008, **64**(7):561–570.
73. Cannon DS, Miller JS, Robison RJ, Villalobos ME, Wahmhoff NK, Allen-Brady K, McMahon WM, Coon H: Genome-wide linkage analyses of two repetitive behavior phenotypes in Utah pedigrees with autism spectrum disorders. *Mol Autism* 2010, **1**(1):3–2392-1-3.

74. Coon H, Villalobos ME, Robison RJ, Camp NJ, Cannon DS, Allen-Brady K, Miller JS, McMahon WM: Genome-wide linkage using the Social Responsiveness Scale in Utah autism pedigrees. *Mol Autism* 2010, **1**(1):8–2392-1-8.
75. Alarcon M, Yonan AL, Gilliam TC, Cantor RM, Geschwind DH: Quantitative genome scan and Ordered-Subsets Analysis of autism endophenotypes support language QTLs. *Mol Psychiatry* 2005, **10**(8):747–757.
76. Chen GK, Kono N, Geschwind DH, Cantor RM: Quantitative trait locus analysis of nonverbal communication in autism spectrum disorder. *Mol Psychiatry* 2006, **11**(2):214–220.
77. Nurmi EL, Dowd M, Tadevosyan-Leyfer O, Haines JL, Folstein SE, Sutcliffe JS: Exploratory subsetting of autism families based on savant skills improves evidence of genetic linkage to 15q11-q13. *J Am Acad Child Adolesc Psychiatry* 2003, **42**(7):856–863.
78. Hu VW, Sarachana T, Kim KS, Nguyen A, Kulkarni S, Steinberg ME, Luu T, Lai Y, Lee NH: Gene expression profiling differentiates autism case-controls and phenotypic variants of autism spectrum disorders: evidence for circadian rhythm dysfunction in severe autism. *Autism Res* 2009, **2**(2):78–97.
79. Hu VW, Addington A, Hyman A: Novel autism subtype-dependent genetic variants are revealed by quantitative trait and subphenotype association analyses of Published GWAS Data. *PLoS One* 2011, **6**(4).
80. Lestrade L, Weber MJ: snoRNA-LBME-db, a comprehensive database of human H/ACA and C/D box snoRNAs. *Nucleic Acids Res* 2006, **34**(Database issue):D158–62.
81. Amaral PP, Clark MB, Gascoigne DK, Dinger ME, Mattick JS: lncRNAdb: a reference database for long noncoding RNAs. *Nucleic Acids Res* 2011, **39**(Database issue):D146–51.
82. Xie J, Zhang M, Zhou T, Hua X, Tang L, Wu W: Sno/scaRNAbase: a curated database for small nucleolar RNAs and cajal body-specific RNAs. *Nucleic Acids Res* 2007, **35**(Database issue):D183–7.

Chapter 9

Targeting Noncoding RNA for Treatment of Autism Spectrum Disorders

Jessica DeWitt and Daniel B. Campbell*

Abstract

An entirely new area of pharmacology is developing from an understanding of endogenous functional long noncoding RNAs (lncRNAs).[1] Potential treatment options for cancer, diabetes and HIV are derived from findings that specific lncRNAs, which regulate networks of genes, can be targeted to treat the disorders. Emerging evidence indicates that specific lncRNAs also contribute to autism spectrum disorder (ASD) risk. Here, we review this evidence and explore possible treatment advances for ASD based on inhibitory RNA-based therapies.

Keywords: Long noncoding RNA; lncRNA; Noncoding RNA; ncRNA; siRNA; Antisense therapy.

Evidence that lncRNAs Contribute to ASD

Autism spectrum disorder (ASD) is an extremely heritable neurodevelopmental disorder distinguished by deficits in social interaction,

*Correspondence to: Zilkha Neurogenetic Institute, 1501 San Pablo Street, University of Southern California Keck School of Medicine, Los Angeles, CA 90089 USA. Emails: Jessica DeWitt: jjdewitt@usc.edu; Daniel B. Campbell: dbcampbe@med.usc.edu

language development, and behavioral flexibility. Although environmental factors contribute to ASD, the estimated heritability of ASD is >70%.[2–4] Thus, genetic approaches offer great promise for determining the factors that contribute to ASD. Recent genome-wide approaches have identified linkage regions likely to harbor rare mutations,[5,6] chromosomal regions with genome-wide significant association,[6–9] rare copy number variations (CNVs),[10–12] and rare mutations of protein-coding genes.[13–18] However, there is very little overlap in the identified genes among these powerful genome-wide approaches, and reliably reproducible candidate genes have proven elusive.

As for all common complex genetic disorders, there is an ongoing debate about the relative contributions of rare and common genetic variants to ASD. The identification of rare *de novo* copy number variants (CNVs) on chromosomes 2p, 7q, 15q, and 16p provide support for rare variants and important biological clues to the etiology of ASD.[10–12,19–21] However, the contribution of these CNVs remains at <6% of cases (with each CNV observed in <1%) despite extensive cohort collection and technological advances that increased CNV detection resolution by 10-fold.[21] It is now clear that CNVs do not account for a large fraction of ASD, either individually or collectively.[22] Further, the implicated CNVs are typically large (~250 kb) and encompass multiple protein-coding genes, making it difficult to determine which gene(s) contribute to the disorder. Genome-wide association studies (GWASs) have proven effective in identifying associated genetic variants in other complex genetic disorders, such as type 2 diabetes, Crohn's disease, and prostate cancer.[23–29] However, application of GWAS to ASD yielded mixed results.[6–9] Because GWAS is the definitive technique for identifying associated common variants of large effect, the lack of obvious protein-coding genes implicated by GWAS pushed the ASD genetics field toward searching for rare causal variants.[13–19,21,30–32] Whole exome sequencing (next-generation sequencing restricted to the exons of protein-coding genes) has proven effective in identifying causal genetic variants in Mendelian disorders, such as Miller syndrome, high myopia, and malformations of cortical development.[33–35] However, application of whole exome sequencing to ASD also yielded mixed results.[13–18] Whole exome sequencing approaches have identified *de novo* mutations that may be specific to individuals with ASD, but the identified *de novo* mutations do not account for a large fraction of cases, either individually or collectively.[13–18] Thus, the definitive experiments have not yet resolved the common variant versus rare variant debate, and suggest that both common variants and rare variants contribute to ASD risk.

Based on a variety of evidence, we consider common variants likely to play a large role in ASD risk. First, the high heritability of ASD suggests that the *de novo* mutations implicated by exome sequencing and CNV detection play a small role in the disorder. Second, sub-clinical ASD traits are present in >50% of family members of affected individuals,[36–39] suggesting that common heritable factors contribute to a large fraction of cases. Third, the heritability of ASD traits is high and indistinguishable between extreme cases and the general population, indicating that these traits are the quantitative extreme of a neurodevelopmental continuum, not a mutation-induced abnormality.[40,41] Fourth, the conclusions from the exome sequencing studies that no single gene is causal and that hundreds of genes will increase risk[13–18] indicate that evidence for association of each genetic variant with the disorder will continue to be of utmost importance. While genome-wide significant association of common markers may drift with additional genotyping, the association is likely to remain significant. In contrast, the nominally significant association of a rare variant can disappear with the positive identification of a mutation in a single control. Although the genetic effect sizes of common variants may be small, similar effect sizes have identified therapeutic targets for Crohn's disease and type 2 diabetes.

The exome sequencing studies[13–18] all concluded that no single gene is causal for ASD and that 400–1000 different genes may increase ASD risk. An alternative interpretation of these data is that *de novo* mutation of protein-coding genes may contribute little to ASD. Instead, genetic variation affecting lncRNA, which are by definition excluded from exome sequencing studies, may contribute substantially. Three lines of evidence support this interpretation. First, while individual common genetic variants contribute small risk to ASD, collectively common genetic variants explain more than 40% of genetic risk for ASD.[42] Second, large-scale whole genome sequencing for high-density lipoprotein (HDL) indicates that non-coding regions contribute as much as protein-coding genes to this complex genetic condition.[43] Third, nearly two-thirds (64%) of all long transcripts in the human brain are lncRNAs.[44] It is also becoming increasingly clear that the number of ncRNAs encoded in a genome increases with developmental complexity,[45] and that genetic variation within human lncRNAs contributes to neurodevelopmental disorders.[46]

Concurrent with these genetic observations, the contribution of lncRNAs became more apparent in other complex genetic disorders, including cancer[47,48] and schizophrenia.[49] Emerging evidence from multiple approaches indicate a contribution of lncRNA to ASD.

The first evidence that lncRNAs may contribute to ASD was derived from a study of X-linked protein-coding genes. Noor *et al.* found that 0.5% of individuals have mutation in the protein-coding *PTCHD1* gene. An additional 0.5% of ASD cases also had deletions of a complex non-coding RNA locus (*PTCHD1AS1*/*PTCHD1AS2*) 5′ to the *PTCHD1* locus that are associated with ASD.[34] Although the deletions are rare and the functional mechanisms of the *PTCHD1AS1* and *PTCHD1AS2* non-coding RNAs have not been explored extensively, they are presumed to regulate expression of the protein-coding gene on the opposite strand, *PTCHD1*.[34]

The second report indicating evidence for lncRNA contribution to ASD was based on a genome-wide significant association in a protein-coding gene-poor region of chromosome 5p14.1. A genome-wide significant association study (GWAS) of ASD indicated the most significant association to date ($P = 10^{-10}$) for rs4307059 on chromosome 5p14.1.[7] The same rs4307059 allele was also associated with social communication phenotypes in a general population sample.[50] However, rs4307059 genotype was not correlated with expression of either of the flanking protein-coding genes, *CDH9* and *CDH10*.[7] We identified a 3.9 kb non-coding RNA that is transcribed directly at the site of the chromosome 5p14.1 ASD GWAS peak.[51] The non-coding RNA is encoded by the opposite (anti-sense) strand of moesin pseudogene 1 (*MSNP1*), and is thus designated *MSNP1AS* (moesin pseudogene 1, anti-sense). *MSNP1AS* is 94% identical and anti-sense to the X chromosome transcript *MSN*, which encodes a protein (moesin) that regulates neuronal architecture and immune response. Expression of *MSNP1AS* in postmortem temporal cortex is increased 12.7-fold in individuals with ASD and increased 22-fold in individuals with the rs4307059 risk allele.[51] The *MSNP1AS* non-coding RNA binds *MSN* and its over-expression in cultured neurons causes significant decreases in *MSN* transcript, moesin protein, neurite number and neurite length. Thus, our discovery reveals a functional lncRNA which, based on the GWAS findings, contributes to ASD risk.

To evaluate whether or not changes in lncRNA expression is part of ASD pathogenesis, 33,000 lncRNAs and 30,000 mRNA transcripts from ASD and control prefrontal cortex and cerebellum postmortem brain tissue were profiled using microarray. Two hundred lncRNAs were differentially expressed in ASD. Previously, it had been reported that mRNA transcription within ASD brains was homogenous.[52,53] Another study also found increased transcriptional homogeneity in lncRNAs in ASD brains.[54] Although the conclusions of this study implicate lncRNA, the sample size was small (2 ASD cases and 2 matched controls). More studies with larger sample sizes will be necessary to determine specific lncRNAs with consistently altered expression in ASD brains.

Therefore, accumulating evidence from multiple experimental approaches suggest that lncRNAs contribute to ASD. The promise of a new pharmacology to target lncRNA suggests possible treatment options.

What is a ncRNA?

The eukaryotic genome is much more complex than previously thought. Protein-coding gene exons comprise <2% of the total genomic material. However, analyses of the transcriptome revealed that the human genome is differentially and dynamically transcribed to produce a spectrum of non-coding RNA (ncRNA) of all shapes and sizes.[55] NcRNAs have been arbitrarily sorted into two categories based on their size: long ncRNA (greater than 200 nucleotides in length) and short ncRNAs (less than 200 nucleotides in length).[56] These categories can be further broken down into subcategories based on other properties. Few ncRNAs have been characterized functionally. However, the functions that have been discovered are obviously an important part of development and organismal maintenance. As key parts of cellular processes, these ncRNAs leave the organism vulnerable to disease when they become dysregulated. Several ncRNAs have been shown to play a role in the pathogenesis of diseases such as cancer, HIV, heart disease, and neurological disorders. Characterizing these cellular components is important to our understanding of these diseases and may provide insight into their pathogenesis and treatment.

Small ncRNA

Small ncRNA, as a category, include RNA such as microRNAs (miRNAs), short interfering RNAs (siRNAs), small nucleolar RNAs (snoRNAs), and PIWI-interacting RNAs (piRNAs). The genes encoding these RNA can be found throughout the genome.[57] Small ncRNA can also be derived from processing lncRNA.[58,59] The main method used by small ncRNA to regulate the genome is by binding to their specific target transcripts via complementary nucleotide sequences, altering downstream events to influence the translation of the target. To read more about small ncRNA subclasses, biogenesis, and forms of genetic regulation, see the following reviews.[57,60–64]

MiRNA is one of the best-studied classes of small ncRNAs and was first discovered in 1993 in *C. elegans*.[65] Since its discovery, miRNA is believed to be relatively conserved among vertebrates.[66–68] MiRNAs can regulate genes post translation by binding to either the 3′-[69] or the 5′-UTR[70] of a target transcript which typically inhibits translation. One particular miRNA can target multiple transcripts and one transcipt can be regulated by multiple miRNA.[71] More than 60% of human protein-coding genes are predicted targets of miRNA, lending credence to the idea that miRNAs are prevalent and important in the human genome.[72] Regulation by miRNA is vital to mammalian developmental processes and, like many other ncRNAs, miRNA expression can be specifically and differentially expressed throughout tissue types and the developmental process as reviewed.[73–75]

Like miRNA, siRNA also binds to complementary sequences on target transcripts and regulates gene expression. The difference is that siRNA complementary sequences have stricter requirements in that they must have close to perfect sequence complementation.[76] Due to this difference, siRNA and miRNA have two separate mechanisms for repression. SiRNA targets with very close complementation undergo direct cleavage, and miRNA targets are destabilized or transcriptionally repressed.[77–79] Endogenous siRNAs (endo-siRNAs) are purported to protect against transposons and dsDNA viruses in somatic tissues in *Drosophila*.[80–82] Endo-siRNAs have also been identified in mice[83–86] and in human cells.[87–89] SiRNA has been extensively researched for use in gene therapy and may be able to correct ncRNA dysregulation in a number of different diseases.[90]

Long ncRNA (lncRNA)

LncRNA is an expansive category that includes many different structural features and mechanisms of action. LncRNA include sense and antisense strands in pseudogenes, in intergenic regions (lincRNA) and in protein-coding genes. After transcription, they may be further processed through splicing and/or the addition of a 5′-methyl-guanosine cap and 3′ poly (A) tail. They also can contain localization signals that direct them to the cytoplasm or the nucleus.[91] LncRNAs have many diverse functional roles as well, such as employing enhancer-like functions, participating in the recruitment of chromatin-modifying complexes, modulating alternative splicing, providing a scaffold for the assembly of protein complexes, and acting as competing endogenous RNAs (ceRNAs).[92–94] XIST, as well as its interaction with the polycomb repressive complex, PRC2, is the most thoroughly studied lncRNA.[95] XIST's antisense gene, TSIX, was discovered in 1999.[96] TSIX serves as a regulatory element of XIST. It is part of a repression pathway along with other pluripotency factors.[97] There is much to be discovered about lncRNA before it can be used to treat disease.

Therapeutic Targeting of ncRNA — A New Type of Pharmacology

The main diseases in which ncRNAs have been used as therapeutic targets are cancer, HIV, and diabetes. Therapies designed to target implicated lncRNAs utilize the lncRNAs in a variety of ways including using them as biomarkers, silencing them, and inducing overexpression. Because many lncRNAs involved in cancer lead to cell proliferation, the primary method of therapy for cancer is inhibition, but overexpression and use as biomarkers are being increasingly investigated for their usefulness in treating other diseases.

Cancer

Dysregulation of lncRNA is a primary feature of cancer because lncRNA governs other cellular processes. LncRNAs are involved in many leukemias and carcinomas, including chronic lymphocytic leukemia, colorectal carcinoma, and hepatocellular carcinoma. Malignant cancer cells have a unique spectrum of expressed ultra-conserved regions when compared with the corresponding normal cells. The data suggests that variations in the expression of translated ultra-conserved regions are involved in cancer formation.[98] Specific genes coding for lncRNAs that have received a lot of attention include MALAT1, HOTAIR, and PTENTP1, among others (Table 1).

Inhibiting miRNAs has been practiced since 2004 when Meister *et al.* used 2′-O-methyl oligoribonucleotides to specifically inactivate miRNA-protein complex RNAi activity.[99] Inhibiting lncRNAs may be more difficult than inhibiting miRNA, possibly because

Table 1. Potential lncRNA therapy targets

Clinical Diagnosis	LncRNA
Autism Spectrum Disorder	PTCHD1AS1/PTCHD1AS2,[53] MSNP1AS[55]
Cancer	MALAT1,[103] HOTAIR,[104–106] PTENP1[107]
Prostate cancer	DD3[108]
Breast cancer	BC1,[108] BC200[108]
Hepatocellular Carcinoma	HULC[108]
Spinocerebellar Ataxia	Mlbn1[109]
Alzheimer's Disease	BACE1/BACE1-AS[110]
Fragile X Syndrome	FMR4[116]
Down's Syndrome	NRON,[117] XIST[118]

they are larger. However, repression of lncRNA with siRNAs is feasible. This method has been used as a therapeutic approach to lncRNAs involved in cancer. Metastasis-associated lung adenocarcinoma transcript 1 (MALAT1) is an lncRNA involved in proliferation and metastasis of many different types of cancer. Guo *et al.* was the first to examine MALAT1's role in cervical cancer pathogenesis. They effectively silenced MALAT1 through the use of siRNAs, demonstrating that the lncRNA could be a therapeutic target. Blocking MALAT1 or other lncRNA involved in cancer progression is a promising therapeutic approach since many lncRNA have been silenced with siRNA.[100]

A second cancer lncRNA that has been studied extensively is HOTAIR, a Hox transcript antisense intergenic RNA. Colorectal cancer patients with increased HOTAIR expression are more likely to have a poor prognosis by increasing cell invasiveness and promoting metastasis.[101] Forcible overexpression of *HOTAIR* in non-small cell lung cancer (NSCLC) cells caused an increase in cancer cell proliferation and migration.[102] Conversely, loss of *HOTAIR* can suppress tumorigenesis and metastasis, which leads to the conclusion that *HOTAIR* is a cancer progression regulator and can drive development and progression. *HOTAIR* helps the

PRC2 complex target specific genes, which trimethylates H3K27 and silences metastasis suppressor genes. *HOTAIR* and PRC2 are interdependent, which means that high HOTAIR expression could be a biomarker for cancers that are sensitive to PRC2 inhibitors. In contrast, tumors with high PRC2 expression might be sensitive to therapies that target *HOTAIR* or prevent *HOTAIR* from interacting with PRC2. These data suggests that reprogramming by lincRNA could be a therapy for any disease with aberrant lincRNA expression.[103]

A pseudogene, PTENP1, that is homologous to PTEN, a tumor suppressor gene, is targeted and silenced by a miRNA. PTENP1 can positively regulate PTEN and de-represses PTEN by binding to PTEN miRNA. SiRNA has also been used to silence PTENP1 and PTEN.[104] Using siRNA to silence lncRNA involved in cancer is an effective method of therapeutic targeting.

LncRNA can be used as biomarkers for cancer based on expression levels alone. DD3 is the most described prostate cancer-specific lncRNA and is a very sensitive marker for cancer cells. BC1 is dysregulated in breast cancer and BC200 is an RNA that is overexpressed in breast cancer with poor prognosis and, therefore, can be used as a biomarker. Highly up-regulated in liver cancer (HULC) is a liver cancer specific gene that can be used as a biomarker for HCC. When HULC is knocked down by siRNA, gene expression of five genes is changed. Its unique expression pattern and the fact that expression can be detected in a blood test and RT-PCR of tissue samples make it a good choice for a HCC biomarker.[105]

Neurodegenerative Disorders

Ataxias

LncRNAs have been associated with several types of spinocerebellar ataxia (SCA), as well. Daughters *et al.* found gain-of-function mutations in lncRNA in SCA type 8 (SCA8) brains. These mutations were characterized by CUG-expansion transcripts that colocalize with muscleblind-like 1 (Mbnl1). They also found that

Mbln1 knockdown mice had decreased motor control, a symptom of SCA8. In mice, humans, and human cell culture, SCA8 CUG-expansion transcripts also increased expression of *GABT4*, a CUGBP1-MBNL1 regulated CNS target, which lead to a loss of GABAergic inhibition in the granular cell layer of the cerebellum. Pathological changes associated with CUG expansion transcripts might be reversed by over-expressing Mbnl1 because its overexpression has been therapeutic in skeletal muscle.[106]

Alzheimer's disease

Alzheimer's disease (AD) causes dementia in much of the aged population by affecting the central nervous system. Currently there are only symptomatic treatments. Therapies that are working toward a cure are based on the amyloid cascade theory, which holds that amyloid β (Aβ) accumulates in the brain and induces a pathway that leads to neurodegeneration. Aβ, the main component of amyloid plaques, is a protein fragment cleaved by β-site APP-cleaving enzyme 1 (BACE1) and γ-secretase from the amyloid precursor protein (APP). BACE1 action is the rate-limiting step in Aβ production and is overexpressed in AD patients, therefore making BACE1 an obvious choice for targeting. Murine knockout of BACE1 appears to have little consequence. Designed BACE1 inhibitors have shown promise in animal models because they have rescued cognitive deficits by reducing amyloid plaque pathology in the brain. The candidate BACE1 inhibitor drugs are well tolerated in Phase I clinical trials and are currently in further trials in AD patients. It has recently been discovered that BACE1 cleaves certain substrates involved in myelination, retinal homeostasis, brain circuitry, and synaptic function, which may point to potential side effects that will have to be resolved before BACE1 inhibitors can be used as a therapy for AD patients.[107]

Faghihi *et al.* suggest that BACE1 is regulated by the lncRNA, *BACE1*-AS. *BACE1*-AS is overexpressed in AD brains and promotes BACE1 stability. Increased expression of *BACE1*-AS is associated with Aβ cell stressors. *BACE1*-AS is elevated in amyloid precursor transgenic mice and Alzheimer's patients indicating that this ncRNA could drive Alzheimer's pathophysiology.[108]

Due to the elevated levels of *BACE1*-AS in AD patients, BACE1 is increased through a feed-forward cycle. *BACE1*-AS masks a miR-485-5p binding site. The resulting BACE1 increase could be due to the lack of repressive miRNA, and this miRNA contributes to Aβ formation and amyloid plaque formation. Amyloid plaque formation is a chronic process and lasts a long time. Even small changes in BACE1 levels could have long term effects on AD pathogenesis.[109,110]

Modarresi *et al.* was also able to use locked nucleic acid siRNA to knockdown either BACE1 or BACE1-AS, reducing the levels of Aβ and promoting accumulation of oligomeric Aβ. Since Aβ is involved in neurogenesis, they believe BACE1-AS could be used either as a novel therapeutic target or a biomarker for early detection of AD.[111]

Neurodevelopmental Disorders

Fragile X syndrome

Autism occurs in 30% of patients with Fragile X syndrome (FXS). FXS is caused by a CGG repeat in the 5′ untranslated region of the fragile mental retardation 1 gene, *FMR1*. Discovering treatment targets for FXS may also be helpful for developing treatments for non-FXS forms of autism.[112] An often over-looked consequence of the CGG repeat is silencing of a lncRNA, *FMR4*, in Fragile X patients. SiRNA knockdown of *FMR4* significantly affected human cell proliferation *in vitro*; knockdown also increased apoptosis and altered the cell cycle. Overexpression of *FMR4* caused an increase in cell proliferation. *FMR4* appears to have an anti-apoptotic function and shows unexpected functional complexity that may contribute to the FXS phenotype.[113]

Down's syndrome

The lncRNA NRON represses NFAT (nuclear factor of activated T cells), a transcription factor that is dysregulated in Down's

Syndrome (DS). When NRON is depleted, T cells show an increase in NFAT-dependent cytokines. The evidence seems to suggest that a complex of lncRNA (NRON) and protein forms a scaffold for a transcription factor (NFAT) and the kinases it regulates, supporting the idea that lncRNAs that bind transcriptional regulators have scaffolding functions.[114]

The normal function of the lncRNA XIST is to trigger X chromosome inactivation. DS is a common chromosomal dosage disorder. Expression of the lncRNA gene, XIST, triggers chromosome 21 repression. It also reverses deficits in neural progenitors and cell proliferation, leading to hypocellularity in brains of DS patients. Upon removal of XIST and doxycycline, heterochromatic silencing is stably maintained. XIST was targeted in non-immortalized fibroblasts from a female DS patient, resulting in the generation of many XIST-carrying cells on chromosome 21. The XIST transgene created by Jiang *et al.* lacked X-chromosome 'counting' sequences, which is compatible with normal female X inactivation. This work reveals DS as a candidate for gene therapy and XIST as a candidate for therapeutic targeting. Before gene therapy can be developed, the gene function, as well as a functional correction, must be revealed. The experiment shows that chromosomal imbalance can be corrected. Advances in X-chromosome biology and lncRNA research opens the way for translational research in DS.[115]

NcRNA in Clinical Trials

NcRNAs are part of many gene regulatory pathways. When lncRNAs are dysregulated by genetic, epigenetic or environmental factors, they can cause dysregulation in other genes of gene networks that contribute to altered development. It would be difficult to alter these dysregulated translated genes that contribute to clinical diagnoses. However, they can be manipulated through the use of their regulatory ncRNAs. The ncRNAs complementary to these translated disease-causing genes can target these genes that are over or under-expressed. The primary challenge is to transport the complementary ncRNA to a position in which they can be effective.

Using ncRNA as therapeutic targets, especially lncRNAs, is a relatively new field of research, but next-generation RNA sequencing is quickly being developed that will help identify the relevant transcripts.[116]

Several types of ncRNAs have been utilized as drugs to suppress or upregulate genes implicated in disease. SiRNA and RNAi are the main candidates for gene suppression and have successfully repressed their target genes in several diseases including cancer as mentioned previously. In addition to siRNA and RNAi, antisense RNAs and ribozymes (RNA that is catalytically active) are also candidates for use in clinical trials.

The first example of RNA therapeutics used in clinical trials is antisense RNA. Antisense RNA usually targets mRNA, but they have also been used to target miRNA as well. These RNA are chemically synthesized oligonucleotides (ON) that are antisense to miRNA (antagomirs) and can inhibit miRNA involved in disease. A locked nucleic acid (LNA)-modified ON, called miravirsen, inhibits miR-122, a liver miRNA involved in Hepatitis C viral (HCV) infection. Miravirsen has completed two Phase 1 clinical trials and is safe. It is currently in Phase IIa clinical trials to test tolerability and efficacy in HCV patients.[117] Twenty-two different siRNA therapies have now reached clinical testing for the treatment of 16 disorders.[118]

SiRNA has enjoyed moderate success with both systemic and local delivery as a drug. SiRNA can specifically knockdown any mRNA and is easily designed and screened. Its silencing effects also last a long time and can retain its catalytic ability in RISC for long periods as well.[118] Many siRNA drugs treating vision loss using local delivery have initiated Phase I clinical trials.[119–122] There are several siRNA and shRNA drugs delivered locally to treat diseases from vision loss to respiratory viruses that are currently in Phase 0 and 1 clinical trials.[123,124]

RNA aptamers are single-stranded nucleic acids with stable three-dimensional shapes that allow them to bind to molecular targets with high affinity and specificity.[125] RNA aptamers are similar to antibodies in affinity and specificity for targets, but they also have certain advantages over antibodies. Since aptamers are formed *in vitro* with SELEX, they are more efficiently, accurately, and economically produced, which makes them a better choice for clinical application. Aptamers are chemically modified, therefore they can target ligands without producing the immune response present with antibodies. The ability of aptamers to be modified is another advantage they hold over antibodies. Aptamers can be joined to ribozymes to make riboswitches or siRNA to make aptamer-siRNA chimeras. They are also smaller than antibodies, which improves tissue penetration and transport. At least six RNA aptamers have been clinically tested. One, a VEGF-specific modified RNA aptamer, treats AMD and is now an FDA approved drug.[126–128]

Ribozymes are RNA molecules that can perform biochemical actions similar to those of an enzyme. Ribozymes have gone under clinical testing as gene therapy-based approaches to treating HIV using CD4+ T cells or CD34+ hematopoietic stem cells (HSCs). CD34+ HSCs differentiate into various hematopoietic lineages including CD4+ T cells.[129–132] ShRNA can work in conjunction with ribozymes. RNA-based treatment was developed for HIV-1 using gene-modified autogolous CD34+ hematopoietic progenitor cells (HPCs) and is now in a Phase 0 clinical study at the City of Hope.[133] The RNA-based treatment involved three RNA components, an shRNA, a Trans-activation-responsive (TAR) decoy, and a hammerhead ribozyme, all encoded in the lentiviral gene vector. The hammerhead ribozyme cleaves the mRNA of the chemokine receptor 5 (CCR5) protein.[134] The CCR5 receptor also is a coreceptor for HIV-1 infection and a subset of CD4+ T cells express it. This receptor is not necessary for normal T cell function and is not prone to mutational escape, therefore it could be a good target for anti-HIV therapy.

Future Directions and Conclusions

Based on recent genetics findings, we anticipate that ncRNAs will account for a large proportion of ASD risk. Our own published and unpublished data indicate that the most significant genetic associations with ASD implicate lncRNAs. The contributions of these lncRNAs to ASD etiology need to be established mechanistically before antisense therapies for ASD can be developed. Antisense therapy has proven effective for other common disorders, including cancer and HIV. Antisense siRNA treatments for HIV and cancer are effective at inhibiting the progression of the diseases in adults. Application of antisense siRNA to treat developmental brain disorders poses unique challenges, both medical and societal. The medical challenges include timing and delivery of antisense therapy to specific brain regions at the correct developmental stages. The societal challenges include the parental decision to alter brain development. However, given the burden of ASD on both families and society, and the lack of effective treatment options, we propose that experiments designed to understand potential biological interventions are worthwhile.

Executive Summary

- Emerging evidence indicates that lncRNAs may contribute to a large fraction of ASD cases
- Long non-coding RNAs (lncRNAs) are more abundant in human brain than protein-coding RNAs
- A new type of pharmacology is emerging to target lncRNAs
- Targeting of lncRNA may be a treatment option for ASD
- Medical and societal challenges await.

References

1. Weinberg MS, Morris KV: Long non-coding RNA targeting and transcriptional de-repression. *Nucleic Acid Ther* 2013, **23**(1):9–14.
2. Folstein S, Rutter M: Genetic influences and infantile autism. *Nature* 1977, **265**(5596):726–728.
3. Ritvo ER, Freeman BJ, Mason-Brothers A, Mo A, Ritvo AM: Concordance for the syndrome of autism in 40 pairs of afflicted twins. *Am J Psychiatry* 1985, **142**(1):74–77.

4. Bailey A, Le Couteur A, Gottesman I, Bolton P, Simonoff E, Yuzda E, Rutter M: Autism as a strongly genetic disorder: evidence from a British twin study. *Psychol Med* 1995, **25**(1):63–77.
5. Szatmari P, Paterson AD, Zwaigenbaum L, Roberts W, Brian J, Liu XQ, Vincent JB, Skaug JL, Thompson AP, Senman L *et al.*: Mapping autism risk loci using genetic linkage and chromosomal rearrangements. *Nat Genet* 2007, **39**(3):319–328.
6. Weiss LA, Arking DE, Daly MJ, Chakravarti A: A genome-wide linkage and association scan reveals novel loci for autism. *Nature* 2009, **461**(7265):802–808.
7. Wang K, Zhang H, Ma D, Bucan M, Glessner JT, Abrahams BS, Salyakina D, Imielinski M, Bradfield JP, Sleiman PM *et al.*: Common genetic variants on 5p14.1 associate with autism spectrum disorders. *Nature* 2009, **459**(7246):528–533.
8. Ma D, Salyakina D, Jaworski JM, Konidari I, Whitehead PL, Andersen AN, Hoffman JD, Slifer SH, Hedges DJ, Cukier HN *et al.*: A genome-wide association study of autism reveals a common novel risk locus at 5p14.1. *Ann Hum Genet* 2009, **73**(Pt 3):263–273.
9. Anney R, Klei L, Pinto D, Regan R, Conroy J, Magalhaes TR, Correia C, Abrahams BS, Sykes N, Pagnamenta AT *et al.*: A genome-wide scan for common alleles affecting risk for autism. *Hum Mol Genet* 2010, **19**(20):4072–4082.
10. Glessner JT, Wang K, Cai G, Korvatska O, Kim CE, Wood S, Zhang H, Estes A, Brune CW, Bradfield JP *et al.*: Autism genome-wide copy number variation reveals ubiquitin and neuronal genes. *Nature* 2009, **459**(7246):569–573.
11. Weiss LA, Shen Y, Korn JM, Arking DE, Miller DT, Fossdal R, Saemundsen E, Stefansson H, Ferreira MA, Green T *et al.*: Association between microdeletion and microduplication at 16p11.2 and autism. *N Engl J Med* 2008, **358**(7):667–675.
12. Bucan M, Abrahams BS, Wang K, Glessner JT, Herman EI, Sonnenblick LI, Alvarez Retuerto AI, Imielinski M, Hadley D, Bradfield JP *et al.*: Genome-wide analyses of exonic copy number variants in a family-based study point to novel autism susceptibility genes. *PLoS Genet* 2009, **5**(6):e1000536.
13. Chahrour MH, Yu TW, Lim ET, Ataman B, Coulter ME, Hill RS, Stevens CR, Schubert CR, Greenberg ME, Gabriel SB *et al.*: Whole-exome sequencing and homozygosity analysis implicate depolarization-regulated neuronal genes in autism. *PLoS Genet* 2012, **8**(4):e1002635.
14. Iossifov I, Ronemus M, Levy D, Wang Z, Hakker I, Rosenbaum J, Yamrom B, Lee YH, Narzisi G, Leotta A *et al.*: De novo gene disruptions in children on the autistic spectrum. *Neuron* 2012, **74**(2):285–299.
15. O'Roak BJ, Deriziotis P, Lee C, Vives L, Schwartz JJ, Girirajan S, Karakoc E, Mackenzie AP, Ng SB, Baker C *et al.*: Exome sequencing in sporadic autism spectrum disorders identifies severe de novo mutations. *Nat Genet* 2011, **43**(6):585–589.

16. O'Roak BJ, Vives L, Girirajan S, Karakoc E, Krumm N, Coe BP, Levy R, Ko A, Lee C, Smith JD *et al.*: Sporadic autism exomes reveal a highly interconnected protein network of de novo mutations. *Nature* 2012, **485**:246–250.
17. Sanders SJ, Murtha MT, Gupta AR, Murdoch JD, Raubeson MJ, Willsey AJ, Ercan-Sencicek AG, Dilullo NM, Parikshak NN, Stein JL *et al.*: De novo mutations revealed by whole-exome sequencing are strongly associated with autism. *Nature* 2012, **485**:237–241.
18. Neale BM, Kou Y, Liu L, Ma'ayan A, Samocha KE, Sabo A, Lin CF, Stevens C, Wang LS, Makarov V *et al.*: Patterns and rates of exonic de novo mutations in autism spectrum disorders. *Nature* 2012, **485**:242–245.
19. Pinto D, Pagnamenta AT, Klei L, Anney R, Merico D, Regan R, Conroy J, Magalhaes TR, Correia C, Abrahams BS *et al.*: Functional impact of global rare copy number variation in autism spectrum disorders. *Nature* 2010, **466**(7304):368–372.
20. Gilman SR, Iossifov I, Levy D, Ronemus M, Wigler M, Vitkup D: Rare de novo variants associated with autism implicate a large functional network of genes involved in formation and function of synapses. *Neuron* 2011, **70**(5):898–907.
21. Sanders SJ, Ercan-Sencicek AG, Hus V, Luo R, Murtha MT, Moreno-De-Luca D, Chu SH, Moreau MP, Gupta AR, Thomson SA *et al.*: Multiple recurrent de novo CNVs, including duplications of the 7q11.23 Williams syndrome region, are strongly associated with autism. *Neuron* 2011, **70**(5):863–885.
22. Schaaf CP, Zoghbi HY: Solving the autism puzzle a few pieces at a time. *Neuron* 2011, **70**(5):806–808.
23. Scott LJ, Mohlke KL, Bonnycastle LL, Willer CJ, Li Y, Duren WL, Erdos MR, Stringham HM, Chines PS, Jackson AU *et al.*: A genome-wide association study of type 2 diabetes in Finns detects multiple susceptibility variants. *Science* 2007, **316**(5829):1341–1345.
24. Zeggini E, Weedon MN, Lindgren CM, Frayling TM, Elliott KS, Lango H, Timpson NJ, Perry JR, Rayner NW, Freathy RM *et al.*: Replication of genome-wide association signals in UK samples reveals risk loci for type 2 diabetes. *Science* 2007, **316**(5829):1336–1341.
25. Hakonarson H, Grant SF: GWAS and its impact on elucidating the etiology of diabetes. *Diabetes Metab Res Rev* 2011.
26. Franke A, Balschun T, Karlsen TH, Hedderich J, May S, Lu T, Schuldt D, Nikolaus S, Rosenstiel P, Krawczak M *et al.*: Replication of signals from recent studies of Crohn's disease identifies previously unknown disease loci for ulcerative colitis. *Nat Genet* 2008, **40**(6):713–715.
27. Franke A, McGovern DP, Barrett JC, Wang K, Radford-Smith GL, Ahmad T, Lees CW, Balschun T, Lee J, Roberts R *et al.*: Genome-wide meta-analysis increases to 71 the number of confirmed Crohn's disease susceptibility loci. *Nat Genet* 2010, **42**(12):1118–1125.

28. Haiman CA, Le Marchand L, Yamamato J, Stram DO, Sheng X, Kolonel LN, Wu AH, Reich D, Henderson BE: A common genetic risk factor for colorectal and prostate cancer. *Nat Genet* 2007, **39**(8):954–956.
29. Haiman CA, Patterson N, Freedman ML, Myers SR, Pike MC, Waliszewska A, Neubauer J, Tandon A, Schirmer C, McDonald GJ *et al.*: Multiple regions within 8q24 independently affect risk for prostate cancer. *Nat Genet* 2007, **39**(5):638–644.
30. Gillis RF, Rouleau GA: The ongoing dissection of the genetic architecture of autistic spectrum disorder. *Mol Autism* 2011, **2**(1):12.
31. Levy D, Ronemus M, Yamrom B, Lee YH, Leotta A, Kendall J, Marks S, Lakshmi B, Pai D, Ye K *et al.*: Rare de novo and transmitted copy-number variation in autistic spectrum disorders. *Neuron* 2011, **70**(5):886–897.
32. Scherer SW, Dawson G: Risk factors for autism: translating genomic discoveries into diagnostics. *Hum Genet* 2011, **130**(1):123–148.
33. Bilguvar K, Ozturk AK, Louvi A, Kwan KY, Choi M, Tatli B, Yalnizoglu D, Tuysuz B, Caglayan AO, Gokben S *et al.*: Whole-exome sequencing identifies recessive WDR62 mutations in severe brain malformations. *Nature* 2010, **467**(7312):207–210.
34. Noor A, Whibley A, Marshall CR, Gianakopoulos PJ, Piton A, Carson AR, Orlic-Milacic M, Lionel AC, Sato D, Pinto D *et al.*: Disruption at the PTCHD1 Locus on Xp22.11 in Autism spectrum disorder and intellectual disability. *Sci Transl Med* 2010, **2**(49):49ra68.
35. Shi Y, Li Y, Zhang D, Zhang H, Li Y, Lu F, Liu X, He F, Gong B, Cai L *et al.*: Exome sequencing identifies ZNF644 mutations in high myopia. *PLoS Genet* 2011, **7**(6):e1002084.
36. Losh M, Adolphs R, Poe MD, Couture S, Penn D, Baranek GT, Piven J: Neuropsychological profile of autism and the broad autism phenotype. *Arch Gen Psychiatry* 2009, **66**(5):518–526.
37. Losh M, Childress D, Lam K, Piven J: Defining key features of the broad autism phenotype: a comparison across parents of multiple- and single-incidence autism families. *Am J Med Genet B Neuropsychiatr Genet* 2008, **147B**(4):424–433.
38. Almasy L, Gur RC, Haack K, Cole SA, Calkins ME, Peralta JM, Hare E, Prasad K, Pogue-Geile MF, Nimgaonkar V *et al.*: A genome screen for quantitative trait loci influencing schizophrenia and neurocognitive phenotypes. *Am J Psychiatry* 2008, **165**(9):1185–1192.
39. Egan MF, Goldberg TE, Gscheidle T, Weirich M, Rawlings R, Hyde TM, Bigelow L, Weinberger DR: Relative risk for cognitive impairments in siblings of patients with schizophrenia. *Biol Psychiatry* 2001, **50**(2):98–107.
40. Robinson EB, Koenen KC, McCormick MC, Munir K, Hallett V, Happe F, Plomin R, Ronald A: Evidence that autistic traits show the same etiology in

the general population and at the quantitative extremes (5%, 2.5%, and 1%). *Arch Gen Psychiatry* 2011, **68**(11):1113–1121.

41. Lundstrom S, Chang Z, Rastam M, Gillberg C, Larsson H, Anckarsater H, Lichtenstein P: Autism spectrum disorders and autisticlike traits: similar etiology in the extreme end and the normal variation. *Arch Gen Psychiatry* 2011, **69**(1):46–52.
42. Devlin B, Melhem N, Roeder K: Do common variants play a role in risk for autism? Evidence and theoretical musings. *Brain Res* 2011, **1380**:78–84.
43. Morrison AC, Voorman A, Johnson AD, Liu X, Yu J, Li A, Muzny D, Yu F, Rice K, Zhu C *et al.*: Whole-genome sequence-based analysis of high-density lipoprotein cholesterol. *Nat Genet* 2013, **45**(8):899–901.
44. Kapranov P, St Laurent G, Raz T, Ozsolak F, Reynolds CP, Sorensen PH, Reaman G, Milos P, Arceci RJ, Thompson JF *et al.*: The majority of total nuclear-encoded non-ribosomal RNA in a human cell is 'dark matter' un-annotated RNA. *BMC Biol* 2010, **8**:149.
45. Mattick JS: The genetic signatures of noncoding RNAs. *PLoS Genet* 2009, **5**(4):e1000459.
46. Qureshi IA, Mattick JS, Mehler MF: Long non-coding RNAs in nervous system function and disease. *Brain Res* 2010, **1338**:20–35.
47. Lin M, Pedrosa E, Shah A, Hrabovsky A, Maqbool S, Zheng D, Lachman HM: RNA-Seq of human neurons derived from iPS cells reveals candidate long non-coding RNAs involved in neurogenesis and neuropsychiatric disorders. *PLoS One* 2011, **6**(9):e23356.
48. Jin G, Sun J, Isaacs SD, Wiley KE, Kim ST, Chu LW, Zhang Z, Zhao H, Zheng SL, Isaacs WB *et al.*: Human polymorphisms at long non-coding RNAs (lncRNAs) and association with prostate cancer risk. *Carcinogenesis* 2011, **32**(11):1655–1659.
49. Ripke S, Sanders AR, Kendler KS, Levinson DF, Sklar P, Holmans PA, Lin DY, Duan J, Ophoff RA, Andreassen OA *et al.*: Genome-wide association study identifies five new schizophrenia loci. *Nat Genet* 2011, **43**(10):969–976.
50. St Pourcain B, Wang K, Glessner JT, Golding J, Steer C, Ring SM, Skuse DH, Grant SF, Hakonarson H, Smith GD *et al.*: Association between a high-risk autism locus on 5p14 and social communication spectrum phenotypes in the general population. *Am J Psychiatry* 2010, **167**(11):1364–1372.
51. Kerin T, Ramanathan A, Rivas K, Grepo N, Coetzee GA, Campbell DB: A noncoding RNA antisense to moesin at 5p14.1 in autism. *Sci Transl Med* 2012, **4**(128):128ra40.
52. Lintas C, Sacco R, Persico AM: Genome-wide expression studies in autism spectrum disorder, Rett syndrome, and Down syndrome. *Neurobiol Dis* 2012, **45**(1):57–68.
53. Voineagu I: Gene expression studies in autism: moving from the genome to the transcriptome and beyond. *Neurobiol Dis* 2012, **45**(1):69–75.

54. Ziats MN, Rennert OM: Aberrant expression of long noncoding RNAs in autistic brain. *J Mol Neurosci: MN* 2013, **49**(3):589–593.
55. Carninci P, Kasukawa T, Katayama S, Gough J, Frith MC, Maeda N, Oyama R, Ravasi T, Lenhard B, Wells C *et al.*: The transcriptional landscape of the mammalian genome. *Science* 2005, **309**(5740):1559–1563.
56. Kapranov P, Cheng J, Dike S, Nix DA, Duttagupta R, Willingham AT, Stadler PF, Hertel J, Hackermüller J, Hofacker IL *et al.*: RNA maps reveal new RNA classes and a possible function for pervasive transcription. *Science* 2007, **316**(5830):1484–1488.
57. Mattick JS, Makunin IV: Small regulatory RNAs in mammals. *Hum Mol Genet* 2005, **14 Spec No 1**:R121–132.
58. Jalali S, Jayaraj GG, Scaria V: Integrative transcriptome analysis suggest processing of a subset of long non-coding RNAs to small RNAs. *Biol Direct* 2012, **7**:25.
59. Keniry A, Oxley D, Monnier P, Kyba M, Dandolo L, Smits G, Reik W: The H19 lincRNA is a developmental reservoir of miR-675 that suppresses growth and Igf1r. *Nat Cell Biol* 2012, **14**(7):659–665.
60. Winter J, Jung S, Keller S, Gregory RI, Diederichs S: Many roads to maturity: microRNA biogenesis pathways and their regulation. *Nat Cell Biol* 2009, **11**(3):228–234.
61. Carthew RW, Sontheimer EJ: Origins and Mechanisms of miRNAs and siRNAs. *Cell* 2009, **136**(4):642–655.
62. Matera AG, Terns RM, Terns MP: Non-coding RNAs: lessons from the small nuclear and small nucleolar RNAs. *Nat Rev Mol Cell Biol* 2007, **8**(3):209–220.
63. Luteijn MJ, Ketting RF: PIWI-interacting RNAs: from generation to transgenerational epigenetics. *Nat Rev Genet* 2013.
64. Kim VN, Han J, Siomi MC: Biogenesis of small RNAs in animals. *Nat Rev Mol Cell Biol* 2009, **10**(2):126–139.
65. Lee RC, Feinbaum RL, Ambros V: The C. elegans heterochronic gene lin-4 encodes small RNAs with antisense complementarity to lin-14. *Cell* 1993, **75**(5): 843–854.
66. Berezikov E, Guryev V, van de Belt J, Wienholds E, Plasterk RH, Cuppen E: Phylogenetic shadowing and computational identification of human microRNA genes. *Cell* 2005, **120**(1):21–24.
67. Altuvia Y, Landgraf P, Lithwick G, Elefant N, Pfeffer S, Aravin A, Brownstein MJ, Tuschl T, Margalit H: Clustering and conservation patterns of human microRNAs. *Nucleic Acids Res* 2005, **33**(8):2697–2706.
68. Ibáñez-Ventoso C, Vora M, Driscoll M: Sequence relationships among C. elegans, D. melanogaster and human microRNAs highlight the extensive conservation of microRNAs in biology. *PLoS One* 2008, **3**(7):e2818.
69. Lai EC: Micro RNAs are complementary to 3′ UTR sequence motifs that mediate negative post-transcriptional regulation. *Nat Genet* 2002, **30**(4):363–364.

70. Lytle JR, Yario TA, Steitz JA: Target mRNAs are repressed as efficiently by microRNA-binding sites in the 5′ UTR as in the 3′ UTR. *Proc Natl Acad Sci USA* 2007, **104**(23):9667–9672.
71. Krek A, Grün D, Poy MN, Wolf R, Rosenberg L, Epstein EJ, MacMenamin P, da Piedade I, Gunsalus KC, Stoffel M *et al.*: Combinatorial microRNA target predictions. *Nat Genet* 2005, **37**(5):495–500.
72. Friedman RC, Farh KK, Burge CB, Bartel DP: Most mammalian mRNAs are conserved targets of microRNAs. *Genome Res* 2009, **19**(1):92–105.
73. Sayed D, Abdellatif M: MicroRNAs in development and disease. *Physiol Rev* 2011, **91**(3):827–887.
74. Wienholds E, Plasterk RH: MicroRNA function in animal development. *FEBS Lett* 2005, **579**(26):5911–5922.
75. Alvarez-Garcia I, Miska EA: MicroRNA functions in animal development and human disease. *Development* 2005, **132**(21):4653–4662.
76. Elbashir SM, Martinez J, Patkaniowska A, Lendeckel W, Tuschl T: Functional anatomy of siRNAs for mediating efficient RNAi in Drosophila melanogaster embryo lysate. *EMBO J* 2001, **20**(23):6877–6888.
77. Guo H, Ingolia NT, Weissman JS, Bartel DP: Mammalian microRNAs predominantly act to decrease target mRNA levels. *Nature* 2010, **466**(7308):835–840.
78. Martinez J, Patkaniowska A, Urlaub H, Lührmann R, Tuschl T: Single-stranded antisense siRNAs guide target RNA cleavage in RNAi. *Cell* 2002, **110**(5):563–574.
79. Hutvágner G, Zamore PD: A microRNA in a multiple-turnover RNAi enzyme complex. *Science* 2002, **297**(5589):2056–2060.
80. Wang XH, Aliyari R, Li WX, Li HW, Kim K, Carthew R, Atkinson P, Ding SW: RNA interference directs innate immunity against viruses in adult Drosophila. *Science* 2006, **312**(5772):452–454.
81. Li H, Li WX, Ding SW: Induction and suppression of RNA silencing by an animal virus. *Science* 2002, **296**(5571):1319–1321.
82. Chung WJ, Okamura K, Martin R, Lai EC: Endogenous RNA interference provides a somatic defense against Drosophila transposons. *Curr Biol* 2008, **18**(11):795–802.
83. Babiarz JE, Ruby JG, Wang Y, Bartel DP, Blelloch R: Mouse ES cells express endogenous shRNAs, siRNAs, and other Microprocessor-independent, Dicer-dependent small RNAs. *Genes Dev* 2008, **22**(20):2773–2785.
84. Watanabe T, Totoki Y, Toyoda A, Kaneda M, Kuramochi-Miyagawa S, Obata Y, Chiba H, Kohara Y, Kono T, Nakano T *et al.*: Endogenous siRNAs from naturally formed dsRNAs regulate transcripts in mouse oocytes. *Nature* 2008, **453**(7194):539–543.
85. Tam OH, Aravin AA, Stein P, Girard A, Murchison EP, Cheloufi S, Hodges E, Anger M, Sachidanandam R, Schultz RM *et al.*: Pseudogene-derived small interfering RNAs regulate gene expression in mouse oocytes. *Nature* 2008, **453**(7194):534–538.

86. Song R, Hennig GW, Wu Q, Jose C, Zheng H, Yan W: Male germ cells express abundant endogenous siRNAs. *Proc Natl Acad Sci USA* 2011, **108**(32): 13159–13164.
87. Yang N, Kazazian HH: L1 retrotransposition is suppressed by endogenously encoded small interfering RNAs in human cultured cells. *Nat Struct Mol Biol* 2006, **13**(9):763–771.
88. Xia J, Joyce CE, Bowcock AM, Zhang W: Noncanonical microRNAs and endogenous siRNAs in normal and psoriatic human skin. *Hum Mol Genet* 2013, **22**(4):737–748.
89. Chan WL, Yuo CY, Yang WK, Hung SY, Chang YS, Chiu CC, Yeh KT, Huang HD, Chang JG: Transcribed pseudogene ψPPM1K generates endogenous siRNA to suppress oncogenic cell growth in hepatocellular carcinoma. *Nucleic Acids Res* 2013, **41**(6):3734–3747.
90. Burnett JC, Rossi JJ: RNA-based therapeutics: current progress and future prospects. *Chem Biol* 2012, **19**(1):60–71.
91. Ponting CP, Oliver PL, Reik W: Evolution and functions of long noncoding RNAs. *Cell* 2009, **136**(4):629–641.
92. Nagano T, Fraser P: No-nonsense functions for long noncoding RNAs. *Cell* 2011, **145**(2):178–181.
93. Rinn JL, Chang HY: Genome regulation by long noncoding RNAs. *Annu Rev Biochem* 2012, **81**:145–166.
94. Wang KC, Chang HY: Molecular mechanisms of long noncoding RNAs. *Mol Cell* 2011, **43**(6):904–914.
95. Brockdorff N: Noncoding RNA and Polycomb recruitment. *RNA (New York, NY)* 2013, **19**(4):429–442.
96. Lee JT, Davidow LS, Warshawsky D: Tsix, a gene antisense to Xist at the X-inactivation centre. *Nat Genet* 1999, **21**(4):400–404.
97. Nesterova TB, Senner CE, Schneider J, Alcayna-Stevens T, Tattermusch A, Hemberger M, Brockdorff N: Pluripotency factor binding and Tsix expression act synergistically to repress Xist in undifferentiated embryonic stem cells. *Epigenetics Chromatin* 2011, **4**(1):17.
98. Calin GA, Liu CG, Ferracin M, Hyslop T, Spizzo R, Sevignani C, Fabbri M, Cimmino A, Lee EJ, Wojcik SE *et al.*: Ultraconserved regions encoding ncRNAs are altered in human leukemias and carcinomas. *Cancer Cell* 2007, **12**(3):215–229.
99. Meister G, Landthaler M, Dorsett Y, Tuschl T: Sequence-specific inhibition of microRNA- and siRNA-induced RNA silencing. *RNA (New York, NY)* 2004, **10**(3):544–550.
100. Guo F, Li Y, Liu Y, Wang J, Li Y, Li G: Inhibition of metastasis-associated lung adenocarcinoma transcript 1 in CaSki human cervical cancer cells suppresses cell proliferation and invasion. *Acta Biochim Biophys Sin* 2010, **42**(3):224–229.
101. Kogo R, Shimamura T, Mimori K, Kawahara K, Imoto S, Sudo T, Tanaka F, Shibata K, Suzuki A, Komune S *et al.*: Long noncoding RNA HOTAIR

regulates polycomb-dependent chromatin modification and is associated with poor prognosis in colorectal cancers. *Cancer Res* 2011, **71**(20):6320–6326.

102. Nakagawa T, Endo H, Yokoyama M, Abe J, Tamai K, Tanaka N, Sato I, Takahashi S, Kondo T, Satoh K: Large noncoding RNA HOTAIR enhances aggressive biological behavior and is associated with short disease-free survival in human non-small cell lung cancer. *Biochem Biophys Res Commun* 2013.
103. Gupta RA, Shah N, Wang KC, Kim J, Horlings HM, Wong DJ, Tsai MC, Hung T, Argani P, Rinn JL *et al.*: Long non-coding RNA HOTAIR reprograms chromatin state to promote cancer metastasis. *Nature* 2010, **464**(7291):1071–1076.
104. Poliseno L, Salmena L, Zhang J, Carver B, Haveman WJ, Pandolfi PP: A coding-independent function of gene and pseudogene mRNAs regulates tumour biology. *Nature* 2010, **465**(7301):1033–1038.
105. Panzitt K, Tschernatsch MM, Guelly C, Moustafa T, Stradner M, Strohmaier HM, Buck CR, Denk H, Schroeder R, Trauner M *et al.*: Characterization of HULC, a novel gene with striking up-regulation in hepatocellular carcinoma, as noncoding RNA. *Gastroenterology* 2007, **132**(1):330–342.
106. Daughters RS, Tuttle DL, Gao W, Ikeda Y, Moseley ML, Ebner TJ, Swanson MS, Ranum LP: RNA gain-of-function in spinocerebellar ataxia type 8. *PLoS Genet* 2009, **5**(8):e1000600.
107. Evin G, Hince C: BACE1 as a therapeutic target in alzheimer's disease: rationale and current status. *Drugs Aging* 2013.
108. Faghihi MA, Modarresi F, Khalil AM, Wood DE, Sahagan BG, Morgan TE, Finch CE, St Laurent G, 3rd, Kenny PJ, Wahlestedt C: Expression of a noncoding RNA is elevated in Alzheimer's disease and drives rapid feed-forward regulation of beta-secretase. *Nat Med* 2008, **14**(7):723–730.
109. Faghihi MA, Zhang M, Huang J, Modarresi F, Van der Brug MP, Nalls MA, Cookson MR, St-Laurent G, 3rd, Wahlestedt C: Evidence for natural antisense transcript-mediated inhibition of microRNA function. *Genome Biol* 2010, **11**(5):R56.
110. Dislich B, Lichtenthaler SF: The membrane-bound aspartyl protease BACE1: molecular and functional properties in alzheimer's disease and beyond. *Front Physiol* 2012, **3**:8.
111. Modarresi F, Faghihi MA, Patel NS, Sahagan BG, Wahlestedt C, Lopez-Toledano MA: Knockdown of BACE1-AS Nonprotein-Coding Transcript Modulates Beta-Amyloid-Related Hippocampal Neurogenesis. *Int J Alzheimers Dis* 2011, **2011**:929042.
112. Hagerman R, Hoem G, Hagerman P: Fragile X and autism: intertwined at the molecular level leading to targeted treatments. *Mol Autism* 2010, **1**(1):12.
113. Khalil AM, Faghihi MA, Modarresi F, Brothers SP, Wahlestedt C: A novel RNA transcript with antiapoptotic function is silenced in fragile X syndrome. *PloS One* 2008, **3**(1):e1486.

114. Sharma S, Findlay GM, Bandukwala HS, Oberdoerffer S, Baust B, Li Z, Schmidt V, Hogan PG, Sacks DB, Rao A: Dephosphorylation of the nuclear factor of activated T cells (NFAT) transcription factor is regulated by an RNA-protein scaffold complex. *Proc Natl Acad Sci USA* 2011, **108**(28):11381–11386.
115. Jiang J, Jing Y, Cost GJ, Chiang JC, Kolpa HJ, Cotton AM, Carone DM, Carone BR, Shivak DA, Guschin DY *et al.*: Translating dosage compensation to trisomy 21. *Nature* 2013.
116. Hu VW: The expanding genomic landscape of autism: discovering the 'forest' beyond the 'trees'. *Future Neurol* 2013, **8**(1):29–42.
117. Lanford RE, Hildebrandt-Eriksen ES, Petri A, Persson R, Lindow M, Munk ME, Kauppinen S, Orum H: Therapeutic silencing of microRNA-122 in primates with chronic hepatitis C virus infection. *Science* 2010, **327**(5962):198–201.
118. Castanotto D, Rossi JJ: The promises and pitfalls of RNA-interference-based therapeutics. *Nature* 2009, **457**(7228):426–433.
119. Dejneka NS, Wan S, Bond OS, Kornbrust DJ, Reich SJ: Ocular biodistribution of bevasiranib following a single intravitreal injection to rabbit eyes. *Mol Vis* 2008, **14**:997–1005.
120. Cho WG, Albuquerque RJ, Kleinman ME, Tarallo V, Greco A, Nozaki M, Green MG, Baffi JZ, Ambati BK, De Falco M *et al.*: Small interfering RNA-induced TLR3 activation inhibits blood and lymphatic vessel growth. *Proc Natl Acad Sci USA* 2009, **106**(17):7137–7142.
121. Kleinman ME, Yamada K, Takeda A, Chandrasekaran V, Nozaki M, Baffi JZ, Albuquerque RJ, Yamasaki S, Itaya M, Pan Y *et al.*: Sequence- and target-independent angiogenesis suppression by siRNA via TLR3. *Nature* 2008, **452**(7187):591–597.
122. Samuel-Abraham S, Leonard JN: Staying on message: design principles for controlling nonspecific responses to siRNA. *FEBS J* 2010, **277**(23):4828–4836.
123. Guo Y, Srinivasula SM, Druilhe A, Fernandes-Alnemri T, Alnemri ES: Caspase-2 induces apoptosis by releasing proapoptotic proteins from mitochondria. *J Biol Chem* 2002, **277**(16):13430–13437.
124. Alvarez R, Elbashir S, Borland T, Toudjarska I, Hadwiger P, John M, Roehl I, Morskaya SS, Martinello R, Kahn J *et al.*: RNA interference-mediated silencing of the respiratory syncytial virus nucleocapsid defines a potent antiviral strategy. *Antimicrob Agents Chemother* 2009, **53**(9):3952–3962.
125. Bouchard PR, Hutabarat RM, Thompson KM: Discovery and development of therapeutic aptamers. *Annu Rev Pharmacol Toxicol* 2010, **50**:237–257.
126. Keefe AD, Pai S, Ellington A: Aptamers as therapeutics. *Nat Rev Drug Discov* 2010, **9**(7):537–550.
127. Sanghvi YS: A status update of modified oligonucleotides for chemotherapeutics applications. *Curr Protoc Nucleic Acid Chem/Edited by Serge L Beaucage [et al.]* 2011, **Chapter 4**:Unit 4. 1. 1–22.

128. Thiel KW, Giangrande PH: Therapeutic applications of DNA and RNA aptamers. *Oligonucleotides* 2009, **19**(3):209–222.
129. Amado RG, Mitsuyasu RT, Rosenblatt JD, Ngok FK, Bakker A, Cole S, Chorn N, Lin LS, Bristol G, Boyd MP *et al.*: Anti-human immunodeficiency virus hematopoietic progenitor cell-delivered ribozyme in a phase I study: myeloid and lymphoid reconstitution in human immunodeficiency virus type-1-infected patients. *Human Gene Ther* 2004, **15**(3):251–262.
130. Macpherson JL, Boyd MP, Arndt AJ, Todd AV, Fanning GC, Ely JA, Elliott F, Knop A, Raponi M, Murray J *et al.*: Long-term survival and concomitant gene expression of ribozyme-transduced CD4+ T-lymphocytes in HIV-infected patients. *J Gene Med* 2005, **7**(5):552–564.
131. Michienzi A, Castanotto D, Lee N, Li S, Zaia JA, Rossi JJ: RNA-mediated inhibition of HIV in a gene therapy setting. *Ann NY Acad Sci* 2003, **1002**:63–71.
132. Wong-Staal F, Poeschla EM, Looney DJ: A controlled, Phase 1 clinical trial to evaluate the safety and effects in HIV-1 infected humans of autologous lymphocytes transduced with a ribozyme that cleaves HIV-1 RNA. *Human Gene Ther* 1998, **9**(16):2407–2425.
133. DiGiusto DL, Krishnan A, Li L, Li H, Li S, Rao A, Mi S, Yam P, Stinson S, Kalos M *et al.*: RNA-based gene therapy for HIV with lentiviral vector-modified CD34(+) cells in patients undergoing transplantation for AIDS-related lymphoma. *Sci Transl Med* 2010, **2**(36):36ra43.
134. Li MJ, Kim J, Li S, Zaia J, Yee JK, Anderson J, Akkina R, Rossi JJ: Long-term inhibition of HIV-1 infection in primary hematopoietic cells by lentiviral vector delivery of a triple combination of anti-HIV shRNA, anti-CCR5 ribozyme, and a nucleolar-localizing TAR decoy. *Mol Ther J Am Soc Gene Ther* 2005, **12**(5):900–909.

Chapter 10

A-to-I RNA Editing in Autism Spectrum Disorder

Alal Eran*, Isaac S. Kohane and Louis M. Kunkel

Abstract

Significant recent progress has been made in deciphering the genetic basis of autism spectrum disorder (ASD), revealing a key role for synaptic gene mutations. However, relevant mechanisms for gene-environment interactions and their contribution to the ASD phenotype remain largely unknown. RNA is emerging as a central substrate for encoding environmental exposures at the molecular level, especially in the nervous system. One effective mechanism for regulating gene function and dosage in response to the changing environment is RNA editing. Recently, we showed that adenosine-to-inosine (A-to-I) RNA editing of synaptic genes is altered in postmortem brains of individuals with ASD, suggesting that this epigenetic mechanism could play a role in ASD. Here we review our current understanding of RNA editing at the molecular level, describe the behavioral phenotypes associated with altered A-to-I editing in animal models, discuss recent discoveries in human A-to-I editing, and speculate about the diagnostic and prognostic potential of A-to-I editing changes in ASD.

Keywords: RNA editing; A-to-I; Autism spectrum disorder; Epigenetics; Gene–environment interactions; Neurodevelopment; Synaptic homeostasis.

---

*Correspondence to: Center for Biomedical Informatics, Harvard Medical School, Boston, MA 02115. Emails: Alal Eran: alal@alum.mit.edu, Isaac S. Kohane: isaac_kohane@harvard.edu, Louis M. Kunkel: kunkel@enders.tch.harvad.edu.

Introduction

Four decades of twin and family studies provide substantial evidence that ASD is one of the most heritable complex disorders.[1–3] Recent advances in autism genetics, brought about with the genomics technology revolution, have revealed its extreme locus heterogeneity.[4–9] Numerous genes and loci have been implicated in ASD, each altered in a small proportion of families, and causing disease in combination with other genomic and environmental factors.[4–9] This has accelerated a growing realization that ASD is comprised of a multitude of etiologies with incompletely overlapping symptomatology and clinical course, which likely converge at the mechanistic or pathway levels.[10–14]

One functional convergence among the various *bona fide* ASD genes is that many are involved in synapse formation and synaptic homeostasis.[13–19] ASD-associated *de novo* mutations,[6,19–21] copy-number variants,[15,22–24] common variants,[25,26] and rare single nucleotide variants[4,7,9,27] tend to commonly predispose to synaptic dysfunction in many cases. Moreover, it seems that there exists a homeostatic ("normal") range of synaptic function, and deviations to either extreme may lead to the common ASD phenotype.[28] For example, both exaggerated and reduced synaptic protein synthesis were shown to cause ASD in animal models.[16] Similarly, both increased and decreased dendritic densities have been associated with syndromic forms of ASD.[29,30] Likewise, mice lacking certain ASD genes show increased excitatory postsynaptic currents while mice missing other autism genes show reduced excitatory postsynaptic currents.[31,32] Therefore, it has been proposed that dysfunctional neuronal homeostasis and altered homeostatic compensation may lead to neuronal circuitries with weakened synaptic flexibility, and ultimately to shared ASD phenotypes.[13,28,33,34]

Both synapse formation and homeostasis are shaped by the interactions of genes with the environment. Although much progress has been made in identifying ASD susceptibility genes, relevant mechanisms for gene-environment interactions remain largely elusive. For example, it is still unknown to what extent mutations in synaptic genes predispose to ASD and in which context. Additionally, environmental factors that influence synapse formation and homeostasis, which could also modulate their impairments in ASD, remain to be discovered. Therefore, identifying mechanisms of gene-environment interactions and their contribution to the observed synaptic dysfunction is needed for both accurate risk assessment and effective intervention in ASD.[28,34,35] A-to-I RNA editing is potentially one such mechanism, linking environmental stimuli with synaptic transmission. We have recently reported that similar to the synaptic phenotypes described above, A-to-I editing also shows both exaggerated and reduced patterns in individuals with ASD. Taken together, understanding the role of A-to-I RNA editing in ASD could improve early diagnosis and ultimately lead to the identification of critical targets for intervention and prevention.

A-to-I RNA Editing Enables Fine-tuned Response to Environmental Cues

A-to-I RNA editing is a context-dependent transcriptional modification, which modulates gene function and dosage in response to the changing environment, mostly in the brain.[36–39] **A**denosine **dea**minase acting on **R**NA (ADAR) enzymes bind double-stranded RNA structures and catalyze the site-specific deamination of adenosine to inosine with precise regional and developmental specificity.[36–39] Editing is enriched in genes that mediate synaptic transmission, including receptors, ion channels, and dendritic microRNA.[36,39,40] *In vitro*, editing levels were shown to dynamically respond to neuronal activity and neuromodulating agents.[41–43] *In vivo*, A-to-I editing changes during neurodevelopment[44–46] and in response to specific environmental exposures,[47,48] learning tasks,[49,50] and pharmacological treatments.[51,52] Canonical signaling pathways acting on the editing enzymes further link A-to-I editing and environmental cues: ADARB1 requires inositol hexakisphosphate to function,[53] and ADAR expression is interferon-inducible.[54] For these reasons, it has been proposed that A-to-I editing constitutes one of the molecular mechanisms connecting environmental exposures and neurophysiological properties.[36,55]

At the molecular level, inosine base pairs with cytidine, and is thus recognized by the cellular machinery as guanosine.[56] Therefore, the impact of an A-to-I edit is equivalent to an A to G mutation affecting a dynamically changing fraction of transcripts. When this A>G change occurs in a coding region it may recode the amino acid sequence of a protein, directly altering protein function.[57–61] A>G changes at splice sites may create or destroy a splice site.[62] A>G changes in structured noncoding RNAs may alter the RNA secondary structure, interfering with typical structure-based biogenesis.[63,64]

Although less than 1% of the human genome codes for proteins, its majority is differentially transcribed throughout development, and thus considered functional.[65] Therefore, most editing events, including those that remain to be discovered, are expected to directly impact non-coding RNA. Nonetheless, most of our knowledge about A-to-I editing stems from a select set of well-characterized recoding events, in which RNA editing changes the amino acid sequence of protein-coding transcripts.[57–61] For example, the G-protein coupled serotonin receptor HTR2C undergoes A-to-I editing in five sites, which dramatically alters its G-protein coupling activity, and therefore the relationship between serotonin levels and postsynaptic signal transduction.[61,66] This editing is regulated by learning, stress, and chronic treatment with antidepressants.[48–50] Another example is the neurodevelopmentally-regulated editing of the AMPA receptors GRIA2, GRIA3 and GRIA4,[45] where arginine to glycine (R/G) recoding of their ligand binding domains leads to faster desensitization recovery.[57] Additional neurodevelopmentally-regulated recoding of glutamine to arginine (Q/R) in the transmembrane domains of the kainate receptors GRIK1 and GRIK2 reduces their calcium permeability.[58] These editing events are dynamically regulated by neuronal activity and neuromodulating agents.[41–43] Since 0 to 100% of mRNA molecules can be edited at any given point,[56] Q/R and R/G editing of ionotropic glutamate receptors provides an efficient means for fine-tuning the glutamatergic synapse in response to the changing environment (Figure 1).

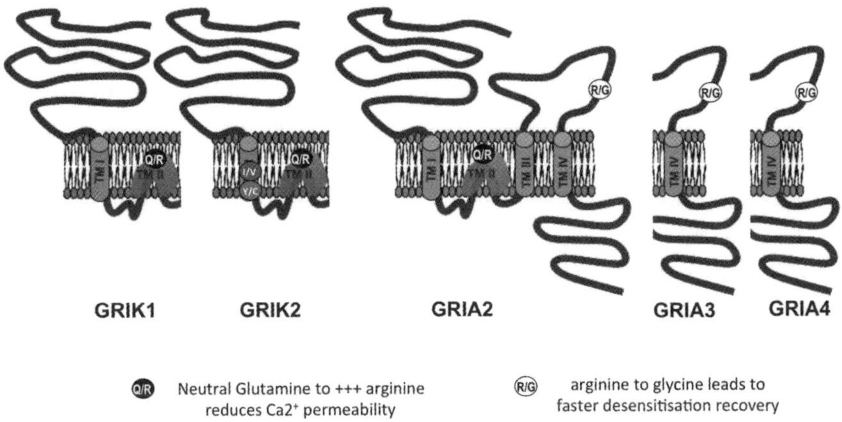

Figure 1. Ionotropic glutamate receptors regulated by A-to-I RNA recoding. Three AMPA receptors and two kainate glutamate receptors are subject to A-to-I RNA editing. Q/R editing of the transmembrane domains of GRIK1, GRIK2, and GRIA2 dramatically alters their calcium permeability,[58,67] while R/G recoding of the intracellular domain of GRIA2, GRIA3, and GRIA4 tunes their kinetics.[57] A-to-I editing of dendritic RNA granules can therefore provide an incredibly efficient means for rapid fine-tuning of synaptic strength and duration in response to environmental cues, contributing to synaptic plasticity.

A-to-I RNA Editing Modulates Complex Behavior in Model Organisms

It was not until recently that the impact of A-to-I editing on animal behavior began to be understood.[55,66,68–75] The extreme phenotypes associated with traditional knockout of the editing enzymes, including seizures, early mortality, and lack of basic motor skills,[76,77] required the development of sophisticated engineering approaches for understanding the relationships between RNA editing and complex behaviors. Several recent breakthrough studies have paved the way for understanding the relationships between environmental exposures, RNA editing, and complex behaviors.[48,49,55,66,68–74,78,79] Studies in flies found that engineering the overall drosophila editing levels results in abnormal circadian patterns, male courtship, and mating behaviors.[55,75] In mice, engineering the editing levels of the serotonin receptor Htr2c leads to increased aggressive-impulsive behaviors in response to a social interaction test,[68] anxiety-like behaviors in the elevated-plus maze paradigm,[66] and a stronger freezing reaction to an innate fear stimulus.[68] These behaviors are similar to those observed in mouse models of *bona fide* autism genes.[31,80] Furthermore, a mouse model of the most common chromosomal alteration in ASD (15q11-13 duplication) shows HTR2C editing alterations, associated with autistic-like behaviors.[81] Perturbing the overall editing levels in mice leads to increased rearing behavior in the open field test, as well as to sensorimotor deficits, including decreased hearing abilities.[70] Conversely, subjecting mice to standard learning paradigms leads to changes in editing levels,[49,68] as does acute stress,[48] and treatment with neuromodulating agents.[48,50,82] Taken together, because different environmental stimuli alter editing levels, and altered editing levels lead to altered complex behavior, it has been postulated that A-to-I RNA editing is one molecular mechanism connecting environmental inputs with behavioral outputs.[36,55]

The Genomics Technology Revolution is Enabling Rapid Discoveries of A-to-I RNA Editing in Humans

Increased editing in humans[83] as compared to other animals,[84] including nonhuman primates,[85] has been proposed to generate the molecular complexity needed for higher-order human cognition.[36] However, little is currently understood about human A-to-I RNA editing, despite its essential roles. The genomics technology revolution has recently enabled the first large-scale editing surveys across various human tissues and conditions.[40,83,86–93] Ultradeep[89,90] or pooled[87,92] RNA-seq can now be effectively used to accurately quantify RNA editing in the presence[89,90] or absence[87,92] of genomic DNA data, via the identification of A/G mixtures on cDNA. Such approaches are becoming increasingly used to clarify the landscape of typical A-to-I editing in humans and its alterations in various forms of abnormal neurodevelopment, including glioblastoma,[94,95] schizophrenia,[89,91] and ASD.[90]

A-to-I RNA Editing in Postmortem Brains of Children with ASD

Several lines of evidence supported the examination of A-to-I editing in ASD. First, model organisms with altered A-to-I editing exhibit maladaptive behaviors characteristic of the human ASD,[55,66,73,75] sometimes with seizures[55] or Prader-Willi-like symptoms,[71] both of which are typically detected in 25% of children with ASD.[96,97] Second, altered editing of the serotonin receptor HTR2C has been detected in a mouse model of autism[81] and in disorders that aggregate in families with ASD,[98] such as schizophrenia and major depression.[89,91] Third, a fly model of Fragile X syndrome, the most common single gene cause of ASD,[99] was recently shown to exhibit significant editing alterations, via a direct interaction between the Fragile X gene dFMR1 and the editing enzyme dADAR.[100] Finally, independent genomic studies have implicated variants in synaptic genes, the most edited type of genes,[40] as a recurring theme in ASD.[10,13–15]

For an initial examination of A-to-I RNA editing in ASD, we recently profiled neurodevelopmentally-regulated editing of synaptic genes in 25 postmortem cerebella from children with ASD and matched neurotypical controls.[90] Using ultra-deep sequencing of RNA and matched genotyping of DNA, we detected A/G mixtures on cDNA that map to homozygous genomic A/A, indicative of A-to-I RNA editing. Editing was measured in postmortem cerebellum, which is one of several brain regions implicated in ASD, by both imaging and autopsy studies.[101] This approach yielded a high confidence dataset of synaptic A-to-I RNA editing from individuals with ASD and matched controls.

Unexpectedly broad distributions of editing levels were found across carefully-matched individuals with ASD and neurotypical controls in all sites previously shown to be neurodevelopmentally regulated in mice.[45] In 20 of 25 examined sites, individuals with ASD were at the extreme of the spectrum of editing seen for that site (Figure 2). Having extreme synaptic editing levels is highly informative of ASD, with a positive predictive value of 78% (i.e. $p(ASD \mid |Z_{editing}| > 2) = 0.78$).

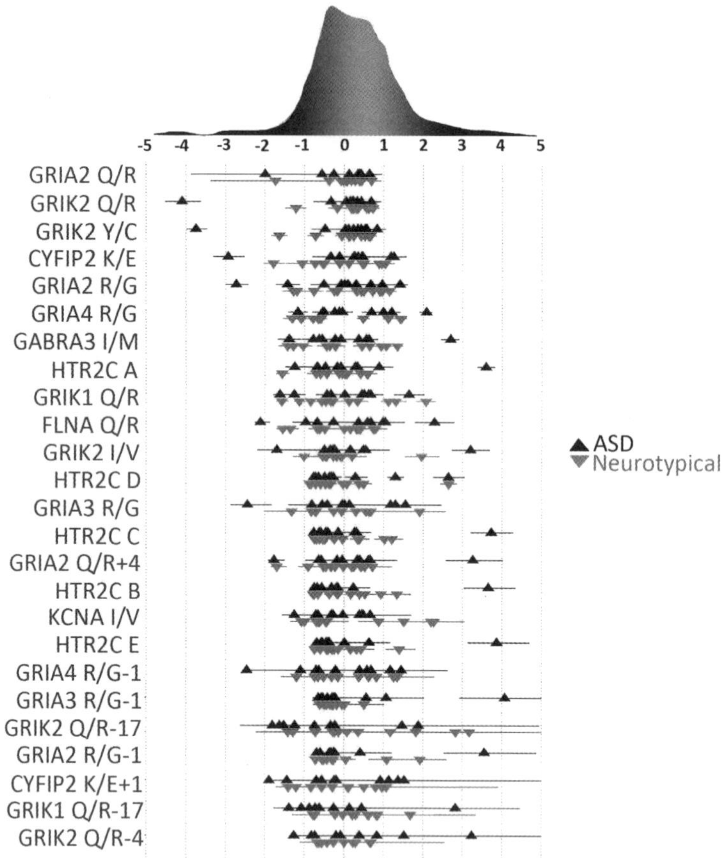

Figure 2. Differences in the landscapes of synaptic editing in individuals with ASD (black) and neurotypical individuals (gray). Shown are the individual standardized deviations from the mean editing level (Z scores) across all sites, demonstrating that the extremes of the population editing distributions tend to be individuals with ASD. Horizontal lines denote the standardized 95% confidence interval of an individual's posterior editing distribution. On top is the summary distribution of all individual Z-scores, showing that extreme editing is specific to individuals with ASD. The positive predictive value of being at the extreme $p(ASD \mid |Z|>2)$ is 0.78.

Outliers at different sites were different individuals with ASD, and editing of more than three standard deviations away from the mean was specific to individuals with ASD. The overall editing variance in ASD was more than two-fold that of neurotypical individuals (Brown-Forsythe p=0.005).

Most of the recoding sites shown in Figure 2 are predominantly edited by ADARB1. To explore the regulatory basis for the observed editing differences, *ADARB1* expression and relative isoform usage were examined in the same cohort. In humans, a dysfunctional *ADARB1* isoform (ADAR2g, NR_027672), resulting from alternative skipping of the exon harboring two double stranded RNA binding domains (dsRBDs), has been found to constitute about 20% of adult cerebellar *ADARB1* mRNA.[102] As such, the presence of this dsRBDs-encoding exon was assayed by both quantitative and semi-quantitative RT-PCR. While individuals with ASD and neurotypical individuals showed similar levels of overall *ADARB1* expression, the relative usage of the inactive *ADARB1* isoform was significantly higher in individuals with ASD (Figure 3, p=0.005, Kolmogorov-Smirnov (KS) test). The relative frequency of the dysfunctional isoform correlated with overall editing levels (Pearson's p=0.03), as measured by the sum of editing Z scores across all sites. Of note, ADARB1 was recently shown to be a direct target of the retinoic acid related orphan receptor alpha (RORA),[103] a sex hormone responsive autism candidate

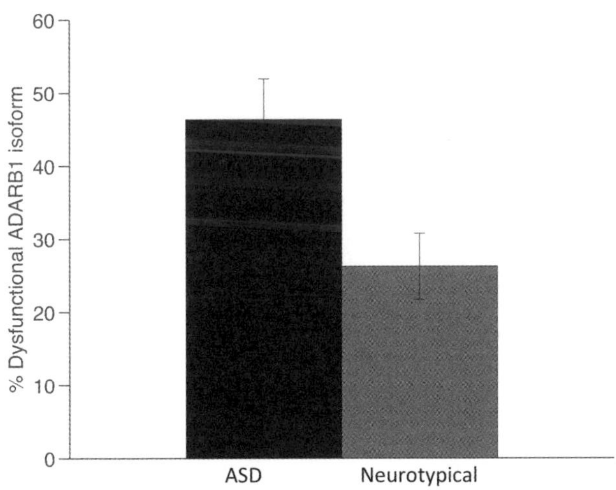

Figure 3. Differences in the relative usage of *ADARB1* inactive form between individuals with ASD and neurotypical individuals (p=0.005, KS test). This isoform lacks the enzyme's double stranded RNA binding domains and is untranslated.[102] ADARB1 is the predominant editing enzyme of the sites shown in Figure 2.

gene,[104] which is reduced in both brain[105] and blood[106] of individuals with ASD.

Future Directions

The recent findings supporting a role for A-to-I RNA editing in ASD raise several key questions and opportunities for exciting future discoveries. First, genome-wide editing patterns in ASD remain to be identified and compared to those of typically developing children. More individuals and more brain regions need to be examined to fully understand the scope and scale of the editing alterations detected in our initial study. The increasing availability of high-quality postmortem brain tissue from individuals with ASD[107] and ongoing large-scale human brain sequencing projects[108,109] should enable a systematic examination of A-to-I editing across populations, individuals, brain regions, cell types, and specific molecular etiologies of ASD.

Second, although A-to-I editing is a brain-centric phenomenon, it may leave traces in peripheral blood, which, if detected, could be highly informative for early diagnosis of ASD. Recent studies suggest that peripheral blood RNA may be informative for understanding nervous system alterations in ASD, with reproducible differential expression at the single gene and pathway levels,[106,10–112] and a blood expression-based diagnostic test in clinical trial.[113] Future work should examine the potential diagnostic utility of detecting RNA editing in blood. One key challenge to overcome is that due to the dynamic range of RNA, traditional RNA-seq cannot accurately quantify the editing levels of rare transcripts in peripheral blood. Therefore, tailored editing detection methods should be employed for blood editotyping, such as the newly developed microfluidics-based multiplex PCR and deep sequencing (mmPCR-Seq[114]).

A final longer-term future direction is based on the anticipated creation of individualized cellular models of ASD. Although the directed differentiation of patient-derived induced pluripotent stem cells (iPSCs) is still a challenge,[115] we anticipate the standardized production of iPSC-derived neurons in the foreseeable future. If and when this is achieved, the relationships between an individual's genomic predisposition to ASD, A-to-I RNA editing, specific environmental stimuli, synaptogenesis, and electrophysiological homeostasis could be explored for the first time. Such a cellular system would also enable personalized *ex vivo* testing of the impact of potential interventions on the complex homeostatic maintenance of affected neurons, ultimately facilitating well-informed treatment decisions in ASD.

Executive Summary

- Adenosine-to-inosine (A-to-I) RNA editing changes the sequence of neuronal transcripts in response to environmental stimuli, fine-tuning gene function and dosage.
- Synaptic function, a major emerging etiology in ASD, is effectively modulated by A-to-I RNA editing, *in vitro* and *in vivo*.
- The behavior of animals with engineered A-to-I editing levels recapitulates several aspects of the human ASD.
- The fragile X mental retardation protein (FMRP), whose loss of function is the most common cause of ASD, directly binds the editing machinery. Flies with no FMRP exhibit significant editing alterations.
- The genomics technology revolution is enabling rapid discoveries of human RNA editing in typical development and disease. Altered editing has been detected in glioblastoma, schizophrenia, and ASD.
- Postmortem brains from individuals with ASD consistently show extreme editing levels of candidate synaptic genes as compared to matched neurotypical controls, suggesting that A-to-I editing may play a role in ASD.
- An inactive form of the editing enzyme ADARB1 is more commonly found in autistic cerebella, and its relative abundance correlates with the overall editing levels.
- Future editing surveys of peripheral blood and patient-derived cellular models of ASD would reveal the diagnostic and prognostic information content of A-to-I editing levels.

Acknowledgements

Our work on RNA editing in ASD has been generously supported by the Nancy Lurie Marks Family Foundation.

References

1. Folstein S, Rutter M: Infantile autism: a genetic study of 21 twin pairs. *J Child Psychol Psychiatry* 1977, **18**(4):297–321.
2. Steffenburg S, Gillberg C, Hellgren L, Andersson L, Gillberg IC, Jakobsson G, Bohman M: A twin study of autism in Denmark, Finland, Iceland, Norway and Sweden. *J Child Psychol Psychiatry* 1989, **30**(3):405–416.
3. Bailey A, Le Couteur A, Gottesman I, Bolton P, Simonoff E, Yuzda E, Rutter M: Autism as a strongly genetic disorder: evidence from a British twin study. *Psychol Med* 1995, **25**(1):63–77.
4. Chahrour MH, Yu TW, Lim ET, Ataman B, Coulter ME, Hill RS, Stevens CR, Schubert CR, Collaboration AAS, Greenberg ME *et al.*: Whole-exome sequencing and homozygosity analysis implicate depolarization-regulated neuronal genes in autism. *PLoS Genet* 2012, **8**(4):e1002635.
5. Marchani EE, Chapman NH, Cheung CY, Ankenman K, Stanaway IB, Coon HH, Nickerson D, Bernier R, Brkanac Z, Wijsman EM: Identification of rare variants from exome sequence in a large pedigree with autism. *Hum Hered* 2012, **74**(3–4):153–164.
6. Neale BM, Kou Y, Liu L, Ma'ayan A, Samocha KE, Sabo A, Lin CF, Stevens C, Wang LS, Makarov V *et al.*: Patterns and rates of exonic de novo mutations in autism spectrum disorders. *Nature* 2012, **485**(7397):242–245.
7. Lim ET, Raychaudhuri S, Sanders SJ, Stevens C, Sabo A, MacArthur DG, Neale BM, Kirby A, Ruderfer DM, Fromer M *et al.*: Rare complete knockouts in humans: population distribution and significant role in autism spectrum disorders. *Neuron* 2013, **77**(2):235–242.
8. Liu L, Sabo A, Neale BM, Nagaswamy U, Stevens C, Lim E, Bodea CA, Muzny D, Reid JG, Banks E *et al.*: Analysis of rare, exonic variation amongst subjects with autism spectrum disorders and population controls. *PLoS Genet* 2013, **9**(4):e1003443.
9. Yu TW, Chahrour MH, Coulter ME, Jiralerspong S, Okamura-Ikeda K, Ataman B, Schmitz-Abe K, Harmin DA, Adli M, Malik AN *et al.*: Using whole-exome sequencing to identify inherited causes of autism. *Neuron* 2013, **77**(2):259–273.
10. Voineagu I, Wang X, Johnston P, Lowe JK, Tian Y, Horvath S, Mill J, Cantor RM, Blencowe BJ, Geschwind DH: Transcriptomic analysis of autistic brain reveals convergent molecular pathology. *Nature* 2011, **474**(7351):380–384.

11. State MW, Levitt P: The conundrums of understanding genetic risks for autism spectrum disorders. *Nat Neurosci* 2011, **14**(12):1499–1506.
12. Pinto D, Pagnamenta AT, Klei L, Anney R, Merico D, Regan R, Conroy J, Magalhaes TR, Correia C, Abrahams BS *et al.*: Functional impact of global rare copy number variation in autism spectrum disorders. *Nature* 2010, **466**(7304):368–372.
13. Toro R, Konyukh M, Delorme R, Leblond C, Chaste P, Fauchereau F, Coleman M, Leboyer M, Gillberg C, Bourgeron T: Key role for gene dosage and synaptic homeostasis in autism spectrum disorders. *Trends Genet* 2010, **26**(8):363–372.
14. Bourgeron T: A synaptic trek to autism. *Curr Opin Neurobiol* 2009, **19**(2):231–234.
15. Gilman SR, Iossifov I, Levy D, Ronemus M, Wigler M, Vitkup D: Rare de novo variants associated with autism implicate a large functional network of genes involved in formation and function of synapses. *Neuron* 2011, **70**(5):898–907.
16. Auerbach BD, Osterweil EK, Bear MF: Mutations causing syndromic autism define an axis of synaptic pathophysiology. *Nature* 2011, **480**(7375):63–68.
17. Bucan M, Abrahams BS, Wang K, Glessner JT, Herman EI, Sonnenblick LI, Alvarez Retuerto AI, Imielinski M, Hadley D, Bradfield JP *et al.*: Genome-wide analyses of exonic copy number variants in a family-based study point to novel autism susceptibility genes. *PLoS Genet* 2009, **5**(6):e1000536.
18. Gai X, Xie HM, Perin JC, Takahashi N, Murphy K, Wenocur AS, D'Arcy M, O'Hara RJ, Goldmuntz E, Grice DE *et al.*: Rare structural variation of synapse and neurotransmission genes in autism. *Mol Psychiatry* 2012, **17**(4):402–411.
19. Iossifov I, Ronemus M, Levy D, Wang Z, Hakker I, Rosenbaum J, Yamrom B, Lee YH, Narzisi G, Leotta A *et al.*: De novo gene disruptions in children on the autistic spectrum. *Neuron* 2012, **74**(2):285–299.
20. Sanders SJ, Murtha MT, Gupta AR, Murdoch JD, Raubeson MJ, Willsey AJ, Ercan-Sencicek AG, DiLullo NM, Parikshak NN, Stein JL *et al.*: De novo mutations revealed by whole-exome sequencing are strongly associated with autism. *Nature* 2012, **485**(7397):237–241.
21. O'Roak BJ, Vives L, Girirajan S, Karakoc E, Krumm N, Coe BP, Levy R, Ko A, Lee C, Smith JD *et al.*: Sporadic autism exomes reveal a highly interconnected protein network of de novo mutations. *Nature* 2012, **485**(7397):246–250.
22. Levy D, Ronemus M, Yamrom B, Lee YH, Leotta A, Kendall J, Marks S, Lakshmi B, Pai D, Ye K *et al.*: Rare de novo and transmitted copy-number variation in autistic spectrum disorders. *Neuron* 2011, **70**(5):886–897.
23. Sanders SJ, Ercan-Sencicek AG, Hus V, Luo R, Murtha MT, Moreno-De-Luca D, Chu SH, Moreau MP, Gupta AR, Thomson SA *et al.*: Multiple recurrent de novo CNVs, including duplications of the 7q11.23 Williams syndrome region, are strongly associated with autism. *Neuron* 2011, **70**(5):863–885.

24. Sakai Y, Shaw CA, Dawson BC, Dugas DV, Al-Mohtaseb Z, Hill DE, Zoghbi HY: Protein interactome reveals converging molecular pathways among autism disorders. *Sci Transl Med* 2011, **3**(86):86ra49.
25. Weiss LA, Arking DE, Gene Discovery Project of Johns H, the Autism C, Daly MJ, Chakravarti A: A genome-wide linkage and association scan reveals novel loci for autism. *Nature* 2009, **461**(7265):802–808.
26. Wang K, Zhang H, Ma D, Bucan M, Glessner JT, Abrahams BS, Salyakina D, Imielinski M, Bradfield JP, Sleiman PM *et al.*: Common genetic variants on 5p14.1 associate with autism spectrum disorders. *Nature* 2009, **459**(7246):528–533.
27. O'Roak BJ, Vives L, Fu W, Egertson JD, Stanaway IB, Phelps IG, Carvill G, Kumar A, Lee C, Ankenman K *et al.*: Multiplex targeted sequencing identifies recurrently mutated genes in autism spectrum disorders. *Science* 2012, **338**(6114):1619–1622.
28. Delorme R, Ey E, Toro R, Leboyer M, Gillberg C, Bourgeron T: Progress toward treatments for synaptic defects in autism. *Nat Med* 2013, **19**(6): 685–694.
29. Chapleau CA, Calfa GD, Lane MC, Albertson AJ, Larimore JL, Kudo S, Armstrong DL, Percy AK, Pozzo-Miller L: Dendritic spine pathologies in hippocampal pyramidal neurons from Rett syndrome brain and after expression of Rett-associated MECP2 mutations. *Neurobiol Dis* 2009, **35**(2):219–233.
30. Henderson C, Wijetunge L, Kinoshita MN, Shumway M, Hammond RS, Postma FR, Brynczka C, Rush R, Thomas A, Paylor R *et al.*: Reversal of disease-related pathologies in the fragile X mouse model by selective activation of GABA(B) receptors with arbaclofen. *Sci Transl Med* 2012, **4**(152):152ra128.
31. Peca J, Feliciano C, Ting JT, Wang W, Wells MF, Venkatraman TN, Lascola CD, Fu Z, Feng G: Shank3 mutant mice display autistic-like behaviours and striatal dysfunction. *Nature* 2011, **472**(7344):437–442.
32. Tabuchi K, Blundell J, Etherton MR, Hammer RE, Liu X, Powell CM, Sudhof TC: A neuroligin-3 mutation implicated in autism increases inhibitory synaptic transmission in mice. *Science* 2007, **318**(5847):71–76.
33. Huguet G, Ey E, Bourgeron T: The genetic landscapes of autism spectrum disorders. *Annu Rev Genomics Hum Genet* 2013, **14**:191–213.
34. Ramocki MB, Zoghbi HY: Failure of neuronal homeostasis results in common neuropsychiatric phenotypes. *Nature* 2008, **455**(7215):912–918.
35. Grafodatskaya D, Chung B, Szatmari P, Weksberg R: Autism spectrum disorders and epigenetics. *J Am Acad Child Adolesc Psychiatry* 2010, **49**(8):794–809.
36. Mattick JS, Mehler MF: RNA editing, DNA recoding and the evolution of human cognition. *Trends Neurosci* 2008, **31**(5):227–233.
37. Nishikura K: Functions and regulation of RNA editing by ADAR deaminases. *Annu Rev Biochem* 2010, **79**:321–349.

38. Maas S: Gene regulation through RNA editing. *Discov Med* 2010, **10**(54):379–386.
39. Rosenthal JJ, Seeburg PH: A-to-I RNA editing: effects on proteins key to neural excitability. *Neuron* 2012, **74**(3):432–439.
40. Li JB, Levanon EY, Yoon JK, Aach J, Xie B, Leproust E, Zhang K, Gao Y, Church GM: Genome-wide identification of human RNA editing sites by parallel DNA capturing and sequencing. *Science* 2009, **324**(5931):1210–1213.
41. Balik A, Penn AC, Nemoda Z, Greger IH: Activity-regulated RNA editing in select neuronal subfields in hippocampus. *Nucleic Acids Res* 2013, **41**(2):1124–1134.
42. Sanjana NE, Levanon EY, Hueske EA, Ambrose JM, Li JB: Activity-dependent A-to-I RNA editing in rat cortical neurons. *Genetics* 2012, **192**(1):281–287.
43. Li B, Zhang S, Zhang H, Hertz L, Peng L: Fluoxetine affects GluK2 editing, glutamate-evoked Ca(2+) influx and extracellular signal-regulated kinase phosphorylation in mouse astrocytes. *J Psychiatry Neurosci* 2011, **36**(5): 322–338.
44. Ekdahl Y, Farahani HS, Behm M, Lagergren J, Ohman M: A-to-I editing of microRNAs in the mammalian brain increases during development. *Genome Res* 2012, **22**(8):1477–1487.
45. Wahlstedt H, Daniel C, Enstero M, Ohman M: Large-scale mRNA sequencing determines global regulation of RNA editing during brain development. *Genome Res* 2009, **19**(6):978–986.
46. Dillman AA, Hauser DN, Gibbs JR, Nalls MA, McCoy MK, Rudenko IN, Galter D, Cookson MR: mRNA expression, splicing and editing in the embryonic and adult mouse cerebral cortex. *Nat Neurosci* 2013, **16**(4): 499–506.
47. Nevo-Caspi Y, Amariglio N, Rechavi G, Paret G: A to I RNA editing is induced upon hypoxia. *Shock* 2011, **35**(6):585–589.
48. Englander MT, Dulawa SC, Bhansali P, Schmauss C: How stress and fluoxetine modulate serotonin 2C receptor pre-mRNA editing. *J Neurosci* 2005, **25**(3):648–651.
49. Du Y, Stasko M, Costa AC, Davisson MT, Gardiner KJ: Editing of the serotonin 2C receptor pre-mRNA: effects of the morris water maze. *Gene* 2007, **391**(1–2):186–197.
50. Iwamoto K, Nakatani N, Bundo M, Yoshikawa T, Kato T: Altered RNA editing of serotonin 2C receptor in a rat model of depression. *Neurosci Res* 2005, **53**(1):69–76.
51. Li B, Dong L, Wang B, Cai L, Jiang N, Peng L: Cell type-specific gene expression and editing responses to chronic fluoxetine treatment in the in vivo mouse brain and their relevance for stress-induced anhedonia. *Neurochem Res* 2012, **37**(11):2480–2495.

52. Caracciolo L, Barbon A, Palumbo S, Mora C, Toscano CD, Bosetti F, Barlati S: Altered mRNA editing and expression of ionotropic glutamate receptors after kainic acid exposure in cyclooxygenase-2 deficient mice. *PLoS One* 2011, **6**(5):e19398.
53. Macbeth MR, Schubert HL, Vandemark AP, Lingam AT, Hill CP, Bass BL: Inositol hexakisphosphate is bound in the ADAR2 core and required for RNA editing. *Science* 2005, **309**(5740):1534–1539.
54. Patterson JB, Samuel CE: Expression and regulation by interferon of a double-stranded-RNA-specific adenosine deaminase from human cells: evidence for two forms of the deaminase. *Mol Cell Biol* 1995, **15**(10):5376–5388.
55. Jepson JE, Savva YA, Yokose C, Sugden AU, Sahin A, Reenan RA: Engineered alterations in RNA editing modulate complex behavior in Drosophila: regulatory diversity of adenosine deaminase acting on RNA (ADAR) targets. *J Biol Chem* 2011, **286**(10):8325–8337.
56. Wulff BE, Nishikura K: Substitutional A-to-I RNA editing. *WIREs RNA* 2010, **1**(1):90–101.
57. Lomeli H, Mosbacher J, Melcher T, Hoger T, Geiger JR, Kuner T, Monyer H, Higuchi M, Bach A, Seeburg PH: Control of kinetic properties of AMPA receptor channels by nuclear RNA editing. *Science* 1994, **266**(5191):1709–1713.
58. Kohler M, Burnashev N, Sakmann B, Seeburg PH: Determinants of Ca^{2+} permeability in both TM1 and TM2 of high affinity kainate receptor channels: diversity by RNA editing. *Neuron* 1993, **10**(3):491–500.
59. Rula EY, Lagrange AH, Jacobs MM, Hu N, Macdonald RL, Emeson RB: Developmental modulation of GABA(A) receptor function by RNA editing. *J Neurosci* 2008, **28**(24):6196–6201.
60. Feldmeyer D, Kask K, Brusa R, Kornau HC, Kolhekar R, Rozov A, Burnashev N, Jensen V, Hvalby O, Sprengel R *et al.*: Neurological dysfunctions in mice expressing different levels of the Q/R site-unedited AMPAR subunit GluR-B. *Nat Neurosci* 1999, **2**(1):57–64.
61. Burns CM, Chu H, Rueter SM, Hutchinson LK, Canton H, Sanders-Bush E, Emeson RB: Regulation of serotonin-2C receptor G-protein coupling by RNA editing. *Nature* 1997, **387**(6630):303–308.
62. Lev-Maor G, Sorek R, Levanon EY, Paz N, Eisenberg E, Ast G: RNA-editing-mediated exon evolution. *Genome Biol* 2007, **8**(2):R29.
63. Kawahara Y, Zinshteyn B, Chendrimada TP, Shiekhattar R, Nishikura K: RNA editing of the microRNA-151 precursor blocks cleavage by the Dicer-TRBP complex. *EMBO Rep* 2007, **8**(8):763–769.
64. Yang W, Chendrimada TP, Wang Q, Higuchi M, Seeburg PH, Shiekhattar R, Nishikura K: Modulation of microRNA processing and expression through RNA editing by ADAR deaminases. *Nat Struct Mol Biol* 2006, **13**(1):13–21.
65. Consortium EP, Dunham I, Kundaje A, Aldred SF, Collins PJ, Davis CA, Doyle F, Epstein CB, Frietze S, Harrow J *et al.*: An integrated encyclopedia of DNA elements in the human genome. *Nature* 2012, **489**(7414):57–74.

66. Mombereau C, Kawahara Y, Gundersen BB, Nishikura K, Blendy JA: Functional relevance of serotonin 2C receptor mRNA editing in antidepressant- and anxiety-like behaviors. *Neuropharmacology* 2010, **59**(6):468–473.
67. Dabiri GA, Lai F, Drakas RA, Nishikura K: Editing of the GLuR-B ion channel RNA *in vitro* by recombinant double-stranded RNA adenosine deaminase. *EMBO J* 1996, **15**(1):34–45.
68. Martin CB, Ramond F, Farrington DT, Aguiar AS, Jr., Chevarin C, Berthiau AS, Caussanel S, Lanfumey L, Herrick-Davis K, Hamon M *et al.*: RNA splicing and editing modulation of 5-HT(2C) receptor function: relevance to anxiety and aggression in VGV mice. *Mol Psychiatry* 2013, **18**(6):656–665.
69. Staber CJ, Gell S, Jepson JE, Reenan RA: Perturbing A-to-I RNA editing using genetics and homologous recombination. *Methods Mol Biol* 2011, **718**:41–73.
70. Horsch M, Seeburg PH, Adler T, Aguilar-Pimentel JA, Becker L, Calzada-Wack J, Garrett L, Gotz A, Hans W, Higuchi M *et al.*: Requirement of the RNA-editing enzyme ADAR2 for normal physiology in mice. *J Biol Chem* 2011, **286**(21):18614–18622.
71. Morabito MV, Abbas AI, Hood JL, Kesterson RA, Jacobs MM, Kump DS, Hachey DL, Roth BL, Emeson RB: Mice with altered serotonin 2C receptor RNA editing display characteristics of Prader-Willi syndrome. *Neurobiol Dis* 2010, **39**(2):169–180.
72. Jepson JE, Reenan RA: Unraveling pleiotropic functions of A-to-I RNA editing in Drosophila. *Fly (Austin)* 2010, **4**(2):154–158.
73. Singh M, Zimmerman MB, Beltz TG, Johnson AK: Affect-related behaviors in mice misexpressing the RNA editing enzyme ADAR2. *Physiol Behav* 2009, **97**(3–4):446–454.
74. Dracheva S, Lyddon R, Barley K, Marcus SM, Hurd YL, Byne WM: Editing of serotonin 2C receptor mRNA in the prefrontal cortex characterizes high-novelty locomotor response behavioral trait. *Neuropsychopharmacology* 2009, **34**(10):2237–2251.
75. Savva YA, Jepson JE, Sahin A, Sugden AU, Dorsky JS, Alpert L, Lawrence C, Reenan RA: Auto-regulatory RNA editing fine-tunes mRNA re-coding and complex behaviour in Drosophila. *Nat Commun* 2012, **3**:790.
76. Palladino MJ, Keegan LP, O'Connell MA, Reenan RA: A-to-I pre-mRNA editing in Drosophila is primarily involved in adult nervous system function and integrity. *Cell* 2000, **102**(4):437–449.
77. Higuchi M, Maas S, Single FN, Hartner J, Rozov A, Burnashev N, Feldmeyer D, Sprengel R, Seeburg PH: Point mutation in an AMPA receptor gene rescues lethality in mice deficient in the RNA-editing enzyme ADAR2. *Nature* 2000, **406**(6791):78–81.
78. Olaghere da Silva UB, Morabito MV, Canal CE, Airey DC, Emeson RB, Sanders-Bush E: Impact of RNA editing on functions of the serotonin 2C receptor *in vivo*. *Front Neurosci* 2010, **4**:26.

79. Doe CM, Relkovic D, Garfield AS, Dalley JW, Theobald DE, Humby T, Wilkinson LS, Isles AR: Loss of the imprinted snoRNA mbii-52 leads to increased 5htr2c pre-RNA editing and altered 5HT2CR-mediated behaviour. *Hum Mol Genet* 2009, **18**(12):2140–2148.
80. Jamain S, Radyushkin K, Hammerschmidt K, Granon S, Boretius S, Varoqueaux F, Ramanantsoa N, Gallego J, Ronnenberg A, Winter D *et al.*: Reduced social interaction and ultrasonic communication in a mouse model of monogenic heritable autism. *Proc Natl Acad Sci USA* 2008, **105**(5): 1710–1715.
81. Nakatani J, Tamada K, Hatanaka F, Ise S, Ohta H, Inoue K, Tomonaga S, Watanabe Y, Chung YJ, Banerjee R *et al.*: Abnormal behavior in a chromosome-engineered mouse model for human 15q11-13 duplication seen in autism. *Cell* 2009, **137**(7):1235–1246.
82. Gurevich I, Tamir H, Arango V, Dwork AJ, Mann JJ, Schmauss C: Altered editing of serotonin 2C receptor pre-mRNA in the prefrontal cortex of depressed suicide victims. *Neuron* 2002, **34**(3):349–356.
83. Peng Z, Cheng Y, Tan BC, Kang L, Tian Z, Zhu Y, Zhang W, Liang Y, Hu X, Tan X *et al.*: Comprehensive analysis of RNA-Seq data reveals extensive RNA editing in a human transcriptome. *Nat Biotechnol* 2012, **30**(3):253–260.
84. Athanasiadis A, Rich A, Maas S: Widespread A-to-I RNA editing of Alu-containing mRNAs in the human transcriptome. *PLoS Biol* 2004, **2**(12):e391.
85. Paz-Yaacov N, Levanon EY, Nevo E, Kinar Y, Harmelin A, Jacob-Hirsch J, Amariglio N, Eisenberg E, Rechavi G: Adenosine-to-inosine RNA editing shapes transcriptome diversity in primates. *Proc Natl Acad Sci U S A* 2010, **107**(27):12174–12179.
86. Tomaselli S, Galeano F, Massimi L, Di Rocco C, Lauriola L, Mastronuzzi A, Locatelli F, Gallo A: ADAR2 editing activity in newly diagnosed versus relapsed pediatric high-grade astrocytomas. *BMC Cancer* 2013, **13**:255.
87. Ramaswami G, Zhang R, Piskol R, Keegan LP, Deng P, O'Connell MA, Li JB: Identifying RNA editing sites using RNA sequencing data alone. *Nat Methods* 2013, **10**(2):128–132.
88. Chen L, Li Y, Lin CH, Chan TH, Chow RK, Song Y, Liu M, Yuan YF, Fu L, Kong KL *et al.*: Recoding RNA editing of AZIN1 predisposes to hepatocellular carcinoma. *Nat Med* 2013, **19**(2):209–216.
89. Zhu H, Urban DJ, Blashka J, McPheeters MT, Kroeze WK, Mieczkowski P, Overholser JC, Jurjus GJ, Dieter L, Mahajan GJ *et al.*: Quantitative analysis of focused a-to-I RNA editing sites by ultra-high-throughput sequencing in psychiatric disorders. *PLoS One* 2012, **7**(8):e43227.
90. Eran A, Li JB, Vatalaro K, McCarthy J, Rahimov F, Collins C, Markianos K, Margulies DM, Brown EN, Calvo SE *et al.*: Comparative RNA editing in autistic and neurotypical cerebella. *Mol Psychiatry* 2013, **18**(9):1041–1048.

91. Silberberg G, Lundin D, Navon R, Ohman M: Deregulation of the A-to-I RNA editing mechanism in psychiatric disorders. *Hum Mol Genet* 2012, **21**(2):311–321.
92. Zhu S, Xiang JF, Chen T, Chen LL, Yang L: Prediction of constitutive A-to-I editing sites from human transcriptomes in the absence of genomic sequences. *BMC Genomics* 2013, **14**(1):206.
93. Park E, Williams B, Wold BJ, Mortazavi A: RNA editing in the human ENCODE RNA-seq data. *Genome Res* 2012, **22**(9):1626–1633.
94. Galeano F, Rossetti C, Tomaselli S, Cifaldi L, Lezzerini M, Pezzullo M, Boldrini R, Massimi L, Di Rocco CM, Locatelli F *et al.*: ADAR2-editing activity inhibits glioblastoma growth through the modulation of the CDC14B/Skp2/p21/p27 axis. *Oncogene* 2013, **32**(8):998–1009.
95. Choudhury Y, Tay FC, Lam DH, Sandanaraj E, Tang C, Ang BT, Wang S: Attenuated adenosine-to-inosine editing of microRNA-376a* promotes invasiveness of glioblastoma cells. *J Clin Invest* 2012, **122**(11):4059–4076.
96. Canitano R: Epilepsy in autism spectrum disorders. *Eur Child Adolesc Psychiatry* 2007, **16**(1):61–66.
97. Veltman MW, Craig EE, Bolton PF: Autism spectrum disorders in Prader-Willi and Angelman syndromes: a systematic review. *Psychiatr Genet* 2005, **15**(4):243–254.
98. Daniels JL, Forssen U, Hultman CM, Cnattingius S, Savitz DA, Feychting M, Sparen P: Parental psychiatric disorders associated with autism spectrum disorders in the offspring. *Pediatrics* 2008, **121**(5):e1357–1362.
99. Hagerman R, Hoem G, Hagerman P: Fragile X and autism: intertwined at the molecular level leading to targeted treatments. *Mol Autism* 2010, **1**(1):12.
100. Bhogal B, Jepson JE, Savva YA, Pepper AS, Reenan RA, Jongens TA: Modulation of dADAR-dependent RNA editing by the Drosophila fragile X mental retardation protein. *Nat Neurosci* 2011, **14**(12):1517–1524.
101. Abrahams BS, Geschwind DH: Advances in autism genetics: on the threshold of a new neurobiology. *Nat Rev Genet* 2008, **9**(5):341–355.
102. Kawahara Y, Ito K, Ito M, Tsuji S, Kwak S: Novel splice variants of human ADAR2 mRNA: skipping of the exon encoding the dsRNA-binding domains, and multiple C-terminal splice sites. *Gene* 2005, **363**:193–201.
103. Sarachana T, Hu VW: Genome-wide identification of transcriptional targets of RORA reveals direct regulation of multiple genes associated with autism spectrum disorder. *Mol Autism* 2013, **4**(1):14.
104. Sarachana T, Xu M, Wu RC, Hu VW: Sex hormones in autism: androgens and estrogens differentially and reciprocally regulate RORA, a novel candidate gene for autism. *PLoS One* 2011, **6**(2):e17116.
105. Nguyen A, Rauch TA, Pfeifer GP, Hu VW: Global methylation profiling of lymphoblastoid cell lines reveals epigenetic contributions to autism

spectrum disorders and a novel autism candidate gene, RORA, whose protein product is reduced in autistic brain. *FASEB J* 2010, **24**(8):3036–3051.

106. Hu VW, Nguyen A, Kim KS, Steinberg ME, Sarachana T, Scully MA, Soldin SJ, Luu T, Lee NH: Gene expression profiling of lymphoblasts from autistic and nonaffected sib pairs: altered pathways in neuronal development and steroid biosynthesis. *PLoS One* 2009, **4**(6):e5775.
107. A priceless resource. *Nature* 2011, **478**(7370):427.
108. Consortium GT: The Genotype-Tissue Expression (GTEx) project. *Nat Genet* 2013, **45**(6):580–585.
109. Kang HJ, Kawasawa YI, Cheng F, Zhu Y, Xu X, Li M, Sousa AM, Pletikos M, Meyer KA, Sedmak G *et al.*: Spatio-temporal transcriptome of the human brain. *Nature* 2011, **478**(7370):483–489.
110. Kong SW, Collins CD, Shimizu-Motohashi Y, Holm IA, Campbell MG, Lee IH, Brewster SJ, Hanson E, Harris HK, Lowe KR *et al.*: Characteristics and predictive value of blood transcriptome signature in males with autism spectrum disorders. *PLoS One* 2012, **7**(12):e49475.
111. Kong SW, Shimizu-Motohashi Y, Campbell MG, Lee IH, Collins CD, Brewster SJ, Holm IA, Rappaport L, Kohane IS, Kunkel LM: Peripheral blood gene expression signature differentiates children with autism from unaffected siblings. *Neurogenetics* 2013, **14**(2):143–152.
112. Luo R, Sanders SJ, Tian Y, Voineagu I, Huang N, Chu SH, Klei L, Cai C, Ou J, Lowe JK *et al.*: Genome-wide transcriptome profiling reveals the functional impact of rare de novo and recurrent CNVs in autism spectrum disorders. *Am J Hum Genet* 2012, **91**(1):38–55.
113. SynapDx Autism Gene Expression Analysis Study (STORY) [http://clinicaltrials.gov/show/NCT01810341]
114. Zhang R, Li X, Ramaswami G, Smith KS, Turecki G, Montgomery SB, Li JB: Quantifying RNA allelic ratios by microfluidic multiplex PCR and sequencing. *Nat Methods* 2014, **11**(1):51–54.
115. Freitas BC, Trujillo CA, Carromeu C, Yusupova M, Herai RH, Muotri AR: Stem cells and modeling of autism spectrum disorders. *Exp Neurol* 2012:Advance Online Publication 2012/10/06.

Chapter 11

Translational Control of Autism and Fragile-X Syndrome

Christos G. Gkogkas,[†] **Argel Aguilar-Valles, Arkady Khoutorsky and Nahum Sonenberg***

Abstract

Control of mRNA translation is a key feature of neurons to respond to experience-dependent changes in brain physiology. Dysregulation of translation has been linked to the etiology of several psychiatric brain disorders with neurodevelopmental origin, such as Autism Spectrum Disorders (ASD). Here, we review the evidence for dysregulated translation in ASD and monogenic disorders such as the Fragile-X syndrome (FXS). Furthermore, we summarize the development of therapies to target the neuronal signaling pathways controlling mRNA translation in these disorders. The alterations in signaling pathways in ASD and FXS have provided a fertile field for the identification of therapeutic agents in treating these conditions.

Keywords: Autism spectrum disorders; Fragile-X syndrome; Translation; Translational control.

---

*Correspondence to: Department of Biochemistry & Goodman Cancer Research Centre, McGill University, Montreal, QC, H3A IA3, Canada.
[†]Patrick Wild Centre and Centre for Integrative Physiology, University of Edinburgh, Hugh Robson Building, George Square, EH8 9XD, Edinburgh, UK.
Emails: Christos G. Gkogkas: christos.gkogkas@ed.ac.uk, Argel Aguilar-Valles: argel.aguilarvalles@mail.mcgill.ca, Arkady Khoutorsky: arkady.khoutorsky@mcgill.ca, Nahum Sonenberg: nahum.sonenberg@mcgill.ca

Introduction

mRNA translation

mRNA translation, also referred to as protein synthesis, is an intricate, multistep process, which is divided temporally into three steps: initiation, elongation and termination, as described below.[1]

Initiation

In eukaryotes, initiation is composed of a series of steps leading to the assembly of the 80S ribosome on the mRNA, at the initiation codon (AUG). Translation initiation is mediated by eIFs (eukaryotic initiation factors). In the first step of initiation, eIF2 associates with GTP and a methionine loaded transfer RNA (the ternary complex) to deliver the latter to the 40S ribosome, aided by other eIFS (i.e. eIF1, eIF1A and eIF3). The 40S ribosome, combined with the ternary complex form the 43S pre-initiation complex (PIC). The 43S PIC is then recruited to the 5′ end cap structure, present in all eukaryotic nuclear-transcribed mRNAs, via the cooperative activities of eIF4F (a complex formed by the RNA helicase eIF4A, the cap-binding eIF4E and the schafold eIF4G), eIF4B and PABP [Poly (A) binding protein]. Finally, the 43S PIC scans the mRNA until it reaches the initiation codon, where the eIF2-bound GTP is hydrolyzed, followed by the joining of the 60S ribosome subunit to form the 80S initiation complex.

Elongation

Elongation consists of repeated cycles facilitated by elongation factors of: (1) aminoacyl-tRNA selection and transfer to the A site of the ribosome, (2) peptide bond formation and (3) translocation. The ribosome continues to translate codons on the mRNA as more aminoacyl-tRNA bind to the A site, until the ribosome reaches a stop codon on an mRNA.

Termination

Recognition of the stop codon, in concert with release factors, causes the destabilization and disassembly of the ribosome and release of the polypeptide chain.

Translational control

Translation is highly regulated and, the majority of regulation is exerted at the level of initiation. Translational control is a powerful means to alter protein amounts in a fast manner, under conditions in which transcription activity is low, such as early development, and in cells where translation is spatially controlled, such as in neurons.[3] The latter aspect of translational control is especially crucial for neuronal physiology, given the key role of local, dendritic translation in neurons.[4] Control of translation initiation is effected via several signaling pathway including the mammalian/mechanistic target of rapamycin (TOR), the mitogen-activated protein kinase (MAPK)/extracellular signal–regulated kinase (ERK) and eIF2α phosphorylation.

mTOR pathway

mTOR is a serine-threonine protein kinase acting as a major sensor that integrates extracellular inputs such as insulin, growth factors and nutrients, and intracellular levels of oxygen, energy levels and aminoacids, to effect outputs including cell growth, proliferation and autophagy.[5] In neurons, mTOR integrates synaptic plasticity related inputs and neuronal activity, downstream of glutamate [e.g. N-methyl-D-aspartate receptor (NMDA) and type I metabotropic glutamate receptors (mGluR)], dopamine [dopamine D1/5], neurotrophin [e.g. Tyrosine kinase receptor B (TrkB)] and insulin receptors [e.g. Insulin-like growth factor (IGF-1)]. mTOR is blocked by rapamycin, an antifungal drug with potent immunosuppressive and antiproliferative properties. mTOR is the catalytic subunit of two distinct complexes: mTORC1, and mTORC2. mTORC1 activity is inhibited by acute rapamycin treatment, while mTORC2 activity is sensitive only to chronic rapamycin treatment.

Upstream of mTORC1, the Tuberous Sclerosis Complex proteins (TSC1 and TSC2) are GTPase activating proteins that inhibit the G-protein Rheb (Ras homolog enriched in brain), which activates mTOR. TSC2 is inactivated when phosphorylated, which alleviates Rheb inhibition to activate mTORC1. PTEN (phosphatase and tensin homolog) is an inhibitor of the mTORC1 pathway, by dephosphorylating phosphatidylinositol (3,4,5)-trisphosphate (PtdIns (3,4,5)*P*3 or PIP3).[5]

The best-characterized mechanism of eIF4F regulation is exerted by members of a family of three small molecular weight proteins, the eIF4E-binding proteins (4E-BPs: 4E-BP1, 4E-BP2 and 4E-BP3).[6] eIF4E is a component of eIF4F that directly binds the 5′-end cap structure of the mRNA to be translated.[2] 4E-BPs specifically inhibit cap-dependent translation initiation by inhibiting the assembly of the eIF4F complex, and subsequently ribosome recruitment to the mRNA.[6,7] 4E-BPs and eIF4G share a common eIF4E-binding site through which they compete for binding to the convex dorsal surface of eIF4E.[8] Non-phosphorylated 4E-BPs bind with high affinity to eIF4E, but following phosphorylation by mTORC1 they dissociate from eIF4E, resulting in stimulation of translation initiation.[9]

mTORC1 also phosphorylates the 40S ribosomal protein S6 kinases (S6Ks: S6K1 and S6K2), which in turn phosphorylate translation initiation factor eIF4B and ribosomal protein S6 to promote translation. S6K1 also targets elongation factor 2 (eEF2), by phosphorylating its kinase (eEF2K).[10] mTORC1 is thought to regulate the phosphorylation of its downstream targets by inhibiting the phosphatase activity of protein phosphatase 2A (PP2A).[11]

Ras/MEK/Erk/Mnk

The MAPK/ERK pathway stimulates protein synthesis. Activated MAPK/ERK phosphorylates and thereby activates MAP kinase-interacting kinases 1 and 2 (Mnk1, 2). Mnks bind to eIF4G and phosphorylate eIF4E on Ser209.[12] Phosphorylation of eIF4E promotes cap-dependent translation.[1]

eIF2α

Phosphorylation of Ser51 of the α subunit of eIF2 converts eIF2 into a dominant inhibitor of eIF2B GEF activity and decreases general translation. However, the translation of a subset of mRNAs, which contain upstream open reading frames (uORFs) in their 5′ UTR, such as *ATF4* mRNA, is paradoxically stimulated. In contrast, dephosphorylation of eIF2α increases general translation and decreases *ATF4* mRNA translation.[1] ATF4 is a repressor of the transcription factor cAMP response element binding protein (CREB)-mediated gene expression, which is required for long-lasting changes in synaptic plasticity and memory.[4]

Neuronal stalled polysomes

Recent work provides data that a significant fraction of dendritic mRNAs are present on stalled polysomes, which are transported in this state to post-synaptic sites.[13] This mode of translation bypasses the rate-limiting initiation step and allows the swift translation of synaptic mRNAs into functional protein upon synaptic signaling.

Autism Spectrum Disorders (ASD)

According to the World Health Organization, the share of the total global burden of disease due to neuropsychiatric disorders is projected to rise from 11% currently to 14.7% by 2020. Despite the prevalence of neuropsychiatric disorders, there is slow progress in understanding the pathophysiology and the development of effective therapies. Thus, the knowledge of the origins of neuropsychiatric disorders, such as schizophrenia, affective disorders (depression and bipolar disorder), and Autism Spectrum Disorders (ASD) is imperative. The understanding and treatment of neurodevelopmental disorders pose many challenges to biomedicine, despite the great advances in preclinical and clinical methodology.

ASD

Impairments in language, social interaction, and behavioral flexibility, all prior to 3 years of age, constitute the triad of symptoms that define ASD. The clinical picture is highly heterogeneous, with large differences in symptom severity among patients.[14,15] The identification of genetic risk factors for ASDs is bolstered by multiple parallel and complementary approaches (Figure 1). Maternally inherited duplications of chromosome 15q11-13, along with other rare chromosomal abnormalities are important causes of ASD. Novel individual genes of major impact (e.g., *NLGN4*, *NRXN1*, and *SHANK3*) were identified by resequencing or array-based methods.[14–16] Linkage studies in ASD have been controversial, as ASD linkage peaks were rarely significant at a genome-wide level and were difficult to replicate across studies. Several large ASD GWAS have identified SNPs, which are significantly associated with ASD

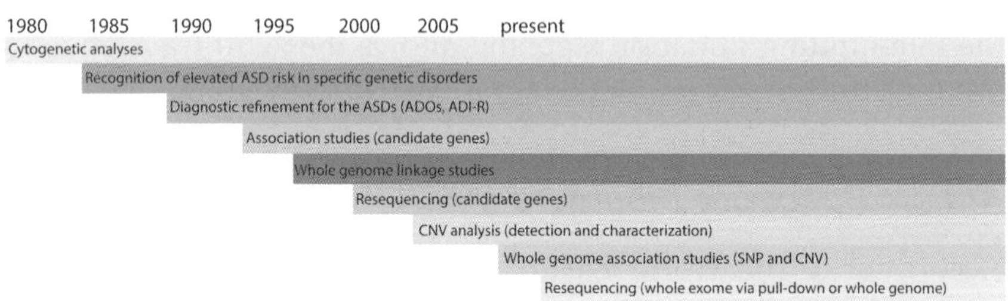

Figure 1. Research in autism spectrum genetics has been accelerated mainly due to the changes in methodology. Large cohorts via international collaborations, in conjunction with array and sequence based technologies allow the genome-wide interrogation of genetic variations. Massive computational infrastructure will be required for the interpretation of these results. ADOS indicates Autism Diagnostic Observation Schedule; ADI-R, Autism Diagnostic Interview–Revised; CNV, copy number variation; and SNP, single-nucleotide polymorphism. Adapted from Ref. [116].

at genome-wide levels. Also, studies have demonstrated a higher incidence of rare *de novo* CNVs in ASD.[17–19] Accordingly, whole-exome sequencing in sporadic ASD has shown presence of rare *de novo* mutations,[20,21] and six ASD genes have been identified: *CHD8*, *DYRK1A*, *GRIN2B*, *KATNAL2*, *POGZ*, and *SCN2A*, which carried recurrent, highly damaging *de novo* mutations.[16] Finally, *de novo* pathogenic SNVs (Single Nucleotide Variants) were found at a higher frequency in ASD cases, as compared to controls.[14,15]

E/I balance

A salient observation in patients and animal models of Schizophrenia and ASD is an imbalance of excitatory to inhibitory (E/I) synaptic transmission ratio.[22–24] Children with ASD exhibit elevated resting-state neuronal activity, consistent with an E/I imbalance as a neurobiological feature of ASD.[25] Modulations of the E/I balance using optogenetics directly influence autism-like behaviors in adult mice.[26] Several ASD mouse models display E/I imbalance due to altered glutamatergic excitation or to altered GABA (γ-aminobutyric acid)-ergic inhibition, leading to an increased[27,28] or decreased[29–31] E/I ratio, respectively. The molecular mechanisms controlling the E/I balance are however not well understood.

Fragile-X Syndrome (FXS)

Genetics of FXS

Fragile-X Syndrome (FXS) is the most common inherited form of intellectual disability (ID) and the most common known monogenic cause of ASD.[32,33] FXS was first described in pedigrees showing ID with X-linkage,[34] and its name stems from the cytogenetic assay showing a characteristic constricted region or fragile site on the long arm of the X chromosome at position Xq27.3.[35] The FXS mutation is an expansion of a trinucleotide repeat in the 5′ untranslated region (UTR) of the *fragile X mental retardation 1* gene (*FMR1*).[36] Expansion occurs in the general population and is considered normal up to 40–50 CGG repeats; a premutation condition is considered when the expansion is approximately 55 to 200 repeats. The full mutation for FXS consists of expansions greater than 200 CGG repeats, as these CpG islands become methylated, transcription of *FMR1* is prevented.[37–39] Only the full mutation causes silencing of the gene and thus FXS.[40]

The causal role of FMRP in FXS has been confirmed by studies demonstrating that other mutations affecting FMRP function or expression are also found in FXS[40] and individuals with FXS-like symptoms.[41,42] The premutation can in turn lead to fragile X-associated premature ovarian insufficiency (FXPOI) in female carriers[43] and an adult-onset neurodegenerative disease, the fragile X-associated tremor/ataxia syndrome (FXTAS).[44] Recently, it was shown that non-expanded (CGG)30–50 and premutation (CGG)59–160 repeats within the *FMR1* gene, are subjected to repeat-associated non-AUG (RAN) translation, in which in the absence of an initiating AUG or cognate AUG codon begins within all reading frames of an expanded (CAG)N repeat,[45] yielding peptides of polyglycine or polyalanine.[46] These polypeptides are prone to aggregation and were proposed to account for the inclusion bodies observed in FXTAS, which contribute to disease pathogenesis.[47] However, it is unclear to what extent the premutation syndromes are due to mRNA-induced toxicity or to the decreased amounts of FMRP.[47]

Clinical phenotype of FXS

Individuals with a fragile-X full mutation and FXS display characteristic physical features and associated medical problems (i.e. macroorchidism, macrocephaly, high arched palate, hyperextensible joints, soft skin, recurrent otitis media and sinusitis, gastroesophageal reflux, strabismus and farsightedness, hyperphagia and obesity, mitral valve prolapse and valvular insufficiency, disrupted sleep pattern, and obstructive sleep apnea), as well as ID and dysfunction in multiple behavioral domains (reviewed in Ref. 32). Syndromes are more severe in males than in females, as heterozygous females express FMRP from the normal X chromosome, the amount of which depends on the extent of X inactivation. Males with FXS have behavioral abnormalities including hyperactivity, impulsivity, attention problems, generalized anxiety with phobias and multiple specific domains of anxiety, mood lability, and autistic features such as poor eye contact, shyness, self-talk, hand-flapping, hand-biting, hyperarousal to sensory stimuli, and substantial perseverative language and behavior.[32] A subset of these problems are common in females, including attention problems, impulsivity, shyness, selective mutism, specific phobias, social anxiety, and social deficits.[32]

The autistic behavioral patterns are very common in boys with FXS (50–90%), particularly avoidant and anxious behaviors, and ~18–36% meet the full criteria for ASD.[48] Despite the overlap with ASD, the patterns of social behavior in FXS often differ from typical autism: individuals with FXS have strong social interest, but high levels of social anxiety.[32] Seizures occur in ~15% of males and 6–8% of females with FXS and appear to be an indicator of an increased risk for autism in FXS.[32]

Dysregulated translational control in ASD

ASD and translation

Several single-gene disorders harboring mutations upstream of mTOR, with high rates of autism, are characterized by increased levels of translation of synaptic proteins and connectivity (Table 1). These disorders include FXS (caused by reduced levels of FMRP);[49] tuberous sclerosis complex, where 85% of affected patients display cognitive impairment tightly linked to autistic features, epilepsy and abnormal or absent speech;[50] PTEN hamartoma syndrome-ASD with macrocephaly, in which a subset of patients show features resembling autism and language impairment are diagnosed.[51] Conversely, in Rett's syndrome, loss of function mutations of MeCP2 (methyl CpG binding protein 2) lead to decreased synaptic protein availability and hypoconnectivity.

A link between eIF4E and autism was first suggested because a *de novo* chromosome translocation between 4q and 5q was mapped

Table 1. Single gene disorders with high aates of autism

Gene	Disorder	Rate of Autism	Rate in Autism	MR	Gene Function
FMR1	Fragile X syndrome	15–30%	2–5%	+	Translational repressor
TSC1/2	Tuberous sclerosis complex	25–60%	1–4%	+	Inhibitor of mTOR
PTEN	PTEN hamartoma syndrome (ASD with macrocephaly)	ND	1%	+	Inhibitor of PI3sK/mTOR signaling
NF1	Neurofibromatosis type 1	4%	0–4%	+	Ras Gap
MECP 2	Rett's syndrome	100%	2%	+	Global transcriptional repressor
UBE3A	Angelman's syndrome	40%	1%	+	E3 ubiquitin ligase
CACNA1C	Timothy's syndrome	60%	<1%	+	L-type voltage-gated calcium channel (Cav1.2)
NLGN3/4	Familial ASD	ND	<1%	+	Synaptic adhesion
NRXN1	Familial ASD	ND	<1%	+	Synaptic adhesion
SHANK3	Familial ASD (22q13 microdeletion syndrome)	ND	<1%	+	PSD scaffolding

to the eIF4E gene in a boy with classic autism.[52] Also, in two unrelated families, two autistic siblings and one of the parents harbored the same single nucleotide insertion at position –25 in the basal element of the eIF4E promoter.[52] This mutation results in increased levels of eIF4E mRNA and most probably higher protein amounts. More recently, a *de novo* mutated eIF4G was reported as a component of a highly interconnected network of proteins linked with sporadic autism.[53] These data suggest that dysregulated eIF4F complex formation could be associated with certain forms of ASD. Interestingly, in sisters with Rett syndrome, harboring the same MeCP2 mutation, and similar levels of X chromosome inactivation [one individual could not speak or walk, and had a profound intellectual deficit (classical Rett syndrome), while the other individual could speak and walk, and had a moderate intellectual disability (Zappella variant)]. Exome sequencing detected PABPC1 mutations in the Rett syndrome patient, but not in the Zappella variant.

Haploinsufficiency of *Pten* or *Tsc2* in mice results in increased brain size,[54–56] which is reminiscent of the high prevalence of macrocephaly amongst children with ASD.[57] Conditional knockout (KO) *Pten* mice display deficits in social interaction and social learning and behavioral abnormalities in response to sensory stimuli, anxiety and learning.[55] These mice exhibit abnormal dendritic and axonal growth and synapse numbers.[55] Similarly, *Tsc2*+/– mice display memory deficits[54] and enlargement of neuronal somata and dendritic spines.[58] Rapamycin prevents and reverses neuronal hypertrophy, resulting in the amelioration of a subset of PTEN-associated abnormal behaviors.[59] Brief treatment of adult *Tsc2*+/– mice with rapamycin rescues not only the synaptic plasticity deficit, but also the behavioral deficits.[54] Recently, two studies demonstrated the role of eIF4E-dependent translation in the development and rescue of ASD-like behaviors in mice.[60,61] Genetic deletion of 4E-BP2 induced exaggerated translation of neuroligins, which engendered an imbalance in the E/I ratio, which was associated with impaired social interactions, repetitive behaviors and defects in communication.[60] Similarly, overexpression of eIF4E induced autism-like behaviors due to exaggerated protein synthesis and an imbalance of E/I.[61] These synaptic and behavioral phenotypes were rescued by inhibiting eIF4F activity (using 4EGI-1, an inhibitor of the eIF4E-eIF4G interaction),[60,61] or by reducing the expression of neuroligin 1.[60]

Dysregulated translational control in FXS

FMRP is a mRNA-binding protein[37] that is thought to play a role in several aspects of activity-dependent mRNA metabolism in neurons, including transport, stability, and regulation of translation.[62] FMRP contains three RNA-binding motifs that interact with specific RNA target elements: two hnRNP K-homology (KH) domains and an RGG box.[62] KH domains are thought to bind "kissing-complex" tertiary motifs in RNA,[63] whilst the RGG box recognizes stem-G-quartet loops, possibly in a methylation-dependent manner.[64]

It has been suggested that FMRP plays a role in transporting mRNA, as the protein is localized in messenger ribonucleoprotein (mRNP) complexes that traffic from the soma to dendrites and axons.[62] FMRP mRNPs have been found in all types of RNA granules (i.e. P-bodies, stress granules and high-density granules), which are believed to be translationally repressed mRNPs.[62] However, FMRP does not appear to be necessary for the localization of the majority of its target mRNAs in dendrites.[62]

FMRP inhibits protein synthesis, but the mechanism by which repression is achieved remains controversial (reviewed in Ref. 62). FMRP can repress translation by blocking elongation, consistent with its binding to the coding sequences of mRNAs on which ribosomes are stalled.[65] Other studies reported a role in control of initiation, as FMRP co-sediments with the monomeric 80S ribosomes and in light mRNP complexes with BC1 (brain cytoplasmic RNA 1), CYFIP1 (cytoplasmic FMRP interacting protein), and translation initiation factors.[62] The differences between these models have been discussed.[66–69]

The mGluR theory

The underlying neural pathophysiology of FXS[32,62,70] has been advanced by studies of the *Fmr1* KO mice, which exhibit increased basal protein synthesis,[71] seizure susceptibility, as well as dendritic[72,73] and behavioral alterations reminiscent of FXS patients. One particular aspect has proven to be highly promising for the development of therapies. The metabotropic glutamate receptors 1/5 (mGluR1/5)-mediated long-term depression (LTD), a specific protein synthesis-dependent form of synaptic plasticity in the hippocampus, is exaggerated,[74] and independent of protein synthesis in *Fmr1* KO mice.[75,76] This is consistent with the idea that FMRP is an inhibitor of synaptic protein synthesis.[77] This led to the formulation of the mGluR theory of FXS that posits that dysregulated mGluR1/5-mediated protein synthesis-dependent forms of synaptic plasticity contribute to the pathology of FXS (Figure 2).[78] As a consequence, mGluR1/5 signaling is considered a promising therapeutic target for FXS. Other mGluR1/5-dependent neuronal mechanisms are also dysregulated in the brain of the mouse model of FXS. In the hippocampus, mGluR1/5 is involved in the prolonged epileptiform discharges, in the increased endocannabinoid synthesis and release and in the alterations found in some forms of LTP (reviewed in Ref. 32). *Fmr1* KO mice exhibit alterations in mGlu1/5 signaling and synaptic plasticity in the anterior cingulate, the barrel cortices, the amygdala and the cerebellum.[32]

Studies in mouse[79,80] and *Drosophila*[81] models have provided strong support for the mGluR theory. Genetic rescue in mice, by generation of *Fmr1* -/y *Mglur5*+/− mice, demonstrated correction of elevated hippocampal protein synthesis, stimulus-induced seizures, exaggerated LTD, increased dendritic spine density and excessive memory extinction.[80] Pharmacological antagonism of mGluR5 has also been proven to ameliorate the alterations resulting from absence of FMRP. 2-methyl-6-(phenylethynyl)-pyridine (MPEP), a mGluR5 negative allosteric modulator reduced the exacerbated protein synthesis and alterations in the polysome and RNA granule,[82,83] as well as the prolonged epileptiform discharges in acute brain slices.[84] These studies provide compelling evidence that altering mGluR5-dependent protein synthesis directly affects the synaptic alterations underlying cognitive and behavioral phenotypes observed in FXS.

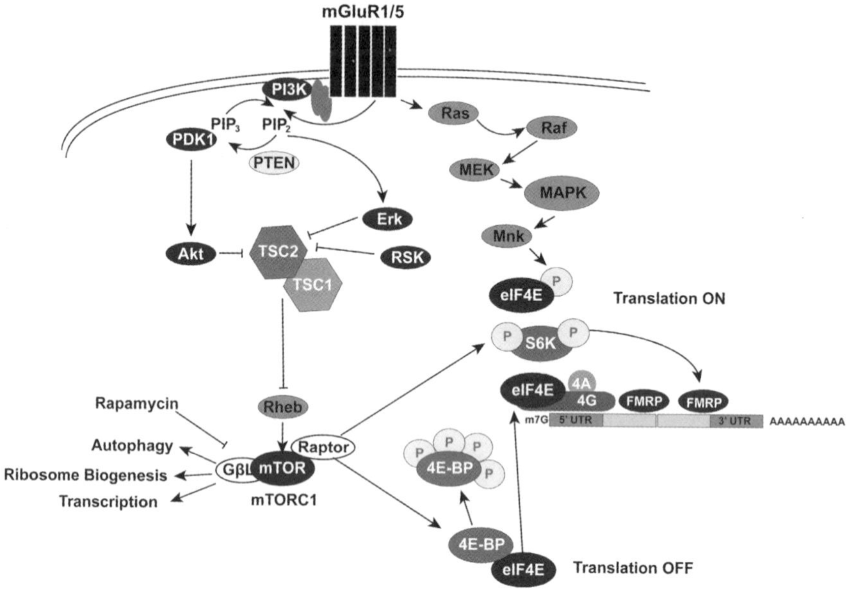

Figure 2. Signaling pathways activated by mGluR1/5 in the control of mRNA translation and are dysfunctional in FXS. mGluR1/5 are G-protein coupled receptors that activate two signaling pathways involved in the control of mRNA translation The PI3K/mTORC1 pathway, which involves the inactivation of TSC1/2 and PTEN, controls translation by phosphorylation of 4E-BPs and S6Ks. In parallel, MAPK/ERK leads to the phosphorylation of eIF4E, also associated to active translation. FMRP is associated to stalled mRNAs, and such inhibition can be relieved by activation of mGluR5. Absence of FMRP results in elevated rates of baseline translation and altered brain physiology downstream of mGluR5, including exacerbated hippocampal LTD. This discovery led to the formulation of the mGluR5 theory of FXS and the development of mGlur5 antagonists now in clinical trials.

FMRP and signaling to translation

The absence or downregulation of FMRP affects several hundred mRNAs,[65,85] which complicates drug development for FXS. However, FMRP phenotypes can be rescued by targeting two signaling pathways that converge to activate mRNA translation initiation, and which have been demonstrated to be dysregulated in FXS. On the one hand, ERK activation is delayed in lymphocytes from FXS-affected individuals.[86] On the other hand, postmortem

FXS brains exhibit enhanced ERK signaling.[87] Accordingly, in *Fmr1* KO mice, inhibition of ERK1/2 corrected the increased basal protein synthesis and reduced audiogenic seizures.[82] In this regard, increased eIF4E phosphorylation (which is a downstream target of the ERK pathway) is observed in mouse models of FXS[88] and confirmed in FXS patients.[89] In parallel, the phosphatidylinositide 3-kinase (PI3K), upstream of mTORC1 activation, is aberrantly activated at synapses in the absence of FMRP, which leads to exacerbated basal protein synthesis and occlusion of agonist-induced protein synthesis.[90,91] As the ERK and PI3K signaling pathways can be activated by several neurotransmitter receptors (i.e. mGluR1/5 (Figure 2), the muscarinic acetylcholine receptors, the dopamine D1/5 receptors and TrkB), their targeting holds promise for the correction of FXS.

Targeting of additional players involved in the control of translation has also been shown to rescue FXS-relevant phenotypes. Genetic reduction of S6K1 in *Fmr1* KO mice reverses FXS-like phenotypes.[92] Moreover administration of lovastatin, a hypolipidemic agent, reversed autism-like phenotypes in the *Fmr1* KO mice by indirectly targeting the ERK pathway.[93] Similarly, genetic deletion of the Cytoplasmic Polyadenylation Element Binding protein 1 (CPEB1) rescued several behavioral and synaptic deficits in *Fmr1* KO mice.[94]

The molecular overlap between FXS and some forms of ASD is based on the dysregulation of genes related to autism in the absence of FMRP.[33] In this regard, some of the proteins whose mRNA translation is controlled directly by FMRP are also associated with autism, including neuroligins, neurexin 1, PTEN, PSD95, MAPK1, and SHANK3.[33]

Future Perspectives

Translation is a "druggable" pathway

We now have a clearer understanding of the mechanisms governing translational control. This led to the synthesis of several new compounds, mainly developed for cancer treatment, some of them in advanced stages of clinical trials.[95] As a result, there are currently direct or indirect translation inhibitors, which function as follows: (1) targeting the regulation of eIF4F complex assembly (Rapamycin analogs or Rapalogs); (2) inhibiting kinases that target mTOR; (3) employing dual specificity PI3K/mTOR inhibitors; (4) targeting eIF4F directly: (i) competing for eIF4E binding to the m^7GpppN cap structure, (ii) uncoupling the eIF4E-eIF4G interaction, (iii) targeting eIF4E synthesis, (iv) blocking eIF4A activity, and (v) inhibiting eIF4E phosphorylation; and (5) blocking the ternary complex formation.

Clinical trials in FXS and ASD

Critical issues concerning testing drugs for ASD are when to initiate therapy and how long to maintain it. In disorders such as ASDs that manifest early in life, this poses the difficult task of testing drugs in children. Exceptionally thorough and rigorous preclinical testing will be required, and the path taken must involve initial trials in adults, with the aim to later move to younger patients.[70] In the case of FXS, one very promising strategy has been to identify and target a pathway that contributes a disproportionate degree to the phenotype, such as the type I mGluRs,[78] PI3K[70] and ERK,[70,96] which converge on control of mRNA translation.

The most studied target to reverse FXS phenotypes has been mGluR5, and a number of group I mGluR antagonists are currently being considered for use in FXS and autism. Early clinical trials in FXS have shown some promise, including encouraging results from an open label trial with fenobam.[97] In this study, a single oral dose was administered to a cohort of 12 adult male and female patients, which showed an improvement in the general anxiety level and in pre-pulse inhibition, without safety concerns. In parallel, based on promising results in the mouse model[98] a randomized double-blind trial with the Novartis compound AFQ056 was performed.[99] This study reported improvements in stereotyped behavior, hyperactivity, and inappropriate speech in individuals with FXS who had fully methylated *FMR1* alleles, whereas individuals with partial *FMR1* gene methylation showed no statistically significant changes in behavior.[99] This outcome was adequate to support ongoing development of AFQ056 for treatment of FXS, with a larger multinational double-blind placebo-controlled three-month trial evaluating the effects of multiple doses of AFQ056 (clinicaltrials.gov).[32] Phase I trials of STX107 (Seaside Therapeutics) have been completed in healthy individuals (no results provided) (clinicaltrials.gov).[32] A double-blind placebo-controlled dose-finding Phase II trial of RO4917523 (Roche) in adult males and females with FXS is underway (clinicaltrials.gov).[32,37]

The GABAergic system is being targeted in clinical trials of the GABAB receptor 1 agonist R-baclofen for FXS and ASD. Expression of the GABABR1 subunit is reduced in the brains of *Fmr1* KO mice,[100] and R-baclofen rescues audiogenic seizures in *Fmr1* null mice.[101] Baclofen is thought to act by inhibiting neuronal presynaptic release of glutamate, which in turn acts to block downstream signaling from mGluR5.[37] Unpublished reports from Seaside Therapeutics suggest that treatment with STX209 (Arbaclofen, R-baclofen) improved sociability and communication in individuals with FXS, while in ASDs patients it improved measures of hyperactivity, inappropriate speech, and repetitive behaviors (www.seasidetherapeutics.com).[33,70] A Phase III randomized double-blind placebo-controlled study examining the safety and efficacy of Arbaclofen treatment on social impairment in adolescents and adults with FXS and a Phase IIb open-label study of social impairments in ASDs have been initiated.[32,70]

Lithium, commonly used to treat bipolar disorder, has recently been tested as a treatment for FXS.[102] Lithium inhibits phospholipase C and glycogen synthase kinase-3β (GSK-3β) and its enzymatic activity is elevated in the *Fmr1* null mouse brain.[103] Lithium treatment of *Fmr1* KO mice reversed some behavioral phenotypes and partially rescued the dendritic spine phenotype.[104–106] It also corrected cognitive impairments in a *Drosophila* model of FXS.[107] The effects of short-term (two months) treatment with lithium were systematically explored in a pilot open-label trial in 15 subjects with FXS.[108] Treatment significantly improved FXS behavioral traits in a Maladaptive Behavior subscore, a parent visual analog scale for target behaviors, and in verbal memory. In addition, lithium treatment corrected the delayed activation of ERK in lymphocytes, suggesting a potential to directly regulate the signaling pathways downstream of mGluR1/5 and upstream of translation. Overall, the results suggest the possibility of using lithium to treat FSX.

The widely used antibiotic, minocycline, also recognized for its neuroprotective and anti-inflammatory effects on autoimmune diseases such as rheumatoid arthritis, has recently been applied to FXS.[109] In a single study, minocycline showed a beneficial effect on dendritic spine morphology and improved behavioral performance in *Fmr1*-deficient mice.[110] The proposed mechanism of action is the reduction of elevated matrix metalloproteinase-9 (MMP-9) activity, which plays a role in the synaptic remodeling associated with plasticity. MMP-9 protein levels are increased by mGlu1/5 stimulation in wild-type mice and under basal conditions in *Fmr1* KO mice.[110] An open-label trial of minocycline in 20 participants with FXS aged ~12 years showed behavioral improvements in several domains affected in FXS.[111] These results were used as the rationale for an in progress double-blind placebo-controlled trial of minocycline in children and adolescents with FXS.[32] A number of side effects of minocycline have been reported (i.e. irreversible yellow/brown discoloration of the permanent teeth, vomiting and/or diarrhea, drug-induced lupus, and pseudotumor cerebri[112]), which diminished the interest in its continued use in clinical trials,[32] however there is still interest in developing more specific drugs to target MMP-9.

Additional targets implicated from studies of the *Fmr1*-deficient mouse include two enzymes, the small GTPase Rac1 and associated p21-activated kinase (PAK), which regulate actin polymerization in dendritic spines and are thought to play a role in plasticity.[113,114] Furthermore, genetic rescue, via the postnatal expression of a dominant-negative form of PAK, reversed some of the FSX relevant phenotypes in *Fmr1* null mice,[113,114] and motivated the development of PAK-specific inhibitors (www.afraxis.com).

Another avenue that has been explored in the treatment of FXS pathophysiology is the targeting the NMDA via the antagonist, memantine.[115] An open-label clinical trial with memantine showed global clinical benefit as measured on the Clinical Global Impression-Improvement subscale over approximately 35 weeks.[115] These promising results have prompted a randomized, double-blind, placebo-controlled study (http://clinicaltrials.gov), although focused on premutation carriers.

Three major therapeutic strategies in FXS and ASD are emerging: (1) strategies targeting upstream extracellular neurotransmitter receptors, such as mGluR5 and GABAB receptors; (2) strategies targeting the intracellular central signal transduction pathways such as ERK1/2 or PI3K; and (3) strategies targeting multiple signaling molecules translated from FMRP target mRNAs (i.e.GSK3b, MMP-9, or PAK). FXS and ASD exhibit a significant overlap in molecular and synaptic pathways, many of which control the process of mRNA translation. The targeted treatments for FXS are likely to target dysregulated mechanisms, such as mRNA translation, in a subgroup of patients with autism who have defects in these pathways.[32] By modulating the protein synthesis mechanisms, we will have the potential to treat a number of neurodevelopmental disorders that arise from dysregulated translation, including FXS, ASD and related ailments.

Executive Summary

- mRNA translation is mainly controlled at the initiation step by signaling pathways, such as ERK, PI3K/mTORC1.
- Dysregulation of these pathways is linked to Autism Spectrum Disorders (ASD) and Fragile X syndrome (FXS).
- Preclinical evidence suggests that targeting these pathways can rescue phenotypes relevant for FXS and ASD.
- Metabotropic glutamate receptor 1/5 (mGluR1/5) controls translation-dependent synaptic plasticity and is involved in FXS pathophysiology.
- Early clinical trials using mGluR5 antagonists have shown lack of adverse effects and signs of improvement in FXS and ASD.
- Additional potential therapies targeting general translation or proteins whose translation is exacerbated are also being considered.

References

1. Sonenberg N, Hinnebusch AG: Regulation of translation initiation in eukaryotes: mechanisms and biological targets. *Cell* 2009, **136**(4):731–745.
2. Sonenberg N, Rupprecht KM, Hecht SM, Shatkin AJ: Eukaryotic mRNA cap binding protein: purification by affinity chromatography on sepharose-coupled m7GDP. *Proc Natl Acad Sci U S A* 1979, **76**(9):4345–4349.
3. Hershey JW, Sonenberg N, Mathews MB: Principles of translational control: an overview. *Cold Spring Harb Perspect Biol* 2012, **4**(12).
4. Costa-Mattioli M, Sossin WS, Klann E, Sonenberg N: Translational control of long-lasting synaptic plasticity and memory. *Neuron* 2009, **61**(1):10–26.
5. Hay N, Sonenberg N: Upstream and downstream of mTOR. *Genes Dev* 2004, **18**(16):1926–1945.
6. Pause A, Belsham GJ, Gingras AC, Donze O, Lin TA, Lawrence Jr. JC, Sonenberg N: Insulin-dependent stimulation of protein synthesis by phosphorylation of a regulator of 5′-cap function. *Nature* 1994, **371**(6500): 762–767.
7. Haghighat A, Mader S, Pause A, Sonenberg N: Repression of cap-dependent translation by 4E-binding protein 1: competition with p220 for binding to eukaryotic initiation factor-4E. *EMBO J* 1995, **14**(22):5701–5709.
8. Marcotrigiano J, Gingras AC, Sonenberg N, Burley SK: Cap-dependent translation initiation in eukaryotes is regulated by a molecular mimic of eIF4G. *Mol Cell* 1999, **3**(6):707–716.
9. Gingras AC, Raught B, Sonenberg N: eIF4 initiation factors: effectors of mRNA recruitment to ribosomes and regulators of translation. *Annu Rev Biochem* 1999, **68**:913–963.
10. Belelovsky K, Elkobi A, Kaphzan H, Nairn AC, Rosenblum K: A molecular switch for translational control in taste memory consolidation. *Eur J Neurosci* 2005, **22**(10):2560–2568.
11. Hoeffer CA, Klann E: mTOR signaling: at the crossroads of plasticity, memory and disease. *Trends Neurosci* 2010, **33**(2):67–75.
12. Furic L, Rong L, Larsson O, Koumakpayi IH, Yoshida K, Brueschke A, Petroulakis E, Robichaud N, Pollak M, Gaboury LA *et al.*: eIF4E phosphorylation promotes tumorigenesis and is associated with prostate cancer progression. *Proc Natl Acad Sci U S A* 2010, **107**(32):14134–14139.
13. Graber TE, Hebert-Seropian S, Khoutorsky A, David A, Yewdell JW, Lacaille JC, Sossin WS: Reactivation of stalled polyribosomes in synaptic plasticity. *Proc Natl Acad Sci USA* 2013, **110**(40):16205–16210.
14. Abrahams BS, Geschwind DH: Advances in autism genetics: on the threshold of a new neurobiology. *Nat Rev Genet* 2008, **9**(5):341–355.
15. Miyauchi S, Voineagu I: Autism susceptibility genes and the transcriptional landscape of the human brain. *Int Rev Neurobiol* 2013, **113**:303–318.

16. Buxbaum JD, Daly MJ, Devlin B, Lehner T, Roeder K, State MW: The autism sequencing consortium: large-scale, high-throughput sequencing in autism spectrum disorders. *Neuron* 2012, **76**(6):1052–1056.
17. Krumm N, O'Roak BJ, Karakoc E, Mohajeri K, Nelson B, Vives L, Jacquemont S, Munson J, Bernier R, Eichler EE: Transmission disequilibrium of small CNVs in simplex autism. *Am J Hum Genet* 2013, **93**(4):595–606.
18. Girirajan S, Johnson RL, Tassone F, Balciuniene J, Katiyar N, Fox K, Baker C, Srikanth A, Yeoh KH, Khoo SJ *et al.*: Global increases in both common and rare copy number load associated with autism. *Hum Mol Genet* 2013, **22**(14):2870–2880.
19. Girirajan S, Dennis MY, Baker C, Malig M, Coe BP, Campbell CD, Mark K, Vu TH, Alkan C, Cheng Z *et al.*: Refinement and discovery of new hotspots of copy-number variation associated with autism spectrum disorder. *Am J Hum Genet* 2013, **92**(2):221–237.
20. O'Roak BJ, Deriziotis P, Lee C, Vives L, Schwartz JJ, Girirajan S, Karakoc E, Mackenzie AP, Ng SB, Baker C *et al.*: Exome sequencing in sporadic autism spectrum disorders identifies severe de novo mutations. *Nat Genet* 2011, **43**(6):585–589.
21. O'Roak BJ, Vives L, Fu W, Egertson JD, Stanaway IB, Phelps IG, Carvill G, Kumar A, Lee C, Ankenman K *et al.*: Multiplex targeted sequencing identifies recurrently mutated genes in autism spectrum disorders. *Science* 2012, **338**(6114):1619–1622.
22. Gogolla N, Leblanc JJ, Quast KB, Sudhof TC, Fagiolini M, Hensch TK: Common circuit defect of excitatory-inhibitory balance in mouse models of autism. *J Neurodev Disord* 2009, **1**(2):172–181.
23. Rubenstein JL, Merzenich MM: Model of autism: increased ratio of excitation/inhibition in key neural systems. *Genes Brain Behav* 2003, **2**(5):255–267.
24. Uhlhaas PJ, Singer W: Neuronal dynamics and neuropsychiatric disorders: toward a translational paradigm for dysfunctional large-scale networks. *Neuron* 2012, **75**(6):963–980.
25. Cornew L, Roberts TP, Blaskey L, Edgar JC: Resting-state oscillatory activity in autism spectrum disorders. *J Autism Dev Disord* 2012, **42**(9):1884–1894.
26. Yizhar O, Fenno LE, Prigge M, Schneider F, Davidson TJ, O'Shea DJ, Sohal VS, Goshen I, Finkelstein J, Paz JT *et al.*: Neocortical excitation/inhibition balance in information processing and social dysfunction. *Nature* 2011, **477**(7363):171–178.
27. Luikart BW, Schnell E, Washburn EK, Bensen AL, Tovar KR, Westbrook GL: Pten knockdown *in vivo* increases excitatory drive onto dentate granule cells. *J Neurosci* 2011, **31**(11):4345–4354.
28. Bateup HS, Takasaki KT, Saulnier JL, Denefrio CL, Sabatini BL: Loss of Tsc1 *in vivo* impairs hippocampal mGluR-LTD and increases excitatory synaptic function. *J Neurosci* 2011, **31**(24):8862–8869.

29. Peca J, Feliciano C, Ting JT, Wang W, Wells MF, Venkatraman TN, Lascola CD, Fu Z, Feng G: Shank3 mutant mice display autistic-like behaviours and striatal dysfunction. *Nature* 2011, **472**(7344):437–442.
30. Schmeisser MJ, Ey E, Wegener S, Bockmann J, Stempel AV, Kuebler A, Janssen AL, Udvardi PT, Shiban E, Spilker C *et al.*: Autistic-like behaviours and hyperactivity in mice lacking ProSAP1/Shank2. *Nature* 2012, **486**(7402):256–260.
31. Won H, Lee HR, Gee HY, Mah W, Kim JI, Lee J, Ha S, Chung C, Jung ES, Cho YS *et al.*: Autistic-like social behaviour in Shank2-mutant mice improved by restoring NMDA receptor function. *Nature* 2012, **486**(7402):261–265.
32. Gross C, Berry-Kravis EM, Bassell GJ: Therapeutic strategies in fragile X syndrome: dysregulated mGluR signaling and beyond. *Neuropsychopharmacology* 2012, **37**(1):178–195.
33. Wang LW, Berry-Kravis E, Hagerman RJ: Fragile X: leading the way for targeted treatments in autism. *Neurotherapeutics* 2010, **7**(3):264–274.
34. Martin JP, Bell J: A Pedigree of Mental Defect Showing Sex-Linkage. *J Neurol Psychiatry* 1943, **6**(3–4):154–157.
35. Sutherland GR: Fragile sites on human chromosomes: demonstration of their dependence on the type of tissue culture medium. *Science* 1977, **197**(4300): 265–266.
36. Verkerk AJ, Pieretti M, Sutcliffe JS, Fu YH, Kuhl DP, Pizzuti A: Identification of a gene (FMR-1) containing a CGG repeat coincident with a breakpoint cluster region exhibiting length variation in fragile X syndrome. *Cell* 1991, **65**:905–914.
37. Healy A, Rush R, Ocain T: Fragile X syndrome: an update on developing treatment modalities. *ACS Chem Neurosci* 2011, **2**(8):402–410.
38. Pieretti M, Zhang FP, Fu YH, Warren ST, Oostra BA, Caskey CT: Absence of expression of the FMR-1 gene in fragile X syndrome. *Cell* 1991, **66**:817–822.
39. Maddalena A, Richards CS, McGinniss MJ, Brothman A, Desnick RJ, Grier RE: Technical standards and guidelines for fragile X: the first of a series of disease-specific supplements to the Standards and Guidelines for Clinical Genetics Laboratories of the American College of Medical Genetics. Quality Assurance Subcommittee of the Laboratory Practice Committee. *Genet Med* 2001, **3**:200–205.
40. De Boulle K, Verkerk AJ, Reyniers E, Vits L, Hendrickx J, Van Roy B: A point mutation in the FMR-1 gene associated with fragile X mental retardation. *Nat Genet* 1993, **3**:31–35.
41. Collins SC, Bray SM, Suhl JA, Cutler DJ, Coffee B, Zwick ME: Identification of novel FMR1 variants by massively parallel sequencing in developmentally delayed males. *Am J Med Genet A* 2010, 152A:2512–2520.
42. Collins SC, Coffee B, Benke PJ, Berry-Kravis E, Gilbert F, Oostra B: Array-based FMR1 sequencing and deletion analysis in patients with a fragile X syndrome-like phenotype. *PLoS One* 2010, **5**:e9476.

43. Sherman SL: Premature ovarian failure in the fragile X syndrome. *Am J Med Genet* 2000, **97**:189–194.
44. Berry-Kravis E, Abrams L, Coffey SM, Hall DA, Greco C, Gane LW: Fragile X-associated tremor/ataxia syndrome: clinical features, genetics, and testing guidelines. *Mov Disord* 2007, **22**:2018–2030.
45. Pearson CE: Repeat associated non-ATG translation initiation: one DNA, two transcripts, seven reading frames, potentially nine toxic entities! *PLoS Genet* 2011, **7**(3):e1002018.
46. Todd PK, Oh SY, Krans A, He F, Sellier C, Frazer M, Renoux AJ, Chen KC, Scaglione KM, Basrur V *et al.*: CGG repeat-associated translation mediates neurodegeneration in fragile X tremor ataxia syndrome. *Neuron* 2013, **78**(3):440–455.
47. Hagerman R, Hagerman P: Advances in clinical and molecular understanding of the FMR1 premutation and fragile X-associated tremor/ataxia syndrome. *Lancet Neurol* 2013, **12**(8):786–798.
48. Wang LW, Berry-Kravis E, Hagerman RJ: Fragile X: leading the way for targeted treatments in autism. *Neurotherapeutics* 2010, **7**:264–274.
49. Levenga J, de Vrij FM, Oostra BA, Willemsen R: Potential therapeutic interventions for fragile X syndrome. *Trends Mol Med* 2010, **16**(11):516–527.
50. Curatolo P, Napolioni V, Moavero R: Autism spectrum disorders in tuberous sclerosis: pathogenetic pathways and implications for treatment. *J Child Neurol* 2010, **25**(7):873–880.
51. Butler MG, Dasouki MJ, Zhou XP, Talebizadeh Z, Brown M, Takahashi TN, Miles JH, Wang CH, Stratton R, Pilarski R *et al.*: Subset of individuals with autism spectrum disorders and extreme macrocephaly associated with germline PTEN tumour suppressor gene mutations. *J Med Genet* 2005, **42**(4):318–321.
52. Neves-Pereira M, Muller B, Massie D, Williams JH, O'Brien PC, Hughes A, Shen SB, Clair DS, Miedzybrodzka Z: Deregulation of EIF4E: a novel mechanism for autism. *J Med Genet* 2009, **46**(11):759–765.
53. O'Roak BJ, Vives L, Girirajan S, Karakoc E, Krumm N, Coe BP, Levy R, Ko A, Lee C, Smith JD *et al.*: Sporadic autism exomes reveal a highly interconnected protein network of de novo mutations. *Nature* 2012, **485**(7397):246–250.
54. Ehninger D, Han S, Shilyansky C, Zhou Y, Li W, Kwiatkowski DJ, Ramesh V, Silva AJ: Reversal of learning deficits in a Tsc2+/− mouse model of tuberous sclerosis. *Nat Med* 2008, **14**(8):843–848.
55. Kwon CH, Luikart BW, Powell CM, Zhou J, Matheny SA, Zhang W, Li Y, Baker SJ, Parada LF: Pten regulates neuronal arborization and social interaction in mice. *Neuron* 2006, **50**(3):377–388.
56. Way SW, McKenna J, 3rd, Mietzsch U, Reith RM, Wu HC, Gambello MJ: Loss of Tsc2 in radial glia models the brain pathology of tuberous sclerosis complex in the mouse. *Hum Mol Genet* 2009, **18**(7):1252–1265.

57. Courchesne E: Brain development in autism: early overgrowth followed by premature arrest of growth. *Ment Retard Dev Disabil Res Rev* 2004, **10**(2):106–111.
58. Tavazoie SF, Alvarez VA, Ridenour DA, Kwiatkowski DJ, Sabatini BL: Regulation of neuronal morphology and function by the tumor suppressors Tsc1 and Tsc2. *Nat Neurosci* 2005, **8**(12):1727–1734.
59. Zhou J, Blundell J, Ogawa S, Kwon CH, Zhang W, Sinton C, Powell CM, Parada LF: Pharmacological inhibition of mTORC1 suppresses anatomical, cellular, and behavioral abnormalities in neural-specific Pten knock-out mice. *J Neurosci* 2009, **29**(6):1773–1783.
60. Gkogkas CG, Khoutorsky A, Ran I, Rampakakis E, Nevarko T, Weatherill DB, Vasuta C, Yee S, Truitt M, Dallaire P *et al.*: Autism-related deficits via dysregulated eIF4E-dependent translational control. *Nature* 2013, **493**(7432):371–377.
61. Santini E, Huynh TN, MacAskill AF, Carter AG, Pierre P, Ruggero D, Kaphzan H, Klann E: Exaggerated translation causes synaptic and behavioural aberrations associated with autism. *Nature* 2013, **493**(7432):411–415.
62. Bhakar AL, Dolen G, Bear MF: The pathophysiology of fragile X (and what it teaches us about synapses). *Annu Rev Neurosci* 2012, **35**:417–443.
63. Darnell JC, Fraser CE, Mostovetsky O, Stefani G, Jones TA, Eddy SR: Kissing complex RNAs mediate interaction between the fragile-X mental retardation protein KH2 domain and brain polyribosomes. *Genes Dev* 2005, **19**:903–918.
64. Blackwell E, Zhang X, Ceman S: Arginines of the RGG box regulate FMRP association with polyribosomes and mRNA. *Hum Mol Genet* 2010, **19**(7):1314–1323.
65. Darnell JC, Van Driesche SJ, Zhang C, Hung KY, Mele A, Fraser CE, Stone EF, Chen C, Fak JJ, Chi SW *et al.*: FMRP stalls ribosomal translocation on mRNAs linked to synaptic function and autism. *Cell* 2011, **146**(2):247–261.
66. Iacoangeli A, Rozhdestvensky TS, Dolzhanskaya N, Tournier B, Schutt J, Brosius J, Denman RB, Khandjian EW, Kindler S, Tiedge H: Reply to Bagni: On BC1 RNA and the fragile X mental retardation protein. *Proc Natl Acad Sci USA* 2008, **105**(22):E29.
67. Iacoangeli A, Rozhdestvensky TS, Dolzhanskaya N, Tournier B, Schutt J, Brosius J, Denman RB, Khandjian EW, Kindler S, Tiedge H: On BC1 RNA and the fragile X mental retardation protein. *Proc Natl Acad Sci USA* 2008, **105**(2):734–739.
68. Stefani G, Fraser CE, Darnell JC, Darnell RB: Fragile X mental retardation protein is associated with translating polyribosomes in neuronal cells. *J Neurosci* 2004, **24**(33):7272–7276.
69. Wang H, Iacoangeli A, Lin D, Williams K, Denman RB, Hellen CU, Tiedge H: Dendritic BC1 RNA in translational control mechanisms. *J Cell Biol* 2005, **171**(5):811–821.

70. Hampson DR, Gholizadeh S, Pacey LK: Pathways to drug development for autism spectrum disorders. *Clin Pharmacol Ther* 2012, **91**(2):189–200.
71. Qin M, Kang J, Burlin TV, Jiang C, Smith CB: Postadolescent changes in regional cerebral protein synthesis: an in vivo study in the FMR1 null mouse. *J Neurosci* 2005, **25**(20):5087–5095.
72. Grossman AW, Aldridge GM, Lee KJ, Zeman MK, Jun CS, Azam HS, Arii T, Imoto K, Greenough WT, Rhyu IJ: Developmental characteristics of dendritic spines in the dentate gyrus of Fmr1 knockout mice. *Brain Res* 2010, **1355**:221–227.
73. Grossman AW, Aldridge GM, Weiler IJ, Greenough WT: Local protein synthesis and spine morphogenesis: Fragile X syndrome and beyond. *J Neurosci* 2006, **26**(27):7151–7155.
74. Huber KM, Gallagher SM, Warren ST, Bear MF: Altered synaptic plasticity in a mouse model of fragile X mental retardation. *Proc Natl Acad Sci U S A* 2002, **99**(11):7746–7750.
75. Nosyreva ED, Huber KM: Metabotropic receptor-dependent long-term depression persists in the absence of protein synthesis in the mouse model of fragile X syndrome. *J Neurophysiol* 2006, **95**:3291–3295.
76. Hou L, Antion MD, Hu D, Spencer CM, Paylor R, Klann E: Dynamic translational and proteasomal regulation of fragile X mental retardation protein controls mGluR-dependent long-term depression. *Neuron* 2006, **51**:441–454.
77. Weiler IJ, Irwin SA, Klintsova AY, Spencer CM, Eberwine J, Greenough WT: Fragile X mental retardation protein is translated near synapses. *Proc Natl Acad Sci USA* 1997, **94**:5395–5400.
78. Bear MF, Huber KM, Warren ST: The mGluR theory of fragile X mental retardation. *Trends Neurosci* 2004, **27**:370–377.
79. Yan QJ, Rammal M, Tranfaglia M, Bauchwitz RP: Suppression of two major Fragile X Syndrome mouse model phenotypes by the mGluR5 antagonist MPEP. *Neuropharmacology* 2005, **49**:1053–1066.
80. Dolen G, Osterweil E, Rao BS, Smith GB, Auerbach BD, Chattarji S: Correction of fragile X syndrome in mice. *Neuron* 2007, **56**:955–962.
81. McBride SM, Choi CH, Wang Y, Liebelt D, Braunstein E, Ferreiro D: Pharmacological rescue of synaptic plasticity, courtship behavior, and mushroom body defects in a Drosophila model of fragile X syndrome. *Neuron* 2005, **45**:753–764.
82. Osterweil EK, Krueger DD, Reinhold K, Bear MF: Hypersensitivity to mGluR5 and ERK1/2 leads to excessive protein synthesis in the hippocampus of a mouse model of fragile X syndrome. *J Neurosci* 2010, **30**:15616–15627.
83. Aschrafi A, Cunningham BA, Edelman GM, Vanderklish PW: The fragile X mental retardation protein and group I metabotropic glutamate receptors regulate levels of mRNA granules in brain. *Proc Natl Acad Sci USA* 2005, **102**(6):2180–2185.

84. Chuang SC, Zhao W, Bauchwitz R, Yan Q, Bianchi R, Wong RK: Prolonged epileptiform discharges induced by altered group I metabotropic glutamate receptor-mediated synaptic responses in hippocampal slices of a fragile X mouse model. *J Neurosci* 2005, **25**:8048–8055.
85. Hagerman R, Hoem G, Hagerman P: Fragile X and autism: Intertwined at the molecular level leading to targeted treatments. *Mol Autism* 2010, **1**(1):12.
86. Weng N, Weiler IJ, Sumis A, Berry-Kravis E, Greenough WT: Early-phase ERK activation as a biomarker for metabolic status in fragile X syndrome. *Am J Med Genet B Neuropsychiatr Genet* 2008, 147B:1253–1257.
87. Wang X, Snape M, Klann E, Stone JG, Singh A, Petersen RB, Castellani RJ, Casadesus G, Smith MA, Zhu X: Activation of the extracellular signal-regulated kinase pathway contributes to the behavioral deficit of fragile x-syndrome. *J Neurochem* 2012, **121**(4):672–679.
88. Gallagher SM, Daly CA, Bear MF, Huber KM: Extracellular signal-regulated protein kinase activation is required for metabotropic glutamate receptor-dependent long-term depression in hippocampal area CA1. *J Neurosci* 2004, **24**(20):4859–4864.
89. Hoeffer CA, Sanchez E, Hagerman RJ, Mu Y, Nguyen DV, Wong H, Whelan AM, Zukin RS, Klann E, Tassone F: Altered mTOR signaling and enhanced CYFIP2 expression levels in subjects with fragile X syndrome. *Genes Brain Behav* 2012, **11**(3):332–341.
90. Gross C, Nakamoto M, Yao X, Chan CB, Yim SY, Ye K: Excess phosphoinositide 3-kinase subunit synthesis and activity as a novel therapeutic target in fragile X syndrome. *J Neurosci* 2010, **30**:10624–10638.
91. Sharma A, Hoeffer CA, Takayasu Y, Miyawaki T, McBride SM, Klann E: Dysregulation of mTOR signaling in fragile X syndrome. *J Neurosci* 2010, **30**:694–702.
92. Bhattacharya A, Kaphzan H, Alvarez-Dieppa AC, Murphy JP, Pierre P, Klann E: Genetic Removal of p70 S6 Kinase 1 Corrects Molecular, Synaptic, and Behavioral Phenotypes in Fragile X Syndrome Mice. *Neuron* 2012, **76**(2):325–337.
93. Osterweil EK, Chuang SC, Chubykin AA, Sidorov M, Bianchi R, Wong RK, Bear MF: Lovastatin corrects excess protein synthesis and prevents epileptogenesis in a mouse model of fragile X syndrome. *Neuron* 2013, **77**(2):243–250.
94. Udagawa T, Farny NG, Jakovcevski M, Kaphzan H, Alarcon JM, Anilkumar S, Ivshina M, Hurt JA, Nagaoka K, Nalavadi VC *et al.*: Genetic and acute CPEB1 depletion ameliorate fragile X pathophysiology. *Nat Med* 2013, **19**(11): 1473–1477.
95. Malina A, Mills JR, Pelletier J: Emerging therapeutics targeting mRNA translation. *Cold Spring Harb Perspect Biol* 2012, **4**(4):a012377.
96. Hampson DR, Adusei DC, Pacey LK: The neurochemical basis for the treatment of autism spectrum disorders and Fragile X Syndrome. *Biochem Pharmacol* 2011, **81**(9):1078–1086.

97. Berry-Kravis E, Hessl D, Coffey S, Hervey C, Schneider A, Yuhas J: A pilot open label, single dose trial of fenobam in adults with fragile X syndrome. *J Med Genet* 2009, **46**:266–271.
98. Levenga J, Hayashi S, de Vrij FM, Koekkoek SK, van der Linde HC, Nieuwenhuizen I, Song C, Buijsen RA, Pop AS, Gomezmancilla B *et al.*: AFQ056, a new mGluR5 antagonist for treatment of fragile X syndrome. *Neurobiol Dis* 2011, **42**(3):311–317.
99. Jacquemont S, Curie A, des Portes V, Torrioli MG, Berry-Kravis E, Hagerman RJ: Epigenetic modification of the FMR1 gene in fragile X syndrome is associated with differential response to the mGluR5 antagonist AFQ056. *Sci Transl Med* 2011, **3**:64ra61.
100. Pacey LK, Doss L, Cifelli C, der Kooy DV, Heximer SP, Hampson DR: Genetic deletion of regulator of G-protein signaling 4 (RGS4) rescues a subset of fragile X related phenotypes in the FMR1 knockout mouse. *Mol Cell Neurosci* 2011, **46**:563–572.
101. Pacey LK, Heximer SP, Hampson DR: Increased GABA(B) receptor-mediated signaling reduces the susceptibility of fragile X knockout mice to audiogenic seizures. *Mol Pharmacol* 2009, **76**(1):18–24.
102. Berry-Kravis E, Sumis A, Hervey C, Nelson M, Porges SW, Weng N, Weiler IJ, Greenough WT: Open-label treatment trial of lithium to target the underlying defect in fragile X syndrome. *J Dev Behav Pediatr* 2008, **29**(4): 293–302.
103. Min WW, Yuskaitis CJ, Yan Q, Sikorski C, Chen S, Jope RS: Elevated glycogen synthase kinase-3 activity in fragile X mice: key metabolic regulator with evidence for treatment potential. *Neuropharmacology* 2009, **56**:463–472.
104. Yuskaitis CJ, Mines MA, King MK, Sweatt JD, Miller CA, Jope RS: Lithium ameliorates altered glycogen synthase kinase-3 and behavior in a mouse model of fragile X syndrome. *Biochem Pharmacol* 2010, **79**:632–646.
105. Liu ZH, Chuang DM, Smith CB: Lithium ameliorates phenotypic deficits in a mouse model of fragile X syndrome. *Int J Neuropsychopharmacol* 2011, **14**:618–630.
106. Mines MA, Yuskaitis CJ, King MK, Beurel E, Jope RS: GSK3 influences social preference and anxiety-related behaviors during social interaction in a mouse model of fragile X syndrome and autism. *PLoS One* 2010, **5**:e9706.
107. Choi CH, McBride SM, Schoenfeld BP, Liebelt DA, Ferreiro D, Ferrick NJ: Age-dependent cognitive impairment in a Drosophila fragile X model and its pharmacological rescue. *Biogerontology* 2010, **11**:347–362.
108. Berry-Kravis E, Sumis A, Hervey C, Nelson M, Porges SW, Weng N: Open-label treatment trial of lithium to target the underlying defect in fragile X syndrome. *J Dev Behav Pediatr* 2008, **29**:293–302.
109. Siller SS, Broadie K: Matrix metalloproteinases and minocycline: therapeutic avenues for fragile X syndrome. *Neural Plast* 2012, **2012**:124548.

110. Bilousova TV, Dansie L, Ngo M, Aye J, Charles JR, Ethell DW: Minocycline promotes dendritic spine maturation and improves behavioural performance in the fragile X mouse model. *J Med Genet* 2009, **46**:94–102.
111. Paribello C, Tao L, Folino A, Berry-Kravis E, Tranfaglia M, Ethell I: Open-label add-on treatment trial of minocycline in fragile X syndrome. *BMC Neurol* 2010, **10**:91.
112. Utari A, Chonchaiya W, Rivera SM, Schneider A, Hagerman RJ, Faradz SM: Side effects of minocycline treatment in patients with fragile X syndrome and exploration of outcome measures. *Am J Intellect Dev Disabil* 2010, **115**:433–443.
113. Hayashi ML, Rao BS, Seo JS, Choi HS, Dolan BM, Choi SY: Inhibition of p21-activated kinase rescues symptoms of fragile X syndrome in mice. *Proc Natl Acad Sci USA* 2007, **104**:11489–11494.
114. Chen LY, Rex CS, Babayan AH, Kramar EA, Lynch G, Gall CM: Physiological activation of synaptic Rac>PAK (p-21 activated kinase) signaling is defective in a mouse model of fragile X syndrome. *J Neurosci* 2010, **30**:10977–10984.
115. Erickson CA, Mullett JE, McDougle CJ: Open-label memantine in fragile X syndrome. *J Autism Dev Disord* 2009, **39**:1629–1635.
116. Abrahams BS, Geschwind DH: Connecting genes to brain in the autism spectrum disorders. *Arch Neurol* 2010, **67**(4):395–399.

Epigenetic, Environmental, and Physiological Contributions to ASD

Chapter 12

Epigenetics in Autism

Matthew Ginsberg and Marvin Natowicz*

Abstract

Despite many gains in our understanding of genetic aspects of autism, the etiologic basis of autism in most affected individuals is unknown and there is a very incomplete understanding about the downstream biology occurring in the autistic brain. Epigenetics has emerged as a promising field that may explain how some cases of autism arise, both as a result of and independently from monogenic and chromosomal mutations. Here we discuss the emerging evidence of how epigenetic dysregulation may play a role in autism pathophysiology, as well as the challenges and potential strategies in studying possible epigenetic processes in autism.

Keywords: Autism; Epigenetics; DNA methylation; Histone, Genetics.

Introduction

There are a number of single gene disorders, chromosomal abnormalities and teratogenic exposures that cause or are major contributing factors in causation of an autistic phenotype. However, chromosomal abnormalities account for only 5–20% of cases of ASDs and other known causes comprise an even smaller portion.[1–4] The majority of individuals having autism are without an etiological

*Correspondence to: Cleveland Clinic Lerner College of Medicine, and Pathology & Laboratory Medicine, Genomic Medicine, Neurological and Pediatrics Institutes, Cleveland Clinic, Cleveland, OH 44195. Email: natowim@ccf.org

basis and are categorized as idiopathic. Even within this group, there is substantial heterogeneity. For example, the classical triad of symptoms that has historically been used to define the ASDs show only moderate covariance in multiple independent studies.[5–7]

Despite this heterogeneity, there is substantial evidence that ASDs are highly heritable. The most convincing evidence comes from twin studies, which show concordance rates of about 90% in monozygotic twins and substantially lower concordance in dizygotic twins, although those numbers are variable depending on whether a broader autistic phenotype is considered in the analysis.[8–12] Yet, despite the discovery of more than 100 monogenic and chromosomal copy number loss/gain risk loci,[13] these mutations do not explain a majority of autism risk. Recently, attention has turned to alternative etiological considerations. Noting the high heritability of autism and the failure to identify loci that together account for most of the etiologic basis of autism, Beaudet suggested that epigenetic factors might be major risk factors for autism.[2]

Epigenetics refers to stable changes of chromatin that affect gene expression, independent of changes in DNA sequence. In this chapter we review basic aspects of epigenetic biology, the role of epigenetic physiology and pathophysiology in the central nervous system, and then discuss evidence for the role of epigenetic factors in autism. We conclude by discussing the implications of current research in autism epigenetics, challenges to the field, and potential pathways forward.

Fundamentals of Epigenetics

An epigenetic signal refers to a stable non-sequence change to chromatin that affects gene regulation (Figure 1). In essence, the epigenome represents a form of regulatory "clay", forming a particular state of a region of chromatin in response to inducers that may persist after the inducer is no longer present. These signals establish gene expression patterns that are essential for establishing commitment to cellular differentiation, the inactivation of X chromosomes in females, genetic imprinting (a process of silencing one allele of a gene from a particular parent), and other important tasks.[14–20]

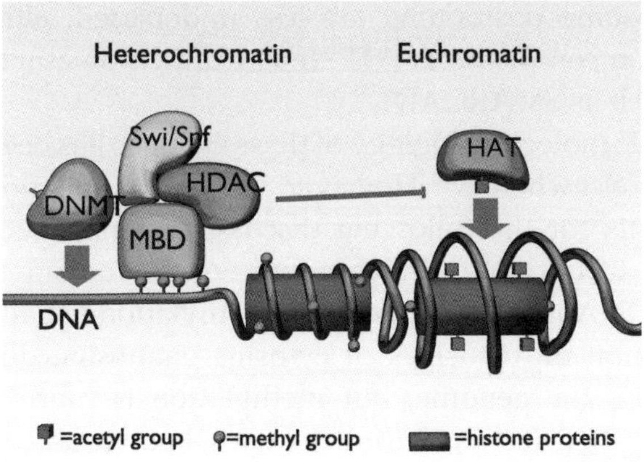

Figure 1. Classical epigenetic mechanisms. DNA forms heterochromatin (repressive) or euchromatin (permissive) in association with histone complexes. DNA methylation by DNA methyltransferases (DNMT) leads to repression of genes by inhibiting binding of transcription factors and promoting binding of Methyl Binding Domain (MBD) proteins such as MECP2. These proteins recruit complexes that include histone deacetylases and chromatin remodeling complexes like Swi/Snf. Euchromatin is associated with lack of methylated DNA and specific histone posttranslational modifications, such as methylation at lysine 4 or acetylation at various residues; HAT = histone acetyltransferase; HDAC = histone deacetylase.

Epigenetic signals can occur as a result of different molecular processes. Epigenetic signals include: (1) direct methylation of cytosine residues in CpG dinucleotides (where a cytosine is directly followed by a guanine); (2) covalent modification of histones; (3) nucleosome positioning/chromosome remodeling (sliding of the histone spools along DNA); and (4) stable, chromatin-associated noncoding RNAs. These signals also require a number of mediators to effectively regulate gene expression, including proteins and noncoding RNAs. Epigenetic marks are often inherited through mitosis; although most of the epigenome is reset during meiosis, there is emerging evidence of inherited epigenomic changes through meiosis as well.[21] DNA methylation typically persists through DNA replication by the action of maintenance DNA methyltransferase, DNMT1. Maintenance of histone post-translational modifications and nucleosome positioning are less understood, although they likely rely on complexes of ATP-dependent chromatin remodeling proteins such as SMARCAD1.[22]

The basic molecular biology of these changes has been reviewed extensively elsewhere.[17,23] However, it is worth mentioning a few salient points for the following discussion. First, methylation of cytosine in CpG dinucleotides occurs via the action of enzymes known as DNA methyltransferases. Methylation usually occurs at the 5′ position on cytosines of cytosine-guanosine dinucleotides (CpG) across the genome, but methylation is more variable in regions enriched for CpG dinucleotides known as CpG islands. These islands often occur in promoters or intragenic noncoding regions, where they regulate the expression of associated genes, but are also found in more distant locations.[24,25]

Histones are proteins that associate with DNA to form chromatin. Octamers of histones H2A, H2B, H3, and H4 associate with DNA in spool-like structures called nucleosomes, and can regulate access of transcription factors and RNA polymerase to this DNA. Covalent modifications of nucleosomal histones cause changes in the structure of the chromatin that can affect methylation and transcription of the associated DNA. Some important modifications include histone methylation and histone acetylation. In general, histone acetylation loosens the association between histones and DNA, allowing for greater transcription. Methylation at particular residues, such as histone 3 lysine 9 (H3K9), is associated with gene repression, while at other sites, such as histone 3 lysine 4 (H3K4), methylation is associated with gene activation.

Chromatin remodeling involves families of ATP-dependent enzymes that act in multi-subunit complexes to position nucleosomes. Although these processes were thought to simply be permissive actions for other regulatory elements such as transcription factors to bind DNA, it is now thought that remodeling complexes are developmentally and tissue specific, and regulate both transcription and chromosomal architecture.[26]

Epigenetic mechanisms play important and unique roles in the nervous system, especially during its early development.[27–30]

Although only a few imprinted regions have been explored in detail, there is now evidence that, in the brain specifically, there may be sexually dimorphic imprinting of a significant number of both X-linked and autosomal genes.[20] This creates a rich and complex developmental regulatory environment with direct relevance for sex-skewed disorders of the nervous system such as autism. Chromatin modifications are also necessary for certain forms of memory, making them relevant for a number of cognitive disorders.[31] Several chromatin modifying enzymes may be involved in low error tolerance regulatory networks important in metabolism and cell cycle control during development, and interact with several autism candidate gene products.[32]

Non-coding RNAs play an important role in mediating epigenetic signals, but they are less understood mechanistically, and are not necessarily as persistent as the other types of epigenetic signals. We will therefore not discuss them in detail, although they may be disrupted in autism and have epigenetic consequences.

Epigenetics in Disease

An epimutation is an abnormal epigenetic signal that leads to abnormal gene expression. Abnormal epigenetic signals have been most studied in the setting of neoplasms, where epimutations arise in somatic cell lineages.[33] Recent work has identified epigenetic pathology in the central nervous system, including in several neuropsychiatric disorders such as depression, addiction, status epilepticus and schizophrenia.[34–39]

There are several useful ways to conceptualize epimutations (Figure 2). One way to classify epimutations is by the manner in which they arise. This type of classification may have relevance in understanding the pathophysiology of a disorder and in an analysis as to how modifiable an epimutation is or how it might be prevented. This developmental classification involves three models: (1) in the *obligatory* model, a particular genotype is always associated with a particular epigenotype [epimutation]; (2) in the *facilitated* model, a particular genotype is permissive

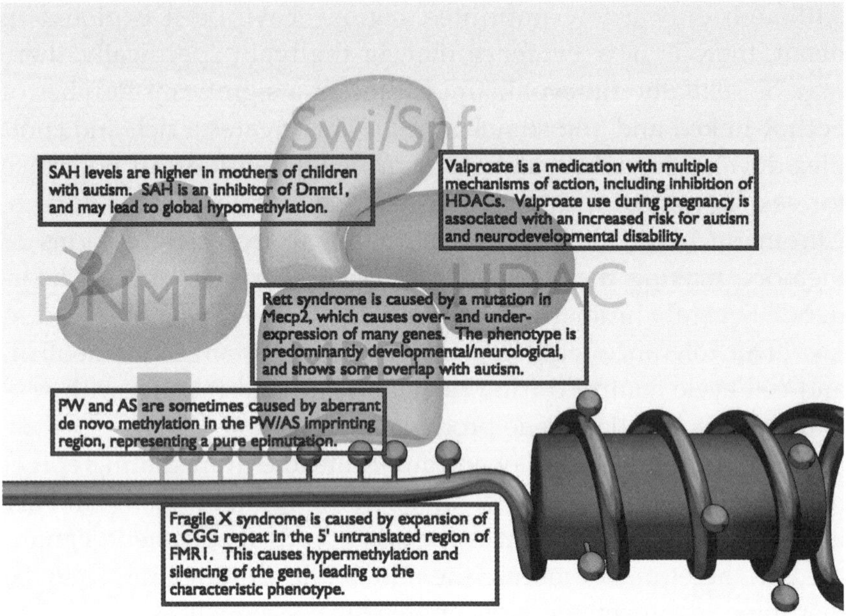

Figure 2. Epigenetic disruption in neurodevelopmental disease. Several neurodevelopmental disorders are caused by disruptions in epigenetic mechanisms. These mechanisms include metabolic or toxic inhibition of key effector proteins, inactivating mutations in the genes encoding these effector proteins, mutations that induce epigenetic inactivation of autism risk genes, and pure epimutations in these autism risk genes. PW = Prader-Willi syndrome; AS = Angelman Syndrome; SAH = s-adenosylhomocysteine; HDAC = histone deacetylase; DNMT = DNA methyltransferase.

but not deterministic of an epimutation; and (3) in the *de novo* model, an epimutation occurs independently of the genotype. The latter two models suggest mechanisms in which the epimutation may be more reversible and more dependent on environmental influences.

Another useful classification system relates to the targets of epimutations and, more specifically, whether an epimutation inactivates a single autism risk gene (analogous to a traditional genetic mutation) or multiple epimutations arise in multiple autism risk genes (and, perhaps, genes that are unrelated to autism) as a result of damage to core epigenomic machinery. This classification scheme speaks to the proximate pathobiology that occurs in epigenetically determined disorders.

Still another classification of epimutations is a mechanistic approach, i.e., a categorization based on which component of epigenesis is dysfunctional, either inherently or on an acquired basis. In this approach, epimutations may be due to abnormalities of DNA methylation, of histone methylation, of histone acetylation, of chromatic remodeling components, *etc.* Within each of these categories, there are obvious subcategories, such as mutations of specific chromatin remodeling genes and abnormalities of different DNA methyltransferases.

We will discuss several syndromic forms of autism that exemplify these models of autism risk, where pathophysiology is better understood. We will then discuss environmental influences on epigenetics and evidence of epigenetic dysregulation in non-syndromic autism, where pathophysiology is less understood, but where epigenetic models in syndromic autism may be helpful in guiding research.

Syndromic Autism

Several genetic syndromes serve as models of epigenetic dysregulation. We will discuss three syndromes, which serve as models of different mechanisms by which epimutations arise and perpetuate an autistic phenotype.

Fragile X syndrome

Fragile X syndrome is an X-linked disorder characterized by mental retardation, dysmorphisms, and a high rate of autistic phenotype. It is caused by a CGG trinucleotide repeat expansion in the 5′ untranslated region of the *FMR1* gene.[40] For example, when the 5′ untranslated region of the *FMR1* gene expands to more than 200 repeats, it is almost always hypermethylated. In turn, expression of the gene and protein are decreased, leading to downstream signaling abnormalities.[41,42] Since the genotype determines the methylation status of the gene, this would be classified as an *obligatory* epimutation that inactivates a single gene (Figure 2).

Recent investigations have probed the use of inhibitors of the mGluR5 metabotropic glutamate receptor to ameliorate symptoms of Fragile X syndrome based on data implicating the pathway in the disorder's pathophysiology. Interestingly, the initial results of a small clinical trial of one such compound were disappointing, but the investigators uncovered a surprising hypothesis in *post hoc* analysis.[43] Subjects of the trial all had full mutations of the *FMR1* gene, but had variable degrees of hypermethylation in the 5′ untranslated region. Only subjects with fully methylated alleles showed significant and consistent improvement in symptoms over the trial period, while those with partial hypermethylation did not demonstrate improvement. This suggests that even in obligatory models of epigenetic diseases, there is variation that may be induced by the environment or even be amenable to therapy.

Rett syndrome

Rett syndrome is a disorder that previously was considered a type of Pervasive Developmental Disorder under the Diagnostic and Statistical Manual IV,[44] although it is not categorized as a specific type of autistic disorder in the DSM V. It is caused by mutations in *MECP2* on the X chromosome, which produces a protein that binds to methylated DNA (Methyl Binding Domain protein) and affects the transcription of multiple genes. The phenotype overlaps significantly with classic autism, including lack of speech and a transient period of social aversion during which ASD criteria may be met. The pathophysiologic basis of the phenotype is the result of abnormal regulation of genes that control the terminal differentiation and maturation of neurons, and that include processes as diverse as cell signaling, metabolism, cell architecture, and neuroprotection.[45,46] MECP2 binds to methylated DNA and recruits additional regulatory elements such as histone deacetylases (HDACs) to inhibit gene expression, but also acts through other mechanisms.[47] Whole-genome analysis revealed that MECP2 actually associates with non-methylated genomic regions and is not only associated with transcriptional repression, but may activate some genes.[48] Thus, Rett syndrome serves as an example of an obligatory model of epigenetic dysregulation affecting many genes. The fact that mutations in *MECP2* cause global dysfunction of methylation-dependent gene regulation and produce a predominantly neurodevelopmental phenotype indicate the importance of epigenetic mechanisms in the formation of the central nervous system. Subsequently, several other MBD proteins[49,50] and other core components of the epigenetic machinery have been shown to be mutated in disorders with an autism phenotype, suggesting that some minority of the autism population may harbor mutations that exert their effects through global epigenetic dysregulation.[51]

Angelman syndrome

Prader-Willi and Angelman syndromes are both caused by abnormalities of imprinted regions on chromosome 15. Angelman syndrome in particular shares phenotypic overlap with classic autism. Prader-Willi syndrome is usually caused by paternally-inherited deletions in the 15q11-13 region, while Angelman syndrome is usually caused by maternally-inherited deletions in this region, as a result of imprinting in the region that silences normal alleles from the non-carrier parent. Although the precise genes involved in Prader-Willi syndrome remain unknown, there is evidence that Angelman syndrome is caused by loss of ubiquitin E3A ligase *UBE3A*.[52,53] Of significance is the fact that these disorders can occasionally be caused by mutations in the Imprinting Center, a genomic region that regulates the imprinting process (either obligatory or facilitated models), or by *de novo* epimutations of the significant genes (Figure 2).[54] In addition, duplications of the 15q11-13 region implicated in PW and AS also cause a syndrome that includes seizures and often an autistic phenotype. Evidence indicates that the duplication causes abnormal methylation in the region and at other imprinted loci across the genome.[55] Importantly, *de novo* epimutations are invisible to traditional genetic techniques, and therefore could represent some fraction of the large number of persons with idiopathic autism after exhaustive genetic investigations. This model also allows for an environmental mechanism of disease.

Environmental Epigenetics

The prevalence of ASDs is currently increasing.[56] There is debate over whether this solely reflects an increase in ascertainment, or also includes a true increase in incidence. However, if there is an increase in the incidence of ASDs, it cannot be due to an increase in frequency of risk-contributing alleles. Additionally, concordance in monozygotic twins is not 100%, indicating some role for environmental factors. Epigenetic mechanisms are an ideal mediator of gene-environment interactions that could be responsible for an increase in ASD incidence (Figure 2).

For example, there is an association between the disorder and abnormalities of methyl-group metabolism. Several studies have linked polymorphisms in folate metabolism genes with autism.[57–60] Folate metabolism is essential for the metabolic maintenance of methyl groups through its assistance in the recycling of homocysteine to methionine. This process is critical for DNA methylation, among other metabolic pathways. In addition, maternal polymorphisms in a folate carrier gene are associated with an increased risk for autism in offspring, and this risk is independent of paternal or proband genotype status, indicating that intrauterine folate dysmetabolism may increase risk for autism.[61] This is also supported by epidemiologic evidence that periconceptional folate supplementation appears to be protective against the development of autism.[62]

Autism has also been linked to prenatal exposure to medications,[63,64] infections,[65,66] and possibly some environmental toxins (such as endocrine disruptors)[67] which may exert their effects through epigenetic mechanisms.[68,69] The above are likely to dysregulate epigenetic proteins or noncoding RNA that affect multiple genes.

The fact that exposure to particular medications can alter epigenetic regulation of genes in the central nervous system also implies that these sites may be potential therapeutic targets. Several medications may be able to positively affect epigenetic aberrations as is already occurring in the treatment of other diseases such as cancer and possibly schizophrenia.[69–71]

Epigenetic Dysregulation in Non-Syndromic Autism

Several recent studies directly support the hypothesis that there are epigenetic contributions to non-syndromic autism, although often we do not know whether the epimutations are a result of underlying genetic variation or whether there are more than one epimutation acting in an individual. In a proof-of-concept study, a person with familial autism showed increased methylation at the promoter of the oxytocin receptor gene (*OXTR*) with concomitant decrease in expression, although no mutation in the gene was identified. Postmortem analysis of a separate set of autistic brains revealed that *OXTR* was significantly more methylated and less expressed in some samples than in control brain samples in the temporal cortex.[72] Oxytocin signaling is a strong candidate for dysregulation in autism given its role in social behavior.[73]

Additional abnormal DNA methylation was subsequently found in a small number of cases at particular autism risk candidate gene loci. For instance, methylation at the promoter of *UBE3A* was observed in an autistic brain.[74] As mentioned previously, UBE3A is involved in several overlapping syndromic forms of autism. Abnormal methylation at the *MECP2* promoter was identified in a small number of autistic subjects as well.[75] These studies were targeted at particular candidate genes, so it is not clear whether these epimutations were part of a broader disruption in the epigenome (except in the case of *MECP2*).

However, studies using high-throughput genome-wide assays of DNA methylation or histone modifications are capable of identifying global patterns of epigenomic disruption. Some investigators have leveraged this technology on discordant monozygotic twins (MZ), who share essentially identical genomes, and offer the opportunity to separate de novo epigenetic events from obligate epigenetic changes induced by genomic variation. In a genome-wide scan of lymphoblastoid cell lines from discordant MZ twins, a number of significant differentially methylated regions were discovered and cross-referenced to gene expression profiling data from the same samples.[76,77] Altered methylation in two genes, *BCL2* and *RORA*, was associated with altered levels of protein product in a small sample of postmortem autistic brains. Another study of twin methylomes, including 6 discordant MZ twin pairs, demonstrated numerous differentially methylated regions (DMRs) between twins, as well as DMRs between autistic and non-autistic twin sets and for specific symptom domains.[78] Although these studies provide evidence for stochastic and non-shared environmental contributions to the autistic phenotype, they are limited in several ways. First, evaluations of MZ discordant twins enrich the sample for epigenetically driven disease, and we cannot necessarily extrapolate results to the broader ASD community. Second, most of these studies have demonstrated methylation differences that are relatively small, given the typical bimodal distribution of DNA methylation. Although such differences may represent biologically relevant changes, the study designs used did not stringently filter for statistically significant results that accounted for multiple hypothesis testing. Thus, the results could reflect spurious associations. Finally, it is unclear that peripheral tissues examined in these studies are representative of epigenetic patterns seen in the brain, an important issue given the tissue-specific epigenetic patterns established during development.

A small number of studies have examined postmortem brain tissue in an attempt to identify epigenetic dysregulation at anatomically relevant sites in a non-syndromic autism population. Our own investigation into genome-wide DNA methylation revealed no global hyper- or hypomethylation in cerebellar hemisphere or occipital cortex.[79] Additionally, there were no significantly DMRs between nine cases and nine matched controls, including at several of the candidate genes *MECP2, OXTR, BCL2*, or *RORA*, although we lacked statistical power to detect subtle differences between groups. Shulha and colleagues analyzed prefrontal cortex in 16 autism-control pairs for abnormal histone modifications.[80] They identified four brains that showed abnormal spreading of histone lysine 4 trimethylation including at multiple autism candidate genes. This study was strengthened by the isolation of neuronal nuclei to isolate epigenetic abnormalities in this unique and most relevant cell type.

Although these studies are methodologically varied and came to differing conclusions, there are some consistent findings relating to the role of epigenetic factors in autism pathogenesis. First, all studies examining genome-wide epigenetic signals have identified no evidence for global epigenetic dysregulation, whether in blood or brain tissue (i.e., no evidence for different overall levels of DNA methylation or histone modfications). This suggests that epigenetic abnormalities are likely to affect relatively small numbers of genes in an individual with ASD. Second, all studies have noted extraordinary heterogeneity within the autism groups. When DMRs have been identified, they are usually present in only one or a few subjects, consistent with the heterogeneity seen at both the genetic and phenotypic level. Third, several genome-wide studies of RNA expression in the autistic brain have been carried out, and have produced more robust results; this may be due to epimutations in autism being targeted, with frequencies similar to genetic mutations.[79,81,82]

Potential Sources of Epigenetic Dysregulation

How could epigenetic dysregulation contribute to the autistic phenotype? First, in a subset of individuals, *de novo* epimutations established in the germline might abnormally silence genes (or de-repress silenced genes) involved in a number of biological pathways implicated in autism pathogenesis, such as synaptic formation and maintenance, neurotransmitter pathways, cell cycle control, and apoptosis.[83] Second, *de novo* epimutations or classical mutations could inactivate or alter the dosage of genes that code for proteins involved with epigenetic machinery, causing more global dysfunctions, such as those seen in Rett syndrome or Rubinstein-Taybi syndrome. In a study of mutations identified by whole exome sequencing of cases of autism, mutated genes were enriched for those related to chromatin maintenance, supporting such a hypothesis.[84] Additionally, non-coding RNAs make up a large portion of the genome, and are key mediators of the epigenomic machinery.[85,86] Given the large proportion of autism cases that are idiopathic despite genetic evaluations that target the exome, and the high heritability of autism, this is an important hypothesis to consider. Third, certain epimutations might even be heritable, caused by incomplete resetting of the methylome during meiosis and transmissible regulatory RNAs.[87,88] A recent study demonstrated that paternal diet could influence gene methylation, expression, and phenotype in F1 offspring, validating epigenetic inheritance in mammals.[21] Fourth, it has recently been shown that non-imprinted, allele-specific methylation is common and is directly related to variation in *cis*.[89–91] Such variants could act as individual "hits" in a multi-hit hypothesis for autism pathogenesis.

Future Directions

There are a number of challenges to investigating epigenetic dysregulation in autism, including both technical and biological limitations.

The advantages and disadvantages of methods that evaluate DNA methylation, posttranslational histone modifications, total RNA, and nucleosome positioning have been reviewed thoroughly.[92–98] One of the most important challenges to future studies will be statistical. High throughput technologies are now allowing rapid, unbiased assaying of genome-wide epigenomic variation. However, such studies need to address the multiple hypothesis testing problem. This may limit statistical power to detect relevant epimutations. For instance, if epimutation X occurs as hypermethylation at a single, key CpG dinucleotide in an autism risk gene, the difference in methylation status between cases and controls will need to be dramatic and consistent to ascertain it in genome-wide analyses. However, autism is a heterogeneous disease; there are already a large number of known rare pathologic copy number variants, single gene disorders, and metabolic disorders that are associated with autism. Relevant epimutations may be just as diverse, making them difficult to identify in case-control studies. There are some potential solutions to this problem. Subgrouping cases in studies into relevant endophenotypes might help to reduce heterogeneity. Another approach is to try and identify novel epimutations indirectly, by assessing their downstream effects. A single epimutation, like a genetic mutation, may give rise to numerous abnormalities at the RNA or protein level. Bioinformatic approaches to studying transcriptomic and proteomic datasets may yield important epigenetic hypotheses.

Another challenge is studying the relevant tissue in epigenetic studies because of tissue specific patterns established during development and differentiation of cell lineages. Thus, for a particular locus, peripheral tissue may normally be methylated and brain demethylated, and in the disease state both might be methylated. Examinations of peripheral tissue would in this instance miss the epimutation. Some or many epimutations might be present in multiple tissues, but any tissue investigated should be validated against brain tissue. Adding complexity to this challenge is that even different regions of the same brain, or particular neuronal populations might have unique epigenetic signatures or epimutations, and such changes may be age-specific. Since acquisition of human brain tissue is challenging, there is a need for animal models that recapitulate epigenetic abnormalities in the autism brain. One approach is to look for common epigenetic abnormalities in different animal models of autism syndromes. Additionally, use of induced pluripotent stem cells from peripheral tissues may be differentiated into neural tissue. However, epigenetic signatures in these cells may or may not faithfully duplicate those in CNS tissue, depending on the source of epigenetic dysregulation and methods used. Efficient use of existing postmortem and surgically-derived human brain tissue will also be essential, along with a better understanding of normal human epigenetic brain development.

Finally, unlike the genome, the epigenome is modifiable, and many psychoactive drugs used by persons with autism may influence the epigenome and bias studies. Epidemiologic investigations may need to exclude, match, or control for medication use, age, and other relevant covariates that might influence results of epigenetic studies.

Conclusions

Identifying the pathogenesis and pathophysiology of ASDs has been extremely challenging. Despite showing high heritability, traditional genetic analyses have not revealed any common mutations that would account for the vast majority of cases. Rather, the data suggests that these disorders are extremely heterogeneous and may have multiple etiologies and involve multiple biological pathways. However, support for an epigenetic contribution to at least some cases of autism stems from evidence that: (1) some neuropsychiatric disorders are mediated by epigenetic mechanisms, (2) several syndromes associated with epigenetic dysregulation show overlapping phenotypes with autism, and (3) preliminary work shows a direct link between abnormal epigenetic regulation at particular genes and autism. It may be that a combination of genetic and epigenetic abnormalities function together to act on multiple biological pathways, and that the particular mutations and epimutations differ in each individual. The implications of this research are not purely academic. Drugs could target pathologic epigenetic modifications; in fact such therapies are already in use in neoplastic and dysplastic diseases.[69] In addition, environmental risk factors for autism may be identified that lead to prevention programs. The identification of unique epigenetic marks in peripheral tissue of autistic persons could also lead to effective biomarkers for diagnosis and subgrouping in autism. Further work should follow up on these leads to validate findings and identify any common pathways that might lead to tools for the diagnosis or treatment of ASDs.

Executive Summary

- Epigenetic regulatory signals are stable, non-sequence changes to chromatin that affect gene expression.
- Epigenetic mechanisms are especially crucial to the developing human brain, and are implicated in neurodevelopmental disorders and autism syndromes.
- There is emerging direct evidence that epigenetic abnormalities may be fundamental to autism pathophysiology in the broader autism population.
- Epigenetic dysregulation can take the form of obligate epimutations in autism risk genes caused by traditional mutations, mutations in core epigenetic regulatory proteins, or pure epimutations that inactivate an autism risk gene.
- Epigenetic regulation provides a mechanism whereby gene by environment interactions may influence autism risk, or even offer opportunities for therapeutic targets.
- Challenges to studying epigenetic factors in autism include technical/statistical limitations, access to tissue, and disease complexity, but there are many promising strategies to address them.

References

1. Abrahams BS, Geschwind DH: Advances in autism genetics: on the threshold of a new neurobiology. *Nat Rev Genet* 2008, **9**:341–355.
2. Beaudet AL: Autism: highly heritable but not inherited. *Nat Med* 2007, **13**:534–536.
3. Rutter M: Chapter 16: Genetic Influences and Autism. In *Handb Autism Pervasive Dev Disord* 3rd edition. Genetic Influences and Autism: Wiley; 2005.
4. Wassink TH, Piven J, Patil SR: Chromosomal abnormalities in a clinic sample of individuals with autistic disorder. *Psychiatr Genet* 2001, **11**:57–63.
5. Hoekstra RA, Happé F, Baron-Cohen S, Ronald A: Association between extreme autistic traits and intellectual disability: insights from a general population twin study. *Br J Psychiatry* 2009, **195**:531–536.
6. Ronald A, Happé F, Bolton P, Butcher LM, Price TS, Wheelwright S, Baron-Cohen S, Plomin R: Genetic heterogeneity between the three components of the autism spectrum: a twin study. *J Am Acad Child Adolesc Psychiatry* 2006, **45**:691–699.

7. Happé F, Ronald A: The "fractionable autism triad": a review of evidence from behavioural, genetic, cognitive and neural research. *Neuropsychol Rev* 2008, **18**:287–304.
8. Steffenburg S, Gillberg C, Hellgren L, Andersson L, Gillberg IC, Jakobsson G, Bohman M: A twin study of autism in Denmark, Finland, Iceland, Norway and Sweden. *J Child Psychol Psychiatry* 1989, **30**:405–416.
9. Bailey A, Le Couteur A, Gottesman I, Bolton P, Simonoff E, Yuzda E, Rutter M: Autism as a strongly genetic disorder: evidence from a British twin study. *Psychol Med* 1995, **25**:63–77.
10. Ritvo ER, Freeman BJ, Mason-Brothers A, Mo A, Ritvo AM: Concordance for the syndrome of autism in 40 pairs of afflicted twins. *Am J Psychiatry* 1985, **142**:74–77.
11. Folstein S, Rutter M: Infantile autism: a genetic study of 21 twin pairs. *J Child Psychol Psychiatry* 1977, **18**:297–321.
12. Rosenberg RE, Law JK, Yenokyan G, McGready J, Kaufmann WE, Law PA: Characteristics and concordance of autism spectrum disorders among 277 twin pairs. *Arch Pediatr Adolesc Med* 2009, **163**:907–914.
13. Betancur C: Etiological heterogeneity in autism spectrum disorders: more than 100 genetic and genomic disorders and still counting. *Brain Res* 2011, **1380**:42–77.
14. Schalkwyk LC, Meaburn EL, Smith R, Dempster EL, Jeffries AR, Davies MN, Plomin R, Mill J: Allelic skewing of DNA methylation is widespread across the genome. *Am J Hum Genet* 2010, **86**:196–212.
15. Dindot SV, Person R, Strivens M, Garcia R, Beaudet AL: Epigenetic profiling at mouse imprinted gene clusters reveals novel epigenetic and genetic features at differentially methylated regions. *Genome Res* 2009, **19**:1374–1383.
16. Weaver ICG, Cervoni N, Champagne FA, D'Alessio AC, Sharma S, Seckl JR, Dymov S, Szyf M, Meaney MJ: Epigenetic programming by maternal behavior. *Nat Neurosci* 2004, **7**:847–854.
17. Law JA, Jacobsen SE: Establishing, maintaining and modifying DNA methylation patterns in plants and animals. *Nat Rev Genet* 2010, **11**:204–220.
18. Shen L, Kondo Y, Guo Y, Zhang J, Zhang L, Ahmed S, Shu J, Chen X, Waterland RA, Issa J-PJ: Genome-wide profiling of DNA methylation reveals a class of normally methylated CpG island promoters. *PLoS Genet* 2007, **3**:2023–2036.
19. Fan G, Hutnick L: Methyl-CpG binding proteins in the nervous system. *Cell Res* 2005, **15**:255–261.
20. Gregg C, Zhang J, Butler JE, Haig D, Dulac C: Sex-specific parent-of-origin allelic expression in the mouse brain. *Science* 2010, **329**:682–685.
21. Ng S-F, Lin RCY, Laybutt DR, Barres R, Owens JA, Morris MJ: Chronic high-fat diet in fathers programs β-cell dysfunction in female rat offspring. *Nature* 2010, **467**:963–966.

22. Mermoud JE, Rowbotham SP, Varga-Weisz PD: Keeping chromatin quiet: how nucleosome remodeling restores heterochromatin after replication. *Cell Cycle* 2011, **10**:4017–4025.
23. Gibney ER, Nolan CM: Epigenetics and gene expression. *Heredity* 2010, **105**:4–13.
24. Maunakea AK, Nagarajan RP, Bilenky M, Ballinger TJ, D'Souza C, Fouse SD, Johnson BE, Hong C, Nielsen C, Zhao Y, Turecki G, Delaney A, Varhol R, Thiessen N, Shchors K, Heine VM, Rowitch DH, Xing X, Fiore C, Schillebeeckx M, Jones SJM, Haussler D, Marra MA, Hirst M, Wang T, Costello JF: Conserved role of intragenic DNA methylation in regulating alternative promoters. *Nature* 2010, **466**:253–257.
25. Flintoft L: DNA methylation: looking beyond promoters. *Nat Rev Genet* 2010, **11**:596.
26. Ho L, Crabtree GR: Chromatin remodelling during development. *Nature* 2010, **463**:474–484.
27. Bagot RC, Meaney MJ: Epigenetics and the biological basis of gene x environment interactions. *J Am Acad Child Adolesc Psychiatry* 2010, **49**:752–771.
28. Dulac C: Brain function and chromatin plasticity. *Nature* 2010, **465**:728–735.
29. Feng J, Fouse S, Fan G: Epigenetic regulation of neural gene expression and neuronal function. *Pediatr Res* 2007, **61**:58R–63R.
30. Ooi L, Wood IC: Regulation of gene expression in the nervous system. *Biochem J* 2008, **414**:327–341.
31. Miller CA, Sweatt JD: Covalent modification of DNA regulates memory formation. *Neuron* 2007, **53**:857–869.
32. Sandhu KS: Systems properties of proteins encoded by imprinted genes. *Epigenetics* 2010, **5**:627–636.
33. Cheung H-H, Lee T-L, Rennert OM, Chan W-Y: DNA methylation of cancer genome. *Birth Defects Res C Embryo Today* 2009, **87**:335–350.
34. Guidotti A, Auta J, Davis JM, Di-Giorgi-Gerevini V, Dwivedi Y, Grayson DR, Impagnatiello F, Pandey G, Pesold C, Sharma R, Uzunov D, Costa E, DiGiorgi Gerevini V: Decrease in reelin and glutamic acid decarboxylase67 (GAD67) expression in schizophrenia and bipolar disorder: a postmortem brain study. *Arch Gen Psychiatry* 2000, **57**:1061–1069.
35. Tremolizzo L, Carboni G, Ruzicka WB, Mitchell CP, Sugaya I, Tueting P, Sharma R, Grayson DR, Costa E, Guidotti A: An epigenetic mouse model for molecular and behavioral neuropathologies related to schizophrenia vulnerability. *Proc Natl Acad Sci USA* 2002, **99**:17095–17100.
36. Huang Y, Doherty JJ, Dingledine R: Altered histone acetylation at glutamate receptor 2 and brain-derived neurotrophic factor genes is an early event triggered by status epilepticus. *J Neurosci* 2002, **22**:8422–8428.

37. Brami-Cherrier K, Valjent E, Hervé D, Darragh J, Corvol J-C, Pages C, Arthur SJ, Simon AJ, Girault J-A, Caboche J: Parsing molecular and behavioral effects of cocaine in mitogen- and stress-activated protein kinase-1-deficient mice. *J Neurosci* 2005, **25**:11444–11454.
38. Tsankova NM, Berton O, Renthal W, Kumar A, Neve RL, Nestler EJ: Sustained hippocampal chromatin regulation in a mouse model of depression and antidepressant action. *Nat Neurosci* 2006, **9**:519–525.
39. Gräff J, Mansuy IM: Epigenetic dysregulation in cognitive disorders. *Eur J Neurosci* 2009, **30**:1–8.
40. Oberle I, Rousseau F, Heitz D, Kretz C, Devys D, Hanauer A, Boue J, Bertheas M, Mandel J: Instability of a 550-base pair DNA segment and abnormal methylation in fragile X syndrome. *Science* 1991, **252**:1097–1102.
41. Mercaldo V, Descalzi G, Zhuo M: Fragile X mental retardation protein in learning-related synaptic plasticity. *Mol Cells* 2009, **28**:501–507.
42. Kelleher RJ, Bear MF: The autistic neuron: troubled translation? *Cell* 2008, **135**:401–406.
43. Jacquemont S, Curie A, des Portes V, Torrioli MG, Berry-Kravis E, Hagerman RJ, Ramos FJ, Cornish K, He Y, Paulding C, Neri G, Chen F, Hadjikhani N, Martinet D, Meyer J, Beckmann JS, Delange K, Brun A, Bussy G, Gasparini F, Hilse T, Floesser A, Branson J, Bilbe G, Johns D, Gomez-Mancilla B: Epigenetic modification of the FMR1 gene in fragile X syndrome is associated with differential response to the mGluR5 antagonist AFQ056. *Sci Transl Med* 2011, **3**:64ra1.
44. Association AP: *Diagnostic and Statistical Manual of Mental Disorders DSM-IV-TR Fourth Edition*. 4th edition. American Psychiatric Publishing, Inc.; 2000.
45. Matarazzo V, Cohen D, Palmer AM, Simpson PJ, Khokhar B, Pan S-J, Ronnett GV: The transcriptional repressor Mecp2 regulates terminal neuronal differentiation. *Mol Cell Neurosci* 2004, **27**:44–58.
46. Kishi N, Macklis JD: MECP2 is progressively expressed in post-migratory neurons and is involved in neuronal maturation rather than cell fate decisions. *Mol Cell Neurosci* 2004, **27**:306–321.
47. Lasalle JM, Yasui DH: Evolving role of MeCP2 in Rett syndrome and autism. *Epigenomics* 2009, **1**:119–130.
48. Yasui DH, Peddada S, Bieda MC, Vallero RO, Hogart A, Nagarajan RP, Thatcher KN, Farnham PJ, Lasalle JM: Integrated epigenomic analyses of neuronal MeCP2 reveal a role for long-range interaction with active genes. *Proc Natl Acad Sci USA* 2007, **104**:19416–19421.
49. Talkowski ME, Mullegama SV, Rosenfeld JA, van Bon BWM, Shen Y, Repnikova EA, Gastier-Foster J, Thrush DL, Kathiresan S, Ruderfer DM, Chiang C, Hanscom C, Ernst C, Lindgren AM, Morton CC, An Y, Astbury C, Brueton LA, Lichtenbelt KD, Ades LC, Fichera M, Romano C, Innis JW, Williams CA, Bartholomew D, Van Allen MI, Parikh A, Zhang L, Wu B-L,

Pyatt RE, *et al.*: Assessment of 2q23.1 microdeletion syndrome implicates MBD5 as a single causal locus of intellectual disability, epilepsy, and autism spectrum disorder. *Am J Hum Genet* 2011, **89**:551–563.

50. Cukier HN, Rabionet R, Konidari I, Rayner-Evans MY, Baltos ML, Wright HH, Abramson RK, Martin ER, Cuccaro ML, Pericak-Vance MA, Gilbert JR: Novel variants identified in methyl-CpG-binding domain genes in autistic individuals. *Neurogenetics* 2010, **11**:291–303.
51. Ronan JL, Wu W, Crabtree GR: From neural development to cognition: unexpected roles for chromatin. *Nat Rev Genet* 2013, **14**:347–359.
52. Matsuura T, Sutcliffe JS, Fang P, Galjaard RJ, Jiang YH, Benton CS, Rommens JM, Beaudet AL: *De novo* truncating mutations in E6-AP ubiquitin-protein ligase gene (UBE3A) in Angelman syndrome. *Nat Genet* 1997, **15**:74–77.
53. Kishino T, Lalande M, Wagstaff J: UBE3A/E6-AP mutations cause Angelman syndrome. *Nat Genet* 1997, **15**:70–73.
54. Buiting K, Gross S, Lich C, Gillessen-Kaesbach G, el-Maarri O, Horsthemke B: Epimutations in Prader-Willi and Angelman syndromes: a molecular study of 136 patients with an imprinting defect. *Am J Hum Genet* 2003, **72**:571–577.
55. Hogart A, Leung KN, Wang NJ, Wu DJ, Driscoll J, Vallero RO, Schanen NC, LaSalle JM: Chromosome 15q11-13 duplication syndrome brain reveals epigenetic alterations in gene expression not predicted from copy number. *J Med Genet* 2009, **46**:86–93.
56. Autism and Developmental Disabilities Network Surveillance Year 2008 Principal Investigators; Centers for Disease Control and Prevention: Prevalence of autism spectrum disorders — autism and developmental disabilities monitoring network, 14 sites, United States, 2008. *MMWR Surveill Summ* 2012, **61**:1–19.
57. Goin-Kochel RP, Porter AE, Peters SU, Shinawi M, Sahoo T, Beaudet AL: The MTHFR 677C-->T polymorphism and behaviors in children with autism: exploratory genotype-phenotype correlations. *Autism Res* 2009, **2**:98–108.
58. Adams M, Lucock M, Stuart J, Fardell S, Baker K, Ng X: Preliminary evidence for involvement of the folate gene polymorphism 19bp deletion-DHFR in occurrence of autism. *Neurosci Lett* 2007, **422**:24–29.
59. Mohammad NS, Jain JMN, Chintakindi KP, Singh RP, Naik U, Akella RRD: Aberrations in folate metabolic pathway and altered susceptibility to autism. *Psychiatr Genet* 2009, **19**:171–176.
60. Paşca SP, Dronca E, Kaucsár T, Craciun EC, Endreffy E, Ferencz BK, Iftene F, Benga I, Cornean R, Banerjee R, Dronca M: One carbon metabolism disturbances and the C677T MTHFR gene polymorphism in children with autism spectrum disorders. *J Cell Mol Med* 2009, **13**:4229–4238.
61. James SJ, Melnyk S, Jernigan S, Pavliv O, Trusty T, Lehman S, Seidel L, Gaylor DW, Cleves MA: A functional polymorphism in the reduced folate carrier gene and DNA hypomethylation in mothers of children with autism. *Am J Med Genet B Neuropsychiatr Genet* 2010, **153B**:1209–1220.

62. Surén P, Roth C, Bresnahan M, Haugen M, Hornig M, Hirtz D, Lie KK, Lipkin WI, Magnus P, Reichborn-Kjennerud T, Schjølberg S, Davey Smith G, Øyen A-S, Susser E, Stoltenberg C: Association between maternal use of folic acid supplements and risk of autism spectrum disorders in children. *JAMA* 2013, **309**:570–577.
63. Rasalam AD, Hailey H, Williams JHG, Moore SJ, Turnpenny PD, Lloyd DJ, Dean JCS: Characteristics of fetal anticonvulsant syndrome associated autistic disorder. *Dev Med Child Neurol* 2005, **47**:551–555.
64. Strömland K, Nordin V, Miller M, Akerström B, Gillberg C: Autism in thalidomide embryopathy: a population study. *Dev Med Child Neurol* 1994, **36**:351–356.
65. Atladóttir HO, Thorsen P, Ostergaard L, Schendel DE, Lemcke S, Abdallah M, Parner ET: Maternal infection requiring hospitalization during pregnancy and autism spectrum disorders. *J Autism Dev Disord* 2010, **40**:1423–1430.
66. Libbey JE, Sweeten TL, McMahon WM, Fujinami RS: Autistic disorder and viral infections. *J Neurovirol* 2005, **11**:1–10.
67. de Cock M, Maas YGH, van de Bor M: Does perinatal exposure to endocrine disruptors induce autism spectrum and attention deficit hyperactivity disorders? Review. *Acta Paediatr* 2012, **101**:811–818.
68. Dolinoy DC, Huang D, Jirtle RL: Maternal nutrient supplementation counteracts bisphenol A-induced DNA hypomethylation in early development. *Proc Natl Acad Sci USA* 2007, **104**:13056–13061.
69. Szyf M: Epigenetics, DNA methylation, and chromatin modifying drugs. *Annu Rev Pharmacol Toxicol* 2009, **49**:243–263.
70. Dong E, Nelson M, Grayson DR, Costa E, Guidotti A: Clozapine and sulpiride but not haloperidol or olanzapine activate brain DNA demethylation. *Proc Natl Acad Sci USA* 2008, **105**:13614–13619.
71. Tremolizzo L, Doueiri M-S, Dong E, Grayson DR, Davis J, Pinna G, Tueting P, Rodriguez-Menendez V, Costa E, Guidotti A: Valproate corrects the schizophrenia-like epigenetic behavioral modifications induced by methionine in mice. *Biol Psychiatry* 2005, **57**:500–509.
72. Gregory SG, Connelly JJ, Towers AJ, Johnson J, Biscocho D, Markunas CA, Lintas C, Abramson RK, Wright HH, Ellis P, Langford CF, Worley G, Delong GR, Murphy SK, Cuccaro ML, Persico A, Pericak-Vance MA: Genomic and epigenetic evidence for oxytocin receptor deficiency in autism. *BMC Med* 2009, **7**:62.
73. McCarthy MM, Altemus M: Central nervous system actions of oxytocin and modulation of behavior in humans. *Mol Med Today* 1997, **3**:269–275.
74. Jiang Y-H, Sahoo T, Michaelis RC, Bercovich D, Bressler J, Kashork CD, Liu Q, Shaffer LG, Schroer RJ, Stockton DW, Spielman RS, Stevenson RE, Beaudet AL: A mixed epigenetic/genetic model for oligogenic inheritance of autism with a limited role for UBE3A. *Am J Med Genet A* 2004, **131**:1–10.

75. Jiang R, Robinson WP, LaSalle JM, Nagarajan RP, Patzel KA, Martin M, Yasui DH, Swanberg SE, Hertz-Picciotto I, Hansen RL, Van de Water J, Pessah IN: MECP2 promoter methylation and X chromosome inactivation in autism. *Autism Res* 2008, **1**:169–178.
76. Hu VW, Frank BC, Heine S, Lee NH, Quackenbush J: Gene expression profiling of lymphoblastoid cell lines from monozygotic twins discordant in severity of autism reveals differential regulation of neurologically relevant genes. *BMC Genomics* 2006, **7**:118.
77. Nguyen A, Rauch TA, Pfeifer GP, Hu VW: Global methylation profiling of lymphoblastoid cell lines reveals epigenetic contributions to autism spectrum disorders and a novel autism candidate gene, RORA, whose protein product is reduced in autistic brain. *FASEB J* 2010, **24**:3036–3051.
78. Wong CCY, Meaburn EL, Ronald A, Price TS, Jeffries AR, Schalkwyk LC, Plomin R, Mill J: Methylomic analysis of monozygotic twins discordant for autism spectrum disorder and related behavioural traits. *Mol Psychiatry* 2013, doi:10.1038/mp.2013.41.
79. Ginsberg MR, Rubin RA, Falcone T, Ting AH, Natowicz MR: Brain transcriptional and epigenetic associations with autism. *PLoS One* 2012, **7**:e44736.
80. Shulha HP, Cheung I, Whittle C, Wang J, Virgil D, Lin CL, Guo Y, Lessard A, Akbarian S, Weng Z: Epigenetic signatures of autism: trimethylated H3K4 landscapes in prefrontal neurons. *Arch Gen Psychiatry* 2012, **69**: 314–324.
81. Voineagu I, Wang X, Johnston P, Lowe JK, Tian Y, Horvath S, Mill J, Cantor RM, Blencowe BJ, Geschwind DH: Transcriptomic analysis of autistic brain reveals convergent molecular pathology. *Nature* 2011, **474**:380–384.
82. Garbett K, Ebert PJ, Mitchell A, Lintas C, Manzi B, Mirnics K, Persico AM: Immune transcriptome alterations in the temporal cortex of subjects with autism. *Neurobiol Dis* 2008, **30**:303–311.
83. Levitt P, Campbell DB: The genetic and neurobiologic compass points toward common signaling dysfunctions in autism spectrum disorders. *J Clin Invest* 2009, **119**:747–754.
84. O'Roak BJ, Vives L, Girirajan S, Karakoc E, Krumm N, Coe BP, Levy R, Ko A, Lee C, Smith JD, Turner EH, Stanaway IB, Vernot B, Malig M, Baker C, Reilly B, Akey JM, Borenstein E, Rieder MJ, Nickerson DA, Bernier R, Shendure J, Eichler EE: Sporadic autism exomes reveal a highly interconnected protein network of de novo mutations. *Nature* 2012, **485**:246–250.
85. Caley DP, Pink RC, Trujillano D, Carter DRF: Long noncoding RNAs, chromatin, and development. *Scientific World Journal* 2010, **10**:90–102.
86. Hawkins PG, Santoso S, Adams C, Anest V, Morris KV: Promoter targeted small RNAs induce long-term transcriptional gene silencing in human cells. *Nucleic Acids Res* 2009, **37**:2984–2995.

87. Rassoulzadegan M, Grandjean V, Gounon P, Vincent S, Gillot I, Cuzin F: RNA-mediated non-mendelian inheritance of an epigenetic change in the mouse. *Nature* 2006, **441**:469–474.
88. Chong S, Whitelaw E: Epigenetic germline inheritance. *Curr Opin Genet Dev* 2004, **14**:692–696.
89. Gibbs JR, van der Brug MP, Hernandez DG, Traynor BJ, Nalls MA, Lai S-L, Arepalli S, Dillman A, Rafferty IP, Troncoso J, Johnson R, Zielke HR, Ferrucci L, Longo DL, Cookson MR, Singleton AB: Abundant quantitative trait loci exist for DNA methylation and gene expression in human brain. *PLoS Genet* 2010, **6**:e1000952.
90. Tycko B: Allele-specific DNA methylation: beyond imprinting. *Hum Mol Genet* 2010, **19**:R210–220.
91. Zhang D, Cheng L, Badner JA, Chen C, Chen Q, Luo W, Craig DW, Redman M, Gershon ES, Liu C: Genetic control of individual differences in gene-specific methylation in human brain. *Am J Hum Genet* 2010, **86**:411–419.
92. Laird PW: Principles and challenges of genome-wide DNA methylation analysis. *Nat Rev Genet* 2010.
93. Park PJ: ChIP-seq: advantages and challenges of a maturing technology. *Nat Rev Genet* 2009, **10**:669–680.
94. Zilberman D, Henikoff S: Genome-wide analysis of DNA methylation patterns. *Development* 2007, **134**:3959–3965.
95. Huang YW, Huang TH, Wang LS: Profiling DNA methylomes from microarray to genome-scale sequencing. *Technol Cancer Res Treat* 2010, **9**:139–147.
96. Collas P: The current state of chromatin immunoprecipitation. *Mol Biotechnol* 2010, **45**:87–100.
97. Gupta R, Nagarajan A, Wajapeyee N: Advances in genome-wide DNA methylation analysis. *BioTechniques* 2010, **49**:iii–xi.
98. Marioni JC, Mason CE, Mane SM, Stephens M, Gilad Y: RNA-seq: an assessment of technical reproducibility and comparison with gene expression arrays. *Genome Res* 2008, **18**:1509–1517

Chapter 13

The Epigenetics of Autism — Running Beyond the Bases

Simon G. Gregory*

Abstract

Previous studies have established that there is a strong genetic component to the development of ASD, but these genetic risks do not account for all of the heritability of the disorder. This raises the prospect that alternative, epigenetic mechanisms may play a role in ASD development. Epigenetic mechanisms facilitate temporal and spatial regulation of gene expression, but are independent of changes to the underlying DNA sequence. Because epigenetic profiles are labile, they represent an intriguing mechanism whereby environmental influences, which are not severe enough to alter the DNA sequence of a cell, may alter gene expression and cellular response and contribute to ASD. In this chapter, we discuss the role of DNA methylation and histone modifications in the development of ASD.

Keywords: Autism; Epigenetics; CpG island; DNA methylation; Chromatin; Histone modification; Environmental influence; Development; Tissue specific.

Introduction

Autism represents a spectrum of disorders that develop from the complex interplay between genetic susceptibility and environmental

*Correspondence to: Duke Molecular Physiology Institute, Duke University Medical Center, Durham, NC 27710, USA. Email: simon.gregory@duke.edu

influence. Significant efforts have been made towards characterizing the strong genetic components that underlie ASD; however, previous studies have established that no single gene accounts for more than 2% of ASD cases.[1] Although the size of genetic studies have increased to detect increasingly smaller genetic effects, it is clear that the risk factors in DNA sequence do not account for all of the heritable elements that contribute to ASD. This raises the interesting prospect that alternative, epigenetic mechanisms may play a role in the development of the disorder.

Epigenetics is a term first described by Waddington in 1942[2] to describe the heritable phenomenon that facilitates temporal and spatial regulation of gene expression, but which is independent of changes to the underlying DNA sequence. Epigenetic profiles are labile and thus represent an intriguing mechanism whereby environmental influences, though not severe enough to alter the DNA sequence of a cell, may alter gene expression and cellular response. Thus epigenetic mechanisms can contribute to the development of disorders by the continual influence of events that may begin *in utero*, persist throughout life, and even be inherited from parents and grandparents. The term epigenetics encompasses several mechanisms of gene regulation and includes non-coding and microRNAs (discussed in Chapters 8 and 9), histone modifications, and DNA methylation. It is the contribution of these latter two to the development of ASD that will be discussed in this chapter.

DNA Methylation

DNA methylation occurs via the addition of a methyl group to the 5th position in the pyrimidine ring of cytosine when it is at the 5′ end of a guanine, with a phospohodiester bond between the two nucleotide bases, thus 'CpG' (Figure 1). The methylation process is carried out by DNA methyltransferases (DNMT) in a context-dependent manner. For example, DNMT1 is involved in maintenance of DNA methylation in daughter strands during replication, while family members DNMT3A and DNMT3B are critical to the

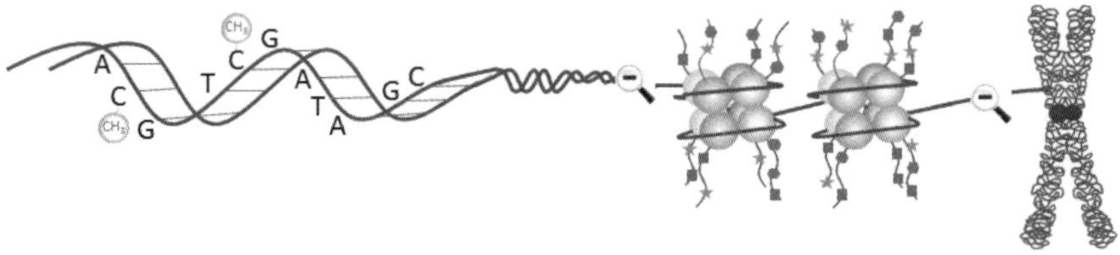

Figure 1. A schematic of DNA methylation, chromatin modification.

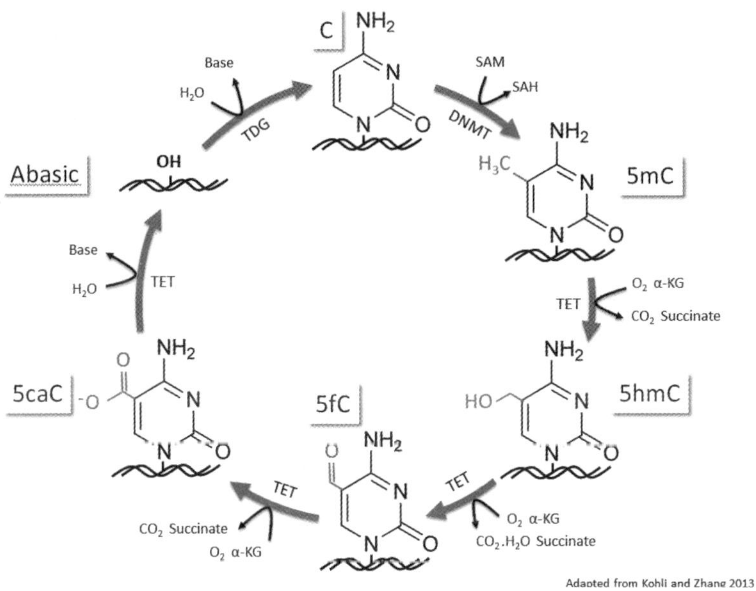

Figure 2. Cytosine methylation and oxidation cycle.

establishment of *de novo* methylation during development. While these *de novo* and maintenance DNA methylation processes are well understood, the demethylation process, until recently, has been less well defined. The discovery of the iterative oxidization of 5-methylcytosine to 5-hydroxymethylcytosine (5hmC), 5-formylcytosine (5fC) and 5-carboxylcytosine (5caC) via the ten-eleven translocation (TET) family of enzymes is a model for dynamic cytosine modification (Figure 2).[3] The potential importance of 5hmC in ASD development is based on its abundance in neuronal cells of the central nervous system.[4,5] The methylation/demethylation of regions within the brain has been shown to be a dynamic process during neuronal development, and the distribution of enriched regions of 5hmC methylation localizes to specific regions of the genome and gene body, as shown in mouse models.[6] Interestingly, there seems to be an interplay between loss and gain of genomic methylation and concomitant histone modifications such as Histone 3, lysine 27 tri-methylation (H3K27me3) in a cooperative manner.[6]

Although DNA methylation is thought of as being primarily methylation of cytosine in the CpG context, mammalian stem cells also contain methylation of CpA and CpT, however, the majority of DNA methylation in the differentiated cell (>99%) occurs at CpG.[7,8] The transitive nature of an epigenetic profile is further exemplified by studies that have shown that DNA methylation profiles change during germ cell specification and development, and throughout early embryonic development.[9] The environment during embryogenesis is particularly important in establishing epigenetic marks because reprogramming occurs shortly after fertilization and during germ cell differentiation. Although the mechanisms of methyl erasure and restoration are an active area of research, it is clear that several factors — including availability of methyl groups, exposure to environmental toxicants, and/or other stressors — can impact this process and can have effects well beyond the developmental period. A number of environmental factors — including nutrition, toxicants, and stress — have been shown to have epigenetic effects.[10] Finally, DNA methylation profiles change normally during aging where these changes are reciprocal; i.e., the aging process can influence the methylome while changes in DNA methylation can also impact aging.[11]

There are approximately 28 million CpG dinucleotides within the genome, with approximately 60% to 90% being methylated.[12,13] The differentiated cell uses methylation of CpG dinucleotides (70% of human cytosines adjacent to guanines (CpG) are methylated) in several key physiological processes, including X chromosome inactivation, imprinting (where gene expression is dependent on the parent-of-origin), silencing of germline-specific genes and sequence repeat elements.[14] However, it is the absence of DNA methylation in the promoter regions of genes that has been known for many years to regulate gene expression by influencing the affinity of the transcription complex to bind to the promoter region of genes.[15] The primary region of regulatory methylation associated with gene expression is the CpG island (CGI). CGIs are genomic regions (>200bp) that contain a GC percentage which is greater than 50%, with an observed-to-expected CpG ratio that is greater than 60%,[16] and are contained in or near ~40% of promoters of mammalian genes.[17] While the CGI has been considered for a number of years to be the prime epigenetic modulator of gene expression, recent interest has focused on the methylation status of CGI shores (up to 2,000 bases on either side of the CpG island). This is because the methylation status of the CGI shore may be more closely regulating gene expression than the CGI itself, both in somatic disease and in neuronal cells where methylation-enriched regions within gene bodies harbor sequence motifs for transcription factor binding.[18,19] Additionally, there is further granularity of methylated regions with the definition of CGI shelves that may also play a role in the regulation of gene expression (Figure 3).[20]

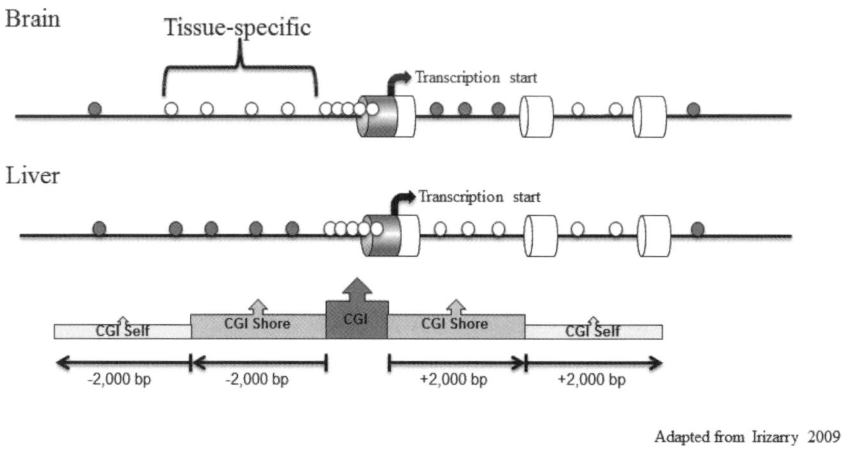

Figure 3. CpG islands, shores and shelves.

Regulation by DNA Methylation

The epigenetic on/off switching of the transcriptional machinery via methylation of CGIs as well as CGI shores and shelves, is not only regulated temporally during development[21] but also spatially between different tissue types. This latter point, the spatial variation of DNA methylation, provides the means whereby a cell, which contains the same genetic material as different cell types in the body, is able to carry out its function in a tissue-specific manner. However, this also means that the methylome of brain tissue, for example, will be different from that of blood. In ASD this is important because the proximal tissue, the brain, is obviously inaccessible from live patients. While autopsy samples represent an available source of material, they are associated with their own potential limitations that may influence their use, such as the mode of death, post-mortem interval, and limited availability. Therefore, there is tremendous impetus to use large, readily available collections of blood samples to report on proximal epigenetic dysregulation in the brain. While DNA methylation profiles in blood do not exactly recapitulate the profile of brain tissue, particularly of different areas of the brain that show differences in methylation, it is possible that epigenetic signatures that are associated with developmental disorders such as ASD occur early during development, possibly *in utero* (see Chapter 14), and thus the dysregulated loci may be manifested in differentiated cell types. For example, high correlations ($r = 0.859$) of methylation between multiple tissues have been documented.[22] Recent research has demonstrated that between-individual differences in methylation are significantly correlated between brain tissues and blood, indicating that any given change in methylation may be detectable in both tissues.[23] Consistent with this correlation, several studies have identified loci in blood which are differentially methylated in the context of neurological diseases.[24] Further, additional studies have pointed towards the blood being a good surrogate for discrimination between ASD cases and controls, either globally[25] or at specific loci[26] for select tissues in the brain.

It is worth noting, however, that regardless of the source, tissues are invariably made up of many different cell types and thus a methylation profile generated from a biological sample represents the average methylation status of a locus from thousands to millions of cells and potentially many different cell types. Therefore the functionally relevant differences in CpG methylation may not be identified because they are being 'lost' in the methylation profiles of other cell types within the tissue. However, recent statistical approaches have been developed to deconvolute homogeneous cell methylation profiles from a heterogeneous mixture of cells such as those represented in blood.[27,28]

Epigenetics and Autism

For a number of years it has been strongly suggested that aberrant epigenetic mechanisms play a role in ASD. Listed below are select syndromic and gene-based examples in which these mechanisms may contribute to the development of ASD.

MECP2

An underlying epigenetic mechanism in ASD development is exemplified by the methyl-DNA-binding protein (MECP2), which is causative of Rett syndrome [MIM:312750].[29] The salience of *MECP2* function in ASD is based on the shared etiology of Rett and ASD, being pervasive developmental disorders, and the identification of *MECP2* mutations, although rare, in individuals with ASD.[30,31] As reviewed by Gonzales and LaSalle,[32] MECP2 is a member of a family of proteins that specifically bind to methylated DNA and translate them functionally into modified gene expression. Although predicted to be a global repressor of gene transcription, other models propose MECP2 as a transcriptional modulator by long-range regulation of chromatin structure from distal methylation sites outside of gene promoters.[33,34] Functionally, it has been shown that 79% of autism cases have decreased *MECP2* expression in the frontal cortex and that an increase in aberrant DNA methylation correlates with this decrease in expression.[35] Thus, MECP2 may mediate ASD via epigenetic mechanisms that regulate transcription, facilitate histone modifications, and inhibit transcription factor binding, chromatin looping, and alternative RNA splicing.[36]

FRAX

Fragile X syndrome (FXS) is caused by a repeat expansion of three nucleotides (CCG) in the 5′ untranslated region of the fragile X mental retardation 1 gene (FMR1) in chromosome Xq27.3. FMR1 encodes a protein (FMRP) that is thought to shuttle RNAs from the nucleus to the cytoplasm,[37] and may be required for the normal processes of maturation in cerebral cortical development[38] and synaptic plasticity and cognition.[39] While FXS individuals are considered under the umbrella of ASD using autism diagnostic criteria [Autism Diagnostic Observation Schedule (ADOS), Autism Diagnostic Interview (ADI), and DSM-IV], interestingly, their ASD risk is dependent on the number of CCG repeats — normal (6 to 44 CCG units), intermediate mutation (45 to 54), premutation (55 to 200), or full mutation (>200).[40] Given that the CCG repeat is located in a CGI within the promoter region of the gene, the expansion of the repeat results in a concomitant increase in DNA methylation and histone deacetylation,[41] resulting in loss of FMR1 expression due to the inability of transcription factors NRF-1 and Sp1 to bind.[42] However, the absence of DNA and histone methylation in a family with a full mutation and no clinical phenotype[43] as well as the observed deletion of the region containing the CGG repeats that did not abolish gene expression,[44] highlight the fact that epigenetic modifications to the gene region are critically important, though not the sole drivers of *FMR1* expression.

Angelman/Prader-Willi syndromes

Imprinted regions of the genome contain genes whose expression is regulated by DNA methylation in a parent-dependent manner.[45] Known as the "parental conflict hypothesis"[46] or "kinship theory",[47] imprinting is essentially a mechanism that balances the interests of each parent in terms of the evolutionary fitness of their genes, broadly, paternally expressed genes tend to be growth-promoting whereas maternally expressed genes tend to be growth-limiting.[46] The first imprinted genetic disorders to be described in humans were the reciprocally-inherited Prader-Willi syndrome (PWS) and Angelman syndrome (AS). Both syndromes are associated with epigenetic dysregulation or chromosomal rearrangement that can include the imprinting control region in 15q11-13 and, interestingly, both have elevated risk of ASD.[48] The genomic interval contains paternally expressed genes, including *SNRPN* and *NDN*, and a maternally expressed gene *UBE3A*. Functional loss (by chromosomal deletion, gain of methylation in the ICR akin to a maternal epigenotype,[49] or by maternal uniparental disomy (UPD)) of the paternally expressed genes is associated with Prader-Willi syndrome, which is characterized by hypotonia, obesity, and hypogonadism. Alternatively, Angelman syndrome (characterized by epilepsy, tremors, and a perpetually smiling facial expression) results from the loss of *UBE3A* by chromosomal deletion or nucleotide mutation, UPD of the paternal chromosome, or imprinting defects such as loss of methylation at the maternal IC, or traditional mutations in the maternal copy of UBE3A.[49]

15q11-13

Genome copy number,[50–52] genetic linkage[53,54] and parent-of-origin effect studies (described above) have consistently implicated human chromosome 15q11-13 in the development of autism and ASD. Further, it is estimated that maternal interstitial duplications of chromosome region 15q11-13 are identified in 1% to 2% of ASD cases,[52] and that maternally inherited isodicentric chromosome 15 have elevated ASD risk.[48] While the dysregulated epigenetic contribution of 15q11-13 region to the development of AS and PWS is well- documented, the effect of such mechanisms on non-imprinted genes in the region, such as the $GABA_A$ receptor subunit genes (*GABRB3*, *GABRA5*, and *GABRG3*) is less well-defined. The GABA genes encode receptor subunits for the major brain inhibitory neurotransmitter gamma-aminobutyric acid (GABA). In a normal state, the GABR genes are biallelically expressed within the brain; however, deficiency of the methyl-DNA-binding protein MECP2, which may act as a chromatin organizer for optimal expression of both alleles of *GABRB3* in neurons, results in reduced expression of *UBE3A* and *GABRB3* in the cortex of individuals with ASD.[36,55]

Chromosome 7

Two distinct regions of chromosome 7, 7q22.1 to 7q22.2 and 7q32.1 to 7q32.3, have been linked with autism[54] and contain clusters of imprinted genes[56] whose expression requires differential DNA methylation.[57] These discrete chromosome 7 loci implicate epigenetic mechanisms in gene regulation as the regions contribute to possible parent-of-origin and sex-specific effects.[58] The autism susceptibility locus on 7q32 contains the maternally-imprinted *MEST* and the non-imprinted coatomer protein complex, subunit gamma 2 (*COPG2*) and centrosomal protein 41kDa (CEP41) genes, while the chromosome 7 cluster also includes two paternally-expressed genes, encoding sarcoglycan epsilon (*SGCE*) and paternally-expressed gene 10 (*PEG10*), both of which were found to be binding targets for Mecp2 in mouse,[33] and paraoxonase 1 gene (*PON1*), which is preferentially expressed from the paternal chromosome in human:mouse somatic cell hybrids.[56]

Methylation pathways

As described in Figure 2, DNA methylation begins with the addition of a methyl group from s-adenosylmethionine (SAM) as part of the remethylation of homocysteine to methionine. However, the availability of methyl groups uses the one-carbon metabolism cycle for which availability of methionine is derived from different dietary micronutrients, such as vitamins B6 and B12 and folate, and enzymes, such as 5-methyltetrahydrofolate reductase (MTHFR), are required to catalyze critical stages of the pathway.[59] Logically, decreased function of enzymes, such as MTHFR, could result in methylation deficits affecting DNA, histones, and gene expression. Therefore, the disruption of MTHFR function by DNA variants in individuals with ASD[60] ties together genetic susceptibility, DNA methylation and dietary intake, the latter being shown to result in metabolic and genetic differences between clinical subtypes of ASDs.[61]

Candidate Genes

As DNA methylation profiling technologies improve (discussed below), the number of non-imprinted ASD genes that show differential methylation in either one or both alleles, and differ spatially between tissue types or temporally during development, will continue to increase. At a global level, DNA methylation has been shown to be hypomethylated within individuals with ASD, while hypermethylation within regions of the genome that regulate the expression of specific genes is also a likely epigenetic mechanism associated with ASD.[25] Listed below are several of the more well-known examples of differentially methylated ASD candidate genes.

OXTR

Deletion of the oxytocin receptor (*OXTR*) was identified in an ASD proband and subsequent DNA methylation screening of the affected sibling and parents identified putatively important regulatory CpG dinucleotides for *OXTR*. Screening of these five CpG dinucleotides in the blood and cortex tissue of two unrelated ASD cohorts established that autism cases had elevated methylation and decreased expression of OXTR.[26] The oxytocin signaling pathway has long been known to play important 'prosocial' roles influencing social behavior and cognitive function in a species-specific manner[62,63]; however, these functions have sexually dimorphic roles in brain function and behavior.[64] Aberrant regulation of oxytocin signaling has already been shown to play a role in the development in high functioning autism spectrum disorder[63,65,66] and genetic linkage[67] as well as association[68–70] studies support a role for the oxytocin signaling pathway in the etiology of autism via OXTR. More recently, an association was identified between the degree of OXTR methylation and brain activity evoked by the perception of animacy as detected by functional MRI.[71]

SHANK3

The SH3 and multiple ankyrin repeat domains 3 gene (*SHANK3*) is a synaptic scaffolding protein that plays an important role in the formation, maturation, and maintenance of synapses. *SHANK3* mutations were identified a small number of ASD families[72] and thus the gene was strongly suspected of being involved in the pathogenesis and neuropathology of ASD. A role for the epigenetic regulation of *SHANK3* has been recently supported by the identification of tissue-specific methylation in one of the five CGIs, within the vicinity of the CpG-island 2 in the mouse brain. Shank3 mutant mice were shown to exhibit autistic-like behaviors, including impaired social interaction and repetitive behaviors.[73]

RELN

Reelin is critical for neuronal migration and synaptogenesis in the cerebral and cerebellar cortex, hippocampus, amygdala, and several brainstem nuclei.[74] The RELN protein and mRNA were significantly reduced in the frontal and cerebellar areas of brains of autistic individuals versus those of controls.[36,75] Akin to the causal trinucleotide repeat length in FXS, the length of the CGG/GCC repeat in the 5′ untranslated region of *RELN* has been genetically associated with ASD.[76–78] However, repeat length-dependent expression of *RELN* in cell lines first hinted at the epigenetic regulation of the gene.[79] Recently, it has been established that promoter methylation levels of *RELN* vary pre- and post-puberty[80] in the temporal cortex of phenotypically normal individuals and further, that 5-hydroxymethylation of the promoter region may influence MeCP2 binding and subsequent expression of RELN in the cerebellum of individuals with ASD.[81]

PRRT1 and *ZFP57*

The advent of high-throughput platforms for identifying differential DNA methylation means that multiple loci can be simultaneously screened in hundreds if not thousands of samples instead of relying on *a priori* hypotheses and separate biological evidence for methylation screening of targeted candidate genes. The genes mentioned here were identified by characterizing the DNA methylation profiles of prefrontal cortex, cerebellum, and temporal cortex from ASD cases, compared to controls, using a SNP based CpG microarray (discussed below). The genes, which were subsequently validated in replication datasets,[82] were identified based on the location of differentially methylated regions (DMRs) that were adjacent to or contained within the gene.

The DMR within proline-rich transmembrane protein 1 (*PPRT1*) is localized to the 3′ UTR of the gene, and overlapping epigenetic data available in the public domain is supportive of the region being functionally important. The gene presents as a reasonable ASD candidate based on the location of its expression,[83] the possible role that the hippocampus may have in ASD[84,85] and the neurological disorders, such as seizures[86,87] and dyskinesia, previously associated with the family of genes.[88,89]

The DMR for the second gene, zinc finger protein 57 (ZFP57), is located about 3.5 kb upstream of the gene. ZFP57 has been shown to be instrumental in maintaining imprinting marks during development[90,91] by providing a mechanism for targeting DNA methyltransferase.[92] The relevance of this mechanism is described above with regard to the importance of MeCP2 in the development of ASD.

Influences of the Environment on the Epigenome

As previously discussed, the epigenome is labile and thus susceptible to environmental influence. It is attractive to consider that these environmental exposures (discussed elsewhere in the book) might mediate epigenomic changes at critical intervals during fetal development.

As reviewed by LaSalle,[93] there are multiple factors that could influence ASD risk, including several that may impact changes to the methylome. As previously discussed, the one-carbon metabolism cycle is critical for providing methionine for DNA methylation. The most notable manipulation of this cycle has been fortification and supplementation with folate (folic acid) to reduce the number of neural tube defects; however, it has also been shown to be protective in ASD.[94,95]

While folate may be protective, there are numerous risk factors that may elevate ASD risk. Although the mechanisms associated with elevated risk from increased paternal,[96] maternal[97] and even grandparental age[98] is likely to include mutation rates and genetic predisposition of older parents,[99] it may be possible that the maintenance of the methylome in the gametes may also play a role.[100]

In utero exposure to the commonly used anti-epileptic valproate has been identified as a risk factor for ASD.[101,102] The epigenetic mechanism is thought to be mediated by enhancement of GABA-mediated neurotransmission, interaction with Wnt/B-catenin and ERK signaling systems, as well as modified inositol and arachidonate metabolism.[36,103]

As reviewed by LaSalle [10], the methylome may also be altered by exposures to environmental toxins, such as arsenic, cadmium, benzene and air pollution, that were found to be associated with reduction of DNA methylation levels at *LINE1* and/or *Alu* repeats in human tissues.[104–107] As previously discussed, individuals with ASD exhibit aberrant global methylation,[25] and these toxin exposures are linked with global hypomethylation changes and targeted hypermethylation at gene loci. While these exposures are unlikely to be solely causal of ASD, epigenomic influences in the context of genetic susceptibility might play a role.

Assisted Reproduction

There is conflicting evidence surrounding elevated risk for the development of ASD in the context of assisted reproductive technologies, such as *in vitro* fertilization.[108] Underlying these conflicting results[109,110] are invariably issues with cohort size, clinical diagnosis, and methodological approaches. While there may be no hard epidemiological evidence for elevated risk or protection, studies have shown that there may be increased frequency of stochastic epigenetic errors in the imprinted genes in *in vitro* cultures of placental tissue from mice,[111] and imprinting disorders are more prevalent after human *in vitro* fertilization or intracytoplasmic sperm injection,[112] while other studies found no such epigenomic evidence.[113]

Technologies for Profiling of DNA Methylation

DNA methylation profiling is undergoing somewhat of a renaissance with the development of microarray and next-generation sequence (NGS)-based profiling methods. The application of these high-throughput approaches is changing the emphasis of DNA methylation profiling from targeted approaches using lower throughput techniques to unbiased discovery approaches, such as epigenome wide association studies (EWAS) that use microarray-based SNP arrays,[114] and enrichment or whole DNA methylome profiling using next-generation sequencing platforms.

As described in the excellent review by Laird,[115] low-throughput techniques include bisulfite Sanger sequencing (which relies on sodium bisulfite conversion of non-methyl cytosine to thymine leaving methylated cytosines as C's followed by PCR and dye-terminator sequencing), pyrosequencing that also uses bisulfite treatment in conjunction with detection of phosphate release upon nucleotide incorporation, and mass spectroscopy measurement of methyl-C and non-methyl-C ratios within native DNA. These targeted approaches have been very useful at establishing the methylation status of genomic regions that have been suggested by some form of *a priori* evidence; however, their main use going forward will be as a screening tool for loci known to be associated with clinical phenotypes, such as neurodevelopmental disorders or cancer, or as validatation of dysregulated methylation loci identified by genome-wide screening approaches, such as microarray or next-generation discovery platforms.

Methyl-microarray approaches represent an attractive intermediate step between the targeted and whole methylome screening approaches.[116] While relatively cheap, they currently sample approximately 480,000 CpG dinucleotides, and methods of array quality control, normalization and methods of analysis and interpretation are fairly well standardized. Although they are restricted to sampling a set number of loci from genomic regions predicted to be important in the regulation of genes, they are already facilitating the discovery of dysregulated loci implicated in ASD.[82]

As previously discussed, there are an approximately 28 million CpGs in the human genome. To date, it is too expensive to generate a whole methylome by bisulfite conversion of DNA from target tissue and generating sequence using an NGS platform. One should consider that the quantitative accuracy of methylation profiling is dependent on the depth of sequence coverage at an individual CpG dinucleotide, which is effectively the number of times you count a C or a bisulfite-converted T at a nucleotide level. Remembering that stochastic differences in homogeneous cells and heterogeneous 'contaminating' tissue types interfere with accurate epigenomic profiling, low coverage of a 3.2 Gb bisulfite- converted genome is inadequate. Thus researchers are faced with generating more than 100-fold coverage of the whole methylome, which is currently cost prohibitive. Therefore, as described by Laird,[115] several approaches have been developed to provide a means by which NGS can be used more economically for quantitative measurement of the methylome. These approaches include targeting 80-100Mb regions of the genome by bait-capture followed by bisulfite sequencing, enrichment of methylated regions by immunoprecipitation using antibodies designed to pull down methyl-binding domain proteins[117] or the methylated DNA itself,[118] or enzyme digestion to generate a reduced representation of the methylome.[119]

However, it should be noted that techniques that rely on the quantitative assessment of CpG methylation following bisulfite conversion cannot distinguish between 5-methyl cytosine and

5-hydroxymethyl cytosine. Although one NGS platform is capable of discriminating between the various methylation states of cytosine,[120] the production of significant amounts of sequence is cost-prohibitive. Therefore, researchers are faced with using recent approaches that screen for 5hmC using oxidative bisulfite sequencing and subtractive alignment to quantitatively measure 5mC and 5hmC.[121]

Histones and Chromatin Modifications

In addition to DNA methylation, gene expression is influenced by a number of chromatin structural modifications. The basic chromatin unit consists of 147 base pairs of DNA wrapped around a histone octamer; this unit is called a nucleosome (Figure 1). Nucleosome structure allows the entire genome to be compacted into the nucleus of a cell and regulates the extent to which transcription machinery has access to the DNA. This mode of transcriptional regulation involves post-translational covalent modification of the histone tails by specific enzymes. The most common modifications are acetylation and methylation, although numerous other modifications have been identified, including phosphorylation, ubiquitination, SUMOylation, citrullination, and adenosine diphosphate ribosylation. In general, acetylation is associated with increased DNA accessibility and transcriptional activity, whereas deacetylation is associated with transcriptional repression; histone methylation may be related to either activation or repression. These modifications alter the charge of the histone tail and therefore alter interactions with the DNA to allow nucleosome mobility and changes in DNA accessibility. Histone modifications can also affect higher-order folding of the chromatin through internucleosome interactions.

Post-translational modifications of core histones, including acetylation, phosphorylation, methylation and ubiquitination, are associated with distinct chromatin states.[122] For example, acetylation of lysines on H3 and H4 tails (H3ac, H4ac), di- and tri-methylation of H3 at Lys 4 (H3K4me2, H3K4me3), define transcriptionally active chromatin.[122–125] H3K4me3 is located at the 5′ ends of active genes, while tri-methylation of lysine 36 on histone H3 (H3K36me3) also strongly marks transcriptional units. Both modifications also mark non-coding transcripts.[126] Conversely, di and tri-methylation at lysine 9 (H3K9me2, H3K9me3) and trimethylation of lysine 27 on H3 (H3K27me3) are associated with transcriptionally silent chromatin.[122,127,128] The different methylation states (mono, di, tri) on the same amino acid residue discriminate between function and functional potential of various chromatin domains. H3K4me2 correlates with chromatin that is "poised" for transcription, while H3K4me3 is associated with transcriptionally active chromatin.[129,130] Histone modifications also coincide with regulatory regions of the genome. For instance, H3K4me2 and H3K4me3 co-localize with DNase hypersensitive sites (HSS), with H3K4me2 correlating with HSS and regulatory regions located proximal and distal to genes.[131] Combinatorial histone modifications, multiple modifications on the same histone tail or on different histone tails within the same nucleosome, create distinctive chromatin states and establish epigenetic signatures or "histone codes" for a genomic region (Figure 4).[132] For example, both H3K9me3 and H3K27me3 define facultative heterochromatin on the inactive X-chromosome in mice and humans.[133–137]

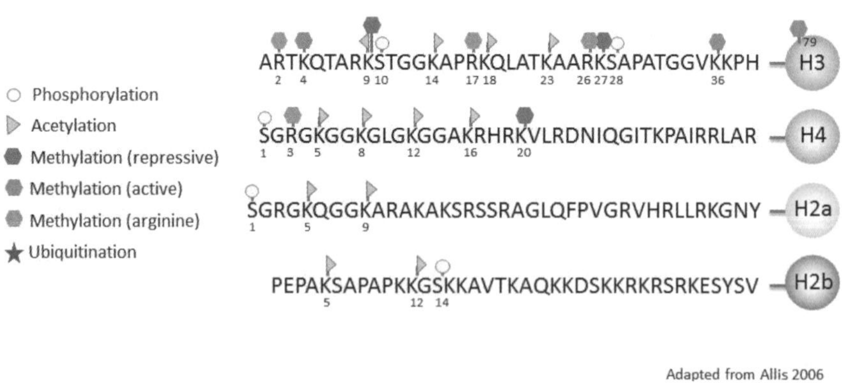

Figure 4. A schematic of the histone code.

Histones and Autism

Although DNA methylation was the first mechanism shown to be important in regulating imprinted genes, subsequent studies linking DNA methylation with the recruitment of histone modifying enzymes and chromatin remodeling complexes have revealed that multiple levels of chromatin regulation are required to establish and maintain epigenetic marks.[138,139] Several paternally-imprinted genes reside within 15q11-13, including SNRPN, NDN, IPW, PWCR1 and MAGEL2, and are expressed from the paternal chromosome while UBE3A is exclusively expressed from the maternal chromosome. In humans, it has been shown that the promoter of SNRPN (15q11.2) is enriched for H3 and H4 acetylation, although the other imprinted genes within the interval do not show significant enrichment of these modifications on either parental chromosome.[140]

In humans, the PWS-IC on the paternal chromosome is enriched for H3K4me2/3 and H3/H4ac while the maternal PWS-IC is enriched for the repressive histone modifications H3K9me2/3 and H4K20me3.[126,141] Conversely, the AS-IC on the maternal chromosomes is enriched for H4ac and H3K4me2/3, but not H3 acetylation.[126]

Future Directions

It is clear that DNA sequence, whether nucleotide mutation/polymorphism or chromosomal loss/gain, does not fully explain the heritability and potential for the environment to impact the development of ASD. The growing weight of evidence suggests that epigenetic mechanisms in the form of dysregulated DNA methylation and/or histone modifications play a role. However, the challenges for the epigenomics field of ASD research are many and involve resolving the following questions. Firstly, are the differences in epigenetic status between cases and controls functionally relevant? That is, do they translate to discernable differences in mRNA and protein expression? Secondly, is the altered epigenetic profile a driver of the phenotype or are these epigenetic changes simply a consequence of some other pathogenic mechanism?

Establishing causal inference will be critical. Additional questions include: how do we identify which cells are sentinel in ASD, how do we overcome tissue heterogeneity, and will stochastic events hamper the usefulness of single cell epigenetic profiling? Finally, what is the required depth of our epidemiological understanding to permit causal connections to be drawn between environmental and maternal exposures and the *in utero* epigenome?

However, there are exciting opportunities ahead. Akin to imputation, methods are being developed to expand the information that can be derived from microarray-based platforms by methyl-CpG imputation; the resolution and throughput of DNA and histone epigenomic profiles is improving; and novel approaches are being developed to profile modifications such as 5hmC. As we generate these data, we should be cognizant that the epigenome does not function in isolation. The field of ASD research will be challenged to explain how dysregulation of the epigenome impacts the transcriptome, the proteome and the metabolome in a systems approach to identify the causes and cures of ASD.

Executive Summary

- DNA Methylation — the mechanism by which a chemical moiety attached to the backbone of DNA regulates gene expression.
- Regulation by DNA Methylation — the hyper- or hypomethylation of DNA that regulates gene expression.
- Epigenetics and Autism — the mechanisms contributing to the development of autism without a change in DNA sequence.
- Candidate Genes — Epigenetically-regulated genes that may play a role in the development of autism.
- Influences of the Environment on the Epigenome — Environmental influences that impact the epigenome and may contribute to ASD.
- Assisted Reproduction — The process of *ex vivo* reproduction that may influence the epigenome.
- Technologies for Profiling of DNA Methylation — Platforms for profiling the epigenome of biological samples.

- Histones and Chromatin Modifications – Chemical modifications of the N-terminal domains of histones that influence gene function.
- Histones and Autism — Histone modifications that play a role in the development of ASD.

Acknowledgements

Dr. Gregory would like to thank Dr. Susan Murphy for adaptation of Figure 3, and Dr. Beth Sullivan for her invaluable advice with the Histone and Chromatin Modifications section.

References

1. Hu VW: The expanding genomic landscape of autism: discovering the 'forest' beyond the 'trees'. *Future Neurol* 2013, **8**(1):29–42.
2. Waddington CH: The epigenotype. 1942. *Int J Epidemiol* 2012, **41**(1):10–13.
3. Kohli RM, Zhang Y: TET enzymes, TDG and the dynamics of DNA demethylation. *Nature* 2013, **502**(7472):472–479.
4. Munzel M, Globisch D, Bruckl T, Wagner M, Welzmiller V, Michalakis S, Muller M, Biel M, Carell T: Quantification of the sixth DNA base hydroxymethylcytosine in the brain. *Angewandte Chemie* 2010, **49**(31):5375–5377.
5. Globisch D, Munzel M, Muller M, Michalakis S, Wagner M, Koch S, Bruckl T, Biel M, Carell T: Tissue distribution of 5-hydroxymethylcytosine and search for active demethylation intermediates. *PloS One* 2010, **5**(12):e15367.
6. Hahn MA, Qiu R, Wu X, Li AX, Zhang H, Wang J, Jui J, Jin SG, Jiang Y, Pfeifer GP *et al.*: Dynamics of 5-hydroxymethylcytosine and chromatin marks in Mammalian neurogenesis. *Cell Reports* 2013, **3**(2):291–300.
7. Ramsahoye BH, Biniszkiewicz D, Lyko F, Clark V, Bird AP, Jaenisch R: Non-CpG methylation is prevalent in embryonic stem cells and may be mediated by DNA methyltransferase 3a. *Proc Nat Acad Sci USA* 2000, **97**(10):5237–5242.
8. Haines TR, Rodenhiser DI, Ainsworth PJ: Allele-specific non-CpG methylation of the Nf1 gene during early mouse development. *Dev Bio* 2001, **240**(2):585–598.
9. Smallwood SA, Kelsey G: *De novo* DNA methylation: a germ cell perspective. *Trends Genet* 2012, **28**(1):33–42.
10. LaSalle JM: A genomic point-of-view on environmental factors influencing the human brain methylome. *Epigenetics* 2011, **6**(7):862–869.
11. Richardson B: Impact of aging on DNA methylation. *Ageing Res Rev* 2003, **2**(3):245–261.

12. Ehrlich M, Gama-Sosa MA, Huang LH, Midgett RM, Kuo KC, McCune RA, Gehrke C: Amount and distribution of 5-methylcytosine in human DNA from different types of tissues of cells. *Nucleic Acids Res* 1982, **10**(8):2709–2721.
13. Tucker KL: Methylated cytosine and the brain: a new base for neuroscience. *Neuron* 2001, **30**(3):649–652.
14. Jones PA, Liang G: Rethinking how DNA methylation patterns are maintained. *Nat Rev Genet* 2009, **10**(11):805–811.
15. Holliday R, Pugh JE: DNA modification mechanisms and gene activity during development. *Science* 1975, **187**(4173):226–232.
16. Gardiner-Garden M, Frommer M: CpG islands in vertebrate genomes. *J Mol Bio* 1987, **196**(2):261–282.
17. Fatemi M, Pao MM, Jeong S, Gal-Yam EN, Egger G, Weisenberger DJ, Jones PA: Footprinting of mammalian promoters: use of a CpG DNA methyltransferase revealing nucleosome positions at a single molecule level. *Nucleic Acids Res* 2005, **33**(20):e176.
18. Irizarry RA, Ladd-Acosta C, Wen B, Wu Z, Montano C, Onyango P, Cui H, Gabo K, Rongione M, Webster M *et al.*: The human colon cancer methylome shows similar hypo- and hypermethylation at conserved tissue-specific CpG island shores. *Nature Genet* 2009, **41**(2):178–186.
19. Kozlenkov A, Roussos P, Timashpolsky A, Barbu M, Rudchenko S, Bibikova M, Klotzle B, Byne W, Lyddon R, Di Narzo AF *et al.*: Differences in DNA methylation between human neuronal and glial cells are concentrated in enhancers and non-CpG sites. *Nucleic Acids Res* 2014, **42**(1):109–127.
20. Sandoval J, Heyn H, Moran S, Serra-Musach J, Pujana MA, Bibikova M, Esteller M: Validation of a DNA methylation microarray for 450,000 CpG sites in the human genome. *Epigenetics* 2011, **6**(6):692–702.
21. Cedar H, Bergman Y: Epigenetic silencing during early lineage commitment. In: *StemBook*. Cambridge (MA); 2008.
22. Byun HM, Siegmund KD, Pan F, Weisenberger DJ, Kanel G, Laird PW, Yang AS: Epigenetic profiling of somatic tissues from human autopsy specimens identifies tissue- and individual-specific DNA methylation patterns. *Hum Mol Genet* 2009, **18**(24):4808–4817.
23. Davies MN, Volta M, Pidsley R, Lunnon K, Dixit A, Lovestone S, Coarfa C, Harris RA, Milosavljevic A, Troakes C *et al.*: Functional annotation of the human brain methylome identifies tissue-specific epigenetic variation across brain and blood. *Genome Bio* 2012, **13**(6):R43.
24. Uddin M, Aiello AE, Wildman DE, Koenen KC, Pawelec G, de Los Santos R, Goldmann E, Galea S: Epigenetic and immune function profiles associated with posttraumatic stress disorder. *Proc Nat Acad Sci USA* 2010, **107**(20):9470–9475.
25. Nguyen A, Rauch TA, Pfeifer GP, Hu VW: Global methylation profiling of lymphoblastoid cell lines reveals epigenetic contributions to autism

spectrum disorders and a novel autism candidate gene, RORA, whose protein product is reduced in autistic brain. *FASEB* 2010, **24**(8):3036–3051.

26. Gregory SG, Connelly JJ, Towers AJ, Johnson J, Biscocho D, Markunas CA, Lintas C, Abramson RK, Wright HH, Ellis P *et al.*: Genomic and epigenetic evidence for oxytocin receptor deficiency in autism. *BMC Med* 2009, **7**:62.
27. Houseman EA, Accomando WP, Koestler DC, Christensen BC, Marsit CJ, Nelson HH, Wiencke JK, Kelsey KT: DNA methylation arrays as surrogate measures of cell mixture distribution. *BMC Bioinformatics* 2012, **13**:86.
28. Adams D, Altucci L, Antonarakis SE, Ballesteros J, Beck S, Bird A, Bock C, Boehm B, Campo E, Caricasole A *et al.*: BLUEPRINT to decode the epigenetic signature written in blood. *Nat Biotechno* 2012, **30**(3):224–226.
29. Amir RE, Van den Veyver IB, Wan M, Tran CQ, Francke U, Zoghbi HY: Rett syndrome is caused by mutations in X-linked MECP2, encoding methyl-CpG-binding protein 2. *Nature Genet* 1999, **23**(2):185–188.
30. Carney RM, Wolpert CM, Ravan SA, Shahbazian M, Ashley-Koch A, Cuccaro ML, Vance JM, Pericak-Vance MA: Identification of MeCP2 mutations in a series of females with autistic disorder. *Pediatric Neurol* 2003, **28**(3):205–211.
31. Coutinho AM, Oliveira G, Katz C, Feng J, Yan J, Yang C, Marques C, Ataide A, Miguel TS, Borges L *et al.*: MECP2 coding sequence and 3¢UTR variation in 172 unrelated autistic patients. *Am J Medi Genet B. Neuropsychiatric Genet* 2007, **144B**(4):475–483.
32. Gonzales ML, LaSalle JM: The role of MeCP2 in brain development and neurodevelopmental disorders. *Curr Psychiatry Rep* 2010, **12**(2):127–134.
33. Horike S, Cai S, Miyano M, Cheng JF, Kohwi-Shigematsu T: Loss of silent-chromatin looping and impaired imprinting of DLX5 in Rett syndrome. *Nat Genet* 2005, **37**(1):31–40.
34. Yasui DH, Peddada S, Bieda MC, Vallero RO, Hogart A, Nagarajan RP, Thatcher KN, Farnham PJ, Lasalle JM: Integrated epigenomic analyses of neuronal MeCP2 reveal a role for long-range interaction with active genes. *Proc Nat Acad Sci USA* 2007, **104**(49):19416–19421.
35. Banine F, Matagne V, Sherman LS, Ojeda SR: Brain region-specific expression of Fxyd1, an Mecp2 target gene, is regulated by epigenetic mechanisms. *J Neurosci Res* 2011, **89**(6):840–851.
36. Grafodatskaya D, Chung B, Szatmari P, Weksberg R: Autism spectrum disorders and epigenetics. *J Am Acad Child Adolescent Psychiatry* 2010, **49**(8):794–809.
37. Tamanini F, Bontekoe C, Bakker CE, van Unen L, Anar B, Willemsen R, Yoshida M, Galjaard H, Oostra BA, Hoogeveen AT: Different targets for the fragile X-related proteins revealed by their distinct nuclear localizations. *Hum Mol Genet* 1999, **8**(5):863–869.
38. Greenough WT, Klintsova AY, Irwin SA, Galvez R, Bates KE, Weiler IJ: Synaptic regulation of protein synthesis and the fragile X protein. *Proc Nat Acad Sci USA* 2001, **98**(13):7101–7106.

39. Aschrafi A, Cunningham BA, Edelman GM, Vanderklish PW: The fragile X mental retardation protein and group I metabotropic glutamate receptors regulate levels of mRNA granules in brain. *Proc Nat Acad Sci USA* 2005, **102**(6):2180–2185.
40. Maddalena A, Richards CS, McGinniss MJ, Brothman A, Desnick RJ, Grier RE, Hirsch B, Jacky P, McDowell GA, Popovich B *et al.*: Technical standards and guidelines for fragile X: the first of a series of disease-specific supplements to the Standards and Guidelines for Clinical Genetics Laboratories of the American College of Medical Genetics. Quality Assurance Subcommittee of the Laboratory Practice Committee. *Genet Med* 2001, **3**(3):200–205.
41. Smith KT, Nicholls RD, Reines D: The gene encoding the fragile X RNA-binding protein is controlled by nuclear respiratory factor 2 and the CREB family of transcription factors. *Nucleic Acids Res* 2006, **34**(4):1205–1215.
42. Crawford DC, Acuna JM, Sherman SL: FMR1 and the fragile X syndrome: human genome epidemiology review. *Genet Med* 2001, **3**(5):359–371.
43. Tabolacci E, Moscato U, Zalfa F, Bagni C, Chiurazzi P, Neri G: Epigenetic analysis reveals a euchromatic configuration in the FMR1 unmethylated full mutations. *Eur J Hum Genet* 2008, **16**(12):1487–1498.
44. Gronskov K, Hjalgrim H, Bjerager MO, Brondum-Nielsen K: Deletion of all CGG repeats plus flanking sequences in FMR1 does not abolish gene expression. *Am J Hum Genet* 1997, **61**(4):961–967.
45. Monk M: Genomic imprinting. Memories of mother and father. *Nature* 1987, **328**(6127):203–204.
46. Moore T, Haig D: Genomic imprinting in mammalian development: a parental tug-of-war. *Trends Genet : TIG* 1991, **7**(2):45–49.
47. Haig D: Parental antagonism, relatedness asymmetries, and genomic imprinting. *Proc Bio Sci R Soc* 1997, **264**(1388):1657–1662.
48. Hogart A, Wu D, LaSalle JM, Schanen NC: The comorbidity of autism with the genomic disorders of chromosome 15q11.2-q13. *Neurobiol Disease* 2010, **38**(2):181–191.
49. Horsthemke B, Buiting K: Imprinting defects on human chromosome 15. *Cytogenetic Genome Res* 2006, **113**(1–4):292–299.
50. Cook EH, Jr., Lindgren V, Leventhal BL, Courchesne R, Lincoln A, Shulman C, Lord C, Courchesne E: Autism or atypical autism in maternally but not paternally derived proximal 15q duplication. *Am J Hum Genet* 1997, **60**(4):928–934.
51. Schroer RJ, Phelan MC, Michaelis RC, Crawford EC, Skinner SA, Cuccaro M, Simensen RJ, Bishop J, Skinner C, Fender D *et al.*: Autism and maternally derived aberrations of chromosome 15q. *Am J Med Genet* 1998, **76**(4):327–336.
52. Abrahams BS, Geschwind DH: Advances in autism genetics: on the threshold of a new neurobiology. *Nat Rev Genet* 2008, **9**(5):341–355.

53. Shao Y, Cuccaro ML, Hauser ER, Raiford KL, Menold MM, Wolpert CM, Ravan SA, Elston L, Decena K, Donnelly SL *et al.*: Fine mapping of autistic disorder to chromosome 15q11-q13 by use of phenotypic subtypes. *Am J Hum Genet* 2003, **72**(3):539–548.
54. Bakker SC, van der Meulen EM, Buitelaar JK, Sandkuijl LA, Pauls DL, Monsuur AJ, van 't Slot R, Minderaa RB, Gunning WB, Pearson PL *et al.*: A whole-genome scan in 164 Dutch sib pairs with attention-deficit/hyperactivity disorder: suggestive evidence for linkage on chromosomes 7p and 15q. *Am J Hum Genet* 2003, **72**(5):1251–1260.
55. Hogart A, Nagarajan RP, Patzel KA, Yasui DH, Lasalle JM: 15q11-13 GABAA receptor genes are normally biallelically expressed in brain yet are subject to epigenetic dysregulation in autism-spectrum disorders. *Human Mol Genet* 2007, **16**(6):691–703.
56. Okita C, Meguro M, Hoshiya H, Haruta M, Sakamoto YK, Oshimura M: A new imprinted cluster on the human chromosome 7q21-q31, identified by human-mouse monochromosomal hybrids. *Genomics* 2003, **81**(6):556–559.
57. Edwards CA, Ferguson-Smith AC: Mechanisms regulating imprinted genes in clusters. *Curr Opin Cell Bio* 2007, **19**(3):281–289.
58. Jorgensen L, Rowsell HC, Dodds WJ, Mustard JF: Myocardial infarction in an old dog. *J Atherosclerosis Res* 1968, **8**(1):97–110.
59. Kofink D, Boks MP, Timmers HT, Kas MJ: Epigenetic dynamics in psychiatric disorders: environmental programming of neurodevelopmental processes. *Neurosci Biobehav Rev* 2013, **37**(5):831–845.
60. Goin-Kochel RP, Porter AE, Peters SU, Shinawi M, Sahoo T, Beaudet AL: The MTHFR 677C-->T polymorphism and behaviors in children with autism: exploratory genotype-phenotype correlations. *Autism Res* 2009, **2**(2):98–108.
61. Pasca SP, Dronca E, Kaucsar T, Craciun EC, Endreffy E, Ferencz BK, Iftene F, Benga I, Cornean R, Banerjee R *et al.*: One carbon metabolism disturbances and the C677T MTHFR gene polymorphism in children with autism spectrum disorders. *J Cell Mol Med* 2009, **13**(10):4229–4238.
62. Insel TR, Shapiro LE: Oxytocin receptor distribution reflects social organization in monogamous and polygamous voles. *Proc Nat Acad Sci USA* 1992, **89**(13):5981–5985.
63. Insel TR, Young L, Wang Z: Central oxytocin and reproductive behaviours. *Rev Reprod* 1997, **2**(1):28–37.
64. Carter CS: Sex differences in oxytocin and vasopressin: implications for autism spectrum disorders? *Behav Brain Res* 2007, **176**(1):170–186.
65. Waterhouse L, Fein D, Modahl C: Neurofunctional mechanisms in autism. *Psychol Rev* 1996, **103**(3):457–489.
66. Bartz JA, Hollander E: The neuroscience of affiliation: forging links between basic and clinical research on neuropeptides and social behavior. *Horm Behav* 2006, **50**(4):518–528.

67. Shao Y, Wolpert CM, Raiford KL, Menold MM, Donnelly SL, Ravan SA, Bass MP, McClain C, von Wendt L, Vance JM *et al.*: Genomic screen and follow-up analysis for autistic disorder. *Am J Med Genet* 2002, **114**(1):99–105.
68. Wu S, Jia M, Ruan Y, Liu J, Guo Y, Shuang M, Gong X, Zhang Y, Yang X, Zhang D: Positive association of the oxytocin receptor gene (OXTR) with autism in the Chinese Han population. *Biol Psychiatry* 2005, **58**(1):74–77.
69. Jacob S, Brune CW, Carter CS, Leventhal BL, Lord C, Cook EH, Jr.: Association of the oxytocin receptor gene (OXTR) in Caucasian children and adolescents with autism. *Neurosci Lett* 2007, **417**(1):6–9.
70. Lerer E, Levi S, Salomon S, Darvasi A, Yirmiya N, Ebstein RP: Association between the oxytocin receptor (OXTR) gene and autism: relationship to Vineland Adaptive Behavior Scales and cognition. *Mol Psychiatry* 2008, **13**(10):980–988.
71. Jack A, Connelly JJ, Morris JP: DNA methylation of the oxytocin receptor gene predicts neural response to ambiguous social stimuli. *Front Hum Neurosci* 2012, **6**:280.
72. Durand CM, Betancur C, Boeckers TM, Bockmann J, Chaste P, Fauchereau F, Nygren G, Rastam M, Gillberg IC, Anckarsater H *et al.*: Mutations in the gene encoding the synaptic scaffolding protein SHANK3 are associated with autism spectrum disorders. *Nat genet* 2007, **39**(1):25–27.
73. Uchino S, Waga C: SHANK3 as an autism spectrum disorder-associated gene. *Brain Dev* 2013, **35**(2):106–110.
74. Tissir F, Goffinet AM: Reelin and brain development. *Nat Rev Neurosci* 2003, **4**(6):496–505.
75. Fatemi SH, Snow AV, Stary JM, Araghi-Niknam M, Reutiman TJ, Lee S, Brooks AI, Pearce DA: Reelin signaling is impaired in autism. *Biol Psychiatry* 2005, **57**(7):777–787.
76. Zhang H, Liu X, Zhang C, Mundo E, Macciardi F, Grayson DR, Guidotti AR, Holden JJ: Reelin gene alleles and susceptibility to autism spectrum disorders. *Mol Psychiatry* 2002, **7**(9):1012–1017.
77. Serajee FJ, Zhong H, Mahbubul Huq AH: Association of Reelin gene polymorphisms with autism. *Genomics* 2006, **87**(1):75–83.
78. Li H, Li Y, Shao J, Li R, Qin Y, Xie C, Zhao Z: The association analysis of RELN and GRM8 genes with autistic spectrum disorder in Chinese Han population. *Am J Med Genet, Neuropsychiatric Genet* 2008, **147B**(2):194–200.
79. Persico AM, Levitt P, Pimenta AF: Polymorphic GGC repeat differentially regulates human reelin gene expression levels. *J Neural Transmission* 2006, **113**(10):1373–1382.
80. Lintas C, Persico AM: Neocortical RELN promoter methylation increases significantly after puberty. *Neuroreport* 2010, **21**(2):114–118.
81. Zhubi A, Chen Y, Dong E, Cook EH, Guidotti A, Grayson DR: Increased binding of MeCP2 to the GAD1 and RELN promoters may be mediated by

an enrichment of 5-hmC in autism spectrum disorder (ASD) cerebellum. *Transl Psychiatry* 2014, **4**:e349.

82. Ladd-Acosta C, Hansen KD, Briem E, Fallin MD, Kaufmann WE, Feinberg AP: Common DNA methylation alterations in multiple brain regions in autism. *Mol Psychiatry* 2013.
83. Fukuoka T, Sumida K, Yamada T, Higuchi C, Nakagaki K, Nakamura K, Kohsaka S, Saito K, Oeda K: Gene expression profiles in the common marmoset brain determined using a newly developed common marmoset-specific DNA microarray. *Neurosci Res* 2010, **66**(1):62–85.
84. Groen W, Teluij M, Buitelaar J, Tendolkar I: Amygdala and hippocampus enlargement during adolescence in autism. *J Am Acad Child Adolesc Psychiatry* 2010, **49**(6):552–560.
85. Agostini G, Mancini J, Chabrol B, Villeneuve N, Milh M, George F, Maurel B, Girard N: [Language disorders in children with morphologic abnormalities of the hippocampus]. *Archives de pediatrie : organe officiel de la Societe francaise de pediatrie* 2010, **17**(7):1008–1016.
86. Heron SE, Ong YS, Yendle SC, McMahon JM, Berkovic SF, Scheffer IE, Dibbens LM: Mutations in PRRT2 are not a common cause of infantile epileptic encephalopathies. *Epilepsia* 2013, **54**(5):e86–89.
87. Marini C, Conti V, Mei D, Battaglia D, Lettori D, Losito E, Bruccini G, Tortorella G, Guerrini R: PRRT2 mutations in familial infantile seizures, paroxysmal dyskinesia, and hemiplegic migraine. *Neurology* 2012, **79**(21):2109–2114.
88. Ono S, Yoshiura K, Kinoshita A, Kikuchi T, Nakane Y, Kato N, Sadamatsu M, Konishi T, Nagamitsu S, Matsuura M *et al.*: Mutations in PRRT2 responsible for paroxysmal kinesigenic dyskinesias also cause benign familial infantile convulsions. *J Human Genet* 2012, **57**(5):338–341.
89. Gardiner AR, Bhatia KP, Stamelou M, Dale RC, Kurian MA, Schneider SA, Wali GM, Counihan T, Schapira AH, Spacey SD *et al.*: PRRT2 gene mutations: from paroxysmal dyskinesia to episodic ataxia and hemiplegic migraine. *Neurology* 2012, **79**(21):2115–2121.
90. Li X, Ito M, Zhou F, Youngson N, Zuo X, Leder P, Ferguson-Smith AC: A maternal-zygotic effect gene, Zfp57, maintains both maternal and paternal imprints. *Deve cell* 2008, **15**(4):547–557.
91. Strogantsev R, Ferguson-Smith AC: Proteins involved in establishment and maintenance of imprinted methylation marks. *Briefings Functional Genomics* 2012, **11**(3):227–239.
92. Zuo X, Sheng J, Lau HT, McDonald CM, Andrade M, Cullen DE, Bell FT, Iacovino M, Kyba M, Xu G *et al.*: Zinc finger protein ZFP57 requires its co-factor to recruit DNA methyltransferases and maintains DNA methylation imprint in embryonic stem cells via its transcriptional repression domain. *J Bio Chem* 2012, **287**(3):2107–2118.

93. LaSalle JM: Epigenomic strategies at the interface of genetic and environmental risk factors for autism. *J Hum Genet* 2013, **58**(7):396–401.
94. Schmidt RJ, Hansen RL, Hartiala J, Allayee H, Schmidt LC, Tancredi DJ, Tassone F, Hertz-Picciotto I: Prenatal vitamins, one-carbon metabolism gene variants, and risk for autism. *Epidemiology* 2011, **22**(4):476–485.
95. Suren P, Roth C, Bresnahan M, Haugen M, Hornig M, Hirtz D, Lie KK, Lipkin WI, Magnus P, Reichborn-Kjennerud T *et al.*: Association between maternal use of folic acid supplements and risk of autism spectrum disorders in children. *J Am Med Assoc* 2013, **309**(6):570–577.
96. Hultman CM, Sandin S, Levine SZ, Lichtenstein P, Reichenberg A: Advancing paternal age and risk of autism: new evidence from a population-based study and a meta-analysis of epidemiological studies. *Mol Psychiatry* 2011, **16**(12):1203–1212.
97. Sandin S, Hultman CM, Kolevzon A, Gross R, MacCabe JH, Reichenberg A: Advancing maternal age is associated with increasing risk for autism: a review and meta-analysis. *J Am Acad Child Adolesc Psychiatry* 2012, **51**(5):477–486 e471.
98. Frans EM, Sandin S, Reichenberg A, Langstrom N, Lichtenstein P, McGrath JJ, Hultman CM: Autism risk across generations: a population-based study of advancing grandpaternal and paternal age. *JAMA Psychiatry* 2013, **70**(5):516–521.
99. Granville-Grossman KL: Parental age and schizophrenia. *Br J Psychiatry* 1966, **112**(490):899–905.
100. Jenkins TG, Aston KI, Cairns BR, Carrell DT: Paternal aging and associated intraindividual alterations of global sperm 5-methylcytosine and 5-hydroxymethylcytosine levels. *Fertility Sterility* 2013, **100**(4):945–951.
101. Bromley RL, Mawer GE, Briggs M, Cheyne C, Clayton-Smith J, Garcia-Finana M, Kneen R, Lucas SB, Shallcross R, Baker GA *et al.*: The prevalence of neurodevelopmental disorders in children prenatally exposed to antiepileptic drugs. *J Neurol, Neurosurgery Psychiatry* 2013, **84**(6):637–643.
102. Christensen J, Gronborg TK, Sorensen MJ, Schendel D, Parner ET, Pedersen LH, Vestergaard M: Prenatal valproate exposure and risk of autism spectrum disorders and childhood autism. *JAMA* 2013, **309**(16):1696–1703.
103. Rosenberg G: The mechanisms of action of valproate in neuropsychiatric disorders: can we see the forest for the trees? *Cellular Mol Life Sci* 2007, **64**(16):2090–2103.
104. Baccarelli A, Wright RO, Bollati V, Tarantini L, Litonjua AA, Suh HH, Zanobetti A, Sparrow D, Vokonas PS, Schwartz J: Rapid DNA methylation changes after exposure to traffic particles. *Am J Respiratory Critical Care Med* 2009, **179**(7):572–578.
105. Bollati V, Baccarelli A, Hou L, Bonzini M, Fustinoni S, Cavallo D, Byun HM, Jiang J, Marinelli B, Pesatori AC *et al.*: Changes in DNA methylation patterns in subjects exposed to low-dose benzene. *Cancer Res* 2007, **67**(3):876–880.

106. Tarantini L, Bonzini M, Apostoli P, Pegoraro V, Bollati V, Marinelli B, Cantone L, Rizzo G, Hou L, Schwartz J *et al.*: Effects of particulate matter on genomic DNA methylation content and iNOS promoter methylation. *Environ Health Perspect* 2009, **117**(2):217–222.
107. Zhao CQ, Young MR, Diwan BA, Coogan TP, Waalkes MP: Association of arsenic-induced malignant transformation with DNA hypomethylation and aberrant gene expression. *Proc Nat Acad Sci USA* 1997, **94**(20):10907–10912.
108. Conti E, Mazzotti S, Calderoni S, Saviozzi I, Guzzetta A: Are children born after assisted reproductive technology at increased risk of autism spectrum disorders? A systematic review. *Hum reprod* 2013, **28**(12):3316–3327.
109. Klemetti R, Sevon T, Gissler M, Hemminki E: Health of children born as a result of *in vitro* fertilization. *Pediatrics* 2006, **118**(5):1819–1827.
110. Maimburg RD, Vaeth M: Do children born after assisted conception have less risk of developing infantile autism? *Hum Reprod* 2007, **22**(7):1841–1843.
111. de Waal E, Mak W, Calhoun S, Stein P, Ord T, Krapp C, Coutifaris C, Schultz RM, Bartolomei MS: *In Vitro* Culture Increases the Frequency of Stochastic Epigenetic Errors at Imprinted Genes in Placental Tissues from Mouse Concepti Produced Through Assisted Reproductive Technologies. *Biol Reprod* 2013.
112. Vermeiden JP, Bernardus RE: Are imprinting disorders more prevalent after human *in vitro* fertilization or intracytoplasmic sperm injection? *Fertility Sterility* 2013, **99**(3):642–651.
113. Oliver VF, Miles HL, Cutfield WS, Hofman PL, Ludgate JL, Morison IM: Defects in imprinting and genome-wide DNA methylation are not common in the *in vitro* fertilization population. *Fertility Sterility* 2012, **97**(1):147–153 e147.
114. Rakyan VK, Down TA, Balding DJ, Beck S: Epigenome-wide association studies for common human diseases. *Nat Rev Genet* 2011, **12**(8):529–541.
115. Laird PW: Principles and challenges of genomewide DNA methylation analysis. *Nat Rev Genet* 2010, **11**(3):191–203.
116. Brunner AL, Johnson DS, Kim SW, Valouev A, Reddy TE, Neff NF, Anton E, Medina C, Nguyen L, Chiao E *et al.*: Distinct DNA methylation patterns characterize differentiated human embryonic stem cells and developing human fetal liver. *Genome Res* 2009, **19**(6):1044–1056.
117. Gebhard C, Schwarzfischer L, Pham TH, Andreesen R, Mackensen A, Rehli M: Rapid and sensitive detection of CpG-methylation using methyl-binding (MB)-PCR. *Nucleic Acids Res* 2006, **34**(11):e82.
118. Weber M, Hellmann I, Stadler MB, Ramos L, Paabo S, Rebhan M, Schubeler D: Distribution, silencing potential and evolutionary impact of promoter DNA methylation in the human genome. *Nat Genet* 2007, **39**(4):457–466.
119. Meissner A, Gnirke A, Bell GW, Ramsahoye B, Lander ES, Jaenisch R: Reduced representation bisulfite sequencing for comparative high-resolution DNA methylation analysis. *Nucleic Acids Res* 2005, **33**(18):5868–5877.

120. Song CX, Clark TA, Lu XY, Kislyuk A, Dai Q, Turner SW, He C, Korlach J: Sensitive and specific single-molecule sequencing of 5-hydroxymethylcytosine. *Nat Methods* 2012, **9**(1):75–77.
121. Booth MJ, Branco MR, Ficz G, Oxley D, Krueger F, Reik W, Balasubramanian S: Quantitative sequencing of 5-methylcytosine and 5-hydroxymethylcytosine at single-base resolution. *Science* 2012, **336**(6083):934–937.
122. Jenuwein T, Allis CD: Translating the histone code. *Science* 2001, **293**(5532):1074–1080.
123. Bannister AJ, Kouzarides T: Reversing histone methylation. *Nature* 2005, **436**(7054):1103–1106.
124. Morris SA, Rao B, Garcia BA, Hake SB, Diaz RL, Shabanowitz J, Hunt DF, Allis CD, Lieb JD, Strahl BD: Identification of histone H3 lysine 36 acetylation as a highly conserved histone modification. *J Bio Chem* 2007, **282**(10):7632–7640.
125. Roh TY, Cuddapah S, Zhao K: Active chromatin domains are defined by acetylation islands revealed by genome-wide mapping. *Genes Deve* 2005, **19**(5):542–552.
126. Mikkelsen TS, Ku M, Jaffe DB, Issac B, Lieberman E, Giannoukos G, Alvarez P, Brockman W, Kim TK, Koche RP *et al.*: Genome-wide maps of chromatin state in pluripotent and lineage-committed cells. *Nature* 2007, **448**(7153):553–560.
127. Bannister AJ, Zegerman P, Partridge JF, Miska EA, Thomas JO, Allshire RC, Kouzarides T: Selective recognition of methylated lysine 9 on histone H3 by the HP1 chromo domain. *Nature* 2001, **410**(6824):120–124.
128. Nakayama J, Rice JC, Strahl BD, Allis CD, Grewal SI: Role of histone H3 lysine 9 methylation in epigenetic control of heterochromatin assembly. *Science* 2001, **292**(5514):110–113.
129. Santos-Rosa H, Schneider R, Bernstein BE, Karabetsou N, Morillon A, Weise C, Schreiber SL, Mellor J, Kouzarides T: Methylation of histone H3 K4 mediates association of the Isw1p ATPase with chromatin. *Mol Cell* 2003, **12**(5):1325–1332.
130. Schneider R, Bannister AJ, Weise C, Kouzarides T: Direct binding of INHAT to H3 tails disrupted by modifications. *J Bio Chem* 2004, **279**(23):23859–23862.
131. Xi H, Shulha HP, Lin JM, Vales TR, Fu Y, Bodine DM, McKay RD, Chenoweth JG, Tesar PJ, Furey TS *et al.*: Identification and characterization of cell type-specific and ubiquitous chromatin regulatory structures in the human genome. *PLoS Genet* 2007, **3**(8):e136.
132. Strahl BD, Allis CD: The language of covalent histone modifications. *Nature* 2000, **403**(6765):41–45.
133. Brinkman AB, Roelofsen T, Pennings SW, Martens JH, Jenuwein T, Stunnenberg HG: Histone modification patterns associated with the human X chromosome. *EMBO Reports* 2006, **7**(6):628–634.

134. Chadwick BP, Willard HF: Multiple spatially distinct types of facultative heterochromatin on the human inactive X chromosome. *Proc Nat Acad Sci USA* 2004, **101**(50):17450–17455.
135. Cohen SM, Brylawski BP, Cordeiro-Stone M, Kaufman DG: Same origins of DNA replication function on the active and inactive human X chromosomes. *J Cellular Biochem* 2003, **88**(5):923–931.
136. Heard E: Delving into the diversity of facultative heterochromatin: the epigenetics of the inactive X chromosome. *Curr Opin Genet Deve* 2005, **15**(5):482–489.
137. Rougeulle C, Chaumeil J, Sarma K, Allis CD, Reinberg D, Avner P, Heard E: Differential histone H3 Lys-9 and Lys-27 methylation profiles on the X chromosome. *Mole Cell Biol* 2004, **24**(12):5475–5484.
138. Jones PL, Veenstra GJ, Wade PA, Vermaak D, Kass SU, Landsberger N, Strouboulis J, Wolffe AP: Methylated DNA and MeCP2 recruit histone deacetylase to repress transcription. *Nature Genet* 1998, **19**(2):187–191.
139. Feng Q, Zhang Y: The MeCP1 complex represses transcription through preferential binding, remodeling, and deacetylating methylated nucleosomes. *Genes Deve* 2001, **15**(7):827–832.
140. Lau JC, Hanel ML, Wevrick R: Tissue-specific and imprinted epigenetic modifications of the human NDN gene. *Nucleic Acids Res* 2004, **32**(11):3376–3382.
141. Xin Z, Tachibana M, Guggiari M, Heard E, Shinkai Y, Wagstaff J: Role of histone methyltransferase G9a in CpG methylation of the Prader-Willi syndrome imprinting center. *J Bio Chem* 2003, **278**(17):14996–15000.

Chapter 14

Genes and Environment in Autism Spectrum Disorders: An Integrated Perspective

Kelly M. Bakulski, Alison B. Singer and M. Daniele Fallin*

Abstract

Autism spectrum disorders (ASDs) are neurodevelopmental disorders characterized by communication and social impairments, as well as repetitive behaviors. ASDs are highly heritable, but environmental contributions to risk are also likely. To date, only a few specific genes or environmental factors have been definitively characterized. Typically, genetic and environmental research in ASD etiology have been distinct pursuits, rather than carried out in a combined gene-environment context. This chapter offers evidence in support of both genetic and environmental etiology perspectives and suggestions for how to integrate the two fields moving forward, highlighting the potential benefits, and how to navigate particular challenges.

Keywords: Environment; Genetics; Epigenetics; Methylation; Interaction.

Etiologic Models for ASD — Genes, Environment, or Both?

Autism spectrum disorders (ASDs) are a complex group of neurodevelopmental disorders characterized by deficits in

*Correspondence to: Departments of Epidemiology, Biostatistics, and Mental Health, John Hopkins Bloomberg School of Public Health, Baltimore, MD 21205. Emails: Kelly M. Bakulski: kbakulsk@jhsph.edu, Alison B. Singer: asinger@jhsph.edu, M. Daniele Fallin: dfallin@jhsph.edu

communication, social impairment, and repetitive behaviors. The most recent report from the Autism and Developmental Disabilities Monitoring (ADDM) Network estimated the prevalence of ASDs in the US as one in 88 among eight-year old children in 2008,[1] with a male-to-female ratio of approximately 4:1.[2] The mechanisms explaining sex differences are not currently understood.[3] Currently, there is no cure or primary prevention strategy for ASDs and ultimately, this will require a thorough understanding of ASD etiology. This chapter presents current evidence for both genetic and environmental perspectives on ASD etiology (Figure 1). We propose that ASD risk is likely a combination of genetic and environmental factors and describe potential mechanisms for gene by environment (GxE) interactions as well as how ASD research can frame questions to integrate these hypotheses.

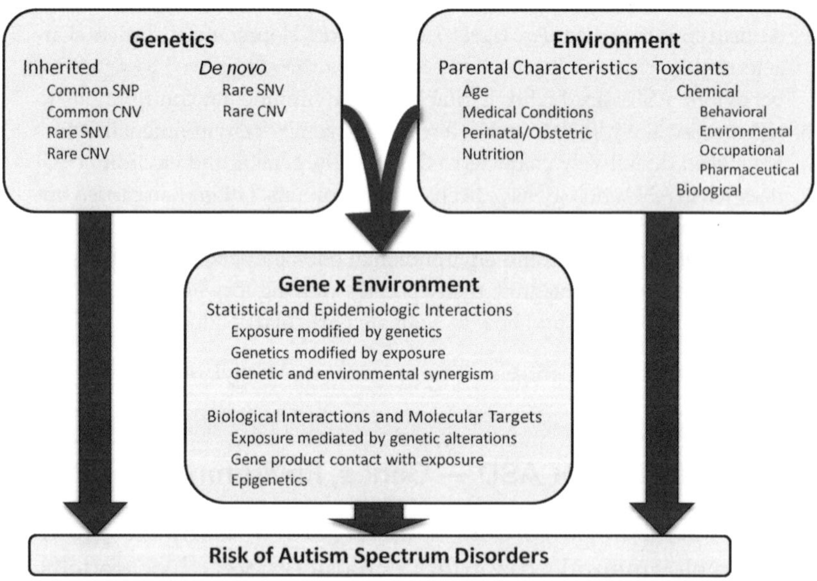

Figure 1. Conceptual paradigm for risk of Autism Spectrum Disorders (ASD), including independent genetic effects, independent environmental effects, and gene-environment interaction effects.

Heritability evidence

Twin studies are a powerful design to estimate the relative contributions of genes and environment to disease.[4] Concordance comparisons between monozygotic (MZ) twins, who share all of their genetic material, and dizygotic (DZ) twins, who share on average half of their genes, can be used to calculate disease heritability. For example, high heritability of ASD was estimated in a US twin study where MZ concordance was 88% and DZ concordance was 31%.[5] This and other twin studies showing high ASD heritability[6,7] catapulted large-scale genetic investigations of ASD; however, as discussed below, many of the findings from these gene discovery efforts have highlighted non-inherited genetic risk factors and have not fully explained the high heritability estimates.

More recent twin studies suggest that shared environment may explain a larger percent of variability in ASDs than previously recognized. In a California Developmental Services study, MZ twin concordance was 58% and DZ concordance was 21%.[8] From these data, shared environmental factors accounted for 58% of the variance in liability, while genetic heritability explained only 38%. Similarly, in the Swedish Twin Registry, MZ twins were 47% concordant and DZ twins were 14% concordant.[9] This study estimated that genetic effects accounted for 80% of the variation in liability of ASD and attributed the remaining 20% to the environment. Importantly, even the most recent twin studies provide evidence for a genetic component to ASD risk, but underscore the role of environment in ASD etiology as well.

Historical perspective on genetic and environmental etiologic work in ASD

On the heels of the first ASD twin concordance case studies,[10,11] researchers used karyotype analysis, where the number and appearance of chromosomes are visualized, to investigate large-scale genetic underpinnings of ASD. Some researchers noted abnormally large Y-chromosomes in some autism cases,[12] while others saw no chromosomal abnormalities.[13] Co-occurrence between the genetic disorder phenylketonuria (PKU) and ASD suggested that smaller scale genetic aberrations might also occur in ASD.[14] Due to the 4:1 male to female sex ratio, special attention has been paid to the role of the X-chromosome. Fragile X syndrome, caused by an increase in trinucleotide repeats (CGG) in the Fragile X mental retardation (*FMR1*) gene, was observed in multiple ASD cases.[15] Early reports showed approximately 30% of Fragile X cases co-occur with ASD,[16] but others report a rate closer to 7–8%.[17,18] However, male-to-male transmission in extended pedigrees seemed to contradict the X-linkage hypothesis.[19] Generally, early genetics research on ASD, which includes the autosomes, resulted in multiple genetic associations that were inconsistent across studies, and there was hope that modern technical advances enabling full genome coverage would lead to discovery of inherited genetic causes.

While this seminal genetic work was being carried out, various case reports suggested a co-occurrence between exposure-induced congenital disorders and ASDs. Chemical and biological agents that were linked to ASDs in case reports and small clinical studies, typically regarding maternal exposure during pregnancy, include valproic acid,[20–23] thalidomide,[24–26] misoprostol,[27] rubella,[28–31] cytomegalovirus,[32,33] measles or mumps,[31] and herpes.[34] These studies provided proof-of-principle that non-genetic factors can influence development of ASDs and have set the groundwork for more recent epidemiologic, or human population-based, studies of autism risk factors.

In the following sections, we will discuss the current body of knowledge on genetic and environmental risk factors for ASDs. We will also discuss the importance of considering a combined genetic and environmental paradigm in autism epidemiology research. We will finish with a discussion of current challenges and future considerations for this growing field of research.

Genetic Risk Factors for ASD

The field of ASD genetics has exploded in the last decade and several reviews have summarized the recent advances.[35–38] The types of genetic variants related to ASDs can be broadly categorized as either inherited, meaning passed from parent to child, or *de novo*, indicating a spontaneous genetic change in the offspring. *De novo* variants are by definition rare (much less than 1% frequency), while inherited variants may be either rare or common (typically >1%).

Inherited genetics in ASD

Given the high heritability of ASDs, there has been a great hope for the discovery of inherited variants that explain the disorders. Genome-wide association studies (GWAS) have examined common single nucleotide polymorphisms (SNPs) across the human genome in family and case-control studies. These studies of common inherited genetic variants have identified associations on chromosome 5p14.1,[39] 5p15.2,[40] and in the *CNTNAP2* gene,[41] although replication across samples has been elusive, and findings from rare variant work have garnered more attention. These GWAS SNP panels also allow examination of copy number variants, where more or less than the expected two copies of a defined sequence is observed in a person, such as a deletion or duplication. Common CNVs have been associated with ASD, notably in the neuroligin *NLGN1* gene, which codes for a neuronal cell surface protein, among several others.[42]

Modern sequencing technology has also allowed detection of rare CNVs and rare single nucleotide variants (SNVs) among ASD cases. Several rare inherited SNVs have been associated with ASDs located in biologically plausible genes[43] such as the X-linked neuroligins, *NLGN3* and *NLGN4*,[44] and *SHANK3* on 22q13, which plays a role in dendritic spine organization and binds neuroligins.[45] Rare CNVs in some of these same genes have been observed. In general, studies consistently find an increased genomic burden of rare CNVs in ASD children, particularly those that intersect genes,[46] although the specific locations of inherited CNVs and associated genes have not been consistent.

De novo genetic change in ASD

Much of the genetic findings in ASD have been related to *de novo* SNVs and CNVs not inherited from a parent. *De novo* missense or nonsense SNVs were observed in 46% of ASD cases in a recent study, and these sites were enriched for ASD-associated genes.[47] In a discordant sibling study, cases had double the number of *de novo* gene disrupting mutations (nonsense, splice site, frame shift) as control siblings,[48] a three-fold increase in number of *de novo* CNVs, and a five-fold increase in large *de novo* CNVs.[49] A similar study showed that SNV locations in cases were enriched for genes associated with intellectual disability and delay.[50] Together, this evidence suggests that rare, *de novo* SNVs and CNVs are associated with ASD, although identification of the particular genes implicated is difficult. Specific genetic regions with recurrent *de novo* ASD-related CNVs include deletions and duplications of 16p11.2, which have been replicated in several studies, and are found in ~1% of cases across several reported samples.[51–53]

Genetics summary

Despite tremendous genome-wide research efforts, to date, observed inherited and *de novo* genetic variation has not explained the majority of ASD cases, and there appears to be great heterogeneity. Identification of single variants with large effects has been difficult, likely due to the polygenic nature of ASD risk and the potential influence of environmental context. Establishing causality of specific variants, especially rare ones, has proved difficult and the general enrichment for *de novo* variation in ASD cases potentially implicates parental, or grandparental, environment or age as a proximal cause. A full, up-to-date summary of the genetic risk factors associated with ASD, and their evidence, can be found at the Simons Foundation Autism Research Initiative Gene 2.0 website (https://gene.sfari.org).

Environmental Risk Factors for ASD

Since genetic background appears to explain only a portion of the variability in ASD etiology, and environmental risk factors have long been implicated in affecting neurodevelopment in general,[54,55] increasing effort has been devoted to examining potential environmental risk factors for ASDs. Below we discuss windows of development that are most relevant for particular environmental stresses to act on ASD risk. We then discuss the current epidemiologic literature regarding environment and ASDs. In this review, we define environmental risk factors broadly, and include discussions of parental characteristics and toxicant exposures.

Windows of susceptibility

Converging evidence implicates the prenatal period as a critical window of susceptibility to ASDs. For example, studies examining the neuroanatomy of autistic brains from children show decreased numbers of Purkinje cells, but no reduction in the number of cells in the inferior olive.[56] Bauman and Kemper[56] suggest that since a connection between the olivary neurons and the Purkinje cells occur at 28–30 weeks of gestation, the insult to Purkinje cells must occur before this period in gestation. Brain stem abnormalities were also described in some cases, implying that the initiation of the disorder may occur around the time of neural tube closure.[57–59] In addition, the critical periods of some teratogens potentially associated with ASD risk, such as thalidomide, are during the early prenatal period.[26,60] The fact that some children with ASDs show behavioral differences[61] and accelerated brain growth[62–64] within the first year of life suggests that the susceptibility window is in the prenatal period or very early life. Finally, exposures to chemicals during the prenatal period have been linked to brain imaging abnormalities[65] and neurodevelopmental outcomes in general.[55,66]

Parental characteristics

Numerous studies report associations between advanced parental age and ASDs, although debate remains regarding whether this is driven by maternal age, paternal age or both.[67–86] Many theories have been proposed regarding the mechanism of this parental age risk, including increased cumulative burden of toxicants, modified maternal *in utero* environment, greater likelihood of using assisted reproductive technology, or higher rates of *de novo* mutations and alterations in epigenetic modifications with parental age.

Parental medical conditions have also been implicated, which may reflect shared genetic susceptibility across different disorders, altered *in utero* environment that may impact the development of the child, or more careful following of children of parents with certain types of medical conditions. These include parental history of mental health conditions,[79,80,87–89] maternal metabolic conditions (type 2 diabetes, gestational diabetes, hypertension, or obesity) during pregnancy,[90] and maternal allergic or autoimmune conditions,[91–93] including rheumatoid arthritis,[94] celiac disease,[94] psoriasis,[95] ulcerative colitis[96] and asthma and allergy during pregnancy.[95] However, these maternal immune associations have not been consistent[95–97] and maternal metabolic conditions are also associated with risk for other developmental delays, suggesting these conditions may not be specific to ASDs.[90] Finally, a meta-analysis on infertility treatment and ASDs found inconsistent results across studies.[98]

A constellation of obstetric complications related to poor birth outcomes, including abnormal presentation, low birth weight, small size for gestational age, low Apgar score, and maternal hemorrhage, are associated with ASDs.[99] However, no single factor appears to drive this association and current interpretations suggest that this reflects co-occurrence of ASD and obstetric complications due to a common cause. Two studies have linked increasing intervals between pregnancies and decreasing odds of autism,[100,101] potentially due to physiological changes in the mother following a previous pregnancy resulting in reduced nutrient availability.[100] This theory corresponds with some recent epidemiologic literature linking prenatal vitamin[102,103] and folic acid intake[103,104] in the preconception and early pregnancy period to ASDs or autistic disorders.

Toxicants

While there is a growing literature on parental characteristics and ASDs, few epidemiology studies have examined the potential impact of environmental toxicants, despite the early evidence from case reports and plausible biological mechanisms. Given the lack of prospective studies for a relatively rare outcome like ASD, research in this area was previously hampered by small sample sizes and imprecise methods for evaluating exposures. In the following section, we will discuss the current knowledge regarding chemical and biological toxicant risks for ASDs and how this field can be expanded.

Chemical toxicants

Chemical toxicants can be grouped by the mode of exposure: behavioral, environmental, occupational, and pharmaceutical.

Behavioral: Associations between prenatal maternal smoking and ASD risk are highly inconsistent, with some inverse associations,[105] some positive associations,[76,106] and null findings[107,108]; moreover, there is some suggestion that positive findings are due to residual confounding by family socioeconomic status.[108] Prenatal alcohol exposure and fetal alcohol syndrome can have autistic-like symptoms[109–111]; however, there is currently a paucity of epidemiology literature looking at fetal alcohol exposure and ASDs.

Environmental: Ubiquitous chemicals that populations are generally exposed to at low levels, including metals, polychlorinated biphenyls (PCBs), organophosphate insecticides, DDT, alcohol, and polybrominated diphenyl ethers (PBDEs), have been implicated in abnormal neurodevelopment in both animal and human studies,[54,55,112,113] but there are currently few published studies of toxicant exposure and ASD risk, though this is a field of growing interest.

Maternal residence near a freeway (≤309 m) at the time of birth is associated with an elevated risk of autism, due to air pollutants.[114] Particular air pollutant compounds such as chlorinated solvents,[115] methylene chloride,[116] heavy metals,[115,116] ozone,[117] particulate matter,[117,118] diesel[116] and nitrogen dioxide[118] have each been associated with ASDs. These associations, however, are not consistent across studies, potentially due to different case or control definitions, variable treatment of potential confounders, or different attention to multiple testing issues. Further, exposure estimation based on data modeling from air monitoring stations, toxicant emission inventories, or pesticide application databases lacks the precision of personal monitors.

Organochlorine pesticide exposure during the prenatal period of central nervous system development (seven days prior to conception through day 49 of gestation) has been associated with ASDs.[119] Chlorpyrifos levels in cord blood[120] and metabolites of organophosphate pesticides in maternal urine from pregnancy[121] were also associated with pervasive developmental disorders (PDDs). While these studies suggest a link between pesticide exposures and ASDs, weaknesses include small sample sizes, broad diagnostic categories, and limited clinical validation. Future environmental epidemiology studies of ASDs should leverage these individual measurements of exposure assessed during appropriate time windows with more precise diagnostic classifications of ASDs.

Occupational: Only two published studies to date looked at the impact of parental occupational exposures on ASDs using occupational histories to estimate workplace exposures, which tend to be higher than environmental exposures. Mothers with an affected child were more likely to have a job with chemical exposures than mothers of typically developing controls.[122] In particular, maternal occupational exposures to disinfectants, and exhaust and combustion products were associated with ASDs.[122] Parental occupational exposure to asphalt and solvents was elevated in ASD cases compared to typically developing controls, but none of the associations were statistically significant after multiple comparison adjustments.[123] More studies are needed to determine if these results can be repeated across different study populations, or if other occupational exposures may be associated with ASDs.

Pharmaceutical: Recent epidemiologic studies found higher risk of ASDs in children of mothers who used antidepressant drugs during pregnancy,[89,124] with one pointing specifically to selective serotonin reuptake inhibitors (SSRIs).[124] At least two studies identified associations between ASD risk and use of drugs during pregnancy that act on the β-2 adrenergic receptor system, such as terbutaline, which is used to treat asthma, but also as a tocolytic (labor repressant) during pregnancy[125,126]; however, most of these results were not statistically significant. Use of valproate, an anticonvulsant drug, during pregnancy was also associated with ASDs.[127] The main criticisms of these findings are that the potential that the underlying condition and reason for taking these drugs may drive the association rather than the pharmaceutical exposure itself, and that attempts to adjust for this concern may not be fully adequate. Additional studies are needed to determine if these prenatal drug exposures are truly risk factors for ASDs.

Biological toxicants

As discussed earlier, case reports link congenital infections and ASDs. Animal studies also demonstrate maternal infections cause neurodevelopmental outcomes in offspring. Further, children with autism may have immune abnormalities. Epidemiologic evidence linking maternal immune activity during pregnancy and ASDs has largely been conflicting. A 2009 meta-analysis did not find an overall association between infections during pregnancy and autism when pooling across 11 studies.[128] However, when including only the four studies that controlled for potential confounders or utilized sibling controls, a statistically significant association between prenatal maternal infections and ASDs was observed.[128] Since publication of the meta-analysis, a number of large studies have found little evidence for associations between maternal infections at any point during pregnancy and risk of ASDs,[129–131] although some trimester-specific associations were observed. The relationship between influenza or fever and autism is mixed.[31,130,132] Numerous factors, however, make comparisons between these studies difficult, including different definitions for the case group and different methodology for collecting exposure information (for example, self-report versus administrative collection after hospitalization). Given the biological evidence from animal models and the mixed support from epidemiologic literature, more investigation of the potential impact of maternal infection during pregnancy on ASDs is needed.

Combined Genetic and Environmental Risk Paradigm

It is evident that genetic as well as environmental factors confer risk to ASD. From twin studies, we understand that ASD is a highly heritable disease, but existing genetic association studies have failed to explain a large proportion of the heritability. One interpretation of this missing heritability is that most of the genetic risk for ASD depends on an environmental context (that previous studies have not considered). There may either be unrecognized genetic and environmental marginal effects, or there may be a direct interaction between genes and environment. The following section discusses the insight that can be gained from gene-environment interaction studies in general and why it is an especially promising area of research for ASD.

The potential applications of gene-environment interaction studies

Joint consideration of genes and environment, and their interactions (GxE) can be incredibly useful in public health. For example, smoking is an established risk factor for lung cancer, though not all smokers develop the disease. Heterogeneity in lung cancer development among smokers can be attributed to genotype.[133–135] There is also a precedence for GxE studies in helping to understand neurotoxicity and neurodevelopment, though GxE research in ASD is just beginning. GxE studies can be used to address diverse epidemiologic research questions ranging from identification and characterization of genetic effects on disease, to establishing a mechanism between exposures and disease, and to potential individualized clinical treatment and public health prevention.

First, GxE studies can be used to recognize and distinguish genetic effects in a genome-wide association study (GWAS) that would have been difficult to find above the background noise in a typical marginal test,[136] because consideration of interaction can improve the ability to detect genes with small signals.[137–139] Similarly, if a gene has already been identified in GWAS to have a relationship with disease, a GxE study can help characterize the nature of the genetic effect across levels of environmental exposure.[136] Thus, GxE research in ASD may identify new genomic locations that are involved in disease, and it may clarify the relationships between known genetic risk factors and ASD that depend on exposures.

Another key role of GxE studies is to provide mechanistic insight between exposures and disease. Environmental epidemiology research often suffers from retrospective exposure assessment and an inability to definitively determine causality in the exposure-disease association. A GxE study can be used to test a functional relationship between exposure and disease.[140] If a given biological pathway is utilized to process a toxicant or to potentially link a toxicant to a disease mechanism, researchers can identify genotypes along the pathways that influence biotransformation or disease manifestation. If the genotype influences the exposure-disease association, there is more confidence in the nature of the exposure-disease relationship. Further, GxE characterization can help determine influential components from complex exposure mixtures, such as diet, air pollution, and cigarette smoke.[140] If the metabolism of a component is altered by a genotype, and disease risk differs by genotype, then it would support a causal role for that component in disease.[140] In the case of ASD, GxE studies would help to infer mechanisms between simple and complex exposures and ASD.

GxE studies may also be influential in a broader public health context and be able to contribute to personalized treatment and prevention. Similar to clinical pharmacogenomics, which is the study of individual response to pharmaceuticals based on their genetic background, it follows that GxE studies could lead to personalized prevention. This idea is promising, but it is unclear whether in the future, genetics will be able to predict individual responses to exposure mixtures.[140] More likely, this research will help to identify susceptible populations that may be at additional risk following exposure. The lower threshold of response in susceptible populations may influence environmental legislation related to population acceptable risk levels. The public health relevance of GxE research will depend on the strength of the effect, the level of exposure in the population, and the allele frequency in the population.[136] In ASD, GxE studies can help identify individuals who are at risk of disease and we may be able to actively prevent or reduce disease severity by eliminating environmental exposures in these sensitive groups. Together, these examples illustrate the widespread utility of GxE studies for research and clinical objectives, and the particular necessity of this research in ASD.

GxE mechanisms

There are multiple possible mechanisms by which genetic and environmental factors may interact with each other to manifest risk of disease. Statisticians and biochemists have very different notions of the nature of GxE interactions; the former characterize interactions based on departures from specific statistical models, while the latter focus on biological interactions between cellular components or molecular targets, which may or may not result in model-based statistical interactions. This next section will describe and provide examples of these two types of GxE mechanisms from the general GxE literature.

Statistical and epidemiological interactions

From an epidemiologic standpoint, an interaction is conceptualized as a scenario where the effect of one exposure differs in the

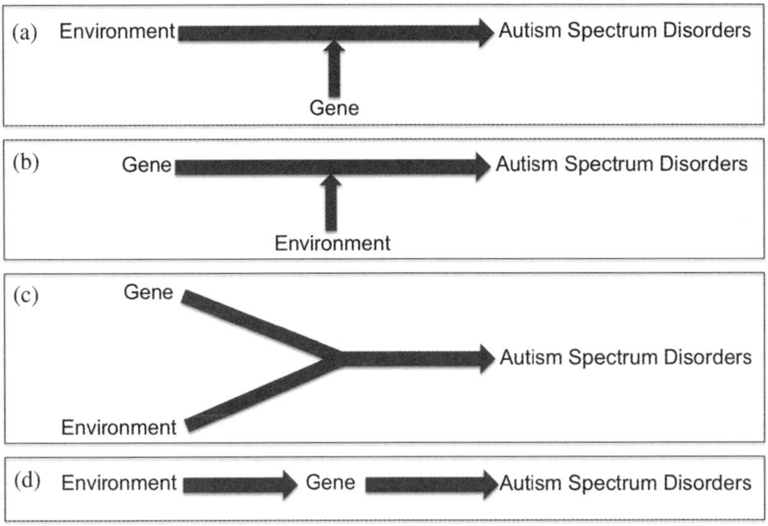

Figure 2. Four potential statistical mechanisms of gene-environment interaction in Autism Spectrum Disorder (ASD): Effect of exposure modified by genetics (a), effect of genetics modified by environment (b), joint effect of both gene and environment (c), effect of environment mediated by genetics (d).

presence or absence of a second exposure. In the case of GxE interactions, we propose three mechanisms: (A) the effect of the exposure on disease is modified by genetic background, (B) the impact of the genetic variant is modified by environment, and (C) synergistic approach whereby only the combined presence of gene and environment is sufficient to cause disease (Figure 2).

This first model of GxE interaction in epidemiologic studies involves an exposure-disease relationship that is dependent on genotype (Figure 2a). For example, the impact of low folate intake during pregnancy on risk for neural tube defects (NTD) is modified by genetic polymorphisms in the one-carbon metabolism pathway that affect folate metabolism, including *MTHFR, MTHFD1, MTR, MTRR,* and *RFC-1.*[141] While in this case, the marginal effect of low folate risk could be observed, this paradigm has great potential for understanding genetic susceptibility and outcomes in children, particularly when the marginal effect is not strong enough and risk effects would only be observed when considering genetic susceptibility. This principle of genetic susceptibility to the effects of exposure on disease is also evident in the asthma, cardiovascular disease, and cancer epidemiology literature.[142–149]

It is also theoretically possible for the effect of genetics on an outcome to be modified by exposure (Figure 2b). A deleterious genotype can lead to disease, but the severity of the disease phenotype may be controlled by expression of the altered gene product. It follows that an exposure that influences gene expression or availability of the deleterious protein would modify the risk. There are a few direct example of this model to date, although imprinting disorders illustrate this point. For example, Angelman syndrome is caused by a genetic mutation in an ubiquitin ligase gene (*UBE3A*), but development of the syndrome depends on the parent of origin of the mutation. Under normal conditions in the brain, the paternal allele of *UBE3A* is silenced via DNA methylation, while the maternal allele is expressed. Thus, the syndrome occurs only when mutations are inherited from the mother where epigenetic mechanisms promote expression. When inherited from the father, the mutation is irrelevant. Thus, the epigenetic "environment" of the genome influences whether a genetic mutation is associated with disease.

A final conceptual model of GxE epidemiological interaction requires genes and environment to have synergistic effects on disease (Figure 2c). For example, Phenylketonuria (PKU) patients have mutations in the phenylalanine hydroxylase gene (*PAH*) that prevents conversion of phenylalanine to tyrosine. Phenylalanine obtained from diet accumulates and affects brain development. Presentation of the disease can be avoided by eating a low phenylalanine diet. Thus, both a mutation in the *PAH* gene and phenylalanine exposure are required to cause disease.

Parsing between models of genetic modification of environmental effect, environmental modification of genetic effect, and synergistic environmental and genetic effects is difficult statistically. How studies choose to structure their models is largely based on conceptual frameworks and theories about causality in that system. Evidence for causality can be provided by biochemical studies of interaction between exposures and gene products.

Biological interactions and molecular targets

Alternative perspectives on GxE interactions focus on the molecular biology of the interaction. There are multiple mechanisms for exposures and genetics or gene products to interact in the cell. Toxicants or toxicant metabolites may directly come in contact with the DNA sequence or DNA epigenetic modifications (Figure 2d). Exposures or by-products may also directly impact gene products (proteins). For most exposures, the mechanisms of toxic action have not been completely worked out, but we have summarized a few examples of biological GxE interactions noted in the literature.

Direct damage to the DNA sequence is a powerful mechanism of action of many well-studied toxicants, which are often also carcinogens. These agents demonstrate a corresponding dose dependent increase in genetic damage and disease risk.[150–153] For example, in healthy individuals, dioxin exposure causes low copy numbers of non-Hodgkin's lymphoma related t(14:18) translocations in lymphocytes, and an increase of t(14:18) positive cells.[154] In addition, ionizing radiation, such as was produced in the Chernobyl and Fukushima nuclear disasters, produces genomic instability and germline DNA mutations in the subsequent generation.[155] In general, *de novo* genetic mutations can be caused by exposures and those mutations can lead to disease, making genes direct mediators of exposure risk. Environmental epigenetics is an emerging area of research that examines exposures which do not influence the underlying DNA sequence, but produce heritable changes in gene expression. Epigenetic changes include DNA methylation, chromatin modifications, and some types of RNA. Many environmental exposures produce epigenetic change either globally or in a gene specific pattern.[156] Epigenetic responses to exposures are non-uniform and genetic factors can modify the epigenetic response to the environment.[157]

Epigenetic mechanisms have been implicated in ASD, and this is a particularly promising area of GxE research for ASD risk and progression.[158–160]

Toxicants and metabolites can directly interact with cellular proteins as well. For example, the heavy metal lead can use proteins for uptake (*DMT1*) or transport (transferrin) in the body, so specific isoforms of these proteins, typically a result of genetic polymorphism, may result in different biological responses and consequences of lead.[161,162] An additional example involves hemoglobin, which is the oxygen carrying molecular of the blood stream. Methemoglobinemia is a blood disorder where an abnormal form of hemoglobin that cannot release oxygen is produced. This can be caused by nitrate exposure, often from farm runoff into drinking wells, or it can be caused by genetic polymorphisms in methemoglobin reductase. When ingested, nitrate is reduced to nitrite, which binds methemoglobin.[163] The effect of nitrate exposure on methemoglobin level is modified by the presence of methemoglobin reductase polymorphisms.[164,165] In these examples, toxicological evidence has shown functional biological links to support statistical associations.

Current GxE autism epidemiology studies

Interactions between genes and exposures have been observed in neurological dysfunction, ranging from the structural development of the nervous system, to childhood and adolescent neurodevelopmental disorders, and neurodegeneration. In addition to the interaction between low folic acid and genetic susceptibility in neural tube defects[166] mentioned earlier, childhood cognitive development, as measured by the mental development index (MDI), is disrupted by an interaction between organophosphate pesticide exposure and paraoxonase 1 (*PON1*) genotype.[167]

To date, there are three GxE findings reported for autism, two regarding risk from maternal nutrition and genetic polymorphisms implicated in folate metabolism. Overall, mothers taking a prenatal vitamin in the three months prior to conception and the first month of pregnancy were less likely to have a child later diagnosed with autism and this was exacerbated among women with genetic variants implicated functionally in the one-carbon metabolism pathway (*MTHFR* 677 T and *CBS* rs234715 GT + TT).[102] The same group showed folic acid intake ≥ 600 ug/day (the US Food & Drug Administration Recommended Daily Allowance) in the first month of pregnancy was protective against ASD,[103] but the effect was driven by mothers with a particular *MTHFR* genotype.[102,103] This research provides support for a mechanism between folate and ASD and it identifies a population that is particularly sensitive to folate intake. The third GxE finding reported an interaction between traffic-related air pollution exposure in pregnancy and *MET* genotypes. Children homozygous for the MET promoter CC genotype at rs1858830 were at high risk for developing an ASD among mothers who lived near high-traffic air pollution during pregnancy, compared to other exposure/genotype combinations.[168]

Finally, Herbert *et al.* took a bioinformatics approach to broadly assess overlap between genetic regions implicated in ASD and genes identified to be "environmentally responsive" in databases such as the NIEHS Environmental Genome Project, the Comparative Toxicogenomics Database, and the SeattleSNPs database, restricted to human immune and inflammatory response. They identified 135 genes that overlapped between the ASD and environmentally responsive gene lists.[169] These are putative risk genes whose effects may vary by environmental exposure and would be good targets for future ASD GxE studies.

Challenges and Suggestions for an Integrated Approach

Genetic measures

Given the recent explosion in genome measurement technology, there are many choices regarding what aspect of "genetics" to study. Early genetics research focused on candidate genes, which was limited at the time by biological knowledge and available coverage of that gene on commercial platforms or via direct genotyping of catalogued genetic markers. Candidate gene studies today are much more informed biologically, and can achieve very high coverage for common variants via multiple SNP genotyping and for rare variants via sequencing. Newer studies can also take an agnostic genome-wide approach using genome-wide or exome-wide SNP genotyping arrays or sequencing. Selection between these methods depends on desired data depth, cost, time, and computing/analytic capabilities. There are also perhaps unintended privacy and ethical consequences that can result following genome-wide genotyping. Discovery-driven GxE platforms will also need to consider the number of additional comparisons relative to the study sample size.

The question of whose genome to measure is also important. If ASD etiology originates *in utero*, maternal genotype may be as relevant as the child's genotype for GxE considerations. Each study will need to determine whether it is feasible or relevant to genotype both the mother and the offspring.[170,171] This particular issue has been explored in ASD with respect to prenatal vitamin and folic acid intake and *MTHFR* C677T genotype.[102,103]

Exposure assessment

In many ways, exposure assessment and analysis is a more complicated endeavor than genotyping. Timing of the exposure assessment and potential for misclassification are critical issues that are of relatively low concern for genetic measures. Especially in studies of environmental risk factors with small magnitudes of effect, misclassification may preclude observation of associations. For ASD research, case-control and case-family studies are more common than prospective cohort designs given the low prevalence of autism that make cohort studies inefficient and extremely expensive. This is particularly problematic when timing of exposure measurement matters, especially given the likelihood that the perinatal window of susceptibility is most relevant. Retrospective exposure assessment is difficult for most exposures and biological or environmental samples from pregnancy or early life may not be available 2–10 years later. Thus, exposure assessment in ASD risk factor epidemiology has often relied on self-reported questionnaires or administrative health records. Both of these methods for exposure classification present specific challenges. When information about exposures is collected through questionnaires administered after case ascertainment, parents of ASD case may recall the prenatal period differently than control parents, potentially biasing the study results. Administrative health and employment records, while likely supplying unbiased information on exposures for cases and controls, may be lacking in terms of detail or accuracy.

When individual biological samples are available, biomarker measurements are considered a preferred method for exposure assessment. However, an understanding of the biological mechanism of the toxicant needs to be incorporated into the analysis. The human body efficiently biotransforms and excretes most toxicants. When a disease or health outcome is identified in the clinic, potentially causative exposures would likely have long since flushed through the body. Thus, particularly for exposures with relatively short biological half-lives, it is critical to select biological samples from developmentally relevant time windows, prior to disease ascertainment and during disease progression, or to choose biomarkers that may be persistent beyond clearance of the original toxicant. Biomarker measurements or environmental sampling provide some advantages for exposure assessment, but rely on the collection of samples at critical time points during the pregnancy. Similar to how genetic studies progressed from candidate gene to genome-wide assessments, the environmental epidemiology research paradigm may shift from candidate exposures to consideration of a person's complete "exposome".[172] Environment-wide studies will make it possible to consider the interaction effects within complex mixtures that we encounter daily.

Phenotype assessment

There is no definitive test for ASDs; diagnosis is based on clinical assessment and diagnostic criteria based on observational and interview information. Thus, measurement error is a critical limitation of ASD research. Further, ASDs are a diverse group of disorders with different potential causes. Dissection and analysis of particular traits or component domains related to ASDs may be more useful in reducing measurement error and increasing detectible effect sizes for particular causal relationships to particular traits. In the GxE setting, all gains of statistical efficiency, via error reduction and/or precise cause-effect paths, will be necessary. Further gains are possible in this regard by considering quantitative "endophenotypes" that have a continuous distribution across the general population, such as social responsiveness scores.[173,174]

Analysis tools

In addition to limitations of availability of G, E, and ASD data, GxE studies in all of epidemiology have been limited by the need for much greater samples sizes than necessary for estimation and testing of main effects when using traditional GxE tests such as cross-product regression terms.[175] Statistical approaches that gain efficiency and allow for more attainable sample sizes have been proposed,[176–183] but most have not been fully extended for use at the genome- and exposome-wide levels of data management and multiplicity. These include case-only designs,[176,177] hybrid case-only/case-control designs and statistical tools,[178,182,183] and testing strategies that only test for joint main and interactive effects, rather than the interaction parameter alone.[179,184] Most of these focus only on a single gene and environment factor in a two-way interaction model. Methods to detect higher-order interactions, and methods that collapse multiple genes or environmental domains into single "G" and "E" factors have also been proposed including applications of hierarchical modeling,[185] random forest approaches,[181] regression tree analyses,[186] Boolean regression tree analysis,[180] and neural networks,[187] as well as brute force combinatorial analyses.[180,188,189] Given the likelihood of GxE interactions, and the likely complexity of true scenarios, these statistical advances will be critical to the discovery and characterization of GxE moving forward.

Study designs and available samples

GxE research related to ASDs has been limited by a lack of available samples with exposures measured in the appropriate neurodevelopmental window, rigorous phenotyping available, and/or adequate numbers to carry out statistical analyses. There are various practical challenges to establishing studies with all three of these characteristics. Studies with large sample sizes are required in order to have power for GxE studies. However, collection of appropriate exposure and detailed phenotyping information may be prohibitively expensive in large samples. Thus, there are benefits to leveraging both smaller studies with more detailed exposure and outcome information, and larger studies with greater power.

Several newly established ASD epidemiology studies with limited sample sizes but detailed exposure assessment and outcome information are now underway that are well positioned to conduct GxE analyses. These studies include case-control studies, such as Childhood Autism Risks from Genetics and Environment (CHARGE)[190] and Study to Explore Early Development (SEED).[191] One potential limitation of case-control studies is that they rely on retrospective exposure assessment. Prospective cohort studies will allow for collection of questionnaire exposure information and biosamples during the pregnancy, but sample sizes in an unselected population cohort would need to be very large to sufficiently power analyses. One approach to combat this power concern is to utilize an enriched pregnancy cohort design in which only families who already have a child with an ASD are enrolled in the cohort.[192,193] Enriched ASD pregnancy cohort studies currently in progress include the Markers of Autism Risk in Babies-Learning Early Signs (MARBLES) study and the Early Autism Risk Longitudinal Investigation (EARLI).[192]

Because sample size is nonetheless a consideration for GxE research and ASDs have a relatively low prevalence limiting the sample size for any single study, even with minimized data collection, it is necessary to combine data across studies. Thus, exposure variables must be harmonizable across studies, requiring common exposure assessment techniques, sample types, and collection timing.[194] The role of a single, smaller study will be to contribute a different and specific genetic background, or environmental exposure level and type, or depth of measurement that allows characterization not possible across the larger group.

Another potential strategy to examine GxE relationships in ASD is to exploit existing studies that were not originally designed for the purposes of studying ASD. Birth cohorts that were unselected for ASD may have a low number of ASD cases, but often have thorough prospective exposure measures and could add clinical assessments to collect continuous phenotypes related to ASD and accrue ASD cases over time that can be pooled across birth cohort studies. Data from health registries could also add important information to the autism GxE literature, largely as a result of the large sample sizes. Studies using health registries will have incomplete information on environmental exposures, but may be highly useful for exposures included in health records, such as medical conditions and pharmaceutical use. Some of these larger cohort studies and registries may also have banked biological samples from the pregnancy that can be utilized to do genotyping and measure exposure biomarkers. Finally, a large consortium is already underway to examine genetic associations with ASD. There is some effort to add exposure data to these consortia with the caveat that this exposure data will be retrospective and subject to bias. Despite the disadvantages in these studies in terms of exposure assessment and phenotyping, the sample sizes will provide great power for GxE analyses.

Summary and Future Directions

There is growing evidence for both genetic and environmental components of ASD risk, with environmental susceptibility most likely occurring in prenatal or early life. While much of the ASD etiologic research has focused on either genes or environment, we advocate the integration of these fields in pursuit of a better understanding of ASDs. While there are great design, measurement, and statistical issues facing both domains, these can be overcome and can inform each other through integrated investigation. Given the potential impact of gene-environment interplay in medicine and health, the field of gene-environment interaction is moving forward rapidly for many disorders, particularly in the fields of epidemiology and biostatistics. ASD research will be able to harness those advances through collaborative efforts, which can set the stage for simultaneous investigation of the genome, epigenome, and exposome. It is therefore a promising time for discovery in ASDs. These GxE characterizations, once established, can inform intervention strategies at the public and clinic levels as well as lead to personalized counseling, prediction, and intervention.

Executive Summary

- There is evidence for both genetic and environmental components to risk for ASDs.
- Environmental susceptibility is most likely to be manifested prenatally or very early in life.
- ASD etiologic research has mostly focused on these two components independently; progress can be made by shifting to a paradigm of GxE integration in ASD research.
- Advances in genotyping technology allow consideration of the whole genome via large-scale SNP or sequencing efforts, but the scope of this measurement will need to take GxE questions into consideration.
- The genome to be considered in GxE risk for ASD should be expanded to include maternal and paternal genomes.
- Timing and accuracy of exposure assessment will be critical, which is a particular challenge for ASD epidemiology, given the common reliance on case-control and case-family designs for a disorder of relatively low prevalence.

- Sample sizes with available and appropriate exposure data, genome data, and phenotype information are currently limited and will continue to be a challenge. Recognition of the immediate value of limited samples with deep information on all three domains must be enhanced, while simultaneously improving the ability to combine data across studies via harmonized, albeit less detailed, information.

References

1. CDC: Prevalence of autism spectrum disorders — Autism and Developmental Disabilities Monitoring Network, 14 sites, United States, 2008. *MMWR Surveill Summ* 2012, **61**(3):1–19.
2. Fombonne E, Quirke S, Hagen A: Prevalence and interpretation of recent trends in rates of pervasive developmental disorders. *Mcgill J Med* 2009, **12**(2):73.
3. Werling DM, Geschwind DH: Sex differences in autism spectrum disorders. *Curr Opin Neurol* 2013, **26**(2):146–153.
4. Ronald A, Hoekstra RA: Autism spectrum disorders and autistic traits: a decade of new twin studies. *Am J Med Genet B Neuropsychiatr Genet* 2011, **156B**(3):255–274.
5. Rosenberg RE, Law JK, Yenokyan G, McGready J, Kaufmann WE, Law PA: Characteristics and concordance of autism spectrum disorders among 277 twin pairs. *Arch Pediatr Adolesc Med* 2009, **163**(10):907–914.
6. Taniai H, Nishiyama T, Miyachi T, Imaeda M, Sumi S: Genetic influences on the broad spectrum of autism: study of proband-ascertained twins. *Am J Med Genet B Neuropsychiatr Genet* 2008, **147B**(6):844–849.
7. Steffenburg S, Gillberg C, Hellgren L, Andersson L, Gillberg IC, Jakobsson G, Bohman M: A twin study of autism in Denmark, Finland, Iceland, Norway and Sweden. *J Child Psychol Psychiatry* 1989, **30**(3):405–416.
8. Hallmayer J, Cleveland S, Torres A, Phillips J, Cohen B, Torigoe T, Miller J, Fedele A, Collins J, Smith K *et al.*: Genetic heritability and shared environmental factors among twin pairs with autism. *Arch Gen Psychiatry* 2011, **68**(11):1095–1102.
9. Lichtenstein P, Carlstrom E, Rastam M, Gillberg C, Anckarsater H: The genetics of autism spectrum disorders and related neuropsychiatric disorders in childhood. *Am J Psychiatry* 2010, **167**(11):1357–1363.
10. Vaillant GE: Twins Discordant for Early Infantile Autism. *Arch Gen Psychiatry* 1963, **9**:163–167.

11. Chapman AH: Early infantile autism in identical twins; report of a case. *AMA Arch Neurol Psychiatry* 1957, **78**(6):621–623.
12. Judd LL, Mandell AJ: Chromosome studies in early infantile autism. *Arch Gen Psychiatry* 1968, **18**(4):450–457.
13. Wolraich M, Bzostek B, Neu RL, Gardner LI: Lack of chromosome aberrations in autism. *N Eng J Med* 1970, **283**(22):1231.
14. Friedman E: The "autistic syndrome" and phenylketonuria. *Schizophrenia* 1968, **1**:249–261.
15. Brown V, Jin P, Ceman S, Darnell JC, O'Donnell WT, Tenenbaum SA, Jin X, Feng Y, Wilkinson KD, Keene JD *et al.*: Microarray identification of FMRP-associated brain mRNAs and altered mRNA translational profiles in fragile X syndrome. *Cell* 2001, **107**(4):477–487.
16. Roberts JE, Mirrett P, Burchinal M: Receptive and expressive communication development of young males with fragile X syndrome. *Am J Ment Retard* 2001, **106**(3):216–230.
17. Li SY, Chen YC, Lai TJ, Hsu CY, Wang YC: Molecular and cytogenetic analyses of autism in Taiwan. *Hum Gen* 1993, **92**(5):441–445.
18. Fombonne E, Du Mazaubrun C, Cans C, Grandjean H: Autism and associated medical disorders in a French epidemiological survey. *J Am Acad Child Adolesc Psychiatry* 1997, **36**(11):1561–1569.
19. Hallmayer J, Spiker D, Lotspeich L, McMahon WM, Petersen PB, Nicholas P, Pingree C, Ciaranello RD: Male-to-male transmission in extended pedigrees with multiple cases of autism. *Am J Med Genet* 1996, **67**(1):13–18.
20. Christianson AL, Chesler N, Kromberg JG: Fetal valproate syndrome: clinical and neuro-developmental features in two sibling pairs. *Dev Med Child Neurol* 1994, **36**(4):361–369.
21. Williams G, King J, Cunningham M, Stephan M, Kerr B, Hersh JH: Fetal valproate syndrome and autism: additional evidence of an association. *Dev Med Child Neurol* 2001, **43**(3):202–206.
22. Moore SJ, Turnpenny P, Quinn A, Glover S, Lloyd DJ, Montgomery T, Dean JC: A clinical study of 57 children with fetal anticonvulsant syndromes. *J Med Genet* 2000, **37**(7):489–497.
23. Williams PG, Hersh JH: A male with fetal valproate syndrome and autism. *Dev Med Child Neurol* 1997, **39**(9):632–634.
24. Rodier PM, Hyman SL: Early environmental factors in autism. *Ment Retard Dev Dis Res Rev* 1998, **4**:121–128.
25. Rodier PM, Ingram JL, Tisdale B, Croog VJ: Linking etiologies in humans and animal models: studies of autism. *Reprod Toxicol* 1997, **11**(2–3):417–422.
26. Stromland K, Nordin V, Miller M, Akerstrom B, Gillberg C: Autism in thalidomide embryopathy: a population study. *Dev Med Child Neurol* 1994, **36**(4):351–356.

27. Bandim JM, Ventura LO, Miller MT, Almeida HC, Costa AE: Autism and Mobius sequence: an exploratory study of children in northeastern Brazil. *Arq Neuropsiquiatr* 2003, **61**(2A):181–185.
28. Chess S: Autism in children with congenital rubella. *J Autism Child Schizophr* 1971, **1**(1):33–47.
29. Chess S: Follow-up report on autism in congenital rubella. *J Autism Child Schizophr* 1977, **7**(1):69–81.
30. Chess S, Fernandez P, Korn S: Behavioral consequences of congenital rubella. *J Pediatrics* 1978, **93**(4):699–703.
31. Deykin EY, MacMahon B: Viral exposure and autism. *Am J Epidemiol* 1979, **109**(6):628–638.
32. Markowitz PI: Autism in a child with congenital cytomegalovirus infection. *J Autism Dev Disord* 1983, **13**(3):249–253.
33. Stubbs EG, Ash E, Williams CP: Autism and congenital cytomegalovirus. *J Autism Dev Disord* 1984, **14**(2):183–189.
34. Ritvo ER, Mason-Brothers A, Freeman BJ, Pingree C, Jenson WR, McMahon WM, Petersen PB, Jorde LB, Mo A, Ritvo A: The UCLA-University of Utah epidemiologic survey of autism: the etiologic role of rare diseases. *Am J Psychiatry* 1990, **147**(12):1614–1621.
35. Buxbaum JD, Daly MJ, Devlin B, Lehner T, Roeder K, State MW, Autism Sequencing C: The autism sequencing consortium: large-scale, high-throughput sequencing in autism spectrum disorders. *Neuron* 2012, **76**(6):1052–1056.
36. Li X, Zou H, Brown WT: Genes associated with autism spectrum disorder. *Brain Res Bull* 2012, **88**(6):543–552.
37. Devlin B, Scherer SW: Genetic architecture in autism spectrum disorder. *Curr Opin Genet Dev* 2012, **22**(3):229–237.
38. Betancur C: Etiological heterogeneity in autism spectrum disorders: more than 100 genetic and genomic disorders and still counting. *Brain Res* 2011, **1380**:42–77.
39. Wang K, Zhang H, Ma D, Bucan M, Glessner JT, Abrahams BS, Salyakina D, Imielinski M, Bradfield JP, Sleiman PM *et al.*: Common genetic variants on 5p14.1 associate with autism spectrum disorders. *Nature* 2009, **459**(7246):528–533.
40. Weiss LA, Arking DE, Daly MJ, Chakravarti A: A genome-wide linkage and association scan reveals novel loci for autism. *Nature* 2009, **461**(7265):802–808.
41. Anney R, Klei L, Pinto D, Almeida J, Bacchelli E, Baird G, Bolshakova N, Bolte S, Bolton PF, Bourgeron T *et al.*: Individual common variants exert weak effects on the risk for autism spectrum disorderspi. *Hum Mol Genet* 2012, **21**(21):4781–4792.
42. Glessner JT, Wang K, Cai G, Korvatska O, Kim CE, Wood S, Zhang H, Estes A, Brune CW, Bradfield JP *et al.*: Autism genome-wide copy number variation reveals ubiquitin and neuronal genes. *Nature* 2009, **459**(7246):569–573.

43. Kenny EM, Cormican P, Furlong S, Heron E, Kenny G, Fahey C, Kelleher E, Ennis S, Tropea D, Anney R *et al.*: Excess of rare novel loss-of-function variants in synaptic genes in schizophrenia and autism spectrum disorders. *Mol Psychiatry* 2013.
44. Jamain S, Quach H, Betancur C, Rastam M, Colineaux C, Gillberg IC, Soderstrom H, Giros B, Leboyer M, Gillberg C *et al.*: Mutations of the X-linked genes encoding neuroligins NLGN3 and NLGN4 are associated with autism. *Nat Genet* 2003, **34**(1):27–29.
45. Durand CM, Betancur C, Boeckers TM, Bockmann J, Chaste P, Fauchereau F, Nygren G, Rastam M, Gillberg IC, Anckarsater H *et al.*: Mutations in the gene encoding the synaptic scaffolding protein SHANK3 are associated with autism spectrum disorders. *Nat Genet* 2007, **39**(1):25–27.
46. Pinto D, Pagnamenta AT, Klei L, Anney R, Merico D, Regan R, Conroy J, Magalhaes TR, Correia C, Abrahams BS *et al.*: Functional impact of global rare copy number variation in autism spectrum disorders. *Nature* 2010, **466**(7304):368–372.
47. Neale BM, Kou Y, Liu L, Ma'ayan A, Samocha KE, Sabo A, Lin CF, Stevens C, Wang LS, Makarov V *et al.*: Patterns and rates of exonic de novo mutations in autism spectrum disorders. *Nature* 2012, **485**(7397):242–245.
48. Iossifov I, Ronemus M, Levy D, Wang Z, Hakker I, Rosenbaum J, Yamrom B, Lee YH, Narzisi G, Leotta A *et al.*: De novo gene disruptions in children on the autistic spectrum. *Neuron* 2012, **74**(2):285–299.
49. Sanders SJ, Ercan-Sencicek AG, Hus V, Luo R, Murtha MT, Moreno-De-Luca D, Chu SH, Moreau MP, Gupta AR, Thomson SA *et al.*: Multiple recurrent de novo CNVs, including duplications of the 7q11.23 Williams syndrome region, are strongly associated with autism. *Neuron* 2011, **70**(5):863–885.
50. O'Roak BJ, Vives L, Girirajan S, Karakoc E, Krumm N, Coe BP, Levy R, Ko A, Lee C, Smith JD *et al.*: Sporadic autism exomes reveal a highly interconnected protein network of de novo mutations. *Nature* 2012, **485**(7397):246–250.
51. Marshall CR, Noor A, Vincent JB, Lionel AC, Feuk L, Skaug J, Shago M, Moessner R, Pinto D, Ren Y *et al.*: Structural variation of chromosomes in autism spectrum disorder. *Am J Hum Genet* 2008, **82**(2):477–488.
52. Kumar RA, KaraMohamed S, Sudi J, Conrad DF, Brune C, Badner JA, Gilliam TC, Nowak NJ, Cook EH, Jr., Dobyns WB *et al.*: Recurrent 16p11.2 microdeletions in autism. *Hum Mol Genet* 2008, **17**(4):628–638.
53. Weiss LA, Shen Y, Korn JM, Arking DE, Miller DT, Fossdal R, Saemundsen E, Stefansson H, Ferreira MA, Green T *et al.*: Association between microdeletion and microduplication at 16p11.2 and autism. *New Eng J Med* 2008, **358**(7):667–675.
54. Landrigan PJ: What causes autism? Exploring the environmental contribution. *Curr Opin Pediatr* 2010, **22**(2):219–225.
55. Bellinger DC: Prenatal Exposures to Environmental Chemicals and Children's Neurodevelopment: An Update. *Saf Health Work* 2013, **4**(1):1–11.

56. Bauman ML, Kemper TL: Neuroanatomic observations of the brain in autism: a review and future directions. *Int J Dev Neurosci* 2005, **23**(2–3):183–187.
57. Rodier PM, Ingram JL, Tisdale B, Nelson S, Romano J: Embryological origin for autism: developmental anomalies of the cranial nerve motor nuclei. *J Comp Neurol* 1996, **370**(2):247–261.
58. Bailey A, Luthert P, Dean A, Harding B, Janota I, Montgomery M, Rutter M, Lantos P: A clinicopathological study of autism. *Brain* 1998, **121**(Pt 5): 889–905.
59. Rodier PM: Converging evidence for brain stem injury in autism. *Dev Psychopathol* 2002, **14**(3):537–557.
60. Miller MT: Thalidomide embryopathy: a model for the study of congenital incomitant horizontal strabismus. *Trans Am Ophthalmol Soc* 1991, **89**:623–674.
61. Zwaigenbaum L, Bryson S, Rogers T, Roberts W, Brian J, Szatmari P: Behavioral manifestations of autism in the first year of life. *Int J Dev Neurosci* 2005, **23**(2–3):143–152.
62. Courchesne E, Carper R, Akshoomoff N: Evidence of brain overgrowth in the first year of life in autism. *Jama* 2003, **290**(3):337–344.
63. Mraz KD, Green J, Dumont-Mathieu T, Makin S, Fein D: Correlates of head circumference growth in infants later diagnosed with autism spectrum disorders. *J Child Neurol* 2007, **22**(6):700–713.
64. Webb SJ, Nalty T, Munson J, Brock C, Abbott R, Dawson G: Rate of head circumference growth as a function of autism diagnosis and history of autistic regression. *J Child Neurol* 2007, **22**(10):1182–1190.
65. Sowell ER, Leow AD, Bookheimer SY, Smith LM, O'Connor MJ, Kan E, Rosso C, Houston S, Dinov ID, Thompson PM: Differentiating prenatal exposure to methamphetamine and alcohol versus alcohol and not methamphetamine using tensor-based brain morphometry and discriminant analysis. *J Neurosci* 2010, **30**(11):3876–3885.
66. Rice D, Barone S, Jr.: Critical periods of vulnerability for the developing nervous system: evidence from humans and animal models. *Environ Health Perspect* 2000, **108** Suppl 3:511–533.
67. Grether JK, Anderson MC, Croen LA, Smith D, Windham GC: Risk of autism and increasing maternal and paternal age in a large north American population. *Am J Epidemiol* 2009, **170**(9):1118–1126.
68. Lampi KM, Hinkka-Yli-Salomaki S, Lehti V, Helenius H, Gissler M, Brown AS, Sourander A: Parental age and risk of autism spectrum disorders in a finnish national birth cohort. *J Autism Dev Dis* 2013.
69. Parner ET, Baron-Cohen S, Lauritsen MB, Jorgensen M, Schieve LA, Yeargin-Allsopp M, Obel C: Parental age and autism spectrum disorders. *Ann Epidemiol* 2012, **22**(3):143–150.
70. van Balkom ID, Bresnahan M, Vuijk PJ, Hubert J, Susser E, Hoek HW: Paternal age and risk of autism in an ethnically diverse, non-industrialized setting: Aruba. *PLoS One* 2012, **7**(9):e45090.

71. Croen LA, Najjar DV, Fireman B, Grether JK: Maternal and paternal age and risk of autism spectrum disorders. *Arch Pediatr Adolesc Med* 2007, **161**(4):334–340.
72. Durkin MS, Maenner MJ, Newschaffer CJ, Lee LC, Cunniff CM, Daniels JL, Kirby RS, Leavitt L, Miller L, Zahorodny W *et al.*: Advanced parental age and the risk of autism spectrum disorder. *Am J Epidemiol* 2008, **168**(11):1268–1276.
73. Glasson EJ, Bower C, Petterson B, de Klerk N, Chaney G, Hallmayer JF: Perinatal factors and the development of autism: a population study. *Arch Gen Psychiatry* 2004, **61**(6):618–627.
74. Golding J, Steer C, Pembrey M: Parental and grandparental ages in the autistic spectrum disorders: a birth cohort study. *PLoS One* 2010, **5**(4):e9939.
75. Hultman CM, Sandin S, Levine SZ, Lichtenstein P, Reichenberg A: Advancing paternal age and risk of autism: new evidence from a population-based study and a meta-analysis of epidemiological studies. *Mol Psychiatry* 2011, **16**(12):1203–1212.
76. Hultman CM, Sparen P, Cnattingius S: Perinatal risk factors for infantile autism. *Epidemiology* 2002, **13**(4):417–423.
77. King MD, Fountain C, Dakhlallah D, Bearman PS: Estimated autism risk and older reproductive age. *Am J Public Health* 2009, **99**(9):1673–1679.
78. Ben Itzchak E, Lahat E, Zachor DA: Advanced parental ages and low birth weight in autism spectrum disorders — rates and effect on functioning. *Res Dev Disabil* 2011, **32**(5):1776–1781.
79. Larsson HJ, Eaton WW, Madsen KM, Vestergaard M, Olesen AV, Agerbo E, Schendel D, Thorsen P, Mortensen PB: Risk factors for autism: perinatal factors, parental psychiatric history, and socioeconomic status. *Am J Epidemiol* 2005, **161**(10):916–925; discussion 926–918.
80. Lauritsen MB, Pedersen CB, Mortensen PB: Effects of familial risk factors and place of birth on the risk of autism: a nationwide register-based study. *J Child Psychol Psychiatry* 2005, **46**(9):963–971.
81. Rahbar MH, Samms-Vaughan M, Loveland KA, Pearson DA, Bressler J, Chen Z, Ardjomand-Hessabi M, Shakespeare-Pellington S, Grove ML, Beecher C *et al.*: Maternal and paternal age are jointly associated with childhood autism in Jamaica. *J Autism Dev Dis* 2012, **42**(9):1928–1938.
82. Reichenberg A, Gross R, Weiser M, Bresnahan M, Silverman J, Harlap S, Rabinowitz J, Shulman C, Malaspina D, Lubin G *et al.*: Advancing paternal age and autism. *Arch Gen Psychiatry* 2006, **63**(9):1026–1032.
83. Sasanfar R, Haddad SA, Tolouei A, Ghadami M, Yu D, Santangelo SL: Paternal age increases the risk for autism in an Iranian population sample. *Mol Autism* 2010, **1**(1):2.
84. Tsuchiya KJ, Matsumoto K, Miyachi T, Tsujii M, Nakamura K, Takagai S, Kawai M, Yagi A, Iwaki K, Suda S *et al.*: Paternal age at birth and high-functioning autistic-spectrum disorder in offspring. *Br J Psychiatry* 2008, **193**(4):316–321.

85. Zhang X, Lv CC, Tian J, Miao RJ, Xi W, Hertz-Picciotto I, Qi L: Prenatal and perinatal risk factors for autism in China. *J Autism Dev Dis* 2010, **40**(11):1311–1321.
86. Frans EM, Sandin S, Reichenberg A, Langstrom N, Lichtenstein P, McGrath JJ, Hultman CM: Autism Risk Across Generations: A Population-Based Study of Advancing Grandpaternal and Paternal Age. *JAMA Psychiatry* 2013:1–6.
87. Jokiranta E, Brown AS, Heinimaa M, Cheslack-Postava K, Suominen A, Sourander A: Parental psychiatric disorders and autism spectrum disorders. *Psychiatry Res* 2013.
88. Daniels JL, Forssen U, Hultman CM, Cnattingius S, Savitz DA, Feychting M, Sparen P: Parental psychiatric disorders associated with autism spectrum disorders in the offspring. *Pediatrics* 2008, **121**(5):e1357–1362.
89. Rai D, Lee BK, Dalman C, Golding J, Lewis G, Magnusson C: Parental depression, maternal antidepressant use during pregnancy, and risk of autism spectrum disorders: population based case-control study. *BMJ* 2013, **346**:f2059.
90. Krakowiak P, Walker CK, Bremer AA, Baker AS, Ozonoff S, Hansen RL, Hertz-Picciotto I: Maternal metabolic conditions and risk for autism and other neurodevelopmental disorders. *Pediatrics* 2012, **129**(5):e1121–1128.
91. Comi AM, Zimmerman AW, Frye VH, Law PA, Peeden JN: Familial clustering of autoimmune disorders and evaluation of medical risk factors in autism. *J Child Neurol* 1999, **14**(6):388–394.
92. Keil A, Daniels JL, Forssen U, Hultman C, Cnattingius S, Soderberg KC, Feychting M, Sparen P: Parental autoimmune diseases associated with autism spectrum disorders in offspring. *Epidemiology* 2010, **21**(6):805–808.
93. Sweeten TL, Bowyer SL, Posey DJ, Halberstadt GM, McDougle CJ: Increased prevalence of familial autoimmunity in probands with pervasive developmental disorders. *Pediatrics* 2003, **112**(5):e420.
94. Atladottir HO, Pedersen MG, Thorsen P, Mortensen PB, Deleuran B, Eaton WW, Parner ET: Association of family history of autoimmune diseases and autism spectrum disorders. *Pediatrics* 2009, **124**(2):687–694.
95. Croen LA, Grether JK, Yoshida CK, Odouli R, Van de Water J: Maternal autoimmune diseases, asthma and allergies, and childhood autism spectrum disorders: a case-control study. *Arch Pediatr Adolesc Med* 2005, **159**(2):151–157.
96. Mouridsen SE, Rich B, Isager T, Nedergaard NJ: Autoimmune diseases in parents of children with infantile autism: a case-control study. *Dev Med Child Neurol* 2007, **49**(6):429–432.
97. Micali N, Chakrabarti S, Fombonne E: The broad autism phenotype: findings from an epidemiological survey. *Autism* 2004, **8**(1):21–37.
98. Hvidtjorn D, Schieve L, Schendel D, Jacobsson B, Svaerke C, Thorsen P: Cerebral palsy, autism spectrum disorders, and developmental delay in children born after assisted conception: a systematic review and meta-analysis. *Arch Pediatr Adolesc Med* 2009, **163**(1):72–83.

99. Gardener H, Spiegelman D, Buka SL: Perinatal and neonatal risk factors for autism: a comprehensive meta-analysis. *Pediatrics* 2011, **128**(2):344–355.
100. Cheslack-Postava K, Liu K, Bearman PS: Closely spaced pregnancies are associated with increased odds of autism in California sibling births. *Pediatrics* 2011, **127**(2):246–253.
101. Gunnes N, Suren P, Bresnahan M, Hornig M, Lie KK, Lipkin WI, Magnus P, Nilsen RM, Reichborn-Kjennerud T, Schjolberg S *et al.*: Interpregnancy interval and risk of autistic disorder. *Epidemiology* 2013, **24**(6):906–912.
102. Schmidt RJ, Hansen RL, Hartiala J, Allayee H, Schmidt LC, Tancredi DJ, Tassone F, Hertz-Picciotto I: Prenatal vitamins, one-carbon metabolism gene variants, and risk for autism. *Epidemiology* 2011, **22**(4):476–485.
103. Schmidt RJ, Tancredi DJ, Ozonoff S, Hansen RL, Hartiala J, Allayee H, Schmidt LC, Tassone F, Hertz-Picciotto I: Maternal periconceptional folic acid intake and risk of autism spectrum disorders and developmental delay in the CHARGE (CHildhood Autism Risks from Genetics and Environment) case-control study. *Am J Clin Nutr* 2012, **96**(1):80–89.
104. Suren P, Roth C, Bresnahan M, Haugen M, Hornig M, Hirtz D, Lie KK, Lipkin WI, Magnus P, Reichborn-Kjennerud T *et al.*: Association between maternal use of folic acid supplements and risk of autism spectrum disorders in children. *JAMA* 2013, **309**(6):570–577.
105. Juul-Dam N, Townsend J, Courchesne E: Prenatal, perinatal, and neonatal factors in autism, pervasive developmental disorder-not otherwise specified, and the general population. *Pediatrics* 2001, **107**(4):E63.
106. Larsson M, Weiss B, Janson S, Sundell J, Bornehag CG: Associations between indoor environmental factors and parental-reported autistic spectrum disorders in children 6–8 years of age. *Neurotoxicology* 2009, **30**(5):822–831.
107. Kalkbrenner AE, Braun JM, Durkin MS, Maenner MJ, Cunniff C, Lee LC, Pettygrove S, Nicholas JS, Daniels JL: Maternal smoking during pregnancy and the prevalence of autism spectrum disorders, using data from the autism and developmental disabilities monitoring network. *Environ Health Perspect* 2012, **120**(7):1042–1048.
108. Lee BK, Gardner RM, Dal H, Svensson A, Galanti MR, Rai D, Dalman C, Magnusson C: Brief report: maternal smoking during pregnancy and autism spectrum disorders. *J Autism Dev Dis* 2012, **42**(9):2000–2005.
109. Nanson JL: Autism in fetal alcohol syndrome: a report of six cases. *Alcohol Clin Exp Res* 1992, **16**(3):558–565.
110. Harris SR, MacKay LL, Osborn JA: Autistic behaviors in offspring of mothers abusing alcohol and other drugs: a series of case reports. *Alcohol Clin Exp Res* 1995, **19**(3):660–665.
111. Fombonne E: Is exposure to alcohol during pregnancy a risk factor for autism? *J Autism Dev Dis* 2002, **32**(3):243.

112. Hertz-Picciotto I, Bergman A, Fangstrom B, Rose M, Krakowiak P, Pessah I, Hansen R, Bennett DH: Polybrominated diphenyl ethers in relation to autism and developmental delay: a case-control study. *Environ Health* 2011, **10**(1):1.
113. National Research Council (U.S.). Committee on Neurotoxicology and Models for Assessing Risk.: Environmental neurotoxicology. Washington, D.C.: National Academy Press; 1992.
114. Volk HE, Hertz-Picciotto I, Delwiche L, Lurmann F, McConnell R: Residential proximity to freeways and autism in the CHARGE study. *Environ Health Perspect* 2011, **119**(6):873–877.
115. Windham GC, Zhang L, Gunier R, Croen LA, Grether JK: Autism spectrum disorders in relation to distribution of hazardous air pollutants in the san francisco bay area. *Environ Health Perspect* 2006, **114**(9):1438–1444.
116. Roberts AL, Lyall K, Hart JE, Laden F, Just AC, Bobb JF, Koenen KC, Ascherio A, Weisskopf MG: Perinatal Air Pollutant Exposures and Autism Spectrum Disorder in the Children of Nurses' Health Study II Participants. *Environ Health Perspect* 2013.
117. Becerra TA, Wilhelm M, Olsen J, Cockburn M, Ritz B: Ambient air pollution and autism in los angeles county, california. *Environ Health Perspect* 2013, **121**(3):380–386.
118. Volk HE, Lurmann F, Penfold B, Hertz-Picciotto I, McConnell R: Traffic-related air pollution, particulate matter, and autism. *JAMA Psychiatry* 2013, **70**(1):71–77.
119. Roberts EM, English PB, Grether JK, Windham GC, Somberg L, Wolff C: Maternal residence near agricultural pesticide applications and autism spectrum disorders among children in the California Central Valley. *Environ Health Perspect* 2007, **115**(10):1482–1489.
120. Rauh VA, Garfinkel R, Perera FP, Andrews HF, Hoepner L, Barr DB, Whitehead R, Tang D, Whyatt RW: Impact of prenatal chlorpyrifos exposure on neurodevelopment in the first 3 years of life among inner-city children. *Pediatrics* 2006, **118**(6):e1845–1859.
121. Eskenazi B, Marks AR, Bradman A, Harley K, Barr DB, Johnson C, Morga N, Jewell NP: Organophosphate pesticide exposure and neurodevelopment in young Mexican-American children. *Environ Health Perspect* 2007, 115(5):792–798.
122. Windham GC, Sumner A, Li SX, Anderson M, Katz E, Croen LA, Grether JK: Use of birth certificates to examine maternal occupational exposures and autism spectrum disorders in offspring. *Autism Res* 2013, **6**(1):57–63.
123. McCanlies EC, Fekedulegn D, Mnatsakanova A, Burchfiel CM, Sanderson WT, Charles LE, Hertz-Picciotto I: Parental occupational exposures and autism spectrum disorder. *J Autism Dev Dis* 2012.
124. Croen LA, Grether JK, Yoshida CK, Odouli R, Hendrick V: Antidepressant use during pregnancy and childhood autism spectrum disorders. *Arch Gen Psychiatry* 2011, **68**(11):1104–1112.

125. Croen LA, Connors SL, Matevia M, Qian Y, Newschaffer C, Zimmerman AW: Prenatal exposure to beta2-adrenergic receptor agonists and risk of autism spectrum disorders. *J Neurodev Disord* 2011, **3**(4):307–315.
126. Connors SL, Crowell DE, Eberhart CG, Copeland J, Newschaffer CJ, Spence SJ, Zimmerman AW: beta2-adrenergic receptor activation and genetic polymorphisms in autism: data from dizygotic twins. *J Child Neurol* 2005, **20**(11):876–884.
127. Christensen J, Gronborg TK, Sorensen MJ, Schendel D, Parner ET, Pedersen LH, Vestergaard M: Prenatal valproate exposure and risk of autism spectrum disorders and childhood autism. *JAMA* 2013, **309**(16):1696–1703.
128. Gardener H, Spiegelman D, Buka SL: Prenatal risk factors for autism: comprehensive meta-analysis. *Br J Psychiatry* 2009, **195**(1):7–14.
129. Atladottir HO, Thorsen P, Ostergaard L, Schendel DE, Lemcke S, Abdallah M, Parner ET: Maternal infection requiring hospitalization during pregnancy and autism spectrum disorders. *J Autism Dev Dis* 2010, **40**(12):1423–1430.
130. Atladottir HO, Henriksen TB, Schendel DE, Parner ET: Autism after infection, febrile episodes, and antibiotic use during pregnancy: an exploratory study. *Pediatrics* 2012, **130**(6):e1447–1454.
131. Buchmayer S, Johansson S, Johansson A, Hultman CM, Sparen P, Cnattingius S: Can association between preterm birth and autism be explained by maternal or neonatal morbidity? *Pediatrics* 2009, **124**(5):e817–825.
132. Zerbo O, Iosif AM, Walker C, Ozonoff S, Hansen RL, Hertz-Picciotto I: Is maternal influenza or fever during pregnancy associated with autism or developmental delays? Results from the CHARGE (CHildhood Autism Risks from Genetics and Environment) study. *J Autism Dev Dis* 2012.
133. Thorgeirsson TE, Geller F, Sulem P, Rafnar T, Wiste A, Magnusson KP, Manolescu A, Thorleifsson G, Stefansson H, Ingason A *et al.*: A variant associated with nicotine dependence, lung cancer and peripheral arterial disease. *Nature* 2008, **452**(7187):638–642.
134. Amos CI, Wu X, Broderick P, Gorlov IP, Gu J, Eisen T, Dong Q, Zhang Q, Gu X, Vijayakrishnan J *et al.*: Genome-wide association scan of tag SNPs identifies a susceptibility locus for lung cancer at 15q25.1. *Nat Genet* 2008, **40**(5):616–622.
135. Hung RJ, McKay JD, Gaborieau V, Boffetta P, Hashibe M, Zaridze D, Mukeria A, Szeszenia-Dabrowska N, Lissowska J, Rudnai P *et al.*: A susceptibility locus for lung cancer maps to nicotinic acetylcholine receptor subunit genes on 15q25. *Nature* 2008, **452**(7187):633–637.
136. Dempfle A, Scherag A, Hein R, Beckmann L, Chang-Claude J, Schafer H: Gene-environment interactions for complex traits: definitions, methodological requirements and challenges. *Eur J Hum Genet* 2008, **16**(10):1164–1172.
137. Kraft P, Yen YC, Stram DO, Morrison J, Gauderman WJ: Exploiting gene-environment interaction to detect genetic associations. *Hum Hered* 2007, **63**(2):111–119.

138. Gauderman WJ, Siegmund KD: Gene-environment interaction and affected sib pair linkage analysis. *Hum Hered* 2001, **52**(1):34–46.
139. Dizier MH, Selinger-Leneman H, Genin E: Testing linkage and gene x environment interaction: comparison of different affected sib-pair methods. *Genet Epidemiol* 2003, **25**(1):73–79.
140. Kraft P, Hunter D: Integrating epidemiology and genetic association: the challenge of gene-environment interaction. *Philos Trans R Soc Lond B Biol Sci* 2005, **360**(1460):1609–1616.
141. Alfarra HY, Alfarra SR, Sadiq MF: Neural tube defects between folate metabolism and genetics. *Indian J Hum Genet* 2011, **17**(3):126–131.
142. Eder W, Klimecki W, Yu L, von Mutius E, Riedler J, Braun-Fahrlander C, Nowak D, Martinez FD: Opposite effects of CD 14/-260 on serum IgE levels in children raised in different environments. *J Allergy Clin Immunol* 2005, **116**(3):601–607.
143. Simpson A, John SL, Jury F, Niven R, Woodcock A, Ollier WE, Custovic A: Endotoxin exposure, CD14, and allergic disease: an interaction between genes and the environment. *Am J Respir Crit Care Med* 2006, **174**(4):386–392.
144. Williams LK, McPhee RA, Ownby DR, Peterson EL, James M, Zoratti EM, Johnson CC: Gene-environment interactions with CD14 C-260T and their relationship to total serum IgE levels in adults. *J Allergy Clin Immunol* 2006, **118**(4):851–857.
145. Park SK, O'Neill MS, Wright RO, Hu H, Vokonas PS, Sparrow D, Suh H, Schwartz J: HFE genotype, particulate air pollution, and heart rate variability: a gene-environment interaction. *Circulation* 2006, **114**(25):2798–2805.
146. Chahine T, Baccarelli A, Litonjua A, Wright RO, Suh H, Gold DR, Sparrow D, Vokonas P, Schwartz J: Particulate air pollution, oxidative stress genes, and heart rate variability in an elderly cohort. *Environ Health Perspect* 2007, **115**(11):1617–1622.
147. Baccarelli A, Cassano PA, Litonjua A, Park SK, Suh H, Sparrow D, Vokonas P, Schwartz J: Cardiac autonomic dysfunction: effects from particulate air pollution and protection by dietary methyl nutrients and metabolic polymorphisms. *Circulation* 2008, **117**(14):1802–1809.
148. Garcia-Closas M, Malats N, Silverman D, Dosemeci M, Kogevinas M, Hein DW, Tardon A, Serra C, Carrato A, Garcia-Closas R *et al.*: NAT2 slow acetylation, GSTM1 null genotype, and risk of bladder cancer: results from the Spanish Bladder Cancer Study and meta-analyses. *Lancet* 2005, **366**(9486):649–659.
149. Marcus PM, Hayes RB, Vineis P, Garcia-Closas M, Caporaso NE, Autrup H, Branch RA, Brockmoller J, Ishizaki T, Karakaya AE *et al.*: Cigarette smoking, N-acetyltransferase 2 acetylation status, and bladder cancer risk: a case-series meta-analysis of a gene-environment interaction. *Cancer Epidemiol Biomarkers Prev* 2000, **9**(5):461–467.

150. Smith SJ, Li Y, Whitley R, Marion MJ, Partilo S, Carney WP, Brandt-Rauf PW: Molecular epidemiology of p53 protein mutations in workers exposed to vinyl chloride. *Am J Epidemiol* 1998, **147**(3):302–308.
151. Jones IM, Galick H, Kato P, Langlois RG, Mendelsohn ML, Murphy GA, Pleshanov P, Ramsey MJ, Thomas CB, Tucker JD *et al.*: Three somatic genetic biomarkers and covariates in radiation-exposed Russian cleanup workers of the chernobyl nuclear reactor 6–13 years after exposure. *Radiation Res* 2002, **158**(4):424–442.
152. Meng Z, Qin G, Zhang B: DNA damage in mice treated with sulfur dioxide by inhalation. *Environ Mole Mutagen* 2005, **46**(3):150–155.
153. Ha M, Yoo KY, Cho SH: Glycophorin A mutant frequency in radiation workers at the nuclear power plants and a hospital. *Mutat Res* 2002, **501**(1–2):45–56.
154. Baccarelli A, Hirt C, Pesatori AC, Consonni D, Patterson DG, Jr., Bertazzi PA, Dolken G, Landi MT: t(14;18) translocations in lymphocytes of healthy dioxin-exposed individuals from Seveso, Italy. *Carcinogenesis* 2006, **27**(10):2001–2007.
155. Charles M: UNSCEAR report 2000: sources and effects of ionizing radiation. United Nations Scientific Comittee on the Effects of Atomic Radiation. *J Radiol Prot* 2001, **21**(1):83–86.
156. Faulk C, Dolinoy DC: Timing is everything: the when and how of environmentally induced changes in the epigenome of animals. *Epigenetics* 2011, **6**(7):791–797.
157. Bjornsson HT, Fallin MD, Feinberg AP: An integrated epigenetic and genetic approach to common human disease. *Trends Genet* 2004, **20**(8):350–358.
158. LaSalle JM: A genomic point-of-view on environmental factors influencing the human brain methylome. *Epigenetics* 2011, **6**(7):862–869.
159. Miyake K, Hirasawa T, Koide T, Kubota T: Epigenetics in autism and other neurodevelopmental diseases. *Adv Exp Med Biol* 2012, **724**:91–98.
160. Schanen NC: Epigenetics of autism spectrum disorders. *Hum Mol Gen* 2006, **15**(Spec No 2):R138–150.
161. Kwong WT, Friello P, Semba RD: Interactions between iron deficiency and lead poisoning: epidemiology and pathogenesis. *Sci Total Environ* 2004, **330**(1–3):21–37.
162. Bannon DI, Abounader R, Lees PS, Bressler JP: Effect of DMT1 knockdown on iron, cadmium, and lead uptake in Caco-2 cells. *Am J Physial Cell Physiol* 2003, **284**(1):C44–50.
163. Ward MH, deKok TM, Levallois P, Brender J, Gulis G, Nolan BT, VanDerslice J, International Society for Environmental E: Workgroup report: Drinking-water nitrate and health — recent findings and research needs. *Environ Health Perspect* 2005, **113**(11):1607–1614.
164. Hodgson E, Levi PE: A textbook of modern toxicology. New York: Elsevier; 1987.

165. Faustman EM, Silbernagel SM, Fenske RA, Burbacher TM, Ponce RA: Mechanisms underlying Children's susceptibility to environmental toxicants. *Environ Health Perspect* 2000, **108**(Suppl 1):13–21.
166. Ross ME: Gene-environment interactions, folate metabolism and the embryonic nervous system. *Wiley Interdiscip Rev Syst Biol Med* 2010, **2**(4):471–480.
167. D'Amelio M, Ricci I, Sacco R, Liu X, D'Agruma L, Muscarella LA, Guarnieri V, Militerni R, Bravaccio C, Elia M *et al.*: Paraoxonase gene variants are associated with autism in North America, but not in Italy: possible regional specificity in gene-environment interactions. *Mol Psychiatry* 2005, **10**(11):1006–1016.
168. Volk HE, Kerin T, Lurmann F, Hertz-Picciotto I, McConnell R, Campbell D: Interaction of the MET receptor tyrosine kinase gene and air pollution exposure in autism spectrum disorder. *Epidemiology* 2013, in press.
169. Herbert MR, Russo JP, Yang S, Roohi J, Blaxill M, Kahler SG, Cremer L, Hatchwell E: Autism and environmental genomics. *Neurotoxicology* 2006, **27**(5):671–684.
170. Ekwo EE, Seals BF, Kim JO, Williamson RA, Hanson JW: Factors influencing maternal estimates of genetic risk. *Am J Med Genet* 1985, **20**(3):491–504.
171. Hammen C, Burge D, Burney E, Adrian C: Longitudinal study of diagnoses in children of women with unipolar and bipolar affective disorder. *Arch Gen Psychiatry* 1990, **47**(12):1112–1117.
172. Wild CP: The exposome: from concept to utility. *Int J Epidemiol* 2012, **41**(1):24–32.
173. Campbell DB, Warren D, Sutcliffe JS, Lee EB, Levitt P: Association of MET with social and communication phenotypes in individuals with autism spectrum disorder. *Am J Med Genet, B Neuropsychiat Genet* 2010, **153B**(2):438–446.
174. Constantino JN: The Social Responsiveness Scale. Los Angeles, California: Western Psychological Services; 2002.
175. Smith PG, Day NE: The design of case-control studies: the influence of confounding and interaction effects. *Int J Epidemiol* 1984, **13**(3):356–365.
176. Piegorsch WW, Weinberg CR, Taylor JA: Non-hierarchical logistic models and case-only designs for assessing susceptibility in population-based case-control studies. *Stat Med* 1994, **13**(2):153–162.
177. Umbach DM, Weinberg CR: Designing and analysing case-control studies to exploit independence of genotype and exposure. *Stat Med* 1997, **16**(15): 1731–1743.
178. Mukherjee B, Chatterjee N: Exploiting gene-environment independence for analysis of case-control studies: an empirical Bayes-type shrinkage estimator to trade-off between bias and efficiency. *Biometrics* 2008, **64**(3):685–694.

179. Kraft P, Yen YC, Stram DO, Morrison J, Gauderman WJ: Exploiting gene-environment interaction to detect genetic associations. *Hum Hered* 2007, **63**(2):111–119.
180. Kooperberg C, Ruczinski I: Identifying interacting SNPs using Monte Carlo logic regression. *Genet Epidemiol* 2005, **28**(2):157–170.
181. Chen X, Liu CT, Zhang M, Zhang H: A forest-based approach to identifying gene and gene gene interactions. *Proc Natl Acad Sci USA* 2007, **104**(49):19199–19203.
182. Murcray CE, Lewinger JP, Gauderman WJ: Gene-environment interaction in genome-wide association studies. *Am J Epidemiol* 2009, **169**(2):219–226.
183. Dai JY, Kooperberg C, Leblanc M, Prentice RL: Two-stage testing procedures with independent filtering for genome-wide gene-environment interaction. *Biometrika* 2012, **99**(4):929–944.
184. Dai JY, Logsdon BA, Huang Y, Hsu L, Reiner AP, Prentice RL, Kooperberg C: Simultaneously testing for marginal genetic association and gene-environment interaction. *Am J Epidemiol* 2012, **176**(2):164–173.
185. Hung RJ, Brennan P, Malaveille C, Porru S, Donato F, Boffetta P, Witte J: Using hierarchical modeling in genetic association studies with multiple markers: Applications to a case-control study of bladder cancer. *Cancer Epidemiol Biomarkers Prev* 2004, **13**:1013–1021.
186. Chen J, Yu K, Hsing A, Therneau TM: A partially linear tree-based regression model for assessing complex joint gene-gene and gene-environment effects. *Genet Epidemiol* 2007, **31**(3):238–251.
187. Gunther F, Pigeot I, Bammann K: Artificial neural networks modeling gene-environment interaction. *BMC Genomics* 2012, **13**(37).
188. Hahn LW, Ritchie MD, Moore JH: Multifactor dimensionality reduction software for detecting gene-gene and gene-environment interactions. *Bioinformatics* 2003, **19**(3):376–382.
189. Ritchie MD, Hahn LW, Roodi N, Bailey LR, Dupont WD, Parl FF, Moore JH: Multifactor-dimensionality reduction reveals high-order interactions among estrogen-metabolism genes in sporadic breast cancer. *Am J Hum Genet* 2001, **69**(1):138–147.
190. Hertz-Picciotto I, Croen LA, Hansen R, Jones CR, van de Water J, Pessah IN: The CHARGE study: an epidemiologic investigation of genetic and environmental factors contributing to autism. *Environ Health Perspect* 2006, **114**(7):1119–1125.
191. Schendel DE, Diguiseppi C, Croen LA, Fallin MD, Reed PL, Schieve LA, Wiggins LD, Daniels J, Grether J, Levy SE *et al.*: The study to explore early development (SEED): a multisite epidemiologic study of autism by the centers for autism and developmental disabilities research and epidemiology (CADDRE) network. *J Autism Dev Dis* 2012.

192. Newschaffer CJ, Croen LA, Fallin MD, Hertz-Picciotto I, Nguyen DV, Lee NL, Berry CA, Farzadegan H, Hess HN, Landa RJ *et al.*: Infant siblings and the investigation of autism risk factors. *J Neurodev Dis* 2012, **4**(1):7.
193. Weinberg CR, Shore DL, Umbach DM, Sandler DP: Using risk-based sampling to enrich cohorts for endpoints, genes, and exposures. *Am J Epidemiol* 2007, **166**(4):447–455.
194. Bookman EB, McAllister K, Gillanders E, Wanke K, Balshaw D, Rutter J, Reedy J, Shaughnessy D, Agurs-Collins T, Paltoo D *et al.*: Gene-environment interplay in common complex diseases: forging an integrative model-recommendations from an NIH workshop. *Genet Epidemiol* 2011.

Chapter 15

The Potential Brain Drain from Environmental Exposures on the Methylome and Genome Across Generations

Janine M. LaSalle* and Keith W. Dunaway

Abstract

While autism spectrum disorders (ASD) have a strong genetic basis, the apparent increase in prevalence in recent generations has raised concerns about environmental exposures and potential interactions with genetic risk factors. DNA methylation is an epigenetic layer on top of DNA sequence at the interface of genetic and environmental factors known to overlap with metabolism and nutrition. Here we review and discuss the evidence from human studies pointing to complex interactions between genetic factors and *in utero* environment. We also examine the impact of chemical exposures on DNA methylation in brain as well as overall effect on development and reproduction. Transgenerational effects of chemical exposures on ASD risk are still an open question, but evidence supporting the need to ask such questions are summarized.

Keywords: Autism; Brain; Epigenetics; Epigenomics; Genome Stability; Toxicology; Neurodevelopment; DNA Methylation.

*Correspondence to: Medical Microbiology and Immunology, MIND Institute, Genome Center, University of California, Davis, Sacramento, CA 95817. Email: jmlasalle@ucdavis.edu

Why Consider a Potential Chemical "Brain Drain" of Generationally Cumulative Environmental Exposures in Autism Risk?

Modern humans live within relatively new environmental conditions, characterized by man-made housing, freeways, manufactured foods, medicines, and consumer products. With these new conditions come novel chemical exposures to our bodies. New manufactured chemicals are usually tested for immediate toxicity; however, the cumulative effects and long-term outcomes of chemical exposures on human health and the health of future generations are often not considered. While a single environmental toxicant is unlikely to be found as the definitive causal factor for autism, the potential for cumulative effects of chemical exposures during pregnancy is concerning. In this chapter, we review the evidence for a diverse array of chemical exposures having complex interactions with the human genome and epigenome as well as the potential for transgenerational effects. While most investigations of etiological risk factors in autism consider genetic versus environmental factors as separate and distinct risks, there is more likely a complex intersection of interactions. Specifically, genes can affect susceptibility to toxins, environmental factors can influence gene expression through epigenetic mechanisms, and epigenetic changes induced by environmental exposures can impact the stability of the human genome.

The Complex Etiology of Autism Spectrum Disorders

Autism spectrum disorders (ASDs) are collectively defined by impaired communication, diminished social interaction, and repetitive behavior.[1] As of March 2012, the Center for Disease Control estimated that 1 in 88 children were afflicted with ASDs in the United States, a rate which has increased from 1 in 150 children just ten years prior.[2] Because the criteria for diagnosis of ASDs has changed multiple times since the disorder was first described by Kanner in 1943, the question of whether there has been a true rise in prevalence or simply improved ascertainment in recent years is not easily answerable.[3,4] The changing diagnostic instruments, lack of definitive laboratory tests, and socioeconomic differences in access to diagnostic centers make the search for etiologic factors for ASDs a decidedly complex problem.

ASDs are considered to be highly heritable, with historic estimates above 90% based on monozygotic twin studies,[5,6] but heritability estimates are lower in more recent studies.[7] A monozygotic versus dizygotic twin study suggested that shared *in utero* environment was a higher risk (30–80%) than genetic risk (14–67%) for ASD.[8] Another strategy for twin studies is to measure autistic traits in the general population to increase sample size by considering ASDs as the extreme on a quantitative trait continuum. Analyses of autistic traits using twin studies from the general population have observed high heritability estimates (53% females, 72% males) and suggest a shared etiology of autistic traits with extreme end ASDs.[9,10] Another general population twin study highlighted evidence for a "female protective effect", as the siblings of female ASD probands scored higher on autistic traits than the siblings of male ASD probands, suggesting a greater genetic contribution to ASD in females.[11] These combined twin studies have suggested a strong genetic component to ASD risk, with additional important roles of shared *in utero* environment and sex differences.

New genomic technologies have provided the tools for many recent insights into the complex genetics of ASDs. While analysis of common inherited variants by genome wide association studies have been extensive, the more recent focus has been to identify rare and *de novo* genetic variants enriched in individuals with ASDs compared to the rest of the population. Rare variants found by copy number variation (CNV) microarray analyses have collectively demonstrated that CNVs occur in 6–8% of autism cases compared to 1–2% of non-autistic individuals.[12–14] Sequencing efforts to identify ASD implicated rare variants in protein coding genes by whole exome sequencing (WES) have led to the identification of mostly non-overlapping candidate mutations.[15–18] In one study, complete knockouts of genes occurred at a two-fold higher rate in

ASD compared to typically developing individuals,[19] however, a different WES study on 1,039 ASD subjects compared to 870 controls found no evidence for enrichment of exonic variants in ASD.[20] These combined studies as well as statistical predictions[16,21] suggest that there are likely hundreds of genes involved in the genetic architecture of ASDs. What is beginning to emerge from the complex genetic architecture of ASDs is that genes found mutated, deleted, or duplicated in rare ASD cases are converging on overlapping pathways. These include synaptic functions,[22] ubiquitin pathways,[23] mammalian target of rapamycin (mTOR) pathways,[24] and chromatin functions.[25]

Some interesting questions that these genomic studies of ASD have raised include: (1) While knockout hits to individual genes within the shared pathways may cause ASD in rare cases, could there be multiple ways of more commonly perturbing the entire pathway of these genes? (2) How do *de novo* mutations in ASD happen? Were *de novo* mutations completely random occurrences or could inherited genetic polymorphisms and environment be interacting in the risk of *de novo* mutations, particularly at common CNV breakpoints? (3) What could explain the missing heritability in ASD risk?

DNA Methylation at the Interface of Genomic Interactions with the Environment

While DNA molecules are primarily defined by their sequential makeup of four basic nucleotides (A,T,G,C), DNA contains additional epigenetic modifications such as DNA methylation. Also, chromosomal DNA is inherited as chromatin, which is defined as the complex of nucleotides and proteins contained within the nucleus. Epigenetics can be defined as modifications to nucleotides or chromatin that do not change DNA sequence but can alter gene expression and eventual phenotype. Epigenetic modifications are considered to exist in a "metastable" state, meaning that they are heritable and/or long-lived but also reversible under the appropriate conditions.[26]

DNA methylation can be considered the first epigenetic layer and can act in concert with noncoding RNAs, the proteome, long-range chromatin interactions, and additional epigenetic layers of histone modifications.[26] Historically, DNA methylation has been considered a silencing mark for promoters of mammalian genes, based on the known promoter methylation and the silent state of X chromosome inactivated and imprinted genes. However, genome-wide analysis of DNA methylation patterns in humans have demonstrated that the human genome is highly saturated for DNA methylation, but CpG islands are predominantly unmethylated in embryonic stem cells and mature tissues such as brain.[27,28] While DNA methylation at promoters negatively correlates with gene expression, gene body methylation shows a positive correlation.[27,29]

DNA methylation is predicted to play a role in genome defense against the rampant expansion of retrotransposons and other repetitive DNA elements throughout mammalian evolution.[30–32] The human genome contains a large fraction of repetitive sequences, together making up roughly half of the 3 GB content. Repeats are found at ribosomal genes, centromeres, telomeres, as well as interspersed throughout chromosomes as transposon-driven repeats, processed pseudogenes, and simple sequence repeats. While most mammals share this collection of common repetitive categories, primates have recently gained a unique category of "low copy" repeats called segmental duplications. These primate-specific segmental duplications are characterized by breakpoints enriched for *Alu* elements that are CpG dense and thus excellent targets for DNA methylation. Interestingly, these Alu-rich segmental duplications coincide with breakpoints of common CNVs observed in human neurodevelopmental disorders and are at higher levels in individuals with autism.[33,34]

The Neuronal Methylome and Its Importance in Autism

Unlike the human genome, which is mostly invariant between cell and tissue types, the methylome is dynamic during early developmental lineages and exhibits differences between tissues in the body. Genomic sequencing analyses at base pair resolution have revealed distinct differences in the methylome landscape between embryonic stem cells and differentiated cell types.[28,35] Differentiated cell lines such as IMR90 fetal lung fibroblasts and SH-SH5Y neuroblastoma cells exhibit the methylome landscape feature of partially methylated domains (PMDs) which do not exist in embryonic stem cells. PMDs encompass 41% of the IMR90 genome, but only 19% of the neuronal SH-SY5Y genome because of the large number of neuronal-specific highly methylated domains (N-HMDs), defined as highly methylated in neuronal cells but PMDs in fetal lung fibroblast cells.[35] The top gene ontology categories for genes within N-HMDs include neuronal differentiation, synaptic transmission, and calcium signaling; the N-HMD genes were also significantly enriched for autism candidate genes.[35]

In addition to 5-methylcytosine (5-mC) of CpG sites, neurons contain relatively high levels of at least two other types of DNA methylation: highly conserved methylation at non-CpG sites (mCH, including CpA, CpC, and CpT) and 5-hydroxymethylcytosine at CpG sites (hmC). A genome-wide sequencing analysis of cell types within the human and mouse brains revealed that neurons have a distinct methylation pattern compared to non-neuronal cells in these types of methylation.[27] Furthermore, widespread changes were observed in the methylation levels between fetal and postnatal brain and neuronally upregulated genes were characterized by high levels of gene body 5-hmC but low levels of mCH.[27] There is still much work ahead in order to understand the dynamics of methylation changes in the mammalian brain, but these first genomic roadmaps of the different methylation types (CpG, CHH, CHG, 5-mC, and 5-hmC) have made clear that neurons are highly saturated with DNA methylation and neuronal genes are differentially marked by methylation patterns.

The critical role of DNA methylation in the mammalian nervous system is revealed when looking at deficiency of enzymes that target DNA for adding or removing methyl groups. Mice with forebrain deficiency in two enzymes that add DNA methyl groups, *Dnmt1* and *Dnmt3a*, show deficits in memory formation and behavioral plasticity.[36,37] Conversion of 5-mC to 5-hmC is catalyzed by a family of TET enzymes that are predicted to play a role in demethylation reactions.[38,39] *Tet1* knockdown diminished the neuronal activity-dependent demethylation of *Bdnf* and *Fgf1* gene promoters, suggesting a role in the dynamic methylation changes in brain.[38] However, global DNA patterns and dynamic changes from fetal to adult brain were only subtly altered in brains from *Tet2*$^{-/-}$ mice compared to wild-type mice,[27] suggesting either that multiple TET enzymes may work in concert to remodel DNA methylation patterns in the brain or that other mechanisms of demethylation may be involved.

Together, these genome-wide maps of the developmental neuronal methylome have suggested an important role for DNA methylation in at least correlating with gene expression patterns. However, it is not clear how much of the methylation differences between fetal and adult brain or neurons versus glia may simply be historic marks of past transcriptional events.[26] Genes specific to neuronal development and synapse formation appear to be kept within PMDs characterized by heterochromatin in non-neuronal dividing cells. Chromatin inaccessibility is the likely explanation for both the presence of PMDs in placenta[29] and non-CpG methylation "deserts" in neurons.[27] Neuronal genes are characterized by high levels of DNA methylation everywhere except CpG island promoters. This suggests a need for CpGs to become methylated in the gene-body and demethylated in promoters during the transcriptional dynamics of the neuronal maturation process. This prediction also raises the question of what factors could impact these complex methylation events during *in utero* and early postnatal brain development, and in the etiology of autism.

Studies showing alterations of DNA methylation patterns in human postmortem brain and blood have provided evidence in support of possible epigenetic ASD etiologies. In postmortem brain studies, candidate gene approaches identified aberrant methylation associated with autism in the *MECP2* promoter and 3′ end of *UBE3A*.[40,41] A genomic analysis of methylation differences in blood-derived cells of monozygotic twins discordant for diagnosis of autism identified altered methylation in the circadian gene *RORA* which was confirmed in postmortem brain.[42] A recent large-scale analysis of methylation changes associated with autism using an Infinium 450k array platform identified commonly altered methylation patterns within several genes in the autism brain.[43]

How Do Environmental Toxins Affect Metabolic Pathways Regulating DNA Methylation?

The human brain genome is clearly saturated with DNA methylation, but what regulates the supply of methyl donors for DNA methylation and the enzymes that methylate and demethylate DNA? While these answers are not entirely known, the intersection of epigenetics with cellular metabolism is likely to provide some of these answers. Furthermore, the intersection of environmental chemicals with the epigenome, particularly the methylome, involves some basic cellular metabolic systems.

The most central metabolic pathway relevant to DNA methylation is the one-carbon or methionine cycle, which serves to supply methyl donors for methylation of DNA, RNA, and proteins (Figure 1). DNA methyltransferases (DNMTs) require methionine to produce S-adenylmethionine (SAM), the methyl donor for DNA methylation. Dietary factors are important for the supply side of the methionine cycle, including choline and B vitamins in the folate cycle that contribute to methionine.[44] Methionine is in turn required in DNA methylation reactions that convert SAM to SAH. Folate and B vitamins are known environmental modifiers of DNA methylation in a genetic A^{vy} mouse model[45,46] and in queen bee determination in honeybees.[47]

In human studies, use of prenatal vitamins containing high folate prior to conception significantly correlated with reduced risk of ASDs.[48–50] For women with a common risk allele in the gene *MTHFR*, encoding the folate cycle enzyme methylenetetrahydrofolate reductase, their child was at a much higher risk for developing autism if she did not take prenatal vitamins prior to conception, demonstrating an important gene by environment interaction.[49]

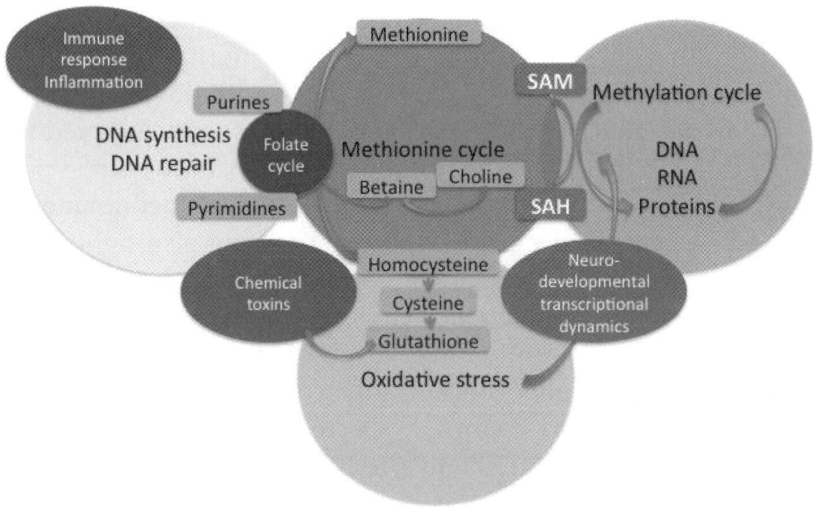

Figure 1. The intersection of metabolic pathways regulating DNA methylation with nutritional, chemical and immune pathways. Methionine levels within the one carbon metabolism cycle are important for the conversion of SAM to SAH in methylation reactions and are also centrally regulated by immune responses, oxidative stress and glutathione pathways that are impacted by a wide variety of environmental toxins. Nutrients such as folate and choline can help counteract the effects of toxins and immune insults by providing an increased supply of methionine.

A meta-analysis of eight case-controlled studies also determined significant risk for ASD associated with the *MTHFR* risk allele, particularly in countries without folate food fortification.[51]

The immune system also plays a role in blocking folate transport by the production of autoantibodies to the folate receptor. In one study, 75% of ASD serum samples were found to be positive for folate receptor autoantibodies that interfere with folate transport across the blood-brain barrier.[52] In a second study of children with infantile autism, folate receptor autoantibodies were found a lower rate of 47%, but still highly significant compared to 3.3% controls without ASD.[53] In addition, ASD-specific maternal autoantibodies that recognize fetal brain antigens have been identified, including lactate dehydrogenase, cypin, stress-induced phosphoprotein 1, collapsing response mediator proteins, and Y-box-binding protein. This suggests the maternal immune system is responding to stress-activated proteins in the fetal brain.[54]

Environmental toxins impact the one-carbon cycle, predominantly through the actions of glutathione and oxidative stress pathways. Methionine synthase and the resulting methionine are reduced by increased oxidative stress. They are also tightly coupled to an increased production of the antioxidant glutathione and a broad reduction in metabolic activity.[55] Children with ASD were found to have significantly lower levels of methionine, SAM, and homocysteine but higher concentrations of SAH and oxidized glutathione compared to controls.[56] Decreased transcript levels of methionine synthase (*MTR*) and alternative splicing patterns were also observed in ASD subjects compared to controls.[55] Furthermore, the increased SAH levels in ASD samples correlated with DNA hypomethylation, showing a direct impact on these metabolic imbalances on methylation levels in blood.[57] Reactive oxygen species from oxidative stress impact mitochondrial function. Evidence for increased mitochondrial damage in children with ASD has been seen in several studies.[58–60]

Evidence for Environmental Exposures Affecting Brain Methylation

DNA methylation is the epigenetic layer with the most support for being impacted by environmental influences. The evidence that environmental exposures affect brain methylation or development and reproduction is summarized in Table 1. Additional evidence that specific environmental exposures affect other tissues in adulthood also exists in abundance, but these studies were not included in Table 1 as they were outside the focus of this chapter.

Heavy metals such as methylmercury can be counteracted by trace dietary metals such as selenium.[61] A small amount of selenium is important for humans because it is required for activation of glutathione peroxidase, an antioxidant pathway.[62] However, an overabundance of selenium has been shown to disrupt embryonic

Table 1. Evidence for DNA methylation effects from chemical exposures in brain, development, or reproduction

Exposure	Methylation Direction	Gene	Species	Tissue Types	Ref.
Selenium	Both (embryonic stage dependent)	global	Zebrafish	embryos (24, 26, and 48 hpf)	[63]
Methylmercury	hyper	*BDNF* (promoter)	mouse	Brain (Hippocampus)	[67]
Methylmercury	hypo	global	rat	Neuronal Stem Cells (NSCs)	[65]
Methylmercury	hypo	global	mink	Brain (cerebrum)	[66]
Methylmercury	no change	glucocorticoid receptor (*GR* promoter)	Sprague–Dawley rats	Hippocampus and liver	[68]
Diethylstilbestrol	hyper	*HOXA10* (intron)	Human	Human endometrial stromal cell line (Uterus-like) and Ishikawa cell line (Uterus)	[71]
Diethylstilbestrol	hyper	c-fos gene (promoter)	CD-1 mice	uterus	[72]
Diethylstilbestrol	hyper	*Mobkl2c*, *BC023892*, *Gm203*, *Ifih1*	C57BL/6 mice	epididymis (testis)	[73]
Diethylstilbestrol	hypo	*Nsbp1* (promoter)	CD-1 mice	uterus	[74]
Vinclozolin	hyper	LPLase gene	Fisher rat	Epididymal sperm	[94]
Bisphenol A	both	global	ICR/Jcl mice	Embryonic forebrain	[75]

(*Continued*)

Table 1. (*Continued*)

Exposure	Methylation Direction	Gene	Species	Tissue Types	Ref.
Bisphenol A	hypo	*Hoxa10* (promoter)	CD-1 mice	Uteri	[76]
Bisphenol A	hypo	*Hoxa10*	ICR mice	right uterine horn of F2 mice	[76]
Bisphenol A	hypo	*Igf2r* and *Peg3*	CD-1 mice	oocytes	[77]
Bisphenol A	hypo	*Igf2r*, *Peg3* and *H19*	Fetal Mouse	Germ cell	[78]
Bisphenol A	hypo	*H19* ICR	Holtzman rats	spermatozoa, embryo	[79]
Bisphenol A	hyper	*Crtc1*	C57BL/6 mice	hippocampus	[80]
Bisphenol A	both (sex specific)	ERα (*Esr1*) (promoter, intergenic)	BALB/c mice	whole brain	[81]
Bisphenol A	hypo		matings between B6 and C7 mice	placenta	[82]
Bisphenol A	hyper	*Stra8*	CD-1 mice	Oocytes	[83]

(*Continued*)

Table 1. *(Continued)*

Exposure	Methylation Direction	Gene	Species	Tissue Types	Ref.
Mixture: BPA, bis(2-ethylhexyl) phthalate (DEHP), and dibutyl phthalate (DBP)	both	global	Sprague Dawley rats	sperm (heads)	[84]
Polychlorinated biphenyl (PCB)	both	*AHR1*	Atlantic Killifish	Liver and brain	[89]
Polychlorinated biphenyl (PCB)	hypo	*LINE-1* (global)	human	blood	[86]
Polychlorinated biphenyl (PCB)	hypo	*LINE-1* (global)	human	Postmortem brain	[87]
Polybrominated diphenyl ethers	hypo	global	mice	brain	[88]

growth and results in improper neuron formation in zebrafish.[63] Methylmercury is made by first introducing mercury to the environment, usually through industrial processes.[64] Then, anaerobic organisms found in aquatic systems form methylmercury by attaching a methyl group to the inorganic mercury. Independent studies have found global hypomethylation correlated with methylmercury exposure,[65,66] while gene specific methylation differences were of variable directions.[67,68]

Historically prescribed medications can have unintended effects on DNA methylation that are only recently being detected. The synthetic nonsteroidal estrogen diethylstilbestrol (DES), classified as an endocrine disruptor, was given to pregnant women to prevent miscarriage for almost 30 years starting in 1940.[69] Doctors incorrectly thought DES would reduce the risk of pregnancy complications but discontinued prescriptions in 1971 when the women of DES pregnancies were shown to have an increased risk of vaginal tumors.[70] While the global effect of DES on DNA methylation has not been determined, increased DNA methylation in multiple genes has been observed in reproductive tissues (uterus and testis).[71–73] There is also evidence for hypomethylation properties at a single gene, Nsbp1.[74]

Bisphenol A (BPA) is a carbon-based synthetic compound, which has been used to make types of plastics and resin since the 1950s. BPA has been shown to affect a wide variety of organs, including brain and reproductive tissues.[75–83] Multiple independent studies have shown impacts on DNA methylation associated with BPA exposure, including both global hypomethylation and gene-specific hypermethylation.[75–83] Furthermore, a mixture of chemicals involving plastics including BPA reduced global DNA methylation levels in sperm.[84]

Polychlorinated biphenyls (PCBs) and polybrominated diphenyl ethers (PBDEs) are persistent organic pollutants with similar structures of two linked aromatic rings with halogens attached to them.[85] These persistent organic pollutants have been associated with global hypomethylation in humans and mouse brain,[86–88] as well as methylation alterations in fish.[89]

A common theme from these studies is that multiple chemicals and mixtures may impact the brain and reproductive methylomes. In analyses of global DNA methylation, reduced levels of methylation is reproducible across species, tissues, and exposures, suggesting a common disruption of the methionine and glutathione pathways discussed in the previous section. Interestingly, gene-specific promoter hypermethylation is often observed in the same tissues or exposures as global hypomethylation, which is an epigenetic hallmark of genes within placental and cancer PMDs.[90,91]

Evidence for Transgenerational Effects of Environmental Exposures

A major question that is raised from the methylation changes due to chemical exposures is whether these epigenetic alterations are inherited in the next generation. Most methylation marks such as those involved in X-inactivation or parental imprinting are erased and re-established in the gametes so methylation changes are unlikely to be passed down to offspring.[92] However, the mammalian genome is highly repetitive, and repeats are partially protected from the global waves of demethylation in gametogenesis, suggesting that some memory of global methylation levels could be heritable.[93] Evidence for transgenerational effects on DNA methylation at the genome-wide level have not been performed, likely due to the expense associated with multigenerational studies. However, there have been some studies that show evidence of environmental exposures contributing to transgenerational effects that are summarized in Table 2.

Vinclozolin is a dicarboximide fungicide used to fight diseases in humans and crops alike. The transgenerational effects of vinclozolin have been studied by independent groups with some contradictory results. Most of the studies showing transgenerational effects of Vinclozolin stem from a single laboratory, showing DNA methylation and/or phenotypic alterations across multiple studies in both rats and mice.[94–99] However, Vandegehuchte *et al.* did show phenotypic and epigenetic effects in planktonic crustacean as an

Table 2. Evidence for transgenerational effects of chemical exposures

Exposure	Effect	Species	Tissue	Ref.
Vinclozolin	Increase in LPLase gene methylation across generations	Fisher rat	Epididymal sperm	[94]
Vinclozolin	Directly contradicts Anway (2005)	Fisher rat	Epididymal sperm	[135]
Vinclozolin	Females three generations removed from the exposure discriminate and prefer males who do not have a history of exposure, whereas males do not exhibit a preference	Sprague–Dawley rats	Behavioral phenotype	[95]
Vinclozolin	Reduction in size (length) in F2 exposed. Also, lowered DNA methylation in F2	Daphnia magna (planktonic crustacean)		[100]
Vinclozolin	Both increased and decreased DNA methylation in various promoters in F3 male sperm of F0 female treated mothers	Harlan Sprague–Dawley rats	sperm	[96]
Vinclozolin	VCZ exposure correlated with hyomethylation of fully methylated *H19* and *Gtl2* alleles but increased those of unmethylated *Peg1*, *Snrpn*, and *Peg3* alleles. The effects of VCZ were transgenerational, but they disappeared gradually from F1 to F3	FVB/N mice	sperm	[97]

(*Continued*)

Table 2. *(Continued)*

Exposure	Effect	Species	Tissue	Ref.
Vinclozolin	Both increase and decreased of DNA methylation at specific sites (assayed by MeDIP) as well as phenotypic effects	gestating 129 inbred and CD1 outbred mice	testis, prostate, kidney and ovary	[98]
Vinclozolin	Both increase and decrease of DNA methylation as well as RNA transcriptional changes (pathway analysis included)	Hsd Sprague Dawley HTMSDHTM female and male rats of an outbred strain (Harlan)	primordial germ cell (PGC)	[99]
Vinclozolin	No significant behavioral or morphological changes (to F3)	Wistar rats	multiple	[101]
Vinclozolin	No significant change morphological changes (to F3)	Wistar rats	whole rat	[102]
Diethylstilbestrol	F2 male hypospadias	human	urethra	[103]
BPA	Social interaction disruption and transcriptional changes (to the F4 generation)	C57BL/6 mice	whole brain	[104]
Mixture: BPA, bis (2-ethylhexyl) phthalate (DEHP), and dibutyl phthalate (DBP)	Both hyper and hypo methylation globally	Sprague Dawley SD (Harlan) rats	sperm (heads)	[84]
Bisphenol A	Hypomethylation of the HOXA10 gene (intron)	ICR mice	right uterine horn of F2	[76]

independent effect.[100] It should be noted that the specific experiments in Anway (2005) were repeated by Inawaka (2009) with no significantly different changes in the Vinclozolin-treated rats. Also, the drug company that produces Vinclozolin, BASF SE, funded two studies authored by Schneider *et al.* showing no significant transgenerational effects of Vinclozolin on a different rat model.[101,102]

Since DES was used to treat pregnant mothers, their daughters' exposure included their oocytes, which could affect the grandchildren of DES exposed women. In a study of human grandsons of DES-exposed pregnant women, one study found a significant increase in hypospadias (birth defect of the urethra).[103]

BPA has been implicated in transgenerational effects in multiple studies. Hypomethylated DNA in sperm was associated with a BPA mixture in rats with transgenerational consequences.[84] Hypomethylation of DNA from the uterine horn as well as increased body weight was observed in the F2 progeny of BPA-treated mice.[76] Furthermore, social effects of BPA- treated mouse progeny have been seen all the way to the F4 generation in one study.[104]

Evidence for Environmental Exposures and Transgenerational Effects in ASD

There is difficulty in finding specific environmental exposures that increase risk for ASD because, similar to genetic mutations and CNVs, they involve complex interactions with other exposures, nutrients, and genes. However, like rare genetic mutations with large effect sizes in ASD, the rare cases of exposures with large effect size serve as proof-of-principle that environmental factors can be involved in ASD. Rubella infection or use of medications such as thalidomide or valproate during pregnancy have been associated with several hundred-fold increase in ASD risk.[105–108] Increased risk for ASD has also been associated with maternal exposure to environmental pollutants such as air pollution, agricultural pesticides, and persistent organic pollutants.[109–118] In an analysis of birth certificate data, mothers of autistic children were twice as likely to work in occupations with chemical exposures compared to mothers of controls. No significant risk was associated with paternal occupation which is consistent with *in utero* exposures being particularly critical.[119] Several maternal pregnancy factors have also been associated with increased ASD risk, including advanced maternal age, maternal obesity, poor fetal growth, later birth order, and shorter inter-pregnancy spacing.[120–124]

Human transgenerational epidemiological studies are only beginning to emerge for ASD. Increased grandpaternal or grandmaternal age was found to be associated with increased ASD risk for grandchildren, suggesting a transgenerational effect.[125,126] A large population-based Nurses Health Study II identified increased risk for ASD associated with maternal exposure to child abuse in early life.[127] Clearly, much more work lies ahead in understanding the transgenerational effect of chemical exposures that may be relevant to ASD etiologies. An intriguing, but unexpected result came in our recent analysis of over 100 postmortem brain samples for distinct congeners of PCB and PBDEs, in which the genetic CNV of maternal 15q11-q13 duplication (Dup15q) was significantly associated with PCB-95 levels, despite the subjects being born after PCBs were banned in the 1970s.[115] In subjects with both PCB-95 and Dup15q, global DNA hypomethylation was observed by LINE-1 pyrosequencing analysis, suggesting an epigenetic impact. These results raise the possibility that chemical exposures in a prior generation could predispose to *de novo* CNVs in the next generation through changes in the methylome leading to instability or by compounding gene by environment (GxE) effects.

Future Directions

In this chapter, we reviewed evidence that ASDs have complex etiologies involving the interface of genetics, metabolism, dietary factors, and chemical exposures. Because the investigation of any one of these risk or protective factors in isolation of the others is likely to give incomplete answers, we propose that future studies include the epigenome, particularly the methylome, and how it is impacted in ASDs (Figure 2). The challenge to these studies is that brains are

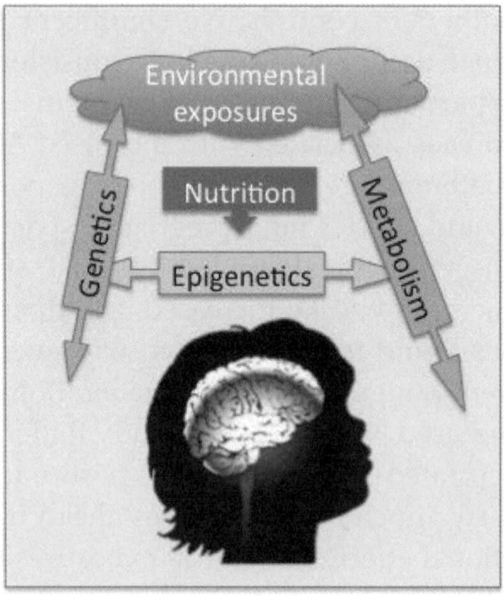

Figure 2. Future investigations of genetic and environmental interactions should be integrative to solve the complex etiologies of ASDs. Environmental exposures and their effects on the human brain can accumulate and persist across generations, so integration of multiple markers of metabolism, nutrition, and genetics with the important interface of epigenetics will be instrumental in improved diagnosis and treatment of ASDs.

only accessible post-mortem, so the path ahead would likely continue to be a combination of human epidemiology studies combined with human postmortem brain studies and animal models.

Human epidemiology studies are beginning to look at GxE interactions, but the number of individuals required to reach significance can be a major limitation. By adding global DNA methylation analyses to epidemiology studies, the sensitivity of detecting seemingly complex interactions between exposures and the genome may be improved. Study designs that incorporate grandparents in addition to parents and investigate both genomic and methylome alterations will likely improve the understanding of the emerging questions surrounding transgenerational effects of environmental exposures in ASDs. Human epigenomic twin studies will also continue to be important in helping to separate genetic from non-genetic influences.[128–130]

Animal studies will continue to be needed to test hypotheses coming from human epidemiologic studies because the genetics and environment are easier to manipulate. Transgenerational studies of environmental exposures need to be pursued further but should include whole genomic analyses for genetic mutations as well as methylation alterations to establish causality of transgenerationally-maintained phenotypes. Carefully designed animal studies are expected to contribute great insight as genomic sequencing becomes standard and less expensive.

But because DNA methylation and gene expression patterns are frequently tissue-specific, human studies using post-mortem brain and human neurons differentiated from stem cells will continue to be important.[131–133] Human neurons in culture have the advantage of being genetic models as well as amenable to high-throughput analyses of many different chemical exposures. However, neurons in a dish are generally not a good model of system-wide disruptions in metabolism or immune imbalances. Furthermore, neuronal cultures do not reach maturity of human neurons in brain, so a continuation of postmortem brain studies as well as growing human neurons with the novel cerebral organoid technology[134] will likely be important for examining environmental exposures on the genome and epigenome of individuals with ASD.

Executive Summary

- While there is a strong genetic component to ASD risk, most genetic mutations identified in ASDs are *de novo*.
- Human twin studies and large population studies have implicated maternal *in utero* environment in ASD risk.
- The epigenetic layer of DNA methylation is at the interface of genetic and environmental risk and protective factors in ASD.
- The developing brain has a dynamic methylome, suggesting that the proper deposition of methyl groups to specific genetic loci at the right time and place is critical for proper brain function.

- Environmental toxins have demonstrated effects on DNA methylation patterns and global DNA methylation levels across organisms.
- Evidence for transgenerational effects of past environmental toxins are emerging in animal models and should be examined more closely for merit in ASD etiology.
- Future investigations should strive to integrate genetic, environmental, and epigenetic analyses in different systems and across generations.

References

1. Volkmar FR, Pauls D: Autism. *Lancet* 2003, **362**(9390):1133–1141.
2. CDC: Prevalence of Autism Spectrum Disorders — Autism and Developmental Disabilities Monitoring Network, 14 Sites, United States, 2008. MMWR. 2012, **61**(SS03):1–19.
3. Wing L, Potter D: The epidemiology of autistic spectrum disorders: is the prevalence rising? *Ment Retard Dev Disabil Res Rev* 2002, **8**(3):151–161.
4. Ratajczak HV: Theoretical aspects of autism: causes — a review. *J Immunotoxicol* 2011, **8**(1):68–79.
5. Steffenburg S, Gillberg C, Hellgren L, Andersson L, Gillberg IC, Jakobsson G, Bohman M: A twin study of autism in Denmark, Finland, Iceland, Norway and Sweden. *J Child Psychol Psychiatry* 1989, **30**(3):405–416.
6. Bailey A, Le Couteur A, Gottesman I, Bolton P, Simonoff E, Yuzda E, Rutter M: Autism as a strongly genetic disorder: evidence from a British twin study. *Psychol Med* 1995, **25**(1):63–77.
7. Ronald A, Hoekstra RA: Autism spectrum disorders and autistic traits: a decade of new twin studies. *Am J Med Genet B Neuropsychiatr Genet* 2011, **156B**(3):255–274.
8. Hallmayer J, Cleveland S, Torres A, Phillips J, Cohen B, Torigoe T, Miller J, Fedele A, Collins J, Smith K *et al.*: Genetic heritability and shared environmental factors among twin pairs with autism. *Archives of General Psychiatry* 2011, **68**(11):1095–1102.
9. Robinson EB, Koenen KC, McCormick MC, Munir K, Hallett V, Happe F, Plomin R, Ronald A: Evidence that autistic traits show the same etiology in the general population and at the quantitative extremes (5%, 2.5%, and 1%). *Archives of General Psychiatry* 2011, **68**(11):1113–1121.
10. Lundstrom S, Chang Z, Rastam M, Gillberg C, Larsson H, Anckarsater H, Lichtenstein P: Autism spectrum disorders and autistic like traits: similar etiology in the extreme end and the normal variation. *Arch Gen Psychiatry* 2012, **69**(1):46–52.

11. Robinson EB, Lichtenstein P, Anckarsater H, Happe F, Ronald A: Examining and interpreting the female protective effect against autistic behavior. *Proc Natl Acad Sci USA* 2013.
12. Sebat J, Lakshmi B, Malhotra D, Troge J, Lese-Martin C, Walsh T, Yamrom B, Yoon S, Krasnitz A, Kendall J *et al.*: Strong association of *de novo* copy number mutations with autism. *Science* 2007, **316**(5823):445–449.
13. Marshall CR, Noor A, Vincent JB, Lionel AC, Feuk L, Skaug J, Shago M, Moessner R, Pinto D, Ren Y *et al.*: Structural variation of chromosomes in autism spectrum disorder. *Am J Hum Genet* 2008, **82**(2):477–488.
14. Girirajan S, Brkanac Z, Coe BP, Baker C, Vives L, Vu TH, Shafer N, Bernier R, Ferrero GB, Silengo M *et al.*: Relative burden of large CNVs on a range of neurodevelopmental phenotypes. *PLoS Genet* 2011, **7**(11):e1002334.
15. O'Roak BJ, Deriziotis P, Lee C, Vives L, Schwartz JJ, Girirajan S, Karakoc E, Mackenzie AP, Ng SB, Baker C *et al.*: Exome sequencing in sporadic autism spectrum disorders identifies severe *de novo* mutations. *Nat Genet* 2011, **43**(6):585–589.
16. Neale BM, Kou Y, Liu L, Ma'ayan A, Samocha KE, Sabo A, Lin CF, Stevens C, Wang LS, Makarov V *et al.*: Patterns and rates of exonic *de novo* mutations in autism spectrum disorders. *Nature* 2012, **485**(7397):242–245.
17. O'Roak BJ, Vives L, Girirajan S, Karakoc E, Krumm N, Coe BP, Levy R, Ko A, Lee C, Smith JD *et al.*: Sporadic autism exomes reveal a highly interconnected protein network of *de novo* mutations. *Nature* 2012, **485**(7397):246–250.
18. Sanders SJ, Murtha MT, Gupta AR, Murdoch JD, Raubeson MJ, Willsey AJ, Ercan-Sencicek AG, DiLullo NM, Parikshak NN, Stein JL *et al.*: *De novo* mutations revealed by whole-exome sequencing are strongly associated with autism. *Nature* 2012, **485**(7397):237–241.
19. Lim ET, Raychaudhuri S, Sanders SJ, Stevens C, Sabo A, MacArthur DG, Neale BM, Kirby A, Ruderfer DM, Fromer M *et al.*: Rare complete knockouts in humans: population distribution and significant role in autism spectrum disorders. *Neuron* 2013, **77**(2):235–242.
20. Liu L, Sabo A, Neale BM, Nagaswamy U, Stevens C, Lim E, Bodea CA, Muzny D, Reid JG, Banks E *et al.*: Analysis of rare, exonic variation amongst subjects with autism spectrum disorders and population controls. *PLoS Genet* 2013, **9**(4):e1003443.
21. Iossifov I, Ronemus M, Levy D, Wang Z, Hakker I, Rosenbaum J, Yamrom B, Lee YH, Narzisi G, Leotta A *et al.*: *De novo* gene disruptions in children on the autistic spectrum. *Neuron* 2012, **74**(2):285–299.
22. Bourgeron T: A synaptic trek to autism. *Curr Opin Neurobiol* 2009, **19**(2):231–234.
23. Glessner JT, Wang K, Cai G, Korvatska O, Kim CE, Wood S, Zhang H, Estes A, Brune CW, Bradfield JP *et al.*: Autism genome-wide copy number variation reveals ubiquitin and neuronal genes. *Nature* 2009, **459**(7246):569–573.
24. Zuko A, Kleijer KT, Oguro-Ando A, Kas MJ, van Daalen E, van der Zwaag B, Burbach JP: Contactins in the neurobiology of autism. *Eur J Pharmacol* 2013.

25. LaSalle J: Autism genes keep turning up chromatin. *OA Autism* 2013, **1**:14.
26. Lasalle JM, Powell WT, Yasui DH: Epigenetic layers and players underlying neurodevelopment. *Trends Neurosci* 2013, **36**(8):460–470.
27. Lister R, Mukamel EA, Nery JR, Urich M, Puddifoot CA, Johnson ND, Lucero J, Huang Y, Dwork AJ, Schultz MD *et al.*: Global epigenomic reconfiguration during mammalian brain development. *Science* 2013, **341**(6146): 1237905.
28. Lister R, Pelizzola M, Dowen RH, Hawkins RD, Hon G, Tonti-Filippini J, Nery JR, Lee L, Ye Z, Ngo QM *et al.*: Human DNA methylomes at base resolution show widespread epigenomic differences. *Nature* 2009.
29. Schroeder DI, Blair JD, Lott P, Yu HO, Hong D, Crary F, Ashwood P, Walker C, Korf I, Robinson WP *et al.*: The human placenta methylome. *Proc Natl Acad Sci USA* 2013, **110**(15):6037–6042.
30. Bestor TH, Tycko B: Creation of genomic methylation patterns. *Nat Genet* 1996, **12**(4):363–367.
31. Buzdin A: Human-specific endogenous retroviruses. *ScientificWorldJournal* 2007, **7**:1848–1868.
32. Zemach A, Zilberman D: Evolution of eukaryotic DNA methylation and the pursuit of safer sex. *Curr Bio: CB* 2010, **20**(17):R780–785.
33. Girirajan S, Dennis MY, Baker C, Malig M, Coe BP, Campbell CD, Mark K, Vu TH, Alkan C, Cheng Z *et al.*: Refinement and discovery of new hotspots of copy-number variation associated with autism spectrum disorder. *Am J Hum Genet* 2013, **92**(2):221–237.
34. Girirajan S, Johnson RL, Tassone F, Balciuniene J, Katiyar N, Fox K, Baker C, Srikanth A, Yeoh KH, Khoo SJ *et al.*: Global increases in both common and rare copy number load associated with autism. *Hum Mol Genet* 2013.
35. Schroeder DI, Lott P, Korf I, LaSalle JM: Large-scale methylation domains mark a functional subset of neuronally expressed genes. *Genome Res* 2011, **21**(10):1583–1591.
36. Feng J, Zhou Y, Campbell SL, Le T, Li E, Sweatt JD, Silva AJ, Fan G: Dnmt1 and Dnmt3a maintain DNA methylation and regulate synaptic function in adult forebrain neurons. *Nat Neurosci* 2010, **13**(4):423–430.
37. LaPlant Q, Vialou V, Covington HE, 3rd, Dumitriu D, Feng J, Warren BL, Maze I, Dietz DM, Watts EL, Iniguez SD *et al.*: Dnmt3a regulates emotional behavior and spine plasticity in the nucleus accumbens. *Nat Neurosci* 2010, **13**(9):1137–1143.
38. Guo JU, Su Y, Zhong C, Ming GL, Song H: Hydroxylation of 5-methylcytosine by TET1 promotes active DNA demethylation in the adult brain. *Cell* 2011, **145**(3):423–434.
39. Iqbal K, Jin SG, Pfeifer GP, Szabo PE: Reprogramming of the paternal genome upon fertilization involves genome-wide oxidation of 5-methylcytosine. *Proc Natl Acad Sci USA* 2011, **108**(9):3642–3647.

40. Jiang YH, Sahoo T, Michaelis RC, Bercovich D, Bressler J, Kashork CD, Liu Q, Shaffer LG, Schroer RJ, Stockton DW *et al.*: A mixed epigenetic/genetic model for oligogenic inheritance of autism with a limited role for UBE3A. *Am J Med Genet* 2004.
41. Nagarajan RP, Patzel KA, Martin M, Yasui DH, Swanberg SE, Hertz-Picciotto I, Hansen RL, Van de Water J, Pessah IN, Jiang R *et al.*: MECP2 promoter methylation and X chromosome inactivation in autism. *Autism Res* 2008, **1**(3):169–178.
42. Nguyen A, Rauch TA, Pfeifer GP, Hu VW: Global methylation profiling of lymphoblastoid cell lines reveals epigenetic contributions to autism spectrum disorders and a novel autism candidate gene, RORA, whose protein product is reduced in autistic brain. *FASEB J* 2010, **24**(8):3036–3051.
43. Ladd-Acosta C, Hansen KD, Briem E, Fallin MD, Kaufmann WE, Feinberg AP: Common DNA methylation alterations in multiple brain regions in autism. *Mol Psychiatr* 2013.
44. Schmidt RJ, LaSalle JM: Interactions between folate, other B vitamins, DNA methylation, and neurodevelopmental disorders. In: *Nutrition, Epigenetic Mechanisms, and Human Disease.* Edited by N. M, Maulik G: CRC Press; 2011: 317–361.
45. Waterland RA, Jirtle RL: Transposable elements: targets for early nutritional effects on epigenetic gene regulation. *Mol Cell Biol* 2003, **23**(15):5293–5300.
46. Dolinoy DC, Huang D, Jirtle RL: Maternal nutrient supplementation counteracts bisphenol A-induced DNA hypomethylation in early development. *Proc Natl Acad Sci U S A* 2007, **104**(32):13056–13061.
47. Kucharski R, Maleszka J, Foret S, Maleszka R: Nutritional control of reproductive status in honeybees via DNA methylation. *Science* 2008, **319**(5871): 1827–1830.
48. Suren P, Roth C, Bresnahan M, Haugen M, Hornig M, Hirtz D, Lie KK, Lipkin WI, Magnus P, Reichborn-Kjennerud T *et al.*: Association between maternal use of folic acid supplements and risk of autism spectrum disorders in children. *JAMA* 2013, **309**(6):570–577.
49. Schmidt RJ, Tancredi DJ, Ozonoff S, Hansen RL, Hartiala J, Allayee H, Schmidt LC, Tassone F, Hertz-Picciotto I: Maternal periconceptional folic acid intake and risk of autism spectrum disorders and developmental delay in the CHARGE (CHildhood Autism Risks from Genetics and Environment) case-control study. *Am J Clin Nutr* 2012, **96**(1):80–89.
50. Schmidt RJ: Maternal folic acid supplements associated with reduced autism risk in the child. *Evid Based Med* 2013.
51. Pu D, Shen Y, Wu J: Association between MTHFR Gene Polymorphisms and the Risk of Autism Spectrum Disorders: A Meta-Analysis. *Autism Res* 2013.
52. Frye RE, Sequeira JM, Quadros EV, James SJ, Rossignol DA: Cerebral folate receptor autoantibodies in autism spectrum disorder. *Mol Psychiatr* 2013, **18**(3):369–381.

53. Ramaekers VT, Quadros EV, Sequeira JM: Role of folate receptor autoantibodies in infantile autism. *Mol Psychiatr* 2013, **18**(3):270–271.
54. Braunschweig D, Krakowiak P, Duncanson P, Boyce R, Hansen RL, Ashwood P, Hertz-Picciotto I, Pessah IN, Van de Water J: Autism-specific maternal autoantibodies recognize critical proteins in developing brain. *Transl Psychiatry* 2013, **3**:e277.
55. Muratore CR, Hodgson NW, Trivedi MS, Abdolmaleky HM, Persico AM, Lintas C, De la Monte S, Deth RC: Age-dependent decrease and alternative splicing of methionine synthase mRNA in human cerebral cortex and an accelerated decrease in autism. *PLoS One* 2013, **8**(2):e56927.
56. James SJ, Cutler P, Melnyk S, Jernigan S, Janak L, Gaylor DW, Neubrander JA: Metabolic biomarkers of increased oxidative stress and impaired methylation capacity in children with autism. *Am J Clin Nutr* 2004, **80**(6):1611–1617.
57. James SJ, Melnyk S, Jernigan S, Hubanks A, Rose S, Gaylor DW: Abnormal transmethylation/transsulfuration metabolism and DNA hypomethylation among parents of children with autism. *J Autism Dev Disord* 2008, **38**(10):1976.
58. Napoli E, Wong S, Giulivi C: Evidence of reactive oxygen species-mediated damage to mitochondrial DNA in children with typical autism. *Mol Autism* 2013, **4**(1):2.
59. Giulivi C, Zhang YF, Omanska-Klusek A, Ross-Inta C, Wong S, Hertz-Picciotto I, Tassone F, Pessah IN: Mitochondrial dysfunction in autism. *JAMA* 2010, **304**(21):2389–2396.
60. Frye RE, Delatorre R, Taylor H, Slattery J, Melnyk S, Chowdhury N, James SJ: Redox metabolism abnormalities in autistic children associated with mitochondrial disease. *Transl Psychiatry* 2013, **3**:e273.
61. Whanger PD: Selenium in the treatment of heavy metal poisoning and chemical carcinogenesis. *J Trace Elem Electrolytes Health Dis* 1992, **6**(4):209–221.
62. Rocourt CR, Cheng WH: Selenium supranutrition: are the potential benefits of chemoprevention outweighed by the promotion of diabetes and insulin resistance? *Nutrients* 2013, **5**(4):1349–1365.
63. Ma Y, Wu M, Li D, Li XQ, Li P, Zhao J, Luo MN, Guo CL, Gao XB, Lu CL *et al.*: Embryonic developmental toxicity of selenite in zebrafish (Danio rerio) and prevention with folic acid. *Food Chem Toxicol* 2012, **50**(8):2854–2863.
64. Ullrich SM, Tanton TW, Abdrashitova SA: Mercury in the aquatic environment: a review of factors affecting methylation. *Cri Rev Env Sci Technol* 2001, **31**(3):241–293.
65. Bose R, Onishchenko N, Edoff K, Janson Lang AM, Ceccatelli S: Inherited effects of low-dose exposure to methylmercury in neural stem cells. *Toxicol Sci* 2012, **130**(2):383–390.
66. Basu N, Head J, Nam DH, Pilsner JR, Carvan MJ, Chan HM, Goetz FW, Murphy CA, Rouvinen-Watt K, Scheuhammer AM: Effects of

methylmercury on epigenetic markers in three model species: mink, chicken and yellow perch. *Comp Biochem Physiol C Toxicol Pharmacol* 2013, **157**(3):322–327.

67. Onishchenko N, Karpova N, Sabri F, Castren E, Ceccatelli S: Long-lasting depression-like behavior and epigenetic changes of BDNF gene expression induced by perinatal exposure to methylmercury. *J Neurochem* 2008, **106**(3):1378–1387.
68. Desaulniers D, Xiao GH, Cummings-Lorbetskie C: Effects of lactational and/or *in utero* exposure to environmental contaminants on the glucocorticoid stress-response and DNA methylation of the glucocorticoid receptor promoter in male rats. *Toxicology* 2013, **308**:20–33.
69. Meyers R: D.E.S., the bitter pill. New York: Seaview/Putnam; 1983.
70. Herbst AL, Ulfelder H, Poskanzer DC: Adenocarcinoma of the vagina. Association of maternal stilbestrol therapy with tumor appearance in young women. *N Engl J Med* 1971, **284**(15):878–881.
71. Bromer JG, Wu J, Zhou Y, Taylor HS: Hypermethylation of homeobox A10 by *in utero* diethylstilbestrol exposure: an epigenetic mechanism for altered developmental programming. *Endocrinology* 2009, **150**(7):3376–3382.
72. Li S, Hansman R, Newbold R, Davis B, McLachlan JA, Barrett JC: Neonatal diethylstilbestrol exposure induces persistent elevation of c-fos expression and hypomethylation in its exon-4 in mouse uterus. *Mol Carcinog* 2003, **38**(2):78–84.
73. Sato K, Fukata H, Kogo Y, Ohgane J, Shiota K, Mori C: Neonatal exposure to diethylstilbestrol alters the expression of DNA methyltransferases and methylation of genomic DNA in the epididymis of mice. *Endocr J* 2006, **53**(3):331–337.
74. Tang WY, Newbold R, Mardilovich K, Jefferson W, Cheng RY, Medvedovic M, Ho SM: Persistent hypomethylation in the promoter of nucleosomal binding protein 1 (Nsbp1) correlates with overexpression of Nsbp1 in mouse uteri neonatally exposed to diethylstilbestrol or genistein. *Endocrinology* 2008, **149**(12):5922–5931.
75. Yaoi T, Itoh K, Nakamura K, Ogi H, Fujiwara Y, Fushiki S: Genome-wide analysis of epigenomic alterations in fetal mouse forebrain after exposure to low doses of bisphenol A. *Biochem Biophys Res Commun* 2008, **376**(3):563–567.
76. Hiyama M, Choi EK, Wakitani S, Tachibana T, Khan H, Kusakabe KT, Kiso Y: Bisphenol-A (BPA) affects reproductive formation across generations in mice. *J Vet Med Sci* 2011, **73**(9):1211–1215.
77. Chao HH, Zhang XF, Chen B, Pan B, Zhang LJ, Li L, Sun XF, Shi QH, Shen W: Bisphenol A exposure modifies methylation of imprinted genes in mouse oocytes via the estrogen receptor signaling pathway. *Histochem Cell Biol* 2012, **137**(2):249–259.

78. Zhang XF, Zhang LJ, Feng YN, Chen B, Feng YM, Liang GJ, Li L, Shen W: Bisphenol A exposure modifies DNA methylation of imprint genes in mouse fetal germ cells. *Mol Biol Rep* 2012, **39**(9):8621–8628.
79. Doshi T, D'Souza C, Vanage G: Aberrant DNA methylation at Igf2-H19 imprinting control region in spermatozoa upon neonatal exposure to bisphenol A and its association with post implantation loss. *Mol Biol Rep* 2013, **40**(8):4747–4757.
80. Jang YJ, Park HR, Kim TH, Yang WJ, Lee JJ, Choi SY, Oh SB, Lee E, Park JH, Kim HP *et al.*: High dose bisphenol A impairs hippocampal neurogenesis in female mice across generations. *Toxicology* 2012, **296**(1–3):73–82.
81. Kundakovic M, Champagne FA: Epigenetic perspective on the developmental effects of bisphenol A. *Brain Behav Immun* 2011, **25**(6):1084–1093.
82. Susiarjo M, Sasson I, Mesaros C, Bartolomei MS: Bisphenol a exposure disrupts genomic imprinting in the mouse. *PLoS Genet* 2013, **9**(4):e1003401.
83. Zhang HQ, Zhang XF, Zhang LJ, Chao HH, Pan B, Feng YM, Li L, Sun XF, Shen W: Fetal exposure to bisphenol A affects the primordial follicle formation by inhibiting the meiotic progression of oocytes. *Mol Biol Rep* 2012, **39**(5):5651–5657.
84. Manikkam M, Tracey R, Guerrero-Bosagna C, Skinner MK: Plastics derived endocrine disruptors (BPA, DEHP and DBP) induce epigenetic transgenerational inheritance of obesity, reproductive disease and sperm epimutations. *PLoS One* 2013, **8**(1):e55387.
85. Hooper K, She J, Sharp M, Chow J, Jewell N, Gephart R, Holden A: Depuration of polybrominated diphenyl ethers (PBDEs) and polychlorinated biphenyls (PCBs) in breast milk from California first-time mothers (primiparae). *Environ Health Perspect* 2007, **115**(9):1271–1275.
86. Rusiecki JA, Baccarelli A, Bollati V, Tarantini L, Moore LE, Bonefeld-Jorgensen EC: Global DNA hypomethylation is associated with high serum-persistent organic pollutants in Greenlandic Inuit. *Environ Health Perspect* 2008, **116**(11):1547–1552.
87. Mitchell MM, Woods R, Chi LH, Schmidt RJ, Pessah IN, Kostyniak PJ, Lasalle JM: Levels of select PCB and PBDE congeners in human postmortem brain reveal possible environmental involvement in 15q11-q13 duplication autism spectrum disorder. *Environ Mol Mutagen* 2012.
88. Woods R, Vallero RO, Golub MS, Suarez JK, Ta TA, Yasui DH, Chi LH, Kostyniak PJ, Pessah IN, Berman RF *et al.*: Long-lived epigenetic interactions between perinatal PBDE exposure and Mecp2308 mutation. *Hum Mol Genet* 2012, **21**(11):2399–2411.
89. Aluru N, Karchner SI, Hahn ME: Role of DNA methylation of AHR1 and AHR2 promoters in differential sensitivity to PCBs in Atlantic Killifish, Fundulus heteroclitus. *Aquat Toxicol* 2011, **101**(1):288–294.

90. Hon GC, Hawkins RD, Caballero OL, Lo C, Lister R, Pelizzola M, Valsesia A, Ye Z, Kuan S, Edsall LE *et al.*: Global DNA hypomethylation coupled to repressive chromatin domain formation and gene silencing in breast cancer. *Genome Res* 2012, **22**(2):246–258.
91. Schroeder DI, Blair JD, Lott P, Yu HO, Hong D, Crary F, Ashwood P, Walker C, Korf I, Robinson WP *et al.*: The human placenta methylome. *Proc Nat Acad Sci U S A* 2013.
92. Razin A, Shemer R: DNA methylation in early development. *Hum Mol Genet* 1995, **4**(Spec No):1751–1755.
93. Hajkova P, Erhardt S, Lane N, Haaf T, El-Maarri O, Reik W, Walter J, Surani MA: Epigenetic reprogramming in mouse primordial germ cells. *Mech Dev* 2002, **117**(1–2):15–23.
94. Anway MD, Cupp AS, Uzumcu M, Skinner MK: Epigenetic transgenerational actions of endocrine disruptors and male fertility. *Science* 2005, **308**(5727):1466–1469.
95. Crews D, Gore AC, Hsu TS, Dangleben NL, Spinetta M, Schallert T, Anway MD, Skinner MK: Transgenerational epigenetic imprints on mate preference. *Proc Natl Acad Sci U S A* 2007, **104**(14):5942–5946.
96. Guerrero-Bosagna C, Settles M, Lucker B, Skinner MK: Epigenetic transgenerational actions of vinclozolin on promoter regions of the sperm epigenome. *PLoS One* 2010, **5**(9).
97. Stouder C, Paoloni-Giacobino A: Transgenerational effects of the endocrine disruptor vinclozolin on the methylation pattern of imprinted genes in the mouse sperm. *Reproduction* 2010, **139**(2):373–379.
98. Guerrero-Bosagna C, Covert TR, Haque MM, Settles M, Nilsson EE, Anway MD, Skinner MK: Epigenetic transgenerational inheritance of vinclozolin induced mouse adult onset disease and associated sperm epigenome biomarkers. *Reprod Toxicol* 2012, **34**(4):694–707.
99. Skinner MK, Haque CG, Nilsson E, Bhandari R, McCarrey JR: Environmentally induced transgenerational epigenetic reprogramming of primordial germ cells and the subsequent germ line. *PLoS One* 2013, **8**(7):e66318.
100. Vandegehuchte MB, Lemiere F, Vanhaecke L, Vanden Berghe W, Janssen CR: Direct and transgenerational impact on Daphnia magna of chemicals with a known effect on DNA methylation. *Comp Biochem Physiol C Toxicol Pharmacol* 2010, **151**(3):278–285.
101. Schneider S, Kaufmann W, Buesen R, van Ravenzwaay B: Vinclozolin — the lack of a transgenerational effect after oral maternal exposure during organogenesis. *Reprod Toxicol* 2008, **25**(3):352–360.
102. Schneider S, Marxfeld H, Groters S, Buesen R, van Ravenzwaay B: Vinclozolin — no transgenerational inheritance of anti-androgenic effects after maternal exposure during organogenesis via the intraperitoneal route. *Reprod Toxicol* 2013, **37**:6–14.

103. Kalfa N, Paris F, Soyer-Gobillard MO, Daures JP, Sultan C: Prevalence of hypospadias in grandsons of women exposed to diethylstilbestrol during pregnancy: a multigenerational national cohort study. *Fertil Steril* 2011, **95**(8):2574–2577.
104. Wolstenholme JT, Edwards M, Shetty SR, Gatewood JD, Taylor JA, Rissman EF, Connelly JJ: Gestational exposure to bisphenol a produces transgenerational changes in behaviors and gene expression. *Endocrinology* 2012, **153**(8):3828–3838.
105. Christianson AL, Chesler N, Kromberg JG: Fetal valproate syndrome: clinical and neuro-developmental features in two sibling pairs. *Dev Med Child Neurol* 1994, **36**(4):361–369.
106. Jensen RA: Autism and the chemical connection. *J Autism Dev Disord* 1994, **24**(6):785–787.
107. Stromland K, Nordin V, Miller M, Akerstrom B, Gillberg C: Autism in thalidomide embryopathy: a population study. *Dev Med Child Neurol* 1994, **36**(4):351–356.
108. Chess S: Autism in children with congenital rubella. *J Autism Child Schizophr* 1971, **1**(1):33–47.
109. Jung CR, Lin YT, Hwang BF: Air pollution and newly diagnostic autism spectrum disorders: a population-based cohort study in Taiwan. *PLoS One* 2013, **8**(9):e75510.
110. Roberts AL, Lyall K, Hart JE, Laden F, Just AC, Bobb JF, Koenen KC, Ascherio A, Weisskopf MG: Perinatal air pollutant exposures and autism spectrum disorder in the children of nurses' health study II participants. *Environ Health Perspect* 2013, **121**(8):978–984.
111. Becerra TA, Wilhelm M, Olsen J, Cockburn M, Ritz B: Ambient air pollution and autism in Los Angeles county, California. *Environ Health Perspect* 2013, **121**(3):380–386.
112. Cheslack-Postava K, Rantakokko PV, Hinkka-Yli-Salomaki S, Surcel HM, McKeague IW, Kiviranta HA, Sourander A, Brown AS: Maternal serum persistent organic pollutants in the Finnish Prenatal Study of Autism: A pilot study. *Neurotoxicol Teratol* 2013, **38**:1–5.
113. Thirtamara Rajamani K, Doherty-Lyons S, Bolden C, Willis D, Hoffman C, Zelikoff J, Chen LC, Gu H: Prenatal and early-life exposure to high-level diesel exhaust particles leads to increased locomotor activity and repetitive behaviors in mice. *Autism Res* 2013, **6**(4):248–257.
114. Volk HE, Lurmann F, Penfold B, Hertz-Picciotto I, McConnell R: Traffic-related air pollution, particulate matter, and autism. *JAMA Psychiatry* 2013, **70**(1):71–77.
115. Mitchell MM, Woods R, Chi LH, Schmidt RJ, Pessah IN, Kostyniak PJ, LaSalle JM: Levels of select PCB and PBDE congeners in human postmortem

brain reveal possible environmental involvement in 15q11-q13 duplication autism spectrum disorder. *Environ Mol Mutag* 2012, **53**(8):589–598.
116. Shelton JF, Hertz-Picciotto I, Pessah IN: Tipping the balance of autism risk: potential mechanisms linking pesticides and autism. *Environ Health Perspect* 2012, **120**(7):944–951.
117. Roberts EM, English PB, Grether JK, Windham GC, Somberg L, Wolff C: Maternal residence near agricultural pesticide applications and autism spectrum disorders among children in the California Central Valley. *Environ Health Perspect* 2007, **115**(10):1482–1489.
118. Volk HE, Hertz-Picciotto I, Delwiche L, Lurmann F, McConnell R: Residential proximity to freeways and autism in the CHARGE study. *Environ Health Perspect* 2011, **119**(6):873–877.
119. Windham GC, Sumner A, Li SX, Anderson M, Katz E, Croen LA, Grether JK: Use of birth certificates to examine maternal occupational exposures and autism spectrum disorders in offspring. *Autism Res* 2013, **6**(1):57–63.
120. Williams EL, Casanova MF: Potential teratogenic effects of ultrasound on corticogenesis: implications for autism. *Med Hypotheses* 2010, **75**(1):53–58.
121. Gardener H, Spiegelman D, Buka SL: Perinatal and neonatal risk factors for autism: a comprehensive meta-analysis. *Pediatrics* 2011, **128**(2):344–355.
122. Lyall K, Pauls DL, Santangelo S, Spiegelman D, Ascherio A: Maternal early life factors associated with hormone levels and the risk of having a child with an autism spectrum disorder in the nurses health study II. *J Autism Dev Disord* 2011.
123. Krakowiak P, Walker CK, Bremer AA, Baker AS, Ozonoff S, Hansen RL, Hertz-Picciotto I: Maternal metabolic conditions and risk for autism and other neurodevelopmental disorders. *Pediatrics* 2012, **129**(5):e1121–1128.
124. Langridge AT, Glasson EJ, Nassar N, Jacoby P, Pennell C, Hagan R, Bourke J, Leonard H, Stanley FJ: Maternal conditions and perinatal characteristics associated with autism spectrum disorder and intellectual disability. *PLoS One* 2013, **8**(1):e50963.
125. Zhang Y, Gao D, Kluetzman K, Mendoza A, Bolivar VJ, Reilly A, Jolly JK, Lawrence DA: The maternal autoimmune environment affects the social behavior of offspring. *J Neuroimmunol* 2013, **258**(1–2):51–60.
126. Golding J, Steer C, Pembrey M: Parental and grandparental ages in the autistic spectrum disorders: a birth cohort study. *PLoS One* 2010, **5**(4):e9939.
127. Roberts A, Lyall K, Rich-Edwards J, Ascherio A, Weisskopf MG: Association of maternal exposure to childhood abuse with elevated risk for autism in offspring. *JAMA Psychiatry* 2013, **doi:10.1001/jamapsychiatry.2013.447**.
128. Miyake K, Yang C, Minakuchi Y, Ohori K, Soutome M, Hirasawa T, Kazuki Y, Adachi N, Suzuki S, Itoh M *et al.*: Comparison of genomic and epigenomic expression in monozygotic twins discordant for rett syndrome. *PLoS One* 2013, **8**(6):e66729.

129. Wong CC, Meaburn EL, Ronald A, Price TS, Jeffries AR, Schalkwyk LC, Plomin R, Mill J: Methylomic analysis of monozygotic twins discordant for autism spectrum disorder and related behavioural traits. *Mol Psychiatry* 2013.
130. Anderson GM: Twin studies in autism: what might they say about genetic and environmental influences. *J Autism Dev Disord* 2012, **42**(7):1526–1527.
131. Chailangkarn T, Acab A, Muotri AR: Modeling neurodevelopmental disorders using human neurons. *Curr Opin Neurobiol* 2012, **22**(5):785–790.
132. DeRosa BA, Van Baaren JM, Dubey GK, Lee JM, Cuccaro ML, Vance JM, Pericak-Vance MA, Dykxhoorn DM: Derivation of autism spectrum disorder-specific induced pluripotent stem cells from peripheral blood mononuclear cells. *Neurosci Lett* 2012, **516**(1):9–14.
133. Vaccarino FM, Stevens HE, Kocabas A, Palejev D, Szekely A, Grigorenko EL, Weissman S: Induced pluripotent stem cells: a new tool to confront the challenge of neuropsychiatric disorders. *Neuropharmacology* 2011, **60**(7–8):1355–1363.
134. Lancaster MA, Renner M, Martin CA, Wenzel D, Bicknell LS, Hurles ME, Homfray T, Penninger JM, Jackson AP, Knoblich JA: Cerebral organoids model human brain development and microcephaly. *Nature* 2013, **501**(7467):373–379.
135. Inawaka K, Kawabe M, Takahashi S, Doi Y, Tomigahara Y, Tarui H, Abe J, Kawamura S, Shirai T: Maternal exposure to anti-androgenic compounds, vinclozolin, flutamide and procymidone, has no effects on spermatogenesis and DNA methylation in male rats of subsequent generations. *Toxicol Appl Pharmacol* 2009, **237**(2):178–187.

Chapter 16

Oxidative Stress and Mitochondrial Dysfunction in ASDs

Feng Gu, Ved Chauhan and Abha Chauhan*

Abstract

Autism spectrum disorders (ASDs) are complex multifactorial disorders involving various genetic vulnerabilities interacting with environmental factors. Although the cause of autism is elusive, numerous studies indicate that oxidative stress and mitochondrial dysfunction are involved in the pathophysiology of ASDs. Several studies have reported increased levels of oxidative damage markers of lipid peroxidation, protein and DNA oxidation as well as impaired antioxidant defense in the plasma, urine and/or brain tissues from autistic subjects. We and other groups have also reported decreased glutathione (GSH) level and GSH/GSSG (oxidized glutathione) ratio, and increased GSSG in the blood and brain samples of autistic subjects, suggesting glutathione redox imbalance in autism. The activities of enzymes involved in GSH homeostasis i.e. glutathione peroxidase (GPx), glutathione-S-transferase (GST) and glutamate cysteine ligase (GCL) are impaired in autism, and association of polymorphisms of GSH, GPx or GST genes with autism has been reported. Mitochondria are not only the source of free radicals, but they are also the target of oxidative damage. Electron transport chain (ETC) in mitochondria is essential for ATP generation. Defects in energy metabolism,

*Correspondence to: NYS Institute for Basic Research in Developmental Disabilities, 1050 Forest Hill Road, Staten Island, NY 10314, USA. Email: abha.chauhan@opwdd.ny.gov.

protein expression and activities of ETC complexes as well as in mitochondrial DNA have been reported in the blood, muscle biopsy and brain samples from autism. All these findings suggest that oxidative stress and mitochondrial dysfunction play critical roles in the pathophysiology of autism.

Keywords: Autism spectrum disorders; Electron transport chain; Glutathione; Mitochondrial dysfunction; Oxidative stress.

Introduction

Autism spectrum disorders (ASDs) are heterogeneous, neurodevelopmental disorders with core symptoms of dysfunctions in social interactions and communication skills, restricted interests, repetitive-stereotypic verbal and non-verbal behaviors. Recent evidence suggests gene-environment interactions in ASDs. The neurobiology and genetics of autism is not clearly understood. Although the cause of autism is elusive, numerous studies indicate that oxidative stress and mitochondrial dysfunction are involved in the pathophysiology of ASDs.[1–10]

Oxidative stress may be defined as an imbalance between free radical production and antioxidant mechanisms of the cell, which results in excess levels of reactive oxygen species (ROS). Under normal conditions, a dynamic equilibrium exists between the production of ROS and the antioxidant capacity of the cell. ROS are generated endogenously during oxidative metabolism and energy production by mitochondria,[11,12] and are neutralized by antioxidant sefense mechanisms. Antioxidant defenses include protective mechanisms by enzymes such as superoxide dismutase (SOD), catalase (CAT) and glutathione peroxidase (GPx), and non-enzymatic antioxidants such as glutathione (GSH), vitamins E and C, metallothionein and phenolic compounds. ROS dramatically increase under certain pathological conditions as well as due to the effect of environmental factors. They are chemically unstable and may attack and damage vital components of the cell, such as polyunsaturated fatty acids, proteins and nucleic acids. An impaired antioxidant system can result in cell membrane damage, alterations in membrane fluidity and permeability, and oxidative stress. Oxidative stress is particularly destructive to the brain because the brain has higher energy requirements, high concentration of polyunsaturated fatty acids and lower reserves of GSH, which is the most important antioxidant defense. Neurons are the first cells to be affected by the increased oxidative stress. Oxidative stress is known to play a vital role in aging and a range of disorders including cardiovascular disease, diabetes, hypertension, psychiatric disorders, neurodegenerative and neurodevelopmental diseases.

The ROS are byproducts of proton cycling in the mitochondria. Mitochondria are very important cellular organelles that generate adenosine triphosphate (ATP), the energy carrier in most mammalian cells by oxidizing glucose and fatty acids. The electron transport chain (ETC) is an essential part of mitochondria, and consists of five multi-subunit complexes of enzymes, i.e., complex I (NADH dehydrogenase), complex II (succinate dehydrogenase), complex III (cytochrome bc1 complex), complex IV (cytochrome c oxidase) and complex V (ATP synthase).[13,14] Complexes I–IV participate in the generation of proton gradient (mitochondrial membrane potential) across the inner membrane of mitochondria, while complex V transports protons from the intermembrane space to the mitochondrial matrix. The generated proton gradient is used by complex V to catalyze the formation of ATP by the phosphorylation of adenosine diphosphate (ADP). Complexes I and III are the main sites of mitochondrial superoxide (O_2^-) production.[15,16] ROS production is highly regulated by complex I.[17–19] Under normal physiological conditions, ROS are neutralized and converted to hydrogen peroxide (H_2O_2) by SOD. However, when the ETC is affected, the excess of electrons can be passed directly to O_2 to generate superoxide anion. In addition to energy generation, mitochondria also play a pivotal role in the maintenance of intracellular calcium homeostasis as well as in amino acid, lipid, and steroid metabolism, thereby regulating developmental processes, including neurite outgrowth, axonal plasticity and synaptic plasticity.[20–24] In particular, neuronal synapses are the area of high energy consumption and therefore, especially rely on mitochondrial function.[21,25] Mitochondria are concentrated in the dendritic and axonal termini where they are involved in ATP production, calcium homeostasis and synaptic plasticity.[26]

Mitochondria have a small amount of their own DNA. Thirteen subunits of ETC complexes I, III, IV and V are coded by 37 mitochondrial DNA (mtDNA) genes,[27] and the other subunits are coded by more than 850 nuclear DNA (nDNA) genes.[28] Therefore, mitochondrial function is under the dual genetic control of both mtDNA and nDNA. The expression, replication and maintenance of mtDNA also require the factors encoded by nuclear genes.[29] Hence, mtDNA and/or nDNA genome mutations may lead to deficiencies of ETC complexes, and subsequently to mitochondrial dysfunction.

Oxidative Stress in Autism

Accumulating evidence from our and other groups suggests that oxidative stress may play a significant role in the development and clinical manifestation of autism.[1–7,9,30–37] Rossignol and Frye performed a comprehensive literature search in ASDs, and reported an association between ASDs and oxidative stress in all of 115 studies that investigated oxidative stress in ASDs.[9]

Increased markers of oxidative damage in autism

Several biomarkers of oxidative stress are reported to be increased in the blood, urine and/or brain samples from the autistic subjects. These include elevated levels of (1) lipid peroxidation markers, i.e., malonyldialdehyde (MDA),[31,35,38,39] F2-isoprostane,[33,40] and 4-hydroxy nonenal (4-HNE)[36] in the blood; (2) protein oxidation/nitration markers, i.e., protein carbonyl in the urine or plasma,[36,39] and 3-nitrotyrosine (3-NT) in the brain[3,41,42] or plasma;[43] and (3) DNA oxidation markers, i.e., 8-hydroxyguanosine (8-OHG) in the brain,[41,44] 8-hydroxydeoxyguanosine (8-OH-dG) in the brain[42] or urine.[33] Analysis of whole blood or polymorphonuclear leukocytes showed that autistic children produced significantly higher levels of ROS.[45] Furthermore, the extent of oxidative stress has been reported to be related with the severity of autism.[36,37]

Decreased antioxidant defense in autism

GSH is the most important endogenous scavenger of environmental toxins and ROS. It is also involved in the maintenance of essential thiol status, storage of cysteine and modulation of cell differentiation, proliferation and apoptosis.[46] Studies from our and other groups have reported decreased levels of total and reduced GSH, while oxidized glutathione (GSSG) levels are increased in the blood and brain samples of subjects with autism.[30,32,37,40,41,43,47–52] Under normal conditions, GSH is the predominant form and GSSG content is less than 1–1.2% of GSH.[53] These findings provide evidence that there is glutathione-redox imbalance in the individuals with autism.

GSH synthesis occurs in the cytosol and includes two consecutive ATP-dependent enzymatic reactions. First, glutamate is coupled with cysteine to form γ-glutamylcysteine (γGC) in a reaction catalyzed by glutamate cysteine ligase (GCL), the key rate-limiting enzyme of GSH biosynthesis. Then, γGC is coupled with glycine to form GSH, and this reaction is catalyzed by glutathione synthetase (GS). In a recent publication, we reported reduced GCL activity and an increased ratio of protein levels of GCLC (a heavy or catalytic subunit of GCL) to GCLM (a light or modulatory subunit of GCL) in the cerebellum of subjects with autism than in that of the control group.[54] Moreover, a positive strong correlation was observed between GCL activity and protein levels of GCLM and GCLC subunits in control subjects but not in autistic subjects, suggesting that regulation of GCL activity is affected in autism.[54]

GSH is oxidized to GSSG by GPx, and GSSG is reversed to GSH when glutathione reductase (GR) catalyzes the reaction. GPx and GR are antioxidant enzymes involved in elimination of ROS. Glutathione-S-transferase (GST) is another antioxidant detoxification enzyme, which is involved in the detoxification of oxidized products by conjugating GSH to electrophilic centers in many toxic substrates to form non-toxic products. Recently, we have reported impaired activities of GPx and GST in the cerebellum of subjects with autism.[54] These results suggest that enzymes involved in GSH homeostasis have impaired activities in the cerebellum in autism, and lower GCL activity in autism may be related to decreased protein expression of GCLM.

The alterations in the activities of antioxidant enzymes GPx,[35,40,55–59] GST,[51] GR,[51] SOD[35,55,57,59–62] and CAT[55,59,60] in the blood from individuals with autism has been reported in many studies and sometimes with conflicting results. Recently, two independent systematic reviews and meta-analyses on oxidative stress-related biomarkers and GSH levels in autism were completed by Main *et al.*[5] and Frustaci *et al.*,[6] which confirmed that plasma GSH levels are lower and GSSG levels are higher in autism. GPx activity in erythrocytes was significantly lower in subjects with ASDs than that in controls, while the activity of SOD in blood failed to demonstrate any significant association with autism.[6] Exogenous antioxidants are also important to prevent the damage of toxins. The levels of vitamin E, A and lycopene are decreased in the blood of autistic individuals.[37,55,63]

Abnormal genes of oxidative stress pathways can increase vulnerability to oxidative stress. Polymorphisms of genes related to GSH, GPx or GST have all been reported in autism.[32,64–66] A polyalanine repeat polymorphism in the GPx gene (GPx-1) was found to be associated with autism.[64] Bowers *et al.*[66] examined genetic variations in 42 genes related to glutathione in 318 families from the Autism Genetic Resource Exchange repository, and results showed a significant association between cystathione gamma lyase, glutaredoxin, glutaredoxin 3 and autism.

Mitochondrial Dysfunction in Autism

Mitochondria are not only the source of free radicals, but they are also the target of oxidative damage. Mitochondria itself cannot synthesize GSH but it transports 10–15% of GSH from the cytosol via an energy-dependent transporter.[67] Therefore, the GSH depletion in mitochondria makes cells more vulnerable to oxidative stress mediated damage. Mitochondrial dysfunction will result in several consequences: decreased ATP production, increased production of

ROS and oxidative damage, alteration of neurotransmission systems, and induction of apoptosis. Mitochondrial dysfunction and oxidative stress have been implicated in a variety of neurodegenerative diseases such as Alzheimer's disease (AD), Parkinson's disease (PD), amyotrophic lateral sclerosis (ALS) and Huntington's disease (HD), as well as neurodevelopmental disorders. An increasing number of studies highlight that mitochondrial dysfunction may have an important role in autism.[8–10,68–78]

Defects in energy metabolism in autism

The prevalence of mitochondrial disease (MD) is 5.0% in the ASDs population,[68] which is much higher than its prevalence of 0.01% in the general population.[79] MD is often caused by the gene mutation or deletion, and the diagnosis of MD is complicated as it is dependent on a significant reduction in ETC complex function and known mtDNA mutations.[80] Only 23% of children with ASDs/MD have a known mtDNA abnormality. Most of children with ASDs/MD do not have a specific genetic mutation to explain the MD. However, brain energy metabolism defects in autism have been suggested by many biochemical, genetic, anatomical and neuroradiographical studies since 1993.[10,75,78,81–85] In 1998, Lombard first presented a hypothesis that mitochondrial dysfunction and defects in neuronal oxidative phosphorylation may be involved in the etiology of autism.[81] Several recent studies have demonstrated alteration of multiple biomarkers of brain energy metabolism, such as high levels of lactate, alanine, lactate to pyruvate ratio, alanine-to-lysine ratio and low carnitine levels in subjects with ASDs, which lend support to the notion that disorders of energy production may affect a significant subset of children with autism.[75,86–90] Pyruvate dehydrogenase (PDH) enzyme activity is significantly reduced in the frontal cortex and lymphocytes from subjects with autism.[73,91] Reduced PDH activity will lead to decreased decarboxylation of pyruvate to acetyl CoA, impaired tricarboxylic acid (TCA) cycle activity, decreased generation of NADH, and increased metabolism of pyruvate to lactate and alanine in autism. Boccuto *et al.*[92]

recently reported that lymphoblastoid cells from the subjects with ASDs had reduced generation of NADH compared to control subjects when tryptophan was the sole energy source. NADH is a critical energy carrier for the ETC, and electrons are transferred from NADH to O_2 through ETC complexes I, III and IV. Therefore, mitochondrial dysfunction may also result from abnormal metabolism of tryptophan, which is a precursor of serotonin, quinolinic acid and kynurenic acid.[92,93]

Defects in activities and protein expression of mitochondrial ETC complexes in autism

Mitochondrial dysfunction results from several causes, and ETC deficiency is an important causative factor among them.[8,10] Several case studies showed that the activities or expression levels of ETC complexes are decreased in autism. For example, a 19-month-old autistic girl with developmental regression had marked reduction in enzymatic activities for complexes I and III,[94] another two children with autism had deficiencies in complexes I-III and CoQ,[95] a 3-year-old child with autism had partial deficiencies of complexes III and IV,[82] and a 12-year-old boy with autism (who had deletion in the 5q14.3 region) had severely decreased activity of complex IV and mildly decreased activity of complex I in a buccal swab sample.[96] For cohort studies, twenty of 25 subjects with ASDs (40% of them demonstrated an unusual pattern of regression) had deficient activity of ETC complexes in tissue ETC or polarographic analysis.[87] The deficiencies of complexes I, III, II and IV were observed in 64%, 20%, 5% and 4% of these autistic subjects respectively, while two subjects had rare mtDNA mutations of likely pathogenicity.[87] Using lymphocytes, Giulivi *et al.*[91] reported that 6 of 10 autistic children had a lower complex I activity, four had a lower complex V activity, and three had a lower complex IV activity compared to the control subjects. On the other hand, the activity of complex IV in the muscle was reported to be elevated in a few cases of ASDs with developmental regression.[97] In a meta-analysis, Rossignol and Frye[68] reported deficiencies of complexes I, III, V, IV and II in 53%,

30%, 23%, 20% and 9% of children with ASDs and concomitant mitochondrial disease (MD), respectively. Multiple complex deficiencies were found in 36% of the children with ASDs/MD.

Using postmortem brain tissue, we reported brain region-specific deficit of protein expression levels of mitochondrial ETC complexes in children with autism.[78] As compared to age-matched control subjects, the levels of complexes III and V in the cerebellum, of complex I in the frontal cortex, and of complexes II, III, and V in the temporal cortex were significantly lower in children with autism (ages: 4–10 years), but there was no change in these ETC complexes in the parietal and occipital cortices.[78] These findings suggest that ETC defect in autism is specific to cerebellum, frontal and temporal lobes. Recently, we also reported defects of complex I, III and V activities in the frontal cortex from autistic subjects.[73] The activities of complexes I and V in autistic group were significantly reduced compared to the control group. When 99% confidence interval (CI) of the control group was taken as a reference range, impaired activities of complexes I, III and V were observed in 43%, 29% and 43% of autistic subjects, respectively.[73] All five ETC complexes showed reduced activities in 14% of autistic cases, and the activities of multiple complexes were decreased in 29% of autistic subjects. Our studies[73] on brain tissues support previous meta-analysis results of Rossignol *et al.*[68] that deficiencies of complex I, III and V are more common in ASDs. In the temporal cortex, Tang *et al.*[70] reported decreased complex I and IV activities, decreased protein expression of SOD as well as increased oxidative DNA damage.

Defects in mtDNA in autism

In our recent study with the brain tissues of subjects with autism, we reported increased copy numbers of three mitochondrial genes, i.e., ND1, ND4 and Cyt B that encode for subunits of mitochondrial ETC complexes I and III. These findings are consistent with the study of Giulivi *et al.*[91] who also reported increased copy number in these mitochondrial genes in the lymphocytes of subjects with autism. Our results suggest that alteration of copy number of mitochondrial genes occurs not only in peripheral tissues but also in the brain of subjects with autism. Oxidative stress has also been shown to increase the mtDNA copy number.[98,99] We also reported mtDNA deletion of ND4 and CytB in the frontal cortex of 44% and 33% of autistic subjects, respectively.[73] Napoli *et al.*[71] recently evaluated the mtDNA deletion in peripheral blood monocytic cells isolated from 2–5 year old children with autism and age-matched typically developing children as well as their parents. They reported that the frequency of mtDNA deletion in children with autism is two-fold higher than in control subjects, and the frequency of deletion is similar to that of their fathers.[71] Higher frequencies of GC to AT transitions and GCT to A transitions in autistic subjects were identical to their mothers.[71] Pons *et al.*[100] reported 72% mtDNA depletion in skeletal muscle in one subject with autism. A3243G mtDNA mutation was observed in tissues from the mothers of two autistic children, but it was not observed in tissues from these children.[100] However, Hadjixenofontos *et al.*[101] did not find any evidence to suggest a major role for mtDNA variation in ASDs susceptibility.

Anitha *et al.*[74] studied the expression of 84 ETC genes in the brain of autistic subjects. Eleven genes of complex I, five genes each of complex III and complex IV, and seven genes of complex V showed brain region-specific reduced expression in autism. ATP5A1 (Complex V), ATP5G3 (Complex V) and NDUFA5 (Complex I) genes showed consistently reduced expression in all the brain regions of autistic subjects. This deficiency in ETC gene expression could have been induced by oxidative stress. They also reported decreased expression of metaxin 2 (MTX2), neurofilament, light polypeptide (NEFL) and solute carrier family 25, member 27 (SLC25A27) genes in all detected brain regions, which relate to diverse functions of mitochondria.[76] ATPase 6 and 8 genes are encoded by mtDNA and are important in energy generation. Piryaei *et al.*[102] reported that 50% of subjects with autism had mutations in ATPase 6 and 8 genes, which also suggests significant correlation between autism and mtDNA variations. All these studies indicate widespread mitochondrial ETC abnormalities in ASDs.

Defects in maintaining calcium homeostasis in autism

Mitochondria play a central role of maintaining Ca^{2+} homeostasis, which is essential for neurotransmitter release and neuronal excitability. Abnormal calcium signaling can also affect the normal mitochondrial function. Mitochondrial aspartate/glutamate carrier (AGC), which is physiologically activated by calcium, plays an important role in energy metabolism and neuron development by transporting glutamate into mitochondria.[77] Palmieri *et al.*[103] reported that AGC transport rates and expression levels of AGC1 protein were significantly higher in the brain of autistic subjects compared with the control subjects. Neocortical calcium levels also increased in these autistic subjects. Excessive calcium levels were responsible for high AGC1 activity and the activation of mitochondrial metabolism in autism.[77,103] Furthermore, the AGC1-encoding SLC25A12 gene is also reported to be associated with autism.[104,105]

Future Perspectives

ASDs are complex multifactorial disorders involving various genetic vulnerabilities interacting with environmental factors. It is difficult to diagnose ASDs before the age of two years because diagnosis is based on behavioral tests designed for older children. Several attempts have been made to find biomarkers for early screening of ASDs, but to date, none have been found. Oxidative stress plays a central role in the pathogenesis of ASDs, and connects with mitochondrial dysfunction, immune abnormalities, inflammation as well as environmental toxicant exposures. Several biomarkers of oxidative stress and mitochondrial dysfunction have been reported to be increased in the blood, urine and/or brain samples from the autistic children. Future studies should attempt to find biomarkers of oxidative stress and/or mitochondrial dysfunction as potential diagnostic or early detection markers of ASDs. Abnormal acyl-carnitine profiles may be one of the potential biomarkers for ASDs/MD children.[106]

Currently, there is no known medication to cure or treat autism. The main method to ameliorate the core behavioral deficits in autistic children is early intensive behavioral and educational interventional therapy. The development of non-pharmacological and pharmacological targets for therapeutic intervention of autism should be the target for future studies. One such approach is to identify agents that can enhance the antioxidant ability and mitochondrial functions in autism. Nutritional supplementation or vitamin supplements have been suggested to decrease oxidative damage and enhance ATP generation in autism.[107] Geier *et al.*[108] reported that L-carnitine therapy administered for three months significantly improved several clinical measurements of ASD severity, which might have been related to improved mitochondrial function. Effective methods to improve glutathione status by enhancing GSH synthesis or inhibiting its degradation may also help in therapeutic intervention in autism. Direct administration of reduced GSH has not been effective in other diseases because of poor absorption, low solubility and stability of GSH. Several glutathione precursors (N-acetyl L-cysteine, L-2-oxothiazolidine-4-carboxylic acid, γ-glutamyl cysteine) as well as the GSH analog YM737 have been successfully developed.[109,110] N-acetyl L-cysteine may provide cystine which is needed for GSH synthesis, and its therapeutic use is in developmental stages (phase II clinical trial).[111] A case report has shown improvement in the symptoms of the autistic child after oral treatment with N-acetyl L-cysteine.[112]

Executive Summary

- The markers of oxidative damage, i.e., lipid peroxidation, protein oxidation and DNA oxidation are increased in the plasma, urine and/or postmortem brain tissues from subjects with autism, whereas antioxidant defense is decreased in autism.
- Several studies have reported glutathione redox imbalance in autism. The levels of GSH and ratio of GSH/GSSG are reduced, while levels of GSSG are increased in the blood and brain samples from autistic subjects. The activities of enzymes involved in GSH homeostasis, i.e., GPx, GST and GCL, are impaired in autism. Furthermore, polymorphisms of GSH, GPx or GST genes have been associated with autism.

- The activities and protein levels or related gene expression of mitochondrial ETC complexes are decreased in the blood, muscle biopsy or brain tissues from subjects with autism.
- Mitochondrial DNA may have deletion, mutation or increased copy number in the subjects with autism. Mitochondrial dysfunction may also result from abnormal tryptophan metabolism.

In conclusion, oxidative stress and mitochondrial dysfunction play important roles in autism, and future studies are needed to understand underlying biochemical mechanisms and to identify potential therapeutic targets in autism.

References

1. Chauhan A, Chauhan V: Oxidative stress in autism. *Pathophysiology* 2006, **13**(3):171–181.
2. Chauhan A, Chauhan V, Brown WT: eds. Autism: *Oxidative Stress, Inflammation and Immune Abnormalities*. 2009, Florida: CRC Press: Taylor and Francis group(pp 1–427).
3. Sajdel-Sulkowska EM, Xu M, McGinnis W, Koibuchi N: Brain region-specific changes in oxidative stress and neurotrophin levels in autism spectrum disorders (ASD). *Cerebellum* 2011, **10**(1):43–48.
4. Ghanizadeh A, Akhondzadeh S, Hormozi M, Makarem A, Abotorabi-Zarchi M, Firoozabadi A: Glutathione-related factors and oxidative stress in autism, a review. *Curr Med Chem* 2012, **19**(23):4000–4005.
5. Main PA, Angley MT, O'Doherty CE, Thomas P, Fenech M: The potential role of the antioxidant and detoxification properties of glutathione in autism spectrum disorders: a systematic review and meta-analysis. *Nutr Metab (Lond)* 2012, **9**:35.
6. Frustaci A, Neri M, Cesario A, Adams JB, Domenici E, Dalla Bernardina B, Bonassi S: Oxidative stress-related biomarkers in autism: systematic review and meta-analyses. *Free Radic Biol Med* 2012, **52**(10):2128–2141.
7. Villagonzalo KA, Dodd S, Dean O, Gray K, Tonge B, Berk M: Oxidative pathways as a drug target for the treatment of autism. *Expert Opin Ther Targets* 2010, **14**(12):1301–1310.
8. Ghanizadeh A, Berk M, Farrashbandi H, Alavi Shoushtari A, Villagonzalo KA: Targeting the mitochondrial electron transport chain in autism, a systematic review and synthesis of a novel therapeutic approach. *Mitochondrion* 2013, **13**(5):515–519.
9. Rossignol DA, Frye RE: A review of research trends in physiological abnormalities in autism spectrum disorders: immune dysregulation,

inflammation, oxidative stress, mitochondrial dysfunction and environmental toxicant exposures. *Mol Psychiatry* 2012, **17**(4):389–401.

10. Chauhan A, Gu F, Chauhan V: Mitochondrial respiratory chain defects in autism and other neurodevelopmental disorders. *J Pediatr Biochem* 2012, **2**(4):213–223.
11. Lenaz G: The mitochondrial production of reactive oxygen species: mechanisms and implications in human pathology. *IUBMB Life* 2001, **52**(3–5):159–164.
12. Cadenas E, Davies KJ: Mitochondrial free radical generation, oxidative stress, and aging. *Free Radic Biol Med* 2000, **29**(3–4):222–230.
13. Boekema EJ, Braun HP: Supramolecular structure of the mitochondrial oxidative phosphorylation system. *J Biol Chem* 2007, **282**(1):1–4.
14. Dudkina NV, Eubel H, Keegstra W, Boekema EJ, Braun HP: Structure of a mitochondrial supercomplex formed by respiratory-chain complexes I and III. *Proc Natl Acad Sci USA* 2005, **102**(9):3225–3229.
15. Barja G: Mitochondrial oxygen radical generation and leak: sites of production in states 4 and 3, organ specificity, and relation to aging and longevity. *J Bioenerg Biomembr* 1999, **31**(4):347–366.
16. Muller FL, Liu Y, Van Remmen H: Complex III releases superoxide to both sides of the inner mitochondrial membrane. *J Biol Chem* 2004, **279**(47):49064–49073.
17. Hansen JM, Go YM, Jones DP: Nuclear and mitochondrial compartmentation of oxidative stress and redox signaling. *Annu Rev Pharmacol Toxicol* 2006, **46**:215–234.
18. Jones DP: Disruption of mitochondrial redox circuitry in oxidative stress. *Chem Biol Interact* 2006, **163**(1–2):38–53.
19. McCord JM: The evolution of free radicals and oxidative stress. *Am J Med* 2000, **108**(8):652–659.
20. Chinnery PF, Schon EA: Mitochondria. *J Neurol Neurosurg Psychiatry* 2003, **74**(9):1188–1199.
21. Mattson MP, Liu D: Energetics and oxidative stress in synaptic plasticity and neurodegenerative disorders. *Neuromolecular Med* 2002, **2**(2):215–231.
22. Szewczyk A, Wojtczak L: Mitochondria as a pharmacological target. *Pharmacol Rev* 2002, **54**(1):101–127.
23. Orth M, Schapira AH: Mitochondria and degenerative disorders. *Am J Med Genet* 2011, **106**:27–36.
24. Davis RE, Williams M: Mitochondrial function and dysfunction: an update. *J Pharmacol Exp Ther* 2012, **342**(3):598–607.
25. Ames A: CNS energy metabolism as related to function. *Brain Res Brain Res Rev* 2000, **34**(1–2):42–68.

26. Li Z, Okamoto K, Hayashi Y, Sheng M: The importance of dendritic mitochondria in the morphogenesis and plasticity of spines and synapses. *Cell* 2004, **119**(6):873–887.
27. Anderson S, Bankier A, Barrell B, de Bruijn M, Coulson A, Drouin J: Sequence and organization of the human mitochondrial genome. *Nature* 1981, **290**(5806):457–465.
28. Cotter D, Guda P, Fahy E, Subramaniam S: MitoProteome: mitochondrial protein sequence database and annotation system. *Nucleic Acids Res* 2004, **32**(Database issue):D463–467.
29. Shadel GS: Expression and maintenance of mitochondrial DNA: new insights into human disease pathology. *Am J Pathol* 2008, **172**(6):1445–1456.
30. Chauhan A, Audhya T, Chauhan V: Brain region-specific glutathione redox imbalance in autism. *Neurochem Res* 2012, **37**(8):1681–1689.
31. Chauhan A, Chauhan V, Brown WT, Cohen I: Oxidative stress in autism: increased lipid peroxidation and reduced serum levels of ceruloplasmin and transferrin — the antioxidant proteins. *Life Sci* 2004, **75**(21):2539–2549.
32. James SJ, Melnyk S, Jernigan S, Cleves MA, Halsted CH, Wong DH, Cutler P, Bock K, Boris M, Bradstreet JJ *et al.*: Metabolic endophenotype and related genotypes are associated with oxidative stress in children with autism. *Am J Med Genet B Neuropsychiatr Genet* 2006, **141B**(8):947–956.
33. Ming X, Stein TP, Brimacombe M, Johnson WG, Lambert GH, Wagner GC: Increased excretion of a lipid peroxidation biomarker in autism. *Prostaglandins Leukot Essent Fatty Acids* 2005, **73**(5):379–384.
34. Deth R, Muratore C, Benzecry J, Power-Charnitsky VA, Waly M: How environmental and genetic factors combine to cause autism: A redox/methylation hypothesis. *Neurotoxicology* 2008, **29**(1):190–201.
35. Meguid NA, Dardir AA, Abdel-Raouf ER, Hashish A: Evaluation of oxidative stress in autism: defective antioxidant enzymes and increased lipid peroxidation. *Biol Trace Elem Res* 2011, **143**(1):58–65.
36. Damodaran LP, Arumugam G: Urinary oxidative stress markers in children with autism. *Redox Rep* 2011, **16**(5):216–222.
37. Adams JB, Audhya T, McDonough-Means S, Rubin RA, Quig D, Geis E, Gehn E, Loresto M, Mitchell J, Atwood S *et al.*: Nutritional and metabolic status of children with autism vs. neurotypical children, and the association with autism severity. *Nutr Metab (Lond)* 2011, **8**(1):34.
38. Ghanizadeh A: Malondialdehyde, Bcl-2, superoxide dismutase and glutathione peroxidase may mediate the association of sonic hedgehog protein and oxidative stress in autism. *Neurochem Res* 2012, **37**(4):899–901.
39. Essa MM, Guillemin GJ, Waly MI, Al-Sharbati MM: Increased markers of oxidative stress in autistic children of the sultanate of Oman. *Biol Trace Elem Res* 2012, **147**(1–3):25–27.

40. Mostafa GA, El-Hadidi ES, Hewedi DH, Abdou MM: Oxidative stress in Egyptian children with autism: relation to autoimmunity. *J Neuroimmunol* 2010, **219**(1–2):114–118.
41. Rose S, Melnyk S, Pavliv O, Bai S, Nick TG, Frye RE, James SJ: Evidence of oxidative damage and inflammation associated with low glutathione redox status in the autism brain. *Transl Psychiatry* 2012, **2**:e134.
42. Sajdel-Sulkowska EM, Xu M, Koibuchi N: Increase in cerebellar neurotrophin-3 and oxidative stress markers in autism. *Cerebellum* 2009, **8**(3):366–372.
43. Frye RE, Delatorre R, Taylor H, Slattery J, Melnyk S, Chowdhury N, James SJ: Redox metabolism abnormalities in autistic children associated with mitochondrial disease. *Transl Psychiatry* 2013, **3**:e273.
44. Chauhan A, Audhya T, Chauhan V: Glutathione redox imbalance and increased DNA oxidation in specific brain regions in autism. *International Meeting for Autism Research (Abstract)* May 2011.
45. Al-Ayadhi LY: Relationship between sonic hedgehog protein, brain-derived neurotrophic factor and oxidative stress in autism spectrum disorders. *Neurochem Res* 2012, **37**(2):394–400.
46. Lu SC: Regulation of glutathione synthesis. *Mol Aspects Med* 2009, **30**(1–2): 42–59.
47. James SJ, Cutler P, Melnyk S, Jernigan S, Janak L, Gaylor DW, Neubrander JA: Metabolic biomarkers of increased oxidative stress and impaired methylation capacity in children with autism. *Am J Clin Nutr* 2004, **80**(6):1611–1617.
48. Geier DA, Kern JK, Garver CR, Adams JB, Audhya T, Geier MR: A prospective study of transsulfuration biomarkers in autistic disorders. *Neurochem Res* 2009, **34**(2):386–393.
49. Geier DA, Kern JK, Garver CR, Adams JB, Audhya T, Nataf R, Geier MR: Biomarkers of environmental toxicity and susceptibility in autism. *J Neurol Sci* 2009, **280**(1–2):101–108.
50. Bertoglio K, Jill James S, Deprey LJ, Brule N, Hendren RL: Pilot study of the effect of methyl B12 treatment on behavioral and biomarker measures in children with autism. *J Altern Complement Med* 2010, **16**(5):555–560.
51. Al-Yafee YA, Al-Ayadhi LY, Haq SH, El-Ansary AK: Novel metabolic biomarkers related to sulfur-dependent detoxification pathways in autistic patients of Saudi Arabia. *BMC Neurol* 2011, **11**:139.
52. James SJ, Rose S, Melnyk S, Jernigan S, Blossom S, Pavliv O, Gaylor DW: Cellular and mitochondrial glutathione redox imbalance in lymphoblastoid cells derived from children with autism. *FASEB J* 2009, **23**(8):2374–2383.
53. Slivka A, Spina MB, Cohen G: Reduced and oxidized glutathione in human and monkey brain. *Neurosci Lett* 1987, **74**(1):112–118.
54. Gu F, Chauhan V, Chauhan A: Impaired synthesis and antioxidant defense of glutathione in the cerebellum of autistic subjects: Alterations in the activities and protein expression of glutathione-related enzymes. *Free Radic Biol Med* 2013, **65C**:488–496.

55. Al-Gadani Y, El-Ansary A, Attas O, Al-Ayadhi L: Metabolic biomarkers related to oxidative stress and antioxidant status in Saudi autistic children. *Clin Biochem* 2009, **42**(10–11):1032–1040.
56. Sogut S, Zoroglu SS, Ozyurt H, Yilmaz HR, Ozugurlu F, Sivasli E, Yetkin O, Yanik M, Tutkun H, Savas HA *et al.*: Changes in nitric oxide levels and antioxidant enzyme activities may have a role in the pathophysiological mechanisms involved in autism. *Clin Chim Acta* 2003, **331**(1–2):111–117.
57. Yorbik O, Sayal A, Akay C, Akbiyik DI, Sohmen T: Investigation of antioxidant enzymes in children with autistic disorder. *Prostaglandins Leukot Essent Fatty Acids* 2002, **67**(5):341–343.
58. Pasca SP, Nemes B, Vlase L, Gagyi CE, Dronca E, Miu AC, Dronca M: High levels of homocysteine and low serum paraoxonase 1 arylesterase activity in children with autism. *Life Sci* 2006, **78**(19):2244–2248.
59. Vergani L, Lanza C, Rivaro P, Abelmoschi M, Genti S, Veneselli E, Minniti G, Grasselli E, Canesi L, Voci A: Metals, metallothioneins and oxidative stress in blood of autistic children. *Res Autism Spect Disord* 2011, **5**(1):286–293.
60. Zoroglu SS, Armutcu F, Ozen S, Gurel A, Sivasli E, Yetkin O, Meram I: Increased oxidative stress and altered activities of erythrocyte free radical scavenging enzymes in autism. *Eur Arch Psychiatry Clin Neurosci* 2004, **254**(3):143–147.
61. Golse B, Debray-Ritzen P, Durosay P, Puget K, Michelson AM: Alterations in two enzymes: superoxide dismutase and glutathione peroxidase in developmental infantile psychosis (infantile autism) (author's transl). *Rev Neurol (Paris)* 1978, **134**(11):699–705
62. Torsdottir G, Hreidarsson S, Kristinsson J, Snaedal J, Johannesson T: Ceruloplasmin, superoxide dismutase and copper in autistic patients. *Basic Clin Pharmacol Toxicol* 2005, **96**(2):146–148.
63. Krajcovicova-Kudlackova M, Valachovicova M, Mislanova C, Hudecova Z, Sustrova M, Ostatnikova D: Plasma concentrations of selected antioxidants in autistic children and adolescents. *Bratisl Lek Listy* 2009, **110**(4):247–250.
64. Ming X, Johnson WG, Stenroos ES, Mars A, Lambert GH, Buyske S: Genetic variant of glutathione peroxidase 1 in autism. *Brain Dev* 2010, **32**(2):105–109.
65. Williams TA, Mars AE, Buyske SG, Stenroos ES, Wang R, Factura-Santiago MF, Lambert GH, Johnson WG: Risk of autistic disorder in affected offspring of mothers with a glutathione S-transferase P1 haplotype. *Arch Pediatr Adolesc Med* 2007, **161**(4):356–361.
66. Bowers K, Li Q, Bressler J, Avramopoulos D, Newschaffer C, Fallin MD: Glutathione pathway gene variation and risk of autism spectrum disorders. *J Neurodev Disord* 2011, **3**(2):132–143.
67. Mari M, Morales A, Colell A, Garcia-Ruiz C, Fernandez-Checa JC: Mitochondrial glutathione, a key survival antioxidant. *Antioxid Redox Signal* 2009, **11**(11):2685–2700.

68. Rossignol DA, Frye RE: Mitochondrial dysfunction in autism spectrum disorders: a systematic review and meta-analysis. *Mol Psychiatry* 2012, **17**(3):290–314.
69. Frye RE, Rossignol DA: Mitochondrial dysfunction can connect the diverse medical symptoms associated with autism spectrum disorders. *Pediatr Res* 2011, **69**(5 Pt 2):41R–47R.
70. Tang G, Gutierrez Rios P, Kuo SH, Akman HO, Rosoklija G, Tanji K, Dwork A, Schon EA, Dimauro S, Goldman J *et al.*: Mitochondrial abnormalities in temporal lobe of autistic brain. *Neurobiol Dis* 2013, **54**:349–361.
71. Napoli E, Wong S, Giulivi C: Evidence of reactive oxygen species-mediated damage to mitochondrial DNA in children with typical autism. *Mol Autism* 2013, **4**(1):2.
72. Guevara-Campos J, Gonzalez-Guevara L, Puig-Alcaraz C, Cauli O: Autism spectrum disorders associated to a deficiency of the enzymes of the mitochondrial respiratory chain. *Metab Brain Dis* 2013, **28**(4):605–612.
73. Gu F, Chauhan V, Kaur K, Brown WT, Lafauci G, Wegiel J, Chauhan A: Alterations in mitochondrial DNA copy number and the activities of electron transport chain complexes and pyruvate dehydrogenase in the frontal cortex from subjects with autism. *Transl Psychiatry* 2013, **3**:e299.
74. Anitha A, Nakamura K, Thanseem I, Matsuzaki H, Miyachi T, Tsujii M, Iwata Y, Suzuki K, Sugiyama T, Mori N: Downregulation of the expression of mitochondrial electron transport complex genes in autism brains. *Brain Pathol* 2013, **23**(3):294–302.
75. Frye RE: Biomarkers of abnormal energy metabolism in children with autism spectrum disorder. *N A J Med Sci* 2012, **5**(3):141–147.
76. Anitha A, Nakamura K, Thanseem I, Yamada K, Iwayama Y, Toyota T, Matsuzaki H, Miyachi T, Yamada S, Tsujii M *et al.*: Brain region-specific altered expression and association of mitochondria-related genes in autism. *Mol Autism* 2012, **3**(1):12.
77. Napolioni V, Persico AM, Porcelli V, Palmieri L: The mitochondrial aspartate/glutamate carrier AGC1 and calcium homeostasis: physiological links and abnormalities in autism. *Mol Neurobiol* 2011, **44**(1):83–92.
78. Chauhan A, Gu F, Essa MM, Wegiel J, Kaur K, Brown WT, Chauhan V: Brain region-specific deficit in mitochondrial electron transport chain complexes in children with autism. *J Neurochem* 2011, **117**(2):209–220.
79. Skladal D, Halliday J, Thorburn DR: Minimum birth prevalence of mitochondrial respiratory chain disorders in children. *Brain* 2003, **126**(Pt 8): 1905–1912.
80. Bernier FP, Boneh A, Dennett X, Chow CW, Cleary MA, Thorburn DR: Diagnostic criteria for respiratory chain disorders in adults and children. *Neurology* 2002, **59**(9):1406–1411.

81. Lombard J: Autism: a mitochondrial disorder? *Med Hypotheses* 1998, **50**(6):497–500.
82. Guevara-Campos J, Gonzalez-Guevara L, Briones P, Lopez-Gallardo E, Bulan N, Ruiz-Pesini E, Ramnarine D, Montoya J: Autism associated to a deficiency of complexes III and IV of the mitochondrial respiratory chain. *Invest Clin* 2010, **51**(3):423–431.
83. Filiano JJ, Goldenthal MJ, Rhodes CH, Marin-Garcia J: Mitochondrial dysfunction in patients with hypotonia, epilepsy, autism, and developmental delay: HEADD syndrome. *J Child Neurol* 2002, **17**(6):435–439.
84. Chugani DC, Sundram BS, Behen M LML, Moore GJ: Evidence of altered energy metabolism in autistic children. *Prog Neuropsychopharmacol Biol Psychiatry* 1999, **23**(4):635–641.
85. Minshew NJ, Goldstein G, Dombrowski SM, Panchalingam K, Pettegrew JW: A preliminary 31P MRS study of autism: evidence for undersynthesis and increased degradation of brain membranes. *Biol Psychiatry* 1993, **33**(11–12):762–773.
86. Mostafa GA, El-Gamal HA, El-Wakkad ASE, El-Shorbagy OE, Hamza MM: Polyunsaturated fatty acids, carnitine and lactate as biological markers of brain energy in autistic children. *Int J Child Neuropsychiatry* 2005, **2**(2):179–188.
87. Weissman JR, Kelley RI, Bauman ML, Cohen BH, Murray KF, Mitchell RL, Kern RL, Natowicz MR: Mitochondrial disease in autism spectrum disorder patients: a cohort analysis. *PLoS One* 2008, **3**(11):e3815.
88. Oliveira G, Diogo L, Grazina M, Garcia P, Ataide A, Marques C, Miguel T, Borges L, Vicente AM, Oliveira CR: Mitochondrial dysfunction in autism spectrum disorders: a population-based study. *Dev Med Child Neurol* 2005, **47**(3):185–189.
89. Correia C, Coutinho AM, Diogo L, Grazina M, Marques C, Miguel T, Ataide A, Almeida J, Borges L, Oliveira C *et al.*: Brief report: High frequency of biochemical markers for mitochondrial dysfunction in autism: no association with the mitochondrial aspartate/glutamate carrier SLC25A12 gene. *J Autism Dev Disord* 2006, **36**(8):1137–1140.
90. Filipek PA, Juranek J, Smith M, Mays LZ, Ramos ER, Bocian M, Masser-Frye D, Laulhere TM, Modahl C, Spence MA *et al.*: Mitochondrial dysfunction in autistic patients with 15q inverted duplication. *Ann Neurol* 2003, **53**(6):801–804.
91. Giulivi C, Zhang YF, Omanska-Klusek A, Ross-Inta C, Wong S, Hertz-Picciotto I, Tassone F, Pessah IN: Mitochondrial dysfunction in autism. *JAMA* 2010, **304**(21):2389–2396.
92. Boccuto L, Chen CF, Pittman AR, Skinner CD, McCartney HJ, Jones K, Bochner BR, Stevenson RE, Schwartz CE: Decreased tryptophan metabolism in patients with autism spectrum disorders. *Mol Autism* 2013, **4**(1):16.

93. Essa MM, Subash S, Braidy N, Al-Adawi S, Lim CK, Manivasagam T, Guillemin GJ: Role of NAD(+), oxidative stress, and tryptophan metabolism in autism spectrum disorders. *Int J Tryptophan Res* 2013, **6**(Suppl 1):15–28.
94. Poling JS, Frye RE, Shoffner J, Zimmerman AW: Developmental regression and mitochondrial dysfunction in a child with autism. *J Child Neurol* 2006, **21**(2):170–172.
95. Tsao CY, Mendell JR: Autistic disorder in 2 children with mitochondrial disorders. *J Child Neurol* 2007, **22**(9):1121–1123.
96. Ezugha H, Goldenthal M, Valencia I, Anderson CE, Legido A, Marks H: 5q14.3 deletion manifesting as mitochondrial disease and autism: case report. *J Child Neurol* 2010, **25**(10):1232–1235.
97. Frye RE, Naviaux RK: Autistic disorder with complex IV overactivity: A new mitochondrial syndrome. *J Ped Neurol* 2011, **9**(4):427–434.
98. Lee HC, Wei YH: Mitochondrial biogenesis and mitochondrial DNA maintenance of mammalian cells under oxidative stress. *Int J Biochem Cell Biol* 2005, **37**(4):822–834.
99. Liu CS, Tsai CS, Kuo CL, Chen HW, Lii CK, Ma YS, Wei YH: Oxidative stress-related alteration of the copy number of mitochondrial DNA in human leukocytes. *Free Radic Res* 2003, **37**(12):1307–1317.
100. Pons R, Andreu AL, Checcarelli N, Vila MR, Engelstad K, Sue CM, Shungu D, Haggerty R, de Vivo DC, DiMauro S: Mitochondrial DNA abnormalities and autistic spectrum disorders. *J Pediatr* 2004, **144**(1):81–85.
101. Hadjixenofontos A, Schmidt MA, Whitehead PL, Konidari I, Hedges DJ, Wright HH, Abramson RK, Menon R, Williams SM, Cuccaro ML *et al.*: Evaluating mitochondrial DNA variation in autism spectrum disorders. *Ann Hum Genet* 2013, **77**(1):9–21.
102. Piryaei F, Houshmand M, Aryani O, Dadgar S, Soheili ZS: Investigation of the mitochondrial ATPase 6/8 and tRNA(Lys) genes mutations in autism. *Cell J* 2012, **14**(2):98–101.
103. Palmieri L, Papaleo V, Porcelli V, Scarcia P, Gaita L: Altered calcium homeostasis in autism-spectrum disorders: evidence from biochemical and genetic studies of the mitochondrial aspartate/glutamate carrier AGC1. *Mol Psychiatry* 2010, **15**(1):38–52.
104. Ramoz N, Reichert JG, Smith CJ, Silverman JM, Bespalova IN, Davis KL, Buxbaum JD: Linkage and association of the mitochondrial aspartate/glutamate carrier SLC25A12 gene with autism. *Am J Psychiatry* 2004, **161**(4):662–669.
105. Segurado R, Conroy J, Meally E, Fitzgerald M, Gill M, Gallagher L: Confirmation of association between autism and the mitochondrial aspartate/glutamate carrier SLC25A12 gene on chromosome 2q31. *Am J Psychiatry* 2005, **162**(11):2182–2184.

106. Frye RE, Melnyk S, Macfabe DF: Unique acyl-carnitine profiles are potential biomarkers for acquired mitochondrial disease in autism spectrum disorder. *Transl Psychiatry* 2013, **3**:e220.
107. Adams JB, Audhya T, McDonough-Means S, Rubin RA, Quig D, Geis E, Gehn E, Loresto M, Mitchell J, Atwood S *et al.*: Effect of a vitamin/mineral supplement on children and adults with autism. *BMC Pediatr* 2011, **11**:111.
108. Geier DA, Kern JK, Davis G, King PG, Adams JB, Young JL, Geier MR: A prospective double-blind, randomized clinical trial of levocarnitine to treat autism spectrum disorders. *Med Sci Monit* 2011, **17**(6):PI15–23.
109. Schulz JB, Lindenau J, Seyfried J, Dichgans J: Glutathione, oxidative stress and neurodegeneration. *Eur J Biochem* 2000, **267**(16):4904–4911.
110. Wu JH, Batist G: Glutathione and glutathione analogues; therapeutic potentials. *Biochim Biophys Acta* 2013, **1830**(5):3350–3353.
111. Kumar B, Prakash A, Sewal RK, Medhi B, Modi M: Drug therapy in autism: a present and future perspective. *Pharmacol Rep* 2012, **64**(6):1291–1304.
112. Ghanizadeh A, Derakhshan N: N-acetylcysteine for treatment of autism, a case report. *J Res Med Sci* 2012, **17**(10):985–987.

Chapter 17

Maternal Autoantibodies in Autism Spectrum Disorder

Karen L. Jones and Judy Van de Water*

Abstract

Despite increasing prevalence rates of autism spectrum disorder (ASD) globally, the underlying causes of the disorder have yet to be identified. However, immune dysregulation has been reported in some individuals with ASD as well as in their family members. In particular, IgG antibodies reactive to fetal brain proteins have been reported in a subset of mothers of children with ASD by several investigators. This chapter provides a summary of the research conducted on these maternal autoantibodies and their potential role in the development of ASD. In this chapter, we will explore the current research on maternal autoantibodies as well as potential therapeutic strategies that may be developed for this subset of ASD.

Keywords: Maternal immunity; Autoantibodies; Fetal brain proteins; Neuroimmunology.

Introduction

Autism spectrum disorders (ASD) are a set of neurodevelopmental disorders classified by deficits in social interaction, social communication, and the presence of restricted and repetitive interests and

*Correspondence to: Department of Rheumatology, Allergy, and Clinical Immunology, M.I.N.D. Institute, University of California, Davis; Sacramento, CA 95817. Email: Karen L. Jones: drkjones@ucdavis.edu and Judy Van de Water: javandewater@ucdavis.edu

behaviors.[1] Currently, the prevalence rate of ASD is thought to be as high as one individual in every 88.[2] While the etiology of ASD is unknown, genetics is thought to play a large role in the disorder.[3] Previous twin studies have shown the monozygotic concordance rate of ASD to be approximately 70%, and dizygotic concordance rates to be approximately 5%.[4,5] While genetics play a large role in the etiology of ASD, evidence suggests that environmental factors also contribute to the disorder. Furthermore, a more recent twin study has suggested that susceptibility to autism has a lesser genetic heritability component and a greater environmental component than previously thought.[6]

One potential non-genetic contributing factor is immune system dysregulation, which has been frequently described in individuals with ASD as well as their family members. For example, ASD has been significantly correlated with increased rates of autoimmunity in family members, particularly mothers.[7–9] Additionally, prenatal immune challenges, such as bacterial or viral infections, have also been related to behavioral disorders in offspring, including autism and schizophrenia.[10,11] Some other irregularities in autism include decreased numbers of B and T cells, reduced lymphocyte responses to stimulation,[12–14] increased numbers of monocytes,[15] and skewed cytokine and chemokine profiles.[16–21] Most notably, mothers of children with ASD have been reported to harbor antibodies reactive to fetal brain proteins.[22] Interestingly, ASD children have also been found to harbor antibodies directed against various brain components.[23–25] This chapter will review the current literature concerning maternal antibodies in ASD and potential therapeutic approaches for this subset of ASD.

Immune Review

Immune system overview

The immune system is intricately linked to every organ system in the body, where it demonstrates great versatility and specialization in different locations. For example, immune activation in the bacterial-filled gastrointestinal tract is very different from that found in the sterile blood stream.

The immune system is made up of "innate" and "adaptive" branches. The innate branch is the first line of bodily defense. It responds rapidly and nonspecifically to injury or infection by recognizing the telltale signs of "danger" that are likely foreign to the host and potentially damage-causing agents. Cells such as neutrophils, dendritic cells and macrophages engulf pathogens, and secrete immune-stimulating molecules including cytokines, chemokines, and reactive oxygen species (ROS). The adaptive immune system is dominated by highly specialized B and T cells that drive a focused and highly efficient response to specific components of the invading agents, known as "antigens." When activated, B cells make target-specific antibodies, while T cells carry out specific helper and cytotoxic functions. Natural killer (NK) cells are specialized members of the innate immune system that respond rapidly to viral infections and tumors by secreting cytokines and cytolytic molecules. These cells have additional importance in stimulating the adaptive immune system, maintaining pregnancy, and regulating autoimmune reactions.[26]

Maternal immunity and the gestational environment

The maternal immune response generates factors that have a direct impact on the gestational environment. Factors including cytokines, chemokines, and antibodies produced during pregnancy can have developmental consequences for the fetus. Under normal conditions, the maternal immune system is uniquely regulated during pregnancy to maintain a pathogen-free, yet non-inflammatory, environment for the developing offspring.[27,28] While the placenta provides a barrier between the mother and the developing fetus, nutrition and certain immune factors are allowed to transfer to the fetus. Among the components of the maternal immune system that enter the fetal compartment, immunoglobulin G (IgG) antibodies transfer at high concentrations beginning around mid-gestation,[29] resulting in circulating IgG levels in the newborn that exceed those in maternal circulation owing to neonatal FC receptor-mediated active transport across the placenta. These IgG antibodies, which are maternal in origin, are known to provide the newborn with a passive defense mechanism against different pathogens. Maternal IgG is also transferred to the newborn during lactation through breast milk (although at much lower levels than IgA), enabling maternal IgG to persist in the newborn through early infancy.[30] However, this physiological process can also have some disadvantages under certain pathological conditions, such as in maternal myasthenia gravis, which can lead to transient neonatal myasthenia gravis and, in rare cases, arthrogryposis multiplex congenital, an often fatal disorder characterized by severe joint contractures in the offspring.[31,32]

Under normal conditions, antibodies are unable to cross the blood-brain barrier (BBB) to access the brain. However, the BBB is not fully mature during early brain development and thus permits maternal antibodies access to the fetal brain.[33] Prenatal exposure to maternal antibodies that bind to fetal brain has been suggested as a trigger for abnormal brain development.[34] The pathology resulting from such brain-reactive maternal antibodies is thought to arise through one of three established mechanisms: (1) blockade or interference with the function of target molecules,[35] (2) acting as an agonist and activating a receptor or signaling cascade,[36] or (3) induction of inflammation and the consequent effects on neurodevelopment.[37] Interestingly, maternal immune activation (MIA) during gestation has been linked with autism in offspring.[10,11] During pregnancy, some maternal cytokines may cross the placenta, as in the case of IL-6,[38–40] or act on placental cells to stimulate the downstream production of immune mediators in the fetal compartment.[41] It is now widely accepted that cytokines are involved in several diverse aspects of normal brain development, including proliferation and differentiation of neural and glial cells.[42] However, fluctuations in the concentrations of these molecules, either increased or decreased, can change the neurodevelopmental trajectory, resulting in altered brain morphology and behavior in the offspring.

Maternal Autoantibodies in Autism

Observational studies

Interest in the association between maternal antibodies and ASD first arose with a study by Warren and colleagues, which found that exposure to fetal lymphocyte antigens elicited an antibody response in some mothers of children with ASD.[43] These maternal antibodies were assumed to be stimulated through exposure to paternal antigens during earlier pregnancy and/or parturition in much the same way in which the production of anti-Rh antibodies are stimulated.

After an intriguing pilot animal study furthered the connection between ASD and maternal antibodies,[44] several studies have expanded this earlier work to more fully characterize these antibodies. Using samples from a large, well-characterized case-control cohort, Braunschweig *et al.* demonstrated antibody reactivity against fetal brain proteins (with molecular weights of 37 and 73 kDa) in plasma from mothers of children with ASD (11.5%) but not in mothers of typically developing children (0%).[45] A similar pilot study by Zimmerman *et al.* looked at 11 mothers of children with autism and 10 mothers of typically developing children, examining maternal serum for autoantibody reactivity to prenatal, postnatal, and adult brain proteins derived from rats.[46] Reactivity to fetal brain tissue was observed in the mothers of children in distinct patterns that were not observed in control mothers. Furthermore, reactivity was not observed in postnatal or adult rat brains, and immunoblotting showed that fetal brain reactivity was not directed at previously identified neuronal targets of autoimmunity.

Continuing upon these findings in a slightly larger sample set, using 100 mothers of children with autism and 100 mothers of neurotypical control children, serum autoantibody reactivity was tested against human and rodent fetal brain tissue.[47] Significantly more mothers of children with autism demonstrated reactivity to a 36 kDa antigen in both human and rodent fetal brain tissues than control mothers.[47] Additionally, maternal reactivity to proteins at 39 kDa in human fetal brain tissue and to proteins at 73 kDa in rodent fetal brain tissue was greater in mothers of children with autism than in control mothers.[47] Interestingly, a significant association between maternal reactivity to fetal brain tissue and having a child with a regressive form of autism was observed in both the Braunschweig and Singer studies.[45,47] Minor variances in band definition between the two studies might be due to different tissue and Western blot protocols used for analyses.

Most recently, researchers have confirmed that mothers of children with autism are significantly more likely to harbor anti-brain antibodies than control women using the largest cohort reported thus far.[48] Additionally, ASD mothers reactive to brain proteins were also more likely to harbor anti-nuclear autoantibodies as well as have an autoimmune disease, suggesting that these brain-reactive antibodies may be associated with autoimmunity. However, it is important to note that the protein bands identified on Western blots in this study were variable, and thus the authors were unable to comment on specific band reactivity.[48]

A central critique of studies linking maternal fetal brain-reactive antibodies and autism has been that specimen collection typically occurs after a child's diagnosis, making it unclear whether antibodies present during collection were present during pregnancy. To address this issue, archived mid-pregnancy blood specimens in a large case-control study were screened for reactivity toward the fetal brain.[49] The gestational presence of maternal fetal brain-reactive antibodies in mothers of children with autism was confirmed in this study, particularly with reactivity to bands at 39 and 73 kDa.

Expanding upon previous findings, a subsequent study by Braunschweig *et al.* looked at an additional 560 mothers of children with autism and of neurotypical control children, measuring the reactivity of maternal antibody to fetal brain proteins derived from Rhesus macaques.[50] Results from this study were consistent with previous observations, with maternal reactivity to bands at 37 and 73 kDa in combination being significantly associated with a diagnosis of autism in the child, and reactivity to the 37/73 kDa band pair remaining exclusive to mothers of children with autism and not control mothers. Interestingly, refinements in immunoblotting techniques led to the observation of another immunoreactive band at approximately 39 kDa. Paired reactivity to 39 and 73 kDa was found to be strongly associated with a broader diagnosis of ASD. Behavioral tests revealed that ASD children of mothers with reactivity to the 37 and 73 kDa bands had significantly impaired expressive language, while ASD children of mothers reactive to both 39 and 73 kDa bands had increased irritability scores.[50] More recently, male ASD children born to mothers reactive to the 37/73 band pair have been found to have more extreme brain enlargement relative to both typically developing and non-37/73 ASD male children.[51] These physical results suggest that reactivity to fetal brain proteins by maternal autoantibodies may play a role in the pathogenesis of ASD for a subset of children.

As the studies mentioned above strongly suggest a role of these maternal antibodies in the etiology of ASD, the identification of the target antigens for maternal autoantibody related (MAR) autism was a critical step in advancing this area of autism research. In the first study of its type, Braunschweig and colleagues successfully determined the identity of several proteins targeted by the candidate antigens in fetal brain tissue.[52] Maternal plasma samples found to react against fetal brain were subsequently used for proteomic antigen identification. Through tandem mass spectrometry peptide sequencing, the autoantigens (or target proteins) were identified as: lactate dehydrogenase (LDH) A and B, guanine deaminase (cypin), stress-induced phosphoprotein 1 (STIP1), collapsin response mediator proteins (CRMPs) 1 and 2, and Y-box binding protein 1 (YBX1).[52] The original 37 kDa antigens were found to be LDH A and LDH B, the 39 kDa antigen to be YBX1, the upper and lower 73 kDa antigens to be STIP1 and CRMP1 respectively, and a 44 kDa protein not noted in previous studies to be guanine deaminase or cypin. Maternal reactivity to any of the antigens, individually or in combination, was highly significantly associated with an outcome of ASD in the child, with an odds ratio of 3.26. The most common pattern of reactivity was to the LDH/STIP1/CRMP1 combination (5% ASD mothers vs 0% control mothers), with several other additional patterns that were only found in the mothers of children with ASD. Additionally, an increase in stereotypical behavior was observed in children of mothers possessing antibodies reactive to LDH in any context, CRMP1, as well as combined reactivity to LDH and STIP1, or LDH/STIP1/CRMP1 in comparison to ASD children of non-reactive mothers. Furthermore, when all antigen patterns were combined, a total of nearly 23% of mothers of ASD children had one of the ASD-specific autoantibody patterns containing two or more of the target proteins in comparison to only 1% of control mothers. Interestingly, each of the identified autoantigens is expressed at significant levels in the human fetal brain and has an established role in neurodevelopment. Thus, the identification of specific IgG reactivity to these identified proteins, which are highly expressed in the developing brain, further supports the potential role of maternal autoantibodies in the etiology of a sub-type of ASD.

Animal models

While observational studies have certainly revealed an association between maternal antibodies reactive to fetal brain proteins and autism, whether these antibodies directly cause autism cannot be ethically demonstrated in human subjects. To gather support in favor of this hypothesis, various animal model studies have been conducted. In an early pilot study, plasma from a mother of a child with autism was injected into pregnant mice, resulting in offspring with altered exploratory behavior and motor coordination.[44] In a separate mouse model, pregnant mice were administered with either IgG from mothers of children with autism or control mother IgG during embryonic days 13 through 18.[37] Offspring exposed to maternal autism IgG displayed abnormal behaviors relative to those exhibited by control offspring, including an increase in anxious-like behavior, hyperactivity, greater magnitude of startle to acoustic stimuli, and impaired social interaction.[37]

A more recent mouse model examined the effects of gestational exposure to a single, intravenous dose of purified brain-reactive

IgG antibodies from mothers of children with autism that display reactivity to both the 37 and 73 kDa fetal brain antigens or from control mothers.[53] Injected human IgG was detected at the highest relative concentration in the fetal brain and liver, supporting a direct mechanism of interaction between maternal IgG and the developing fetal brain. Offspring exposed to maternal IgG positive for 37/73 kDa bands displayed slower motor and sensory development during pre-weaning periods. Additionally, behavioral testing of juvenile offspring revealed that these offspring displayed altered social and anxious-like behaviors in comparison to control offspring.[53] In a recent study designed to examine the direct effects of the maternal antibodies on neuronal development, researchers administered mid-gestational intracerebroventricular injections of purified IgG of both 37/73 reactive ASD mothers and control mothers in embryonic mice. Fetal brains exposed to ASD maternal autoantibodies displayed increased proliferation in the subventricular zone of the embryonic cerebral cortex, increased total brain size and weight, increased cortical neuron size, and altered stereotypical and social behaviors in comparison to controls.[54] The results from each of these models demonstrate that gestational exposure of maternal IgG reactive to the 37 and 73 kDa bands produces physiological and behavioral effects in offspring, further supporting the role of these maternal antibodies in the etiology of ASD.

While there are many benefits to studying mouse models of ASD, the core features of ASD are difficult to assess in such model systems. In contrast, non-human primate models offer unique translational advantages, as they demonstrate many features of human physiology, anatomy, and behavior that mouse models are unable to fully recapitulate. In the first non-human primate study of maternal antibodies in autism, Martin *et al.* administered purified pooled IgG from mothers of children with autism or control mothers three separate times during gestation via intravenous injection to pregnant rhesus macaques.[55] While gestational exposure to the ASD maternal antibodies did not result in any profound differences in social behavior, these offspring demonstrated hyperactivity and dramatic whole-body stereotypies, such as repetitive back flipping and pacing, in comparison to animals exposed to control IgG.

In a follow-up study, pregnant rhesus monkeys were administered purified IgG from either mothers of children with autism with antibody reactivity specific to 37 and 73 kDa fetal brain proteins or from control mothers.[56] In analyzing offspring behavior, monkeys prenatally exposed to the 37/73 kDa antibodies displayed aberrant social behavior in comparison to control offspring. To study potential changes in brain development in the exposed offspring, longitudinal magnetic resonance imaging (MRI) was also conducted with emphasis on the male offspring. At two years of age, the 37/73 kDa antibody exposed male offspring had significantly greater total brain, frontal lobe, and occipital lobe volumes than that of control male offspring. Further, these offspring had significantly more total white matter volume than control offspring.[56] These results mirror those found in the male children of mothers with antibodies to the 37/73 kDa proteins described above.

The recapitulation of an autistic-like phenotype in both mouse and non-human primate models of MAR ASD suggests that these maternal antibodies play role in the pathology of ASD. How these antibodies arise and the mechanisms through which they interfere with neurodevelopment remain to be elucidated. While it is possible that binding of these maternal antibodies to fetal brain proteins directly result in altered neurodevelopment, maternal antibodies may alternatively serve as a biomarker of cell destruction. Despite the lack of mechanistic animal studies of MAR ASD, current animal models offer translatable findings to further support the role of these antibodies in the etiology of ASD.

Maternal antibodies and genetic susceptibility

As genetics is thought to play a large role in the etiology of ASD, certain genetic polymorphisms related to immune functioning may contribute to the potential mechanism through which these maternal antibodies arise. One such genetic risk factor is a functional polymorphism in the tyrosine kinase MET receptor. In addition to its roles in brain development and gastrointestinal repair, MET serves as a key regulator of immune responsiveness and influences self-tolerance in a dose-dependent fashion.[57,58] A functional variant in the 5′ promoter of the *MET* gene has been found to be significantly associated with ASD in several independent genetic studies.[59–62] This variant, rs1858830, is a common G-to-C single-nucleotide polymorphism (SNP); in ASD, the 'C' allele is inherited by individuals with ASD more often than predicted by chance and is more common in individuals with ASD than in the general population. In a recent study by Heuer *et al.*, the 'C' allele was also found to be significantly associated with the presence of maternal autoantibodies to fetal brain proteins in mothers of children with autism in comparison to control mothers.[63] Additionally, the authors found that the 'C' allele was associated with decreased MET protein expression and decreased levels of IL-10.[63] A decrease in IL-10, an immunosuppressive cytokine, may result in a breakdown of maternal immune tolerance and thus could predispose to the production of autoantibodies. Currently, no other studies have addressed the relationship between these ASD-specific maternal antibodies and genetic risk factors.

Future Directions

While it appears that maternal antibodies reactive to fetal brain proteins play a role in the etiology of a subset of ASD, future studies are necessary to further this area of research. For example, while the initial identification of the specific targeted proteins is promising, studies are needed to validate the true incidence of maternal antibodies to these proteins and their relevance to ASD. Additionally, further animal experiments are necessary to better understand if maternal autoantibody reactivity to these identified proteins produces behavioral and physiological alterations in offspring similar to those seen in individuals with ASD. These studies would also allow researchers to investigate the underlying mechanism of MAR ASD, leading to the future development of targeted therapeutics for this subset of ASD.

Unfortunately, there are no biological therapies currently available for MAR ASD. Prior to the development of antibody-inhibiting therapies, however, an antibody-based screening test needs to be developed so that women with these reactive antibodies can be identified. While it is uncertain whether these maternal antibodies are directly mediating a pathological effect, animal studies of MAR ASD support the notion that they are pathologically significant, as discussed above. Therefore, therapeutics that may prevent these pathogenic antibodies from accessing the fetal compartment would be ideal. To prevent their access, antibodies may be removed from circulation, either specifically or non-specifically. While we are a long way from developing any type of therapeutic for MAR autism, we hope that through careful clinical studies, a thorough understanding of the pathologic mechanisms related to these antibodies, and a clinically relevant animal model will move this concept towards that ultimate goal.

Conclusions

In conclusion, numerous studies support the role of maternal autoantibodies reactive to fetal brain proteins in the etiology of a subset of ASD (Table 1). Further, these autoantibodies seem to be reactive specifically to proteins at approximately 36–39 and 73 kDa, with researchers initially identifying these proteins as LDH A and B, STIP1, CRMP 1 and 2, guanine deaminase (cypin), and YBX1. However, future experiments are necessary to determine whether these antibodies play a direct role in the pathogenesis of MAR ASD as well as to elucidate the potential mechanisms through which they may act. Understanding the potential pathogenic role of these antibodies will be critical in the development of future therapeutic strategies for use in this subset of ASD.

Table 1. Compiled results of observational and behavioral studies associating maternal antibodies with autism

Antibody Target	Study Description	Year Published
Observational Studies		
Child/parental lymphocytes[43]	Mother of ASD subjects demonstrated increased antibody reactivity against lymphocytes of their ASD children	1990
Rat fetal/juvenile/adult brain[46]	Mothers of ASD subjects were shown to have IgG reactivity to rat fetal brain proteins	2007
Human fetal brain[45]	Plasma from mothers of ASD subjects showed IgG reactivity to human fetal brain proteins at 37 and 73 kDa, unlike mothers of controls	2007
Fetal/adult human and rat brain[47]	Mothers of ASD subjects were confirmed to have IgG reactivity to human fetal brain proteins at 36, 39 and 73 kDa	2008
Human fetal brain[49]	Serum collected during the second trimester from mothers of ASD subjects showed increased IgG reactivity to human fetal brain proteins at 39 and 73 kDa, unlike control mother serum	2008
Mouse fetal/adult brain[48]	Mothers of ASD subjects were confirmed to have IgG reactivity to fetal brain antibodies as well as anti-nuclear autoantibodies, unlike mothers of controls	2013
Fetal human/monkey brain[50]	Mothers of AU and ASD subjects were confirmed to have IgG reactivity to fetal brain proteins at 37/73 kDa (AU mothers) and as 39/73 kDa (ASD mothers) pairs	2012

(*Continued*)

Table 1. (*Continued*)

Antibody Target	Study Description	Year Published
Fetal monkey brain[63]	Mothers of ASD subjects reactive to fetal brain proteins at 37/73 kDa were found to inherit a functional variant of the *MET* promoter C allele more often than non-reactive ASD mothers or controls	2011
Fetal monkey brain[51]	Male ASD children of mothers reactive to 37/73 kDa fetal brain proteins were shown to have more extreme brain enlargement relative to control children	2013
Fetal monkey brain[52]	Identity of the fetal brain proteins were reported as: LDH A and B, cypin, STIP1, YBX1, and CRMP 1 and 2	2013

Species	Study Description	Year Published
Animal Models		
Mouse[44]	Serum from a mother of an ASD subject was injected into pregnant mice. Resulting offspring displayed altered exploratory behavior and motor coordination in comparison to controls.	2003
Mouse[37]	Pregnant mice were injected with IgG from mothers of ASD subjects or of controls. ASD IgG exposed offspring displayed a variety of autistic-like behaviors, unlike control offspring.	2009
Mouse[55]	Purified pooled IgG from mothers of ASD subjects or of control subjects was injected into pregnant rhesus monkeys. ASD IgG exposed offspring displayed profound differences in social and stereotypical behaviors.	2008

(*Continued*)

Table 1. *(Continued)*

Species	Study Description	Year Published
Mouse[53]	Pregnant mice were injected with purified IgG from either ASD mothers reactive to fetal brain proteins at 37/73 kDa bands or control mothers. 37/73 IgG offspring displayed impaired motor and sensory development as well as altered social and anxiety behaviors.	2012
Mouse[54]	Purified IgG of 37/73 ASD mothers or control mothers was injected into the ventricles of embryonic mice during gestation. Offspring prenatally exposed to ASD maternal antibodies displayed altered brain morphology and behaviors compared to control injected offspring.	2013

Executive Summary

- Immune dysregulation has been noted in both individuals with ASD as well as their family members.
- Gestational exposure to maternal antibodies reactive to fetal brain proteins may be involved in some cases of autism.
- While the exact reasons women develop these antibodies remains to be determined, research suggests that genetic susceptibility factors may combine with anti-brain antibodies to result in ASD in the offspring.
- The exact mechanisms underlying maternal antibodies in autism remain ambiguous, and are an active area of investigation.

References

1. APA: Diagnostic and statistical manual of mental disorders: DSM-IV text revision. *American Psychiatric Association* 2000, Washington, DC.
2. Centers for Disease Control and Prevention: Prevalence of Autism Spectrum Disorders — Autism and Developmental Disabilities Monitoring Network, 14 Sites, United States, 2008. *MMWR Surveill Summ* 2012, **61**(SS03):1–19.
3. Fombonne E: Epidemiology of autistic disorder and other pervasive developmental disorders. *J Clin Psychiatry* 2005, **66**(Suppl) **10**:3–8.
4. Bailey A, Le Couteur A, Gottesman I, Bolton P, Simonoff E, Yuzda E, Rutter M: Autism as a strongly genetic disorder: evidence from a British twin study. *Psychol Med* 1995, **25**(1):63–77.
5. Folstein S, Rutter M: Infantile autism: a genetic study of 21 twin pairs. *J Child Psychol Psychiatry* 1977, **18**(4):297–321.
6. Hallmayer J, Cleveland S, Torres A, Phillips J, Cohen B, Torigoe T, Miller J, Fedele A, Collins J, Smith K *et al.*: Genetic heritability and shared environmental factors among twin pairs with autism. *Arch Gen Psychiatry* 2011, **68**(11):1095–1102.
7. Atladottir HO, Pedersen MG, Thorsen P, Mortensen PB, Deleuran B, Eaton WW, Parner ET: Association of family history of autoimmune diseases and autism spectrum disorders. *Pediatrics* 2009, **124**(2):687–694.
8. Croen LA, Grether JK, Yoshida CK, Odouli R, Van de Water J: Maternal autoimmune diseases, asthma and allergies, and childhood autism spectrum disorders: a case-control study. *Arch Pediatr Adolesc Med* 2005, **159**(2):151–157.
9. Sweeten TL, Bowyer SL, Posey DJ, Halberstadt GM, McDougle CJ: Increased prevalence of familial autoimmunity in probands with pervasive developmental disorders. *Pediatrics* 2003, **112**(5):e420.

10. Atladottir HO, Thorsen P, Ostergaard L, Schendel DE, Lemcke S, Abdallah M, Parner ET: Maternal infection requiring hospitalization during pregnancy and autism spectrum disorders. *J Autism Dev Disord* 2010, **40**(12):1423–1430.
11. Shi L, Fatemi SH, Sidwell RW, Patterson PH: Maternal influenza infection causes marked behavioral and pharmacological changes in the offspring. *J Neurosci* 2003, **23**(1):297–302.
12. Stubbs EG, Crawford ML: Depressed lymphocyte responsiveness in autistic children. *J Autism Child Schizophr* 1977, **7**(1):49–55.
13. Plioplys AV, Greaves A, Kazemi K, Silverman E: Lymphocyte function in autism and Rett syndrome. *Neuropsychobiology* 1994, **29**(1):12–16.
14. Warren RP, Foster A, Margaretten NC: Reduced natural killer cell activity in autism. *J Am Acad Child Adolesc Psychiatry* 1987, **26**(3):333–335.
15. Sweeten TL, Posey DJ, McDougle CJ: High blood monocyte counts and neopterin levels in children with autistic disorder. *Am J Psychiatry* 2003, **160**(9):1691–1693.
16. Ashwood P, Krakowiak P, Hertz-Picciotto I, Hansen R, Pessah I, Van de Water J: Elevated plasma cytokines in autism spectrum disorders provide evidence of immune dysfunction and are associated with impaired behavioral outcome. *Brain Behav Immun* 2011, **25**(1):40–45.
17. Ashwood P, Krakowiak P, Hertz-Picciotto I, Hansen R, Pessah IN, Van de Water J: Associations of impaired behaviors with elevated plasma chemokines in autism spectrum disorders. *J Neuroimmunol* 2011, **232**(1–2):196–199.
18. Goines PE, Croen LA, Braunschweig D, Yoshida CK, Grether J, Hansen R, Kharrazi M, Ashwood P, Van de Water J: Increased mid-gestational IFN-gamma, IL-4, and IL-5 in women giving birth to a child with autism: a case-control study. *Mol Autism* 2011, **2**(1):13–40.
19. Abdallah MW, Larsen N, Grove J, Bonefeld-Jorgensen EC, Norgaard-Pedersen B, Hougaard DM, Mortensen EL: Neonatal chemokine levels and risk of autism spectrum disorders: findings from a Danish historic birth cohort follow-up study. *Cytokine* 2013, **61**(2):370–376.
20. Abdallah MW, Larsen N, Grove J, Norgaard-Pedersen B, Thorsen P, Mortensen EL, Hougaard DM: Amniotic fluid inflammatory cytokines: Potential markers of immunologic dysfunction in autism spectrum disorders. *World J Biol Psychiatry 2013,* **14**(7): 528–538.
21. Abdallah MW, Larsen N, Grove J, Norgaard-Pedersen B, Thorsen P, Mortensen EL, Hougaard DM: Amniotic fluid chemokines and autism spectrum disorders: an exploratory study utilizing a Danish Historic Birth Cohort. *Brain Behav Immun* 2012, **26**(1):170–176.
22. Braunschweig D, Van de Water J: Maternal autoantibodies in autism. *Arch Neurol* 2012, **69**(6):693–699.
23. Connolly AM, Chez M, Streif EM, Keeling RM, Golumbek PT, Kwon JM, Riviello JJ, Robinson RG, Neuman RJ, Deuel RM: Brain-derived neurotrophic factor and autoantibodies to neural antigens in sera of children with autistic

spectrum disorders, Landau-Kleffner syndrome, and epilepsy. *Biol Psychiatry* 2006, **59**(4):354–363.

24. Goines P, Haapanen L, Boyce R, Duncanson P, Braunschweig D, Delwiche L, Hansen R, Hertz-Picciotto I, Ashwood P, Van de Water J: Autoantibodies to cerebellum in children with autism associate with behavior. *Brain Behav Immun* 2011, **25**(3):514–523.
25. Wills S, Cabanlit M, Bennett J, Ashwood P, Amaral DG, Van de Water J: Detection of autoantibodies to neural cells of the cerebellum in the plasma of subjects with autism spectrum disorders. *Brain Behav Immun* 2009, **23**(1):64–74.
26. Perricone R, Perricone C, De Carolis C, Shoenfeld Y: NK cells in autoimmunity: a two-edg'd weapon of the immune system. *Autoimmun Rev* 2008, **7**(5):384–390.
27. Wegmann TG, Lin H, Guilbert L, Mosmann TR: Bidirectional cytokine interactions in the maternal-fetal relationship: is successful pregnancy a TH2 phenomenon? *Immunol Today* 1993, **14**(7):353–356.
28. Chaouat G: The Th1/Th2 paradigm: still important in pregnancy? *Semin Immunopathol* 2007, **29**(2):95–113.
29. Garty BZ, Ludomirsky A, Danon YL, Peter JB, Douglas SD: Placental transfer of immunoglobulin G subclasses. *Clin Diagn Lab Immunol* 1994, **1**(6):667–669.
30. Van de Perre P: Transfer of antibody via mother's milk. *Vaccine* 2003, **21**(24):3374–3376.
31. O'Carroll P, Bertorini TE, Jacob G, Mitchell CW, Graff J: Transient neonatal myasthenia gravis in a baby born to a mother with new-onset anti-MuSK-mediated myasthenia gravis. *J Clin Neuromuscul Dis* 2009, **11**(2):69–71.
32. Hoff JM, Daltveit AK, Gilhus NE: Artrogryposis multiplex congenita — a rare fetal condition caused by maternal myasthenia gravis. *Acta Neurol Scand Suppl* 2006, **183**:26–27.
33. Saunders NR, Liddelow SA, Dziegielewska KM: Barrier mechanisms in the developing brain. *Front Pharmacol* 2012, **3**:46.
34. Diamond B, Honig G, Mader S, Brimberg L, Volpe BT: Brain-reactive antibodies and disease. *Annu Rev Immunol* 2013, **31**:345–385.
35. Cole RN, Ghazanfari N, Ngo ST, Gervasio OL, Reddel SW, Phillips WD: Patient autoantibodies deplete postsynaptic muscle-specific kinase leading to disassembly of the ACh receptor scaffold and myasthenia gravis in mice. *J Physiol* 2010, **588**(Pt 17):3217–3229.
36. Lee JY, Huerta PT, Zhang J, Kowal C, Bertini E, Volpe BT, Diamond B: Neurotoxic autoantibodies mediate congenital cortical impairment of offspring in maternal lupus. *Nat Med* 2009, **15**(1):91–96.
37. Singer HS, Morris C, Gause C, Pollard M, Zimmerman AW, Pletnikov M: Prenatal exposure to antibodies from mothers of children with autism produces neurobehavioral alterations: A pregnant dam mouse model. *J Neuroimmunol* 2009, **211**(1–2):39–48.

38. Zaretsky MV, Alexander JM, Byrd W, Bawdon RE: Transfer of inflammatory cytokines across the placenta. *Obstet Gynecol* 2004, **103**(3):546–550.
39. Ashdown H, Dumont Y, Ng M, Poole S, Boksa P, Luheshi GN: The role of cytokines in mediating effects of prenatal infection on the fetus: implications for schizophrenia. *Mol Psychiatry* 2006, **11**(1):47–55.
40. Samuelsson AM, Jennische E, Hansson HA, Holmang A: Prenatal exposure to interleukin-6 results in inflammatory neurodegeneration in hippocampus with NMDA/GABA(A) dysregulation and impaired spatial learning. *Am J Physiol Regul Integr Comp Physiol* 2006, **290**(5):R1345–1356.
41. Hauguel-de Mouzon S, Guerre-Millo M: The placenta cytokine network and inflammatory signals. *Placenta* 2006, **27**(8):794–798.
42. Mehler MF, Kessler JA: Cytokines in brain development and function. *Adv Protein Chem* 1998, **52**:223–251.
43. Warren RP, Cole P, Odell JD, Pingree CB, Warren WL, White E, Yonk J, Singh VK: Detection of maternal antibodies in infantile autism. *J Am Acad Child Adolesc Psychiatry* 1990, **29**(6):873–877.
44. Dalton P, Deacon R, Blamire A, Pike M, McKinlay I, Stein J, Styles P, Vincent A: Maternal neuronal antibodies associated with autism and a language disorder. *Ann Neurol* 2003, **53**(4):533–537.
45. Braunschweig D, Ashwood P, Krakowiak P, Hertz-Picciotto I, Hansen R, Croen LA, Pessah IN, Van de Water J: Autism: maternally derived antibodies specific for fetal brain proteins. *Neurotoxicology* 2007, **29**(2):226–231.
46. Zimmerman AW, Connors SL, Matteson KJ, Lee LC, Singer HS, Castaneda JA, Pearce DA: Maternal antibrain antibodies in autism. *Brain Behav Immun* 2007, **21**:351–357.
47. Singer HS, Morris CM, Gause CD, Gillin PK, Crawford S, Zimmerman AW: Antibodies against fetal brain in sera of mothers with autistic children. *J Neuroimmunol* 2008, **194**(1–2):165–172.
48. Brimberg L, Sadiq A, Gregersen PK, Diamond B: Brain-reactive IgG correlates with autoimmunity in mothers of a child with an autism spectrum disorder. *Mol Psychiatry* 2013, **18**(11): 1171–1177.
49. Croen LA, Braunschweig D, Haapanen L, Yoshida CK, Fireman B, Grether JK, Kharrazi M, Hansen RL, Ashwood P, Van de Water J: Maternal mid-pregnancy autoantibodies to fetal brain protein: the early markers for autism study. *Biol Psychiatry* 2008, **64**(7):583–588.
50. Braunschweig D, Duncanson P, Boyce R, Hansen R, Ashwood P, Pessah IN, Hertz-Picciotto I, Van de Water J: Behavioral correlates of maternal antibody status among children with autism. *J Autism Dev Disord* 2012, **42**(7):1435–1445.
51. Nordahl CW, Braunschweig D, Iosif AM, Lee A, Rogers S, Ashwood P, Amaral DG, Van de Water J: Maternal autoantibodies are associated with abnormal brain enlargement in a subgroup of children with autism spectrum disorder. *Brain Behav Immun* 2013, **30**:61–65.

52. Braunschweig D, Krakowiak P, Duncanson P, Boyce R, Hansen RL, Ashwood P, Hertz-Picciotto I, Pessah IN, Van de Water J: Autism-specific maternal autoantibodies recognize critical proteins in developing brain. *Transl Psychiatry* 2013, **3**:e277.
53. Braunschweig D, Golub MS, Koenig CM, Qi L, Pessah IN, Van de Water J, Berman RF: Maternal autism-associated IgG antibodies delay development and produce anxiety in a mouse gestational transfer model. *J Neuroimmunol* 2012, **252**(1–2):56–65.
54. Martínez-Cerdeño V: Prenatal Exposure to Autism-Specific Maternal Autoantibodies Impairs Proliferation of Neural Precursor Cells, Neuronal Morphology and Animal Behavior. In: *International Meeting for Autism Research (IMFAR)*. San Sebastián, Spain, Kursaal Centre; 2013.
55. Martin LA, Ashwood P, Braunschweig D, Cabanlit M, Van de Water J, Amaral DG: Stereotypies and hyperactivity in rhesus monkeys exposed to IgG from mothers of children with autism. *Brain Behav Immun* 2008, **22**:806–816.
56. Bauman MD, Iosif AM, Ashwood P, Braunschweig D, Lee A, Schumann CM, Van de Water J, Amaral DG: Maternal antibodies from mothers of children with autism alter brain growth and social behavior development in the rhesus monkey. *Transl Psychiatry* 2013, **3**:e278.
57. Rutella S, Bonanno G, Procoli A, Mariotti A, de Ritis DG, Curti A, Danese S, Pessina G, Pandolfi S, Natoni F *et al.*: Hepatocyte growth factor favors monocyte differentiation into regulatory interleukin (IL)-10++IL-12low/neg accessory cells with dendritic-cell features. *Blood* 2006, **108**(1):218–227.
58. Futamatsu H, Suzuki J, Mizuno S, Koga N, Adachi S, Kosuge H, Maejima Y, Hirao K, Nakamura T, Isobe M: Hepatocyte growth factor ameliorates the progression of experimental autoimmune myocarditis: a potential role for induction of T helper 2 cytokines. *Circ Res* 2005, **96**(8):823–830.
59. Campbell DB, Sutcliffe JS, Ebert PJ, Militerni R, Bravaccio C, Trillo S, Elia M, Schneider C, Melmed R, Sacco R *et al.*: A genetic variant that disrupts MET transcription is associated with autism. *Proc Natl Acad Sci USA* 2006, **103**(45):16834–16839.
60. Campbell DB, Li C, Sutcliffe JS, Persico AM, Levitt P: Genetic evidence implicating multiple genes in the MET receptor tyrosine kinase pathway in autism spectrum disorder. *Autism Res* 2008, **1**(3):158–168.
61. Jackson PB, Boccuto L, Skinner C, Collins JS, Neri G, Gurrieri F, Schwartz CE: Further evidence that the rs1858830 C variant in the promoter region of the MET gene is associated with autistic disorder. *Autism Res* 2009, **2**(4):232–236.
62. Thanseem I, Nakamura K, Miyachi T, Toyota T, Yamada S, Tsujii M, Tsuchiya KJ, Anitha A, Iwayama Y, Yamada K *et al.*: Further evidence for the role of MET in autism susceptibility. *Neurosci Res* 2010, **68**(2):137–141.
63. Heuer L, Braunschweig D, Ashwood P, Van de Water J, Campbell DB: Association of a MET genetic variant with autism-associated maternal

autoantibodies to fetal brain proteins and cytokine expression. *Transl Psychiatry* 2011, **1**:e48.
64. Schröder A, Linker RA, Gold R: Plasmapheresis for neuroinflammatory disorders. *Clin Exp Neuroimmunol* 2010, **1**(3):95–102.
65. Rönspeck W, Brinckmann R, Egner R, Gebauer F, Winkler D, Jekow P, Wallukat G, Müller J, Kunze R: Peptide based adsorbers for therapeutic immunoadsorption. *Ther Apher Dial* 2003, **7**(1):91–97.
66. Hershko AY, Naparstek Y: Removal of pathogenic autoantibodies by immunoadsorption. *Ann N Y Acad Sci* 2005, **1051**(1):635–646.
67. Schwab I, Nimmerjahn F: Intravenous immunoglobulin therapy: how does IgG modulate the immune system? *Nat Rev Immunol* 2013, **13**(3):176–189.
68. Schroder A, Linker RA, Gold R: Plasmapheresis for neurological disorders. *Expert Rev Neurother* 2009, **9**(9):1331–1339.
69. Kuo TT, Aveson VG: Neonatal Fc receptor and IgG-based therapeutics. *mAbs* 2011, **3**(5):422–430.
70. Vaccaro C, Zhou J, Ober RJ, Ward ES: Engineering the Fc region of immunoglobulin G to modulate *in vivo* antibody levels. *Nat Biotechnol* 2005, **23**(10):1283–1288.
71. Bloom O, Cheng KF, He M, Papatheodorou A, Volpe BT, Diamond B, Al-Abed Y: Generation of a unique small molecule peptidomimetic that neutralizes lupus autoantibody activity. *Proc Natl Acad Sci USA* 2011, **108**(25):10255–10259.
72. Knittelfelder R, Riemer AB, Jensen-Jarolim E: Mimotope vaccination--from allergy to cancer. *Expert Opin Biol Ther* 2009, **9**(4):493–506.
73. Sun Y, Fong KY, Chung MC, Yao ZJ: Peptide mimicking antigenic and immunogenic epitope of double-stranded DNA in systemic lupus erythematosus. *Int Immunol* 2001, **13**(2):223–232.
74. Sathasivam S: Steroids and immunosuppressant drugs in myasthenia gravis. *Nat Clin Pract Neurol* 2008, **4**(6):317–327.
75. Gomez AM, Willcox N, Molenaar PC, Buurman W, Martinez-Martinez P, De Baets MH, Losen M: Targeting plasma cells with proteasome inhibitors: possible roles in treating myasthenia gravis? *Ann N Y Acad Sci* 2012, **1274**:48–59.
76. Neubert K, Meister S, Moser K, Weisel F, Maseda D, Amann K, Wiethe C, Winkler TH, Kalden JR, Manz RA *et al.*: The proteasome inhibitor bortezomib depletes plasma cells and protects mice with lupus-like disease from nephritis. *Nat Med* 2008, **14**(7):748–755.
77. Ichikawa HT, Conley T, Muchamuel T, Jiang J, Lee S, Owen T, Barnard J, Nevarez S, Goldman BI, Kirk CJ *et al.*: Beneficial effect of novel proteasome inhibitors in murine lupus via dual inhibition of type I interferon and autoantibody-secreting cells. *Arthritis Rheum* 2012, **64**(2):493–503.

Chapter 18

Why is Autism More Common in Males?

Simon Baron-Cohen*, Michael V. Lombardo, Bonnie Auyeung, Emma Ashwin, Bhismadev Chakrabarti and Rebecca Knickmeyer

Abstract

Autism Spectrum Conditions (ASC) are much more common in males, a bias that may offer clues to the etiology of this condition. We argue that it occurs because ASC are an extreme manifestation of the male brain. The extreme male brain (EMB) theory, first proposed in 1997, is an extension of the Empathizing-Systemizing (E-S) theory of typical sex differences that proposes that females on average have a stronger drive to empathize while males on average have a stronger drive to systemize. In this chapter, we describe some of the evidence relating to the EMB theory of ASC and consider how typical sex differences in brain structure may be relevant to ASC. One possible biological mechanism to account for the male bias is the effect of fetal testosterone (fT). The weight of evidence in favor of the fT theory is growing from converging sources (longitudinal studies from gestation to age 10 years old based on fT in amniotic fluid, current hormone studies, and genetic association studies of SNPs in genes involved in the sex steroid biosynthetic pathways). The strong test of the fT theory will come from the large collaborative study with the Danish biobank that is adequately powered to test if fT and the related fetal steroids in the Δ4 pathway are elevated in children who go on to develop ASC.

*Correspondence to: Autism Research Centre, Department of Psychiatry, University of Cambridge, UK. Email: sb205@cam.ac.uk

Keywords: Sex bias; Extreme male brain (EMB) theory; Emphatizing-Systemizing (E-S) theory; Fetal testosterone; Amniotic fluid; Sex hormones; Sex steriod biosynthetic pathways; Fetal androgens; Sexual dimorphism in human brain.

The Male Bias in Autism

The diagnosis of classic autism and Asperger syndrome (AS), known as Autism Spectrum Conditions (ASC), rests on difficulties in reciprocal social interaction and communication, alongside strongly repetitive behavior and unusually narrow interests.[1] The prevalence of ASC is estimated to be 1%.[2,3] A diagnosis of classic autism, unlike AS, also requires the presence of additional learning difficulties and language delay. ASC are neurobiological, evidenced by atypical brain development in structure and function.[4] ASC are also genetic[5,6] though not without some interaction with environmental influences. ASC is strongly biased towards males,[7] with ratios of 4:1 (male:female) for classic autism[8] and as high as 11:1 in individuals with AS.[9] The specific factors responsible for the higher male prevalence in ASC remain unclear. ASC is not the only neurodevelopmental condition more common among males — a greater prevalence in males versus females is also seen in Attention Deficit and Hyperactivity Disorder (ADHD), dyslexia, conduct disorder (CD), specific language impairment, Tourette Syndrome, and Learning Difficulties.[10] However, the male bias is much more pronounced in ASC, especially in the case of AS.

This male bias could simply reflect the difficulty of diagnosing AS in females. Though classic autism would not be missed in females, AS could be if it presented as some other condition, such as anorexia[11] or borderline personality disorder,[12] both of which involve the exercise of excessive control over the environment or other people, and a certain degree of self-centeredness. Equally, AS in females could be under-diagnosed if females are more motivated to learn to conform socially or have better imitation skills that allow them to "pretend to be normal".[13] Finally, this male bias might reflect the inability of the widely used diagnostic instruments (the Autism Diagnostic Observation Schedule (ADOS) or

Autism Diagnostic Interview-Revised (ADI-R)) to detect the more subtle ways in which AS may present in females.

While these explanations of mis- or under-diagnosis of females may explain part of the male bias, there may also be *biological* reasons for the male bias in ASC. We argue that the bias can be understood as an extreme expression of the psychological and physiological attributes of the male brain; that is, males need only slight psychological and physiological changes to exhibit ASC while females would require more, thus making ASC rarer in females. What factors might favor overdevelopment of male characteristics? One possible biological mechanism could be the masculinizing effect of fetal testosterone (fT). Here, we lay out some of the evidence for this theory in explaining the male bias in ASC.

Are ASC an Extreme Expression of the Male Brain?

The Extreme Male Brain (EMB) theory of autism extends the Empathizing-Systemizing (E-S) theory of typical sex differences,[14] which proposes that females on average have a stronger drive to *empathize* (to identify with another person's thoughts and feelings and to respond to these with an appropriate emotion), while males on average have a stronger drive to *systemize* (to analyze or construct rule-based systems). Whilst sociologists still debate if there are any sex differences at all and, if so, whether these are purely the result of cultural conditioning, biologists have long known from animal research that sex differences in behavior exist in all primates and are influenced by biology as well as the environment. On the Empathy Quotient (EQ)[15] typical females score higher than typical males who score higher than those with ASC.[15] On the Systemizing Quotient (SQ), individuals with ASC score higher than typical males who score higher than typical females.[16–18] Additional psychological evidence (summarized in Table 1) shows that — irrespective of the direction of sex difference — people with autism show an extreme of the male profile. Note that the EMB theory does not state that all psychological sex differences will be exaggerated in ASC — only those relating to empathy and systemizing.

Table 1. A summary of the psychological evidence for the Extreme Male Brain (EMB) theory

Psychological Measure	Autism > Male > Female	Female > Male > Autism	Key References
Adolescent AQ	✓		[56]
Adult Autism Spectrum Quotient (AQ)	✓		[57–61]
Adult Systemizing Quotient (SQ)	✓		[16]
Child AQ	✓		[62]
Child SQ	✓		[63]
Childhood Autism Spectrum Test (CAST)	✓		[64–67]
Embedded Figures Test	✓		[68, 69]
Intuitive Physics Test	✓		[70, 71]
Social Responsiveness Scale	✓		[72, 73]
Quantitative Checklist for Autism in Toddlers (Q-CHAT)	✓		[74]
Adult Empathy Quotient (EQ)		✓	[15]
Child EQ		✓	[63]
Faux Pas Test		✓	[75]
Friendship and Relationship Questionnaire (FQ)		✓	[76]
Reading the Mind in the Eyes		✓	[77]
Social Stories Questionnaire (SSQ)		✓	[70]

Sexual Dimorphism in the Human Brain

Additional support for the EMB theory of ASC comes from evidence of neural sexual dimorphism across development. Some key examples of typical sexual dimorphism reveal an extreme of the typical male profile in the neurodevelopment of ASC.[19] However, one caveat to keep in mind is that just as all *psychological* sex differences do not constitute an exaggerated form of maleness in ASC, neither do all *neural* differences. Indeed, given that the EMB theory is defined at the psychological level, we should expect only a narrow set of neural sex differences will be involved in such hypermasculinization in ASC. A key finding supporting this prediction is that infant males on average have a larger brain than females[20] and children with autism have even larger brains early in life right around the time they would typically receive a diagnosis (2–4 years).[21] In addition, independent of global differences in brain size, the amygdala in typical males tends to be larger than in females,[22] and early in development the amygdala in autism is even more enlarged than that observed in typical males.[23–25] In addition to such *structural* sexual dimorphism in the brain, exaggeration of neural sexual dimorphism extends to brain *function* and corroborates predictions from the EMB theory (see Table 2 for fuller discussion).[26–29]

Table 2. A summary of the evidence consistent with the EMB theory at the neural level

Brain Region	Autism > Male > Female	Female > Male > Autism	Key References
Structure			
Total brain volume	✓		[20, 78–80]
Amgydala	✓		[22–25, 81–87]
Corpus callosum		✓	[88, 89]
Perisylvian language areas (Heschl's gyrus/planum temporale)		✓	[22, 90–93]
L>R asymmetry in planum temporale		✓	[22, 91, 94–97]
Lateral fronto-parietal cortex		✓	[81, 82, 84, 87, 93, 98–102]
Function			
Default Mode Network Connectivity		✓	[103, 104]
Embedded Figures fMRI		✓	[27–29, 105]
Reading the Mind in the Eyes task fMRI		✓	[26, 28]

The set of striking findings of hyper-masculinization in ASC at three simultaneous levels (cognitive, neuroanatomy, and neural function) raises the question as to which biological mechanism(s) are involved. A plausible mechanism that could give rise to sexual dimorphism, hyper-masculinization, and/or the absence of typical sexual dimorphism at the levels of brain, cognition, and behavior is the "organizing" effects of fetal testosterone (fT).[30–32] We review this interesting hypothesis. This is not proposed as a complete explanation for ASC, since ASC is recognized to be multi-factorial, but it may form an important part of the explanation.

The Fetal Testosterone (fT) Theory

Fetal androgens affect the brain: Evidence from animal and human studies

Animal studies, especially in rodents, confirm that early exposure to androgens (such as testosterone) acts on the brain to produce sex differences in behavior, cognition, brain structure, and function.[31–33] It is widely accepted that fT exposure also affects brain development and behavior in humans. Human males experience a surge in fT between weeks 8 to 24 of gestation,[34–36] reaching almost pubertal levels. There is also a second surge soon after birth (here called "neonatal testosterone," or nT). Usually the levels remain high and then drop to barely detectable levels by 4–6 months,[37] until the third surge at puberty. Whilst the third surge is understood to be controlling the onset of puberty, the function of the first surge (fT) is believed to control brain masculinization. While direct manipulation of hormones as has been conducted in animal studies is unquestionably unethical in human fetuses and infants, alternative research strategies include relating individual variation in amniotic fT exposure to later development,[38] or studying people in whom — for medical reasons — the sex hormones are higher or lower than expected for a person's sex,[39] and using proxy measures of fT exposure. Here we review evidence from studies of cognitive traits relevant to ASC and their relationship with amniotic fT.

Fetal androgens affect ASC traits: Evidence from amniotic fluid testosterone

fT can be measured in amniotic fluid, obtained during routine amniocentesis. Because amniocentesis is typically performed during the second trimester of pregnancy (usually 14–20 weeks of gestation), when serum testosterone peaks in male fetuses, it offers a unique opportunity to compare fT with ASC traits. There is a well-documented large sex difference in amniotic androgen levels.[40–44] The origin of androgens in amniotic fluid appears to be the fetus itself, and testosterone obtained in amniotic fluid is thought to be a good reflection of the levels in the fetus.[38] In the Cambridge Fetal Testosterone Project, initiated by our group in 1998, children whose mothers had amniocentesis during pregnancy (but who were otherwise developing normally) have been followed up after birth every year or two and are now approximately 11 years of age.[34]

Evidence that amniotic fT affects individual differences in cognitive development in typically developing children (but with clear relevance to ASC) includes the following: fT is *inversely* associated with frequency of eye contact at 12 months old[45] and with size of vocabulary development at 18 and 24 months.[46] fT is also *inversely* associated with quality of social relationships at 48 months[47] and with empathy at 48 and 96 months.[48,49] In contrast, amniotic fT is *positively* associated with narrow interests at 48 months,[47] with "systemizing" at 96 months,[18] and with performance on the Embedded Figures Test (EFT) as a measure of attention to detail at 96 months.[50] These are all behaviors that show sexual dimorphism, but critically, these fT effects are often found within one sex as well as when analyzing the sexes combined. The finding of a consistent inverse correlation between fT and social domains, and a consistent positive correlation between fT and non-social domains, across development, is striking and suggests these are real effects which substantiate the notion that fT plays an "organizational" role in development.

In the first study to directly assess if fT affects not just human cognition but also human brain structure, we found that increasing levels of fT are associated with increasing rightward asymmetry in the thickness of one subsection of the corpus callosum, the isthmus.[51] This is interesting since the isthmus projects to posterior parietal and superior temporal cortices, which are integral for language and visuospatial ability and are known to be sexually dimorphic in lateralization, structure, and function. We also found that fT predicts local brain volume within the right temporo-parietal junction (RTPJ), right planum temporale/parietal operculum, and posterior lateral orbitofrontal cortex. These areas were all found to be sexually dimorphic in an independent cohort of age-matched individuals, and as predicted the directionality of the association with fT was congruent with the directionality of the sexual dimorphism (e.g., positive correlation between fT and RTPJ and a Male > Female sexual dimorphism).[52] This is the first evidence to show that, like work in non-human species, fT has real long-lasting effects for sexual differentiation of human brain structure as well. Functionally, these regions are important for autism and other forms of developmental psychopathology, as they are involved in cognitive domains such as mentalizing/theory of mind, language, and self-regulation. Moving further into directly assaying fT's influence on human brain function, using fMRI we also found that increased fT predicted enhanced selectivity of reward system (e.g., striatal) response to positive compared to negatively valenced stimuli. Furthermore, this fT influence on enhancing reward system sensitivity to positive stimuli mediated fT's influence on behavioral approach tendencies (risk taking, impulsivity, reward sensitivity) as measured by the Sensitivity to Punishment and Sensitivity to Rewards questionnaire.[53] Thus, in addition to being relevant to the phenotype of autism, fT's influence on human brain function also extends to influencing neural sensitivity and phenotypic domains that are relevant for many other neuropsychiatric conditions with an asymmetry between the sexes, aberrant reward system function, and extremes of behavioral approach tendencies.

All of the above behavioral domains (eye contact, language development, quality of social relationships, narrow interests, empathy, systemizing, and embedded figures/attention to detail) and brain structure show sexual dimorphism and appear hypermasculinized in ASC, raising the possibility that fT may play a role in the development of ASC itself. Three recent experiments have confirmed a positive correlation between fT levels and the number of autistic traits a child shows in toddlerhood[54] and in preschool.[55]

In the present article, we have considered studies that suggest fetal testosterone is involved in sex differences in key areas of behavior and cognition in the general population (in social development, language development, empathy, systemizing, and attention to detail), as well as in influencing brain structure, and the number of autistic traits an individual possesses. Understanding the relationship between empathy and systemizing will require more research because presenting them as independent measures ignores the fact that both are related to fT. Nor can we yet extrapolate the fT results to individuals with an ASC diagnosis since this will require much larger collections of amniotic samples than has been possible to date. Strengthening a role for fT in ASC is the recent genetic evidence in which SNPs in key sex steroid genes are associated with either diagnosed AS and/or autistic traits.

Future Directions

The Cambridge Fetal Testosterone Project has too few children (currently $n = 635$ are enrolled) to test whether fT is elevated in those who later are diagnosed with ASC, but testing for a *direct* association between fT levels and diagnosed ASC will be possible in our ongoing collaboration with the Danish Biobank, which has tens of thousands of amniotic samples, with adequate power to test this hypothesis. Using a different line of evidence, a number of studies have found also *current* androgen dysregulation in ASC or in their relatives, or androgen-related genes being associated with ASC (see Table 3 for a summary of the evidence for the fT/androgen theory). Although some studies have failed to support a role for testosterone in ASC (and most of these have not been able to study fT specifically), the studies reported above suggest that fT is implicated in the biased sex ratio seen in ASC. Finally, whilst it may be that the psychiatric classification system is "carving nature at its joints," it is also possible that some of the underlying hormonal

Table 3. Evidence for the effect of sex steroids in autism

Evidence	Key References
From typically developing children	
Eye contact is inversely related to fT	[45]
Quality of social relationships are inversely related to fT	[47]
Vocabulary size is inversely related to fT	[46]
Empathy is inversely related to fT	[48, 49]
Autistic traits are positively associated with fT	[54, 55]
Restricted interests are positively associated with fT	[47]
Systemizing is positively associated with fT	[18]
Rightward asymmetry in the isthmus of the corpus callosum is positively associated with fT	[51]
Variation in sexually dimorphic brain regions is predicted by fT	[52]
Variation in reward system sensitivity and behavioral approach tendencies are predicted by fT	[53]
From people with ASC	
10 genes involved in sex steroid synthesis, transport, and/or metabolism associated with AS or AQ or empathy: *HSD11B1, LHCGR, CYP17A1, CYP19A1, SCP2, CYP11B1, ESR1, ESR2, HSD17B4, HSD17B2*	[106]
Timing of puberty: Boys with ASC enter puberty earlier. Girls with ASC enter puberty later	[107–109]
Testosterone related medical conditions in women with ASC and their mothers (e.g., PCOS, breast and ovarian cancers, acne)	[109]
Testosterone related characteristics in women with ASC and their mothers	[109, 110]
Lower 2D:4D ratio in ASC, and parents	[111–113]
SRD5A1 and AR genes associated with ASC	[114, 115]
Decreased expression of RORA gene and aromatase in post-mortem frontal and cerebellar tissue	[116, 117]
Females with Congenital Adrenal Hyperplasia (CAH) have elevated AQ	[118]
Testosterone levels are elevated in ASC	[119]
Androstenedione levels are elevated in ASC	[120]

and genetic mechanisms are involved not just in ASC but are relevant to a broader category of neurodevelopmental conditions.

Executive Summary

- Autism Spectrum Conditions (ASC) are much more common in males.
- Typical females on average have a stronger drive to empathize while typical males on average have a stronger drive to systemize.
- People with ASC appear to fit an extreme of the typical male cognitive profile: below average empathy and intact or above average systemizing.
- Fetal testosterone (fT) is associated with individual differences in empathy, systemizing and autistic traits.
- The large collaborative study with the Danish biobank is adequately powered to test if fT (and the related fetal steroids in the Δ4 pathway) are elevated in children who go on to develop ASC.

Acknowledgments

This chapter is adapted from our work published in *PLOS Biology* (2011). The authors were supported by the MRC, the Wellcome Trust, the British Academy, and the Autism Research Trust, during the period of this work.

References

1. A.P.A: *DSM-IV Diagnostic and Statistical Manual of Mental Disorders*, 4th Edition. Washington DC: American Psychiatric Association; 1994.
2. Baird G, Simonoff E, Pickles A, Chandler S, Loucas T, Meldrum D, Charman T: Prevalence of disorders of the autism spectrum in a population cohort of children in South Thames: the Special Needs and Autism Project (SNAP). *Lancet* 2006, **368**:210–215.
3. Baron-Cohen S, Scott FJ, Allison C, Williams JG, Bolton P, Matthews FE, Brayne C: Autism Spectrum Prevalence: a school-based U.K. population study. *Brit J Psychiatry* 2009, **194**:500–509.
4. Bauman ML, Kemper TL: Neuroanatomic observations of the brain in autism: a review and future directions. *Int J Dev Neurosci* 2005, **23**:183–187.

5. Stodgell CJ, Ingram JI, Hyman SL: The role of candidate genes in unravelling the genetics of autism. *Int Rev Res Mental Retard* 2001, **23**:57–81.
6. Geschwind DH: Autism: many genes, common pathways? *Cell* 2008, **135**:391–395.
7. Fombonne E: The changing epidemiology of autism. *J App Res Intellectual Disabilities* 2005, **18**:281–294.
8. Chakrabarti S, Fombonne E: Pervasive Developmental Disorders in preschool children. *J Am Med Ass* 2001, **285**:3093–3099.
9. Gillberg C, Cederlund M, Lamberg K, Zeijlon L: Brief report: "the autism epidemic". The registered prevalence of autism in a Swedish urban area. *J Autism Dev Disord* 2006, **36**:429–435.
10. Rutter M, Caspi A, Moffitt TE: Using sex differences in psychopathology to study causal mechanisms: unifying issues and research strategies. *J Child Psychology Psychiatry* 2003, **44**:1092–1115.
11. Treasure JL: Getting beneath the phenotype of anorexia nervosa: the search for viable endophenotypes and genotypes. *Can J Psychiatry* 2007, **52**:212–219.
12. New AS, Triebwasser J, Charney DS: The case for shifting borderline personality disorder to Axis I. *Biol Psychiatry* 2008, **64**:653–659.
13. Holliday-Willey L: Pretending to be normal. London: Jessica Kingsley Ltd; 1999.
14. Baron-Cohen S: *The Essential Difference: Men, Women and the Extreme Male Brain*. Penguin: London; 2003.
15. Baron-Cohen S, Wheelwright S: The Empathy Quotient (EQ). An investigation of adults with Asperger Syndrome or High Functioning Autism, and normal sex differences. *J Autism Dev Disord* 2004, **34**:163–175.
16. Baron-Cohen S, Richler J, Bisarya D, Gurunathan N, Wheelwright S: The Systemising Quotient (SQ): An investigation of adults with Asperger Syndrome or High Functioning Autism and normal sex differences. *Phil Trans R Soc B Bio Sci* 2003, **358**:361–374.
17. Wheelwright S, Baron-Cohen S, Goldenfeld N, Delaney J, Fine D, Smith R, Weil L, Wakabayashi A: Predicting Autism Spectrum Quotient (AQ) from the Systemizing Quotient-Revised (SQ-R) and Empathy Quotient (EQ). *Brain Res* 2006, **1079**:47–56.
18. Auyeung B, Baron-Cohen S, Chapman E, Knickmeyer R, Taylor K, Hackett G: Foetal testosterone and the Child Systemizing Quotient (SQ-C). *Eur J Endrocrinol* 2006, **155**:123–130.
19. Baron-Cohen S, Knickmeyer R, Belmonte M: Sex differences in the brain: implications for explaining autism. *Science* 2005, **310**:819–823.
20. Gilmore JH, Lin W, Prastawa MW, Looney CB, Vetsa YS, Knickmeyer RC, Evans DD, Smith JK, Hamer RM, Lieberman JA *etal.*: Regional gray matter growth, sexual dimorphism, and cerebral asymmetry in the neonatal brain. *Neurosci* 2007, **27**:1255–1260.

21. Courchesne E, Campbell K, Solso S: Brain growth across the life span in autism: Age-specific changes in anatomical pathology. *Brain Res* 2010, **1380**:138–145.
22. Good CD, Johnsrude I, Ashburner J, Henson RN, Friston KJ, Frackowiak RS: Cerebral asymmetry and the effects of sex and handedness on brain structure: a voxel-based morphometric analysis of 465 normal adult human brains. *Neuroimage* 2001, **14**:685–700.
23. Schumann CM, Hamstra J, Goodlin-Jones BL, Lotspeich LJ, Kwon H, Buonocore MH, Lammers CR, Reiss AL, Amaral DG: The amygdala is enlarged in children but not adolescents with autism; the hippocampus is enlarged at all ages. *J Neurosci* 2004, **24**:6392–6401.
24. Schumann CM, Barnes CC, Lord C, Courchesne E: Amygdala enlargement in toddlers with autism related to severity of social and communication impairments. *Bio Psychiatry* 2009, **66**:942–949.
25. Mosconi MW, Cody-Hazlett H, Poe MD, Gerig G, Gimpel-Smith R, Piven J: Longitudinal study of amygdala volume and joint attention in 2- to 4-year-old children with autism. *Arch Gen Psychiatry* 2009, **66**:509–516.
26. Baron-Cohen S, Ring H, Wheelwright S, Bullmore ET, Brammer MJ, Simmons A, Williams S: Social intelligence in the normal and autistic brain: an fMRI study. *Eur J Neurosci* 1999, **11**:1891–1898.
27. Ring H, Baron-Cohen S, Williams S, Wheelwright S, Bullmore E, Brammer M, Andrew C: Cerebral correlates of preserved cognitive skills in autism. A functional MRI study of Embedded Figures task performance. *Brain* 1999, **122**:1305–1315.
28. Baron-Cohen S, Ring H, Chitnis X, Wheelwright S, Gregory L, Williams S, Brammer M, Bullmore E: fMRI of parents of children with Asperger Syndrome: a pilot study. *Brain Cognition* 2006, **61**:122–130.
29. Manjaly ZM, Bruning N, Neufang S, Stephan KE, Brieber S, Marshall JC, Kamp-Becker I, Remschmidt H, Herpertz-Dahlmann B, Konrad K *et al.*: Neurophysiological correlates of relatively enhanced local visual search in autistic adolescents. *Neuroimage* 2007, **35**:283–291.
30. Geschwind N, Galaburda AM: Cerebral lateralization: biological mechanisms, associations and pathology. III. A hypothesis and a program for research. *Arch Neurol* 1985, **42**:634–654.
31. Phoenix CH, Goy RW, Gerall AA, Young WC: Organizing action of prenatally administered testosterone propionate on the tissues mediating mating behavior in the female guinea pig. *Endocrinology* 1959, **65**:369–382.
32. Arnold AP, Breedlove SM: Organizational and activational effects of sex steroids on brain and behavior: a reanalysis. *Hormones Behav* 1985, **19**:469–498.
33. De Vries G, Simerley RB: Anatomy, development and function of sexually dimorphic neural circuits in the mammalian brain. In: *Hormones, Brain and Behaviour: Development of Hormone-Dependent Neuronal Systems.* Edited by

Pfaff DW, Arnold AP, Etgen AM, Fahrbach SE, Moss RL, Rubin RT, vol. IV. San Diego: Academic Press; 2002: 137–191.

34. Baron-Cohen S, Lutchmaya S, Knickmeyer R: Prenatal testosterone in mind: amniotic fluid studies. Cambridge, USA: MIT/Bradford Books; 2004.
35. Collaer M, Hines M: Human behavioural sex differences: a role for gonadal hormones during early development? *Psychological Bulletin* 1995, **118**:55–107.
36. Hines M: Brain gender. New York: Oxford University Press; 2004.
37. Smail PJ, Reyes FI, Winter JSD, Fairman C: The fetal hormonal environment and its effect on the morphogenesis of the genital system. In: *Pediatric Andrology*. Edited by Kogan SJ, Hafez ESE. The Hague: Martinus Nijhoff; 1981: 9–19.
38. van de Beek C, Thijssen JHH, Cohen-Kettenis PT, van Goozen SHM, Buitelaar JK: Relationships between sex hormones assessed in amniotic fluid, and maternal and umbilical cord serum: what is the best source of information to investigate the effects of fetal hormonal exposure? *Hormones Behav* 2004, **46**:663–669.
39. Money J, Ehrhardt AA: Man and Woman, Boy and Girl. Baltimore: Johns Hopkins University Press; 1972.
40. Dawood MY: Hormones in amniotic fluid. *Am J Obstetrics Gynecol* 1977, **128**:576–583.
41. Finegan JA, Bartleman B, Wong PY: A window for the study of prenatal sex hormone influences on postnatal development. *J Genet Psychol* 1989, **150**(1): 101–112.
42. Judd HL, Robinson JD, Young PE, Jones OW: Amniotic fluid testosterone levels in midpregnancy. *Obstetrics Gynecol* 1976, **48**:690–692.
43. Nagami M, McDonough P, Ellegood J, Mahesh V: Maternal and amniotic fluid steroids throughout human pregnancy. *Am J Obstetrics Gynaecol* 1979, **134**:674–680.
44. Robinson JD, Judd HL, Young PE, Jones OW, Yen SS: Amniotic fluid androgens and estrogens in midgestation. *J Clin Endocrinol Metabolism* 1977, **45**:755–761.
45. Lutchmaya S, Baron-Cohen S, Raggatt P: Foetal testosterone and eye contact in 12 month old infants. *Infant Behav Dev* 2002, **25**:327–335.
46. Lutchmaya S, Baron-Cohen S, Raggatt P: Foetal testosterone and vocabulary size in 18- and 24-month-old infants. *Infant Behav Dev* 2002, **24**:418–424.
47. Knickmeyer R, Baron-Cohen S, Raggatt P, Taylor K: Foetal testosterone, social cognition, and restricted interests in children. *J Child Psychol Psychiatry* 2005, **46**:198–210.
48. Chapman E, Baron-Cohen S, Auyeung B, Knickmeyer R, Taylor K, Hackett G: Foetal testosterone and empathy: evidence from the Empathy Quotient (EQ) and the 'Reading the Mind in the Eyes' Test'. *Social Neurosci* 2006, **1**:135–148.

49. Knickmeyer R, Baron-Cohen S, Raggatt P, Taylor K, Hackett G: Foetal testosterone and empathy. *Hormones Behav* 2006, **49**:282–292.
50. Auyeung B, Ashwin E, Knickmeyer R, Taylor K, Hackett G, Baron-Cohen S: Effects of fetal testosterone on visiospatial ability. *Unpublished MS, University of Cambridge* submitted.
51. Chura LR, Lombardo MV, Ashwin E, Auyeung B, Chakrabarti B, Bullmore ET, Baron-Cohen S: Organizational effects of fetal testosterone on human corpus callosum size and asymmetry. *Psychoneuroendocrinology* 2010, **35**:122–132.
52. Lombardo MV, Ashwin E, Auyeung B, Chakrabarti B, Taylor K, Hackett G, Bullmore ET, Baron-Cohen S: Fetal testosterone influences sexually dimorphic gray matter in the human brain. *J Neurosci* 2012, **32**(2):674–680.
53. Lombardo MV, Ashwin E, Auyeung B, Chakrabarti B, Lai MC, Taylor K, Hackett G, Bullmore ET, Baron-Cohen S: Fetal programming effects of testosterone on the reward system and behavioral approach tendencies in humans. *Biol Psychiatry* 2012, **72**(10):839–847.
54. Auyeung B, Taylor K, Hackett G, Baron-Cohen S: Foetal testosterone and autistic traits in 18 to 24-month-old children. *Mol Autism* 2010, **1**:11.
55. Auyeung B, Baron-Cohen S, Ashwin E, Knickmeyer R, Taylor K, Hackett G: Fetal testosterone and autistic traits. *Brit J Psychol* 2009, **100**:1–22.
56. Baron-Cohen S, Hoekstra RA, Knickmeyer R, Wheelwright S: The Autism-Spectrum Quotient (AQ)-Adolescent version. *J Autism Dev Disord* 2006, **36**:343–350.
57. Wakabayashi A, Baron-Cohen S, Wheelwright S: The Autism Spectrum Quotient (AQ) Japanese version: evidence from high-functioning clinical group and normal adults. *Jap J Psychol* 2004, **75**:78–84.
58. Wakabayashi A, Baron-Cohen S, Wheelwright S, Tojo Y: The Autism-Spectrum Quotient (AQ) in Japan: a cross-cultural comparison. *J Autism Dev Disord* 2006, **36**:263–270.
59. Wakabayashi A, Baron-Cohen S, Uchiyama T, Yoshida Y, Tojo Y, Kuroda M, Wheelwright S: The Autism-Spectrum Quotient (AQ) children's version in Japan: a cross-cultural comparison. *J Autism Dev Disord* 2007, **37**:491–500.
60. Hoekstra R, Bartels M, Cath DC, Boomsma DI: Factor structure, reliability and criterion validity of the Autism-Spectrum Quotient (AQ): a study in Dutch population and patient groups. *J Autism Dev Disord* 2008, **38**:1555–1566.
61. Baron-Cohen S, Wheelwright S, Skinner R, Martin J, Clubley E: The Autism Spectrum Quotient (AQ): evidence from Asperger Syndrome/High Functioning Autism, Males and Females, Scientists and Mathematicians. *J Autism Dev Disord* 2001, **31**:5–17.
62. Auyeung B, Baron-Cohen S, Wheelwright S, Allison C: The Autism Spectrum Quotient: Children's Version (AQ-Child). *J Autism Dev Disord* 2008, **38**:1230–1240.

63. Auyeung B, Wheelwright S, Allison C, Atkinson M, Samarawickrema N, Baron-Cohen S: The children's Empathy Quotient and Systemizing Quotient: sex differences in typical development and in autism spectrum conditions. *J Autism Dev Disord* 2009, **39**:1509–1521.
64. Scott F, Baron-Cohen S, Bolton P, Brayne C: Prevalence of autism spectrum conditions in children aged 5–11 years in Cambridgeshire, UK. *Autism* 2002, **6**:231–237.
65. Scott F, Baron-Cohen S, Bolton P, Brayne C: The CAST (Childhood Asperger Syndrome Test): preliminary development of UK screen for mainstream primary-school children. *Autism* 2002, **6**:9–31.
66. Williams J, Allison C, Scott F, Bolton P, Baron-Cohen S, Matthews F, Brayne C: The Childhood Autism Spectrum Test (CAST): Sex Differences. *J Autism Dev Disord* 2008, **38**.
67. Williams J, Scott FJ, Allison C, Bolton P, Baron-Cohen S, Brayne C: The CAST (Childhood Asperger Syndrome Test): test accuracy. *Autism* 2005, **9**:45–68.
68. Shah A, Frith U: An islet of ability in autism: a research note. *J Child Psychol Psychiatry* 1983, **24**:613–620.
69. Jolliffe T, Baron-Cohen S: Are people with autism or Asperger's Syndrome faster than normal on the Embedded Figures Task? *J Child Psychol Psychiatry* 1997, **38**:527–534.
70. Lawson J, Baron-Cohen S, Wheelwright S: Empathising and systemising in adults with and without Asperger Syndrome. *J Autism Dev Disord* 2004, **34**:301–310.
71. Baron-Cohen S, Wheelwright S, Scahill V, Lawson J, Spong A: Are intuitive physics and intuitive psychology independent? *J Dev Learning Disord* 2001, **5**:47–78.
72. Constantino JN, Todd RD: Autistic traits in the general population. *Arch Gen Psychiatry* 2003, **60**:524–530.
73. Constantino JN, Todd RD: Intergenerational transmission of subthreshold autistic traits in the general population. *Bio Psychiatry* 2005, **57**:655–660.
74. Allison C, Baron-Cohen S, Wheelwright S, Charman T, Richler J, Pasco G, Brayne C: The Q-CHAT (Quantitative CHecklist for Autism in Toddlers): a normally distributed quantitative measure of autistic traits at 18–24 months of age: preliminary report. *J Autism Dev Disord* 2008, **38**:1414–1425.
75. Baron-Cohen S, O'Riordan M, Jones R, Stone V, Plaisted K: A new test of social sensitivity: detection of faux pas in normal children and children with Asperger syndrome. *J Autism Dev Disord* 1999, **29**:407–418.
76. Baron-Cohen S, Wheelwright S: The Friendship Questionnaire (FQ): an investigation of adults with Asperger Syndrome or High Functioning Autism, and normal sex differences. *J Autism Dev Disord* 2003, **33**:509–517.
77. Baron-Cohen S, Jolliffe T, Mortimore C, Robertson M: Another advanced test of theory of mind: evidence from very high functioning adults with autism or Asperger Syndrome. *J Child Psychol Psychiatry* 1997, **38**:813–822.

78. Redcay E, Courchesne E: When is the brain enlarged in autism? A meta-analysis of all brain size reports. *Bio Psychiatry* 2005, **58**:1–9.
79. Hazlett HC, Poe M, Gerig G, Smith RG, Provenzale J, Ross A, Gilmore J, Piven J: Magnetic resonance imaging and head circumference study of brain size in autism: birth through age 2 years. *Arch Gen Psychiatry* 2005, **62**:1366–1376.
80. Courchesne E, Carper R, Akshoomoff NA: Evidence of Brain Overgrowth in the First Year of Life in Autism. *J Am Med Assoc* 2003, **290**:337–344.
81. Cheng Y, Chou KH, Decety J, Chen IY, Hung D, Tzeng OJ, Lin CP: Sex differences in the neuroanatomy of human mirror-neuron system: a voxel-based morphometric investigation. *Neuroscience* 2009, **158**:713–720.
82. Yamasue H, Abe O, Suga M, Yamada H, Rogers MA, Aoki S, Kato N, Kasai K: Sex-linked neuroanatomical basis of human altruistic cooperativeness. *Cerebal Cortex* 2008, **18**:2331–2340.
83. Giedd JN, Viatuzis AC, Hamburger SD, Lange N, Rajapakse JC, Kaysen D, Vauss YC, Rapoport JL: Quantatitive MRI of the temporal lobe, amygdala and hippocampus in normal human development: ages 4–18 years. *J Comparative Neurol* 1996, **366**:223–230.
84. Goldstein JM, Seidman LJ, Horton NJ, Makris N, Kennedy DN, Caviness Jnr VS, Faraone SV, Tsuang MT: Normal sexual dimorphism of the adult human brain assessed by *In Vivo* Magnetic Resonance Imaging. *Cerebral Cortex* 2001, **11**:490–497.
85. Wilke M, Krageloh-Mann I, Holland SK: Global and local development of gray and white matter volume in normal children and adolescents. *Exp Brain Res* 2007, **178**:296–307.
86. Peper JS, Brouwer RM, Schnack HG, van Baal GC, van Leeuwen M, van den Berg SM, Delemarre-Van de Waal HA, Janke AL, Collins DL, Evans AC *et al.*: Cerebral white matter in early puberty is associated with luteinizing hormone concentrations. *Psychoneuroendocrinology* 2008, **33**:909–915.
87. Chen X, Sachdev PS, Wen W, Anstey KJ: Sex differences in regional gray matter in healthy individuals aged 44–48 years: a voxel-based morphometric study. *Neuroimage* 2007, **36**:691–699.
88. Lenroot RK, Gogtay N, Greenstein DK, Wells EM, Wallace GL, Clasen LS, Blumenthal JD, Lerch J, Zijdenbos AP, Evans AC *et al.*: Sexual dimorphism of brain developmental trajectories during childhood and adolescence. *Neuroimage* 2007, **36**:1065–1073.
89. Frazier TW, Hardan AY: A meta-analysis of the corpus callosum in autism. *Biol Psychiatry* 2009, **66**:935–941.
90. Rojas DC, Bawn SD, Benkers TL, Reite ML, Rogers SJ: Smaller left hemisphere planum temporale in adults with autistic disorder. *Neurosci Lett* 2002, **328**:237–240.
91. Rojas DC, Camou SL, Reite ML, Rogers SJ: Planum temporale volume in children and adolescents with autism. *J Autism Dev Disord* 2005, **35**:479–486.

92. Witelson SF, Glezer II, Kigar DL: Women have greater density of neurons in posterior temporal cortex. *J Neurosci* 1995, **15**:3418–3428.
93. Sowell ER, Peterson BS, Kan E, Woods RP, Yoshii J, Bansal R, Xu D, Zhu H, Thompson PM, Toga AW: Sex differences in cortical thickness mapped in 176 healthy individuals between 7 and 87 years of age. *Cerebral Cortex* 2007, **17**:1550–1560.
94. Witelson SF, Kigar DL: Sylvian fissure morphology and asymmetry in men and women: bilateral differences in relation to handedness in men. *J Comparative Neurol* 1992, **323**:326–340.
95. Herbert MR, Ziegler DA, Deutsch CK, O'Brien LM, Kennedy DN, Filipek PA, Bakardjiev AI, Hodgson J, Takeoka M, Makris N *et al.*: Brain asymmetries in autism and developmental language disorder: a nested whole-brain analysis. *Brain* 2005, **128**:213–226.
96. Wada JA, Clarke R, Hamm A: Cerebral hemispheric asymmetry in humans. Cortical speech zones in 100 adults and 100 infant brains. *Arch Neurol* 1975, **32**:239–246.
97. Gage NM, Juranek J, Filipek PA, Osann K, Flodman P, Isenberg AL, Spence MA: Rightward hemispheric asymmetries in auditory language cortex in children with autistic disorder: an MRI investigation. *J Neurodev Disord* 2009, **1**:205–214.
98. Im K, Lee JM, Lee J, Shin YW, Kim IY, Kwon JS, Kim SI: Gender difference analysis of cortical thickness in healthy young adults with surface-based methods. *Neuroimage* 2006, **31**:31–38.
99. Luders E, Narr KL, Thompson PM, Rex DE, Woods RP, Deluca H, Jancke L, Toga AW: Gender effects on cortical thickness and the influence of scaling. *Hum Brain Mapping* 2006, **27**:314–324.
100. Brun CC, Lepore N, Luders E, Chou YY, Madsen SK, Toga AW, Thompson PM: Sex differences in brain structure in auditory and cingulate regions. *Neuroreport* 2009, **20**:930–935.
101. Hadjikhani N, Joseph RM, Snyder J, Tager-Flusberg H: Anatomical differences in the mirror neuron system and social cognition network in autism. *Cerebal Cortex* 2006, **16**:1276–1282.
102. McAlonan GM, Cheung V, Suckling J, Lam GY, Tai KS, Yip L, Murphy DG, Chua SE: Mapping the brain in autism: a voxel based MRI study of volumetric differences and intercorrelations in autism. *Brain* 2005, **128**:268–276.
103. Biswal BB, Mennes M, Zuo XN, Gohel S, Kelly C, Smith SM, Beckmann CF, Adelstein JS, Buckner RL, Colcombe S *et al.*: Toward discovery science of human brain function. *Proc Nat Acad Sci USA* 2010, **107**:4734–4739.
104. Kennedy DP, Courchesne E: The intrinsic functional organization of the brain is altered in autism. *Neuroimage* 2008, **39**:1877–1885.
105. Lee PS, Foss-Feig J, Henderson JG, Kenworthy LE, Gilotty L, Gaillard WD, Vaidya CJ: Atypical neural substrates of Embedded Figures Task performance in children with Autism Spectrum Disorder. *Neuroimage* 2007, **38**:184–193.

106. Chakrabarti B, Dudbridge F, Kent L, Wheelwright S, Hill-Cawthorne G, Allison C, Banerjee-Basu S, Baron-Cohen S: Genes related to sex steroids, neural growth, and social-emotional behavior are associated with autistic traits, empathy, and Asperger syndrome. *Autism Res* 2009, **2**:157–177.
107. Tordjman A, Ferrari P, Sulmont V, Duyme M, Roubertoux P: Androgenic activity in autism. *Am J Psychiatry* 1997, **154**:1626–1627.
108. Knickmeyer R, Baron-Cohen S: Foetal Testosterone and sex differences in typical social development and in Autism. *J Child Neurol* 2006, **21**:825845–845845.
109. Ingudomnukul E, Baron-Cohen S, Knickmeyer R, Wheelwright S: Elevated rates of testosterone-related disorders in a sample of women with autism spectrum conditions. *Hormones Behav* 2007, **51**:597–604.
110. Knickmeyer R, Wheelwright S, Baron-Cohen SB: Sex-typical play: masculinization/defeminization in girls with an autism spectrum condition. *J Autism Dev Disord* 2008, **38**:1028–1035.
111. Manning J, Baron-Cohen S, Wheelwright S, Sanders G: Autism and the ratio between 2nd and 4th digit length. *Dev Med Child Neurol* 2001, **43**:160–164.
112. Milne E, White S, Campbell R, Swettenham J, Hansen P, Ramus F: Motion and form coherence detection in autistic spectrum disorder: relationship to motor control and 2:4 digit ratio. *J Autism Dev Disord* 2006, **36**:1–13.
113. de Bruin EI, Verheij F, Wiegman T, Ferdinand RF: Differences in finger length ratio between males with autism, pervasive developmental disorder-not otherwise specified, ADHD, and anxiety disorders. *Dev Med Child Neurol* 2006, **48**:962–965.
114. Henningsson S, Jonsson L, Ljunggren E, Westberg L, Gillberg C, Rastam M, Anckarsater H, Nygren G, Landen M, Thuresson K *et al.*: Possible association between the androgen receptor gene and autism spectrum disorder. *Psychoneuroendocrinology* 2009, **34**:752–761.
115. Hu VW, Nguyen A, Kim KS, Steinberg ME, Sarachana T, Scully MA, Soldin SJ, Luu T, Lee NH: Gene expression profiling of lymphoblasts from autistic and nonaffected sib pairs: altered pathways in neuronal development and steroid biosynthesis. *PLoS One* 2009, **4**:e5775.
116. Sarachana T, Xu M, Wu RC, Hu VW: Sex hormones in autism: androgens and estrogens differentially and reciprocally regulate RORA, a novel candidate gene for autism. *PLoS One* 2011, **6**(2):e17116.
117. Nguyen A, Rauch TA, Pfeifer GP, Hu VW: Global methylation profiling of lymphoblastoid cell lines reveals epigenetic contributions to autism spectrum disorders and a novel autism candidate gene, RORA, whose protein product is reduced in autistic brain. *FASEB J* 2010, **24**:3036–3051.
118. Knickmeyer R, Baron-Cohen S, Fane BA, Wheelwright S, Mathews GA, Conway GS, Brook CG, Hines M: Androgens and autistic traits: a study of individuals with congenital adrenal hyperplasia. *Hormones Behav* 2006, **50**:148–153.

119. Schmidtova E, Kelemenova S, Celec P, Ficek A, Ostatnikova D: Polymorphisms in genes involved in testosterone metabolism in slovak autistic boys. *The Endocrinologist* 2010, **20**:245–249.
120. Ruta L, Ingudomnukul E, Taylor K, Chakrabarti B, Baron-Cohen S: Increased serum androstenedione in adults with autism spectrum conditions. *Psychoneuroendocrinology* 2011.

Moving Towards Personalized Treatment of ASD and Lifespan Issues

Chapter 19

Future Directions in Psychopharmacology of Autism Spectrum Disorder

Deepali Mankad and Evdokia Anagnostou*

Abstract

Autism spectrum disorder (ASD) is a neurodevelopmental disorder characterized by social communication deficits and repetitive behaviors/restrictive interests. Current approaches for development of therapeutics have been based on the assumption that phenotypic spectra suggest neurobiological spectra and, as such, medications approved for conditions with overlapping symptoms may be of use in ASD. The approach has not been completely fruitless, but has produced no medications that target core symptom domains. With advances in our understanding of the basic neuroscience of ASD, attempts have been made to identify the various brain pathways and molecular targets that may be responsible for core deficits and may be amenable to pharmacologic interventions. In this chapter, we review such candidate pathways and promising approaches.

Keywords: Autism spectrum disorder; Psychopharmacology; Oxytocin; Immune dysfunction; Synaptic dysfunction.

*Correspondence to: Bloorview Research Institute, Department of Pediatrics, University of Toronto, Toronto, Ontario, Canada. Emails: Deepali Mankad: dmankad@hollandbloorview.ca, Evdokia Anagnostou: eanagnostou@hollandbloorview.ca

Introduction

Autism spectrum Disorder (ASD) refers to a group of neurodevelopmental disorders characterized by social communication deficits and repetitive behaviors. The word "autism," comes from the Greek word "autos," meaning "self." The term was used by Leo Kanner in 1943 to describe children with limited interaction — hence, an isolated self.[1] The two constructs of social communication impairment and repetitive behaviors have been extensively described.[2] Impairment in social communication includes deficits in basic social cognitive processes, such as eye contact, recognition of faces and emotions, and theory of mind among others as well as difficulties with social emotional reciprocity and ultimate failure to develop peer, age-appropriate relationships. Retrospective studies and systematic reviews of videotapes recorded at birthdays have shown that abnormalities with eye contact, joint attention, and imitative behaviors can be identified as early as 12 months of age.[3,4] Restrictive interests and repetitive behaviors together make up the second category to make the diagnosis of ASD. Severity of repetitive behaviors at age 2 predicts whether or not a child would meet diagnostic criteria for autism at age 9.[5]

Therapeutics in autism has been plagued by the lack of molecular targets. Most medications tested in ASD have been chosen based on the fact that they may be efficacious in another neuropsychiatric/neurodevelopmental disorder with seemingly overlapping symptoms. The approach has been somewhat successful for treating associated domains, e.g. Risperidone and Aripiprazole for irritability[6] but has produced no medications for core symptom domains. As the neuroscience of ASD matures, new approaches, based on targets related to the pathophysiology of ASD are emerging. Still several challenges remain. Genomic findings have challenged the common disease–common variant model in favor for a common disease/rare variant model, suggesting not only phenotypic heterogeneity in ASD but also biologic heterogeneity.[7] As such, it is unlikely that one treatment would fit all.

Several approaches have been proposed to address this issue:

1. Enhance biomarker discovery to facilitate stratification of populations entering trials into those more likely to respond.
2. Translate emerging neurobiology of ASD into novel therapeutics targeting the underlying pathophysiology.
3. Translate evidence from systems neuroscience that identifies target systems, which although may not be directly involved in the pathophysiology of ASD, may modify severity of core and associated symptoms
4. Combine pharmaco-therapeutics targeting the biology of ASD with existing and evolving psycho-educational interventions to facilitate skill acquisition.

Stratification Based on Biomarkers

Depending on methodology used, up to 40% of individuals with ASD carry a rare mutation.[8] Several cluster analyses have been attempted to date[8] suggesting that several of the genes involved map onto synaptogenesis and synaptic plasticity, neuronal motility and axon targeting.[9] Examples include the neuroligin family which are important for the formation and consolidation of both inhibitory and excitatory synapses as well as SHANK3, which codes for a synaptic scaffolding protein. Variation in these genes is likely to affect the excitation to inhibition ratio in the brain, a fact that reveals a potential target for therapeutics, to be discussed later.

However, a significant portion of participants entering the studies still have no available genomic information. There is a need for downstream biomarkers that will assist with stratifying patients in the absence of a known mutation. Gene expression studies, imaging, proteomics/metabolomics have all been suggested, in addition to detailed phenotyping, to serve this role and hold promise, although data confirming the feasibility of this approach is still pending. Cross-species platforms are being developed to translate the neurobiology from known animal models to humans, and iPS cells are being explored as a potential system for testing cellular effects of drugs directly on patient cells. For examples of such initiatives, see www.eu-aims.com and http://pond-network.ca/.

Medications Targeting Emerging Pathophysiology of ASD

We chose two areas to discuss which emerge from genomics/proteomics and neuropathology. This is by no means an exhaustive list and is meant to illustrate the process.

Drugs altering the excitation to inhibition (E:I) balance

As previously mentioned, several rare mutations co-localize in pathways responsible for synaptic formation and synaptic plasticity of glutamatergic and GABAergic synapses, suggesting that E:I balance is disturbed. The phenomenon was also suggested much before the CNV data was available.

There have been several studies suggesting activation of the glutamatergic system, which could play a role in brain maturation alterations in autism, and evidence comes from biochemistry as well as neuropathology. Patients with autism or Asperger's syndrome were found to have raised levels of glutamate in plasma compared to healthy controls in multiple studies.[10–12] Glutamic acid decarboxylase protein, the enzyme responsible for normal conversion of glutamate to GABA in the brain, has been noted to be reduced in the brain of autistic subjects in postmortem studies, suggesting increased levels of glutamate or transporter receptor density in autistic brains.[13] Genetic studies have so far revealed a mutation in the glutamate receptor gene GRIK2 in 6q21 that is present in 8% of autistic subjects and 4% of controls and seems to be transmitted from mothers to autistic males.[14] Furthermore, evidence for the presence of a susceptibility mutation in linkage disequilibrium with variants in the metabotropic glutamate receptor 8 gene has been found on chromosome 7.[15] A single nucleotide polymorphism of the GRIN2A subunit of the NMDA receptor has been reported to be associated with autism.[16] Recent understanding of the genetics/proteomics of Fragile X, a syndrome with prevalent ASD symptoms associated with dysregulated mGluR signalling,[17] highlights the importance of this system in the pathophysiology of autistic-like symptoms. There are few glutamatergic/GABA modulators studied in autism: Riluzole, Memantine, Amantadine, D-Cyclosporine, Dextromethorphan, and emerging data on Arbaclofen. Data on several others including mGLUR-5 inhibitors is pending.

Riluzole

This is a glutamatergic modulator with neuroprotective and neuroplasticity enhancing properties[18] that has an important role in the treatment of Amyotropic Lateral Sclerosis. There is emerging evidence that Riluzole may be useful in the treatment of OCD, bipolar disorder and generalized anxiety disorder. Wink *et al.*, in 2011, reported the use of Riluzole in three cases of ASD. Improvements were noted in repetitive behaviors, self-injury and aggression.[19] A case report by Veenstra-Vander Weele in a 13-year-old girl treated with 50 mg BID of Riluzole suggested improvements in compulsivity and aggression.[20] Lastly, a recent small randomized controlled trial suggested that Riluzole is effective as an adjunct to Risperidone for irritability in children with autistic disorder.[21] Mechanisms of action of interest include inhibition of glutamate release, enhancing glutamate transporter uptake, increasing AMPA trafficking, stimulating synthesis of BDNF and inhibition of Protein Kinase C resulting in reduced oxidative stress.[22] It is important to investigate this drug further using better study designs to evaluate the potential efficacy and safety of the medication for core and associated symptom domains of autism.

Memantine

This drug blocks the pathological tonic activation of NMDA receptors and suppresses synaptic noise but allows the relevant physiological synaptic signal to be detected. This provides both neuroprotection and symptomatic restoration of synaptic plasticity.[23] This medication has shown positive clinical benefits on memory in patients with Alzheimer's disease.[24,25] There have been several open label studies reporting on the effect of memantine on ASD. Owley *et al.*[26] enrolled 13 children, ages 3–12 years, in an 8 week study. Patients showed significant changes in irritability, lethargy, hyperactivity, inappropriate speech, stereotypy. and memory. A retrospective review reported by Erikson *et al.*[27] described administration of memantine between 2.5 and 20 mg per day for a mean of 19 weeks as part of clinical care. Eleven out of 18 patients were reported to be responders and showed significant improvements in levels of hyperactivity. Chez *et al.*[28] also reported on 151 children with ASD treated with memantine as part of clinical care. Clinical improvements in language, behavior and repetitive movements were noted in 70% of patients. Niederhofer *et al.*,[29] reported on a series of four patients with ASD who received memantine for four weeks. Significant improvements were noted in their irritability, hyperactivity, and inappropriate speech. Lastly, Ghaleiha *et al.* (2013) reported on a randomized controlled trial of memantine used as adjunctive treatment to Risperidone and reported statistically significant improvements on irritability.[30] Several randomized trials are currently registered at clinical trials.gov.

Amantadine

Amantadine has NMDA non-competitive inhibitor activity at routine doses. A double-blind placebo-controlled trial of Amantadine was carried out in autistic children. While it was tolerated, it was noted to have at best modest effect on irritability and hyperactivity.[31] However, amantadine has very low affinity for the NMDA receptor so that 10- to 20-fold higher doses than memantine may need to be used.[32]

D-Cycloserine

D-Cycloserine is a partial agonist of the NMDA receptor. A single blind trial of this medication in autistic children resulted in significant improvement on the CGI and social withdrawal subscale of the Aberrant Behavior Checklist and it was tolerated at most of the doses used.[33]

Dextromethorphan

Dextromerthophan is a NMDA receptor antagonist. In a series of case studies and single subject design studies of children with autism, it has been reported to improve problem behaviors, such as tantrums and self-injurious behavior (SIB), as well as anxiety, motor planning, socialization and language.[34,35]

GABA-B agonists

An alternative approach to improving E:I ratios is to enhance GABA activity. Early open-label data of a selective GABA-B agonist with preliminary evidence of benefit in fragile X syndrome (Arbaclofen) suggests improvements in social responsiveness, irritability and global function,[36] whereas a small randomized controlled trial of this compound in patients with fragile X has shown promise for the social communication domain.[37]

In summary, manipulation of the glutamate/GABA systems may offer a new molecular target for the treatment of symptoms associated with ASD and further work is urgently needed in this area.

Immune Modulators

Abnormalities in immune and inflammatory responses as well as in oxidative stress have been proposed to play a role in at least some of the individuals with autism.

A series of studies have demonstrated an association between autism and family history of autoimmune disorders,[38] although the link with specific disorders remains inconsistent across studies. In fact, the data is most consistent with global autoimmunity during pregnancy being associated with increased risk for ASD. One hypothesis is that anti-fetal brain antibodies present in mothers may bind to neuronal targets during development. Animal models have been supportive of such a hypothesis[39] although the exact mechanisms remain unclear.

Immune differences in children with ASD have also been well-documented. A postmortem study of the brain tissue of individuals with autism showed active and ongoing neuroinflammatory process in the cerebral cortex, white matter, and the cerebellum. Both microglia and astroglia activation was noted.[40] Activation of neuroglial cells can produce significant neuronal and synaptic changes that are likely to contribute to CNS dysfunction. Increase in cerebrospinal fluid (CSF) pro-inflammatory cytokines, including IL-6, interferon (IFN)-gamma, and MCP-1, have also been reported [40–42] as well as decreased plasma TGFb1 and IL-10 (an important counter-regulatory cytokine) -producing T cells.[43–45] All these changes in turn suggest that there is a dysregulated immune response and elevated levels of pro-inflammatory cytokines in some patients with ASD.[46] There is evidence suggesting an association between altered cytokine levels and behavioral changes in animals and humans. Direct administration of IL-1, IL-2 and IL-6 in mice, for example, results in profound behavioral changes, including digging, exploration, rearing and grooming,[46] and other increases in stereotypic motor behaviour.

Lastly several neurotransmitters shown to be associated with ASD have immunomodulatory properties. For example, glutamate receptors are differentially expressed in naïve vs. activated T cells and are thought to participate in T cell regulation.[47] GABA receptors may increase pathogen clearance by macrophages and act in an inhibitory fashion in a subset of lymphocytes.[48,49] Both receptors may be targets of antibodies, although this has not been established in ASD. It has however been established in some seizure and motor disorders;[38] thus, dysregulation of neurotransmitters may be responsible for some immune differences observed in ASD.

The issue of whether the immune system could be a therapeutic target remains controversial. To date, only small studies of steroids, intravenous (IV) immunoglobulin, and antibiotics are in the literature, presenting contradictory findings from small samples.[50–53]

Other potential compounds may be those modulating cytokines (e.g., minocycline, omega-3 fatty acids),[54,55] those providing antioxidant protection (e.g. N-acetycycteine [NAC] or L-carnosine),[56,57] or those having effects on microglia activation (e.g., pioglitazone, omega-3 fatty acids). The data for all of the above is at best mixed, with the exception of NAC, which has been shown to improve irritability in randomized controlled trials. However, NAC is also an NMDA inhibitor, and it is not clear what the contribution of the antioxidant mechanism is to its effect.

We will focus on two drugs, Omega-3 fatty acids and Pioglitazone, as examples of approaches to immune modulation that may be useful in ASD.

Omega-3 fatty acids

Omega-3 fatty acids are a special class of poly-unsaturated fatty acids, and include ALA (alpha-linoleic acid), DHA (docosahexaenoic acid) and EPA (eicosapentaenoic acid). DHA and EPA are derived from fish oil. There is accumulating data to support that these two fatty acids are important structural and functional elements of the brain and have been suggested to have therapeutic effects in multiple neurodevelopmental disorders including mood disorders, schizophrenia, ADHD, and autism. During the last trimester of fetal life and the first two years of childhood, the brain undergoes a period of rapid growth. DHA is one nutrient required for the development of the sensory, perceptual, cognitive, and motor neural systems during the brain growth spurt.[58,59] While it is DHA that matters the most in the structure of neuronal membranes, EPA nonetheless plays a significant role in brain function, and its derivatives are key regulators of immune, endocrine, and cardiovascular function.[60] The supplementation of omega-3 fatty acids has been shown to have some efficacy in treating many neurodevelopmental disorders, although the data is mixed. Many children with ADHD and dyslexia have responded to oral supplementation of these nutrients.[60,61] A couple of studies have reported that Omega-3 fatty acids also reduce aggression and impulsivity in university students.[62,63] The use of omega-3 supplements has also established its role as a mood stabilizer in adults and children with a bipolar disorder.[64–67] Omega-3 fatty acids have been shown to have anti-inflammatory effects that may include decreasing pro-inflammatory substances[54] and lower microglia activation, thus acting as an efficient immune modulator.

There have been few studies that have suggested potential efficacy of omega-3s in children with autism.[68,69] Amminger *et al.*[70] reported effect sizes above d=0.7 in improvement of hyperactivity and stereotypy. In 2011, Bent *et al.*[71] published the results of the pilot randomized controlled trial which was conducted to determine the feasibility and initial safety and efficacy of omega-3 fatty acids for the treatment of hyperactivity in 27 children (ages 3–8) with ASD. After 12 weeks, hyperactivity, as measured by the Aberrant Behavior Checklist, improved 2.7 (± 4.8) points in the omega-3 group compared to 0.3 (± 7.2) points in the placebo group ($p = 0.40$; effect size = 0.38). Correlations were found between decreases in five fatty acid levels and decreases in hyperactivity, and the treatment was well-tolerated. However, no effects have been documented against core symptom domains yet, and this remains an active area of research.

Pioglitazone

Pioglitazone is an agonist of the peroxisome proliferator-activated receptor (PPAR)-gamma, a nuclear hormone receptor which modulates insulin sensitivity. Agonists of the PPAR-gamma include omega-3 and omega-6 fatty acids, non-steroidal anti-inflammatory drugs (NSAIDs), some prostaglandins, and thiazolidinediones (TZDs), which include pioglitazone and rosiglitazone. This class of medications was originally designed to treat type II diabetes, given their insulin-modulating effects. However, a series of basic science studies have shown anti-inflammatory effects and, as such, they are being considered for the treatment of inflammatory conditions such as atherosclerosis,[72] psoriasis, and inflammatory bowel disease.[73–75] The majority of studies have focused on the effects on monocyte/macrophages and endothelial cells since PPAR-gamma may modulate the production of inflammatory cytokines by these cells and might control immune cell differentiation and function.[76] A recently published trial of pioglitazone in multiple sclerosis reported a significant reduction in grey matter atrophy in participants randomized to pioglitazone versus those randomized to placebo.[77] Pioglitazone is also known to exert several PPAR-gamma-independent effects [78,79] such as NMDAR inhibition, which is of particular interest given the recent model suggesting hyperplasticity in autism.

Pioglitazone has been shown *in vitro* and in early *in vivo* studies to have a variety of effects of interest in the pathophysiology of autism. Those include inhibition of expression of inflammatory cytokines (IL-1b, IL-6 and IL-8),[80] reduction in TNF-α,[81] inhibition of microglial activation, reduced NMDAR mediated currents (potentially having a role to play in neuroprotection but also in modulation of plasticity), and reduction in oxidative stress and mitochondrial biogenesis.[82]

There has been some data in the literature reporting benefits in children with autism when treated with pioglitazone. Improvements were noted in various domains with very minimal safety concern;[83] however the dose in this study exceeded the maximum recommended dose in adults and, as such, further research into this compound is warranted. Additionally, concerns regarding bladder toxicity and heart failure need to be taken into consideration.

In summary, immune modulation remain an attractive therapeutic target, but the pathophysiology of the immune difference in ASD is not yet clarified and early studies of immune-modulating agents have provided mixed results at best.

Medications that Target Neurocircuitry which May Modify Severity of Core Symptoms

Systems neuroscience has been suggesting several circuitries and molecular targets which may be useful in modulating severity, although not necessarily involved in the pathophysiology of the disorder. Examples are MET polymorphisms,[84] CNTNAP polymorphisms[85] and the Oxytocin and Vasopressin systems. We will briefly review the data supporting the choice of oxytocin and oxytocin receptors as molecular targets for ASD.

Oxytocin (OXT) is a peptide synthesized in the paraventricular and supraoptic nucleus of the hypothalamus. The central release of oxytocin and its closely related peptide arginine vasopressin (AVP) have also been found to be involved in aspects of social cognition and functioning, including social recognition, social memory, separation distress, and other aspects of social attachment.

The two hormones have been extensively studied in both animal models and typical humans due to their strong influence on social interaction and repetitive behaviors. In animal models, oxytocin has been known to affect social processing, recognition, bonding, and stereotyped behaviors such as exaggerated grooming;[5,86,87] moreover, OXT knockout mice suffer deficits in social recognition.[88] There are significant challenges in administering oxytocin in humans. OXT is metabolized by chymotrypsin in the gastrointestinal tract and thus is not administered orally. Intravenous administration of this medication can cause both peripheral as well as central behavioral effects, and is not practical for treatment studies. Thus, the administration of intranasal oxytocin has been attempted in human studies. Domes *et al.*[89] found that a single dose of intranasal OXT facilitated performance on Reading the Mind in the Eyes Test, which assesses the ability to infer the mental states of others or theory of mind (ToM), and Guastella *et al.*[90] found that OXT increases gaze to the eye region of human faces. These studies are particularly relevant to autism because individuals with autism have deficits in ToM and in orienting gaze to the eye region. Savaskan *et al.* also reported improved facial identity recognition memory with 20 IU of OXT (administered post-learning) versus placebo.[91] A number of researchers have reported abnormalities in OXT levels in ASD. Modahl *et al.* reported that children with autism have lower average levels of blood OXT levels in comparison to typically developing children matched for age.[92] Other studies provide evidence, albeit weak, of a genetic association between the OXT receptor gene and autism.[93–98]

Based on the evidence from basic science research and early human investigations, early clinical trials were undertaken in the last 10–15 years. In 2003, Hollander *et al.*[99] reported a significant reduction in repetitive behaviors on administration of IV OXT. The same group in 2007[100] found that IV administration of OXT facilitated retention of cognition in adults with ASD. A six-week feasibility trial exploring safety and efficacy issues related to the use of IN-OXT on social cognition/functioning and repetitive behaviors was reported by Anagnostou *et al. in* 2012.[101] Analysis of primary outcome measures suggested IN-OXT treatment vs. placebo resulted in improvements in social cognition/functioning and repetitive behaviors, and global functioning. Several larger clinical trials are currently underway to examine the efficacy of intranasal oxytocin in children, adolescents, and adults with ASD. The question still remains whether administration of oxytocin is the best way to manipulate the oxytocin system, and several companies are working on compounds that are either analogs of oxytocin or may result in increased oxytocin secretion in the CNS.

Future Directions: Combining Pharmaco-therapeutics Targeting the Biology of ASD with Behavioral Interventions

Ultimately, novel pharmacological agents affecting synaptic plasticity or social learning or neuroinflammation will have to be used in combination with successful psychoeducational interventions in order to facilitate their effect. Multimodal interventions will be necessary for maximum effect as, at the end of the day, ASD is in part a disorder of skill acquisition. Such studies are cumbersome and expensive and require significant commitments for funding agencies and institutions. However, the feasibility of the approach has started to be demonstrated.[102]

Conclusion

Emerging evidence from genomics, proteomics/metabolomics, neuropathology and systems neuroscience are providing us with a series of novel molecular targets that will hopefully allow for the development of pharmacologic agents targeting the specific "lesion(s)" of ASD and, as such, hold promise for the treatment of core symptom domains. With improved translation capacity from basic science and the emergence of novel compounds based on such translation, the hope is that combination of novel pharmacologic agents with behavioral approaches targeting learning will improve long-term outcomes for individuals with ASD.

Executive Summary

- Most agents currently available for the pharmacologic treatment of symptoms associated with ASD have not come through the translational route. As such, they have had minimal impact on core symptom domains of ASD.
- Emerging evidence from genomics, proteomics, neuropathology and systems neuroscience are revealing a series of molecular targets that are candidates for drug development.
- Combining such therapeutics with behavioral/psychosocial learning interventions is likely to have significant impact on long term outcomes of individuals with ASD.

References

1. Kanner L: Autistic disturbances of affective contact. *Nervous Child* 1943, **2**:217–250.
2. American Psychiatric Association: *Diagnostic and Statistical Manual of Mental Disorders: DSM-5.* 5th ed. Washington, D.C.: American Psychiatric Association; 2013.
3. Dahlgren SO, Gillberg C: Symptoms in the first two years of life. A preliminary population study of infantile autism. *Eur Arch Psychiatry Neurological Sci* 1989, **238**(3):169–174.
4. Osterling J, Dawson G: Early recognition of children with autism: a study of first birthday home videotapes. *J Autism Dev Disord* 1994, **24**(3):247–257.
5. Lord C, Risi S, DiLavore PS, Shulman C, Thurm A, Pickles A: Autism from 2 to 9 years of age. *Arch Gen Psychiatry* 2006, **63**(6):694–701.
6. McPheeters ML, Warren Z, Sathe N, Bruzek JL, Krishnaswami S, Jerome RN, Veenstra-Vanderweele J: A systematic review of medical treatments for children with autism spectrum disorders. *Pediatrics* 2011, **127**(5):e1312–21.

7. Betancur C: Etiological heterogeneity in autism spectrum disorders: more than 100 genetic and genomic disorders and still counting. *Brain Res* 2011, **1380**:42–77.
8. Jiang YH, Yuen RK, Jin X, Wang M, Chen N, Wu X, Ju J, Mei J, Shi Y, He M, Wang G, Liang J, Wang Z, Cao D, Carter MT, Chrysler C, Drmic IE, Howe JL, Lau L, Marshall CR, Merico D, Nalpathamkalam T, Thiruvahindrapuram B, Thompson A, Uddin M, Walker S, Luo J, Anagnostou E, Zwaigenbaum L, Ring RH, Wang J, Lajonchere C, Wang J, Shih A, Szatmari P, Yang H, Dawson G, Li Y, Scherer SW: Detection of clinically relevant genetic variants in autism spectrum disorder by whole-genome sequencing. *Am J Hum Genet* 2013, **93**(2):249–263.
9. Pinto D, Pagnamenta AT, Klei L, Anney R, Merico D, Regan R, Conroy J, Magalhaes TR, Correia C, Abrahams BS, Almeida J, Bacchelli E, Bader GD, Bailey AJ, Baird G, Battaglia A, Berney T, Bolshakova N, Bolte S, Bolton PF, Bourgeron T, Brennan S, Brian J, Bryson SE, Carson AR, Casallo G, Casey J, Chung BH, Cochrane L, Corsello C, Crawford EL, Crossett A, Cytrynbaum C, Dawson G, de Jonge M, Delorme R, Drmic I, Duketis E, Duque F, Estes A, Farrar P, Fernandez BA, Folstein SE, Fombonne E, Freitag CM, Gilbert J, Gillberg C, Glessner JT, Goldberg J, Green A, Green J, Guter SJ, Hakonarson H, Heron EA, Hill M, Holt R, Howe JL, Hughes G, Hus V, Igliozzi R, Kim C, Klauck SM, Kolevzon A, Korvatska O, Kustanovich V, Lajonchere CM, Lamb JA, Laskawiec M, Leboyer M, Le Couteur A, Leventhal BL, Lionel AC, Liu XQ, Lord C, Lotspeich L, Lund SC, Maestrini E, Mahoney W, Mantoulan C, Marshall CR, McConachie H, McDougle CJ, McGrath J, McMahon WM, Merikangas A, Migita O, Minshew NJ, Mirza GK, Munson J, Nelson SF, Noakes C, Noor A, Nygren G, Oliveira G, Papanikolaou K, Parr JR, Parrini B, Paton T, Pickles A, Pilorge M, Piven J, Ponting CP, Posey DJ, Poustka A, Poustka F, Prasad A, Ragoussis J, Renshaw K, Rickaby J, Roberts W, Roeder K, Roge B, Rutter ML, Bierut LJ, Rice JP, Salt J, Sansom K, Sato D, Segurado R, Sequeira AF, Senman L, Shah N, Sheffield VC, Soorya L, Sousa I, Stein O, Sykes N, Stoppioni V, Strawbridge C, Tancredi R, Tansey K, Thiruvahindrapduram B, Thompson AP, Thomson S, Tryfon A, Tsiantis J, Van Engeland H, Vincent JB, Volkmar F, Wallace S, Wang K, Wang Z, Wassink TH, Webber C, Weksberg R, Wing K, Wittemeyer K, Wood S, Wu J, Yaspan BL, Zurawiecki D, Zwaigenbaum L, Buxbaum JD, Cantor RM, Cook EH, Coon H, Cuccaro ML, Devlin B, Ennis S, Gallagher L, Geschwind DH, Gill M, Haines JL, Hallmayer J, Miller J, Monaco AP, Nurnberger JI,Jr, Paterson AD, Pericak-Vance MA, Schellenberg GD, Szatmari P, Vicente AM, Vieland VJ, Wijsman EM, Scherer SW, Sutcliffe JS, Betancur C: Functional impact of global rare copy number variation in autism spectrum disorders. *Nature* 2010, **466**(7304):368–372.
10. Rolf LH, Haarmann FY, Grotemeyer KH, Kehrer H: Serotonin and amino acid content in platelets of autistic children. *Acta Psychiatr Scand* 1993, **87**(5):312–316.

11. Moreno-Fuenmayor H, Borjas L, Arrieta A, Valera V, Socorro-Candanoza L: Plasma excitatory amino acids in autism. *Invest Clin* 1996, **37**(2):113–128.
12. Aldred S, Moore KM, Fitzgerald M, Waring RH: Plasma amino acid levels in children with autism and their families. *J Autism Dev Disord* 2003, **33**(1):93–97.
13. Fatemi SH: The role of Reelin in pathology of autism. *Mol Psychiatry* 2002, **7**(9):919–920.
14. Philippe A, Martinez M, Guilloud-Bataille M, Gillberg C, Rastam M, Sponheim E, Coleman M, Zappella M, Aschauer H, Van Maldergem L, Penet C, Feingold J, Brice A, Leboyer M: Genome-wide scan for autism susceptibility genes. Paris Autism Research International Sibpair Study. *Hum Mol Genet* 1999, **8**(5):805–812.
15. Serajee FJ, Zhong H, Nabi R, Huq AH: The metabotropic glutamate receptor 8 gene at 7q31: partial duplication and possible association with autism. *J Med Genet* 2003, **40**(4):e42.
16. Barnby G, Abbott A, Sykes N, Morris A, Weeks DE, Mott R, Lamb J, Bailey AJ, Monaco AP, International Molecular Genetics Study of Autism, Consortium: Candidate-gene screening and association analysis at the autism-susceptibility locus on chromosome 16p: evidence of association at GRIN2A and ABAT. *Am J Hum Genet* 2005, **76**(6):950–966.
17. Gross C, Berry-Kravis EM, Bassell GJ: Therapeutic strategies in fragile X syndrome: dysregulated mGluR signaling and beyond. *Neuropsychopharmacology* 2012, **37**(1):178–195.
18. Zarate CA,Jr, Manji HK: The role of AMPA receptor modulation in the treatment of neuropsychiatric diseases. *Exp Neurol* 2008, **211**(1):7–10.
19. Wink LK, Erickson CA, Stigler KA, McDougle CJ: Riluzole in autistic disorder. *J Child Adolesc Psychopharmacol* 2011, **21**(4):375–379.
20. Veenstra-VanderWeele J: Increase in valproic acid levels during riluzole treatment in an adolescent with autism. *J Child Adolesc Psychopharmacol* 2010, **20**(2):163–165.
21. Ghaleiha A, Mohammadi E, Mohammadi M-, Farokhnia M, Modabbernia A, Yekehtaz H, Ashrafi M, Hassanzadeh E, Akhondzadeh S: Riluzole as an adjunctive therapy to risperidone for the treatment of irritability in children with autistic disorder: A double-blind, placebo-controlled, randomized trial. *Pediatric Drugs* 2013, **15**(6):505–514.
22. Zarate C,Jr, Machado-Vieira R, Henter I, Ibrahim L, Diazgranados N, Salvadore G: Glutamatergic modulators: the future of treating mood disorders? *Harv Rev Psychiatry* 2010, **18**(5):293–303.
23. Danysz W, Parsons CG: The NMDA receptor antagonist memantine as a symptomatological and neuroprotective treatment for Alzheimer's disease: preclinical evidence. *Int J Geriatr Psychiatry* 2003, **18**(Suppl 1):S23–32.

24. Winblad B, Poritis N: Memantine in severe dementia: results of the 9M-Best Study (Benefit and efficacy in severely demented patients during treatment with memantine). *Int J Geriatr Psychiatry* 1999, **14**(2):135–146.
25. Reisberg B, Doody R, Stoffler A, Schmitt F, Ferris S, Mobius HJ: Memantine in moderate-to-severe Alzheimer's disease. *N Engl J Med* Apr 2003, **348**(14):1333–1341.
26. Owley T, Salt J, Guter S, Grieve A, Walton L, Ayuyao N, Leventhal BL, Cook EH,Jr: A prospective, open-label trial of memantine in the treatment of cognitive, behavioral, and memory dysfunction in pervasive developmental disorders. *J Child Adolesc Psychopharmacol* 2006, **16**(5):517–524.
27. Erickson CA, Posey DJ, Stigler KA, Mullett J, Katschke AR, McDougle CJ: A retrospective study of memantine in children and adolescents with pervasive developmental disorders. *Psychopharmacology (Berl)* 2007, **191**(1):141–147.
28. Chez MG, Burton Q, Dowling T, Chang M, Khanna P, Kramer C: Memantine as adjunctive therapy in children diagnosed with autistic spectrum disorders: an observation of initial clinical response and maintenance tolerability. *J Child Neurol* 2007, **22**(5):574–579.
29. Niederhofer H: Glutamate antagonists seem to be slightly effective in psychopharmacologic treatment of autism. *J Clin Psychopharmacol* 2007, **27**(3):317–318.
30. Ghaleiha A, Asadabadi M, Mohammadi MR, Shahei M, Tabrizi M, Hajiaghaee R, Hassanzadeh E, Akhondzadeh S: Memantine as adjunctive treatment to risperidone in children with autistic disorder: a randomized, double-blind, placebo-controlled trial. *Int J Neuropsychopharmcol* 2013, **16**(4):783–789.
31. King BH, Wright DM, Handen BL, Sikich L, Zimmerman AW, McMahon W, Cantwell E, Davanzo PA, Dourish CT, Dykens EM, Hooper SR, Jaselskis CA, Leventhal BL, Levitt J, Lord C, Lubetsky MJ, Myers SM, Ozonoff S, Shah BG, Snape M, Shernoff EW, Williamson K, Cook EH,Jr: Double-blind, placebo-controlled study of amantadine hydrochloride in the treatment of children with autistic disorder. *J Am Acad Child Adolesc Psychiatry* 2001, **40**(6):658–665.
32. Kornhuber J, Weller M, Schoppmeyer K, Riederer P: Amantadine and memantine are NMDA receptor antagonists with neuroprotective properties. *J Neural Transm Suppl* 1994, **43**(Supplementum. 43):91–104.
33. Posey DJ, Kem DL, Swiezy NB, Sweeten TL, Wiegand RE, McDougle CJ: A pilot study of D-cycloserine in subjects with autistic disorder. *Am J Psychiatry* 2004, **161**(11):2115–2117.
34. Woodard C, Groden J, Goodwin M, Shanower C, Bianco J: The treatment of the behavioral sequelae of autism with dextromethorphan: a case report. *J Autism Dev Disord* 2005, **35**(4):515–518.
35. Philippe A, Martinez M, Guilloud-Bataille M, Gillberg C, Rastam M, Sponheim E, Coleman M, Zappella M, Aschauer H, Van Maldergem L,

Penet C, Feingold J, Brice A, Leboyer M: Genome-wide scan for autism susceptibility genes. Paris Autism Research International Sibpair Study. *Hum Mol Genet* 1999, **8**(5):805–812.
36. Erickson CA, VeenstraVanderweele JM, Melmed RD, McCracken JT, Ginsberg LD, Sikich L, Scahill L, Cherubini M, Zarevics P, WaltonBowen K, Carpenter RL, Bear MF, Wang PP, King BH: Stx209 (arbaclofen) for autism spectrum disorders: An 8-week open-label study. *J Autism Dev Disord* 2013, (Nov):No Pagnaton Sefe.
37. Berry-Kravis EM, Hessl D, Rathmell B, Zarevics P, Cherubini M, Walton-Bowen K, Mu Y, Nguyen DV, Gonzalez-Heydrich J, Wang PP, Carpenter RL, Bear MF, Hagerman RJ: Effects of STX209 (arbaclofen) on neurobehavioral function in children and adults with fragile X syndrome: a randomized, controlled, phase 2 trial. *Sci Transl Med* 2012, **4**(152):152ra127.
38. Careaga M, Van de Water J, Ashwood P: Immune dysfunction in autism: a pathway to treatment. *Neurotherapeutics* 2010, **7**(3):283–292.
39. Brimberg L, Sadiq A, Gregersen PK, Diamond B: Brain-reactive IgG correlates with autoimmunity in mothers of a child with an autism spectrum disorder. *Mol Psychiatry* 2013, **18**(11):1171–1177.
40. Vargas DL, Nascimbene C, Krishnan C, Zimmerman AW, Pardo CA: Neuroglial activation and neuroinflammation in the brain of patients with autism. *Ann Neurol* 2005, **57**(1):67–81.
41. Pardo CA, Vargas DL, Zimmerman AW: Immunity, neuroglia and neuroinflammation in autism. *Int Rev Psychiatry* 2005, **17**(6):485–495.
42. Zimmerman AW, Connors SL, PardoVillamizar CA: Neuroimmunology and Neurotransmitters in Autism. In *Autism: A neurological disorder of early brain development*. Edited by Tuchman R, Rapin I. London NW3 5RN, England: Mac Keith Press; England; 2006:141–159.
43. Ashwood P, Anthony A, Torrente F, Wakefield AJ: Spontaneous mucosal lymphocyte cytokine profiles in children with autism and gastrointestinal symptoms: mucosal immune activation and reduced counter regulatory interleukin-10. *J Clin Immunol* 2004, **24**(6):664–673.
44. Ashwood P, Wakefield AJ: Immune activation of peripheral blood and mucosal CD3+ lymphocyte cytokine profiles in children with autism and gastrointestinal symptoms. *J Neuroimmunol* Apr 2006, **173**(1–2):126–134.
45. Ashwood P, Enstrom A, Krakowiak P, Hertz-Picciotto I, Hansen RL, Croen LA, Ozonoff S, Pessah IN, Van de Water J: Decreased transforming growth factor beta1 in autism: a potential link between immune dysregulation and impairment in clinical behavioral outcomes. *J Neuroimmunol* 2008, **204**(1–2):149–153.
46. Ashwood P, Anthony A, Torrente F, Wakefield AJ: Spontaneous mucosal lymphocyte cytokine profiles in children with autism and gastrointestinal symptoms: mucosal immune activation and reduced counter regulatory interleukin-10. *J Clin Immunol* 2004, **24**(6):664–673.

47. Pacheco R, Ciruela F, Casado V, Mallol J, Gallart T, Lluis C, Franco R: Group I metabotropic glutamate receptors mediate a dual role of glutamate in T cell activation. *J Biol Chem* 2004, **279**(32):33352–33358.
48. Alam S, Laughton DL, Walding A, Wolstenholme AJ: Human peripheral blood mononuclear cells express GABAA receptor subunits. *Mol Immunol* 2006, **43**(9):1432–1442.
49. Lubick K, Radke M, Jutila M: Securinine, a GABAA receptor antagonist, enhances macrophage clearance of phase II C. burnetii: Comparison to TLR agonists (Journal of Leukocyte Biology (2007) 82, 5, (1062–1069) DOI:10.1189/jlb.0407255). *J Leukoc Biol* 2008, **83**(4):1068.
50. Buitelaar JK, Van Engeland H, Van Ree JM, De Wied D: Behavioral effects of Org 2766, a synthetic analog of the adrenocorticotrophic hormone (4–9), in 14 outpatient autistic children. *J Autism Dev Disord* 1990, **20**(4):467–478.
51. Dalakas MC: Intravenous immune globulin therapy for neurologic diseases. *Ann Intern Med* 1997, **126**(9):721–730.
52. Gupta S, Plioplys AV: Treatment of children with autism with intravenous immunoglobulin (multiple letters). *J Child Neurol* 1999, **14**(3):203–205.
53. Sandler RH, Finegold SM, Bolte ER, Buchanan CP, Maxwell AP, Vaisanen ML, Nelson MN, Wexler HM: Short-term benefit from oral vancomycin treatment of regressive-onset autism. *J Child Neurol* 2000, **15**(7):429–435.
54. Li H, Ruan XZ, Powis SH, Fernando R, Mon WY, Wheeler DC, Moorhead JF, Varghese Z: EPA and DHA reduce LPS-induced inflammation responses in HK-2 cells: evidence for a PPAR-gamma-dependent mechanism. *Kidney Int* 2005, **67**(3):867–874.
55. Pardo CA, Buckley A, Thurm A, Lee L-, Azhagiri A, Neville DM, Swedo SE: A pilot open-label trial of minocycline in patients with autism and regressive features. *J Neurodev Disord* 2013, **5**(1):1–9.
56. Hardan AY, Fung LK, Libove RA, Obukhanych TV, Nair S, Herzenberg LA, Frazier TW, Tirouvanziam R: A randomized controlled pilot trial of oral N-acetylcysteine in children with autism. *Biol Psychiatry* 2012, **71**(11):956–961.
57. Chez MG, Buchanan CP, Aimonovitch MC, Becker M, Schaefer K, Black C, Komen J: Double-blind, placebo-controlled study of L-carnosine supplementation in children with autistic spectrum disorders. *J Child Neurol* 2002, **17**(11):833–837.
58. McCann JC, Ames BN: Is docosahexaenoic acid, an n-3 long-chain polyunsaturated fatty acid, required for development of normal brain function? An overview of evidence from cognitive and behavioral tests in humans and animals. *Am J Clin Nutr* 2005, **82**(2):281–295.
59. SanGiovanni JP, Parra-Cabrera S, Colditz GA, Berkey CS, Dwyer JT: Meta-analysis of dietary essential fatty acids and long-chain polyunsaturated fatty acids as they relate to visual resolution acuity in healthy preterm infants. *Pediatrics* 2000, **105**(6):1292–1298.

60. Richardson AJ: Omega-3 fatty acids in ADHD and related neurodevelopmental disorders. *Int Rev Psychiatry* 2006, **18**(2):155–172.
61. Harding KL, Judah RD, Gant C: Outcome-based comparison of Ritalin versus food-supplement treated children with AD/HD. *Altern Med Rev* 2003, **8**(3):319–330.
62. Hamazaki T, Sawazaki S, Itomura M, Asaoka E, Nagao Y, Nishimura N, Yazawa K, Kuwamori T, Kobayashi M: The effect of docosahexaenoic acid on aggression in young adults. A placebo-controlled double-blind study. *J Clin Invest* 1996, **97**(4):1129–1133.
63. Itomura M, Hamazaki K, Sawazaki S, Kobayashi M, Terasawa K, Watanabe S, Hamazaki T: The effect of fish oil on physical aggression in schoolchildren — a randomized, double-blind, placebo-controlled trial. *J Nutr Biochem* 2005, **16**(3):163–171.
64. Stoll AL, Severus WE, Freeman MP, Rueter S, Zboyan HA, Diamond E, Cress KK, Marangell LB: Omega 3 fatty acids in bipolar disorder: a preliminary double-blind, placebo-controlled trial. *Arch Gen Psychiatry* 1999, **56**(5):407–412.
65. Keck PE,Jr, Mintz J, McElroy SL, Freeman MP, Suppes T, Frye MA, Altshuler LL, Kupka R, Nolen WA, Leverich GS, Denicoff KD, Grunze H, Duan N, Post RM: Double-blind, randomized, placebo-controlled trials of ethyl-eicosapentanoate in the treatment of bipolar depression and rapid cycling bipolar disorder. *Biol Psychiatry* 2006, **60**(9):1020–1022.
66. Marangell LB, Suppes T, Ketter TA, Dennehy EB, Zboyan H, Kertz B, Nierenberg A, Calabrese J, Wisniewski SR, Sachs G: Omega-3 fatty acids in bipolar disorder: clinical and research considerations. *Prostaglandins Leukot Essent Fatty Acids* 2006, **75**(4–5):315–321.
67. Wozniak J, Biederman J, Mick E, Waxmonsky J, Hantsoo L, Best C, Cluette-Brown JE, Laposata M: Omega-3 fatty acid monotherapy for pediatric bipolar disorder: a prospective open-label trial. *Eur Neuropsychopharmacol* 2007, **17**(6–7):440–447.
68. Bell JG, Sargent JR, Tocher DR, Dick JR: Red blood cell fatty acid compositions in a patient with autistic spectrum disorder: a characteristic abnormality in neurodevelopmental disorders? *Prostaglandins Leukot Essent Fatty Acids* 2000, **63**(1–2):21–25.
69. Politi P, Cena H, Comelli M, Marrone G, Allegri C, Emanuele E, Ucelli di Nemi S: Behavioral effects of omega-3 fatty acid supplementation in young adults with severe autism: an open label study. *Arch Med Res* 2008, **39**(7):682–685.
70. Amminger GP, Berger GE, Schafer MR, Klier C, Friedrich MH, Feucht M: Omega-3 fatty acids supplementation in children with autism: a double-blind randomized, placebo-controlled pilot study. *Biol Psychiatry* 2007, **61**(4):551–553.

71. Bent S, Bertoglio K, Ashwood P, Bostrom A, Hendren RL: A pilot randomized controlled trial of omega-3 fatty acids for autism spectrum disorder. *J Autism Dev Disord* 2011, **41**(5):545–554.
72. Bongartz T, Coras B, Vogt T, Scholmerich J, Muller-Ladner U: Treatment of active psoriatic arthritis with the PPARgamma ligand pioglitazone: an open-label pilot study. *Rheumatology* 2005, **44**(1):126–129.
73. Gniadecki R, Calverley MJ: Emerging drugs in psoriasis. *Expert Opin Emerg Drugs* 2002, **7**(1):69–90.
74. Katayama K, Wada K, Nakajima A, Mizuguchi H, Hayakawa T, Nakagawa S, Kadowaki T, Nagai R, Kamisaki Y, Blumberg RS, Mayumi T: A novel PPAR gamma gene therapy to control inflammation associated with inflammatory bowel disease in a murine model. *Gastroenterology* 2003, **124**(5):1315–1324.
75. Wada K, Nakajima A, Blumberg RS: PPARgamma and inflammatory bowel disease: a new therapeutic target for ulcerative colitis and Crohn's disease. *Trends Mol Med* 2001, **7**(8):329–331.
76. Aoun P, Watson DG, Simpkins JW: Neuroprotective effects of PPARgamma agonists against oxidative insults in HT-22 cells. *Eur J Pharmacol* 2003, **472**(1–2):65–71.
77. Kaiser CC, Shukla DK, Stebbins GT, Skias DD, Jeffery DR, Stefoski D, Katsamakis G, Feinstein DL: A pilot test of pioglitazone as an add-on in patients with relapsing remitting multiple sclerosis. *J Neuroimmunol* 2009, **211**(1–2):124–130.
78. Cunard R, Ricote M, DiCampli D, Archer DC, Kahn DA, Glass CK, Kelly CJ: Regulation of cytokine expression by ligands of peroxisome proliferator activated receptors. *J Immunol* 2002, **168**(6):2795–2802.
79. Argmann CA, Cock TA, Auwerx J: Peroxisome proliferator-activated receptor gamma: the more the merrier? *Eur J Clin Invest* 2005, **35**(2):82–92; discussion 80.
80. Schutz B, Reimann J, Dumitrescu-Ozimek L, Kappes-Horn K, Landreth GE, Schurmann B, Zimmer A, Heneka MT: The oral antidiabetic pioglitazone protects from neurodegeneration and amyotrophic lateral sclerosis-like symptoms in superoxide dismutase-G93A transgenic mice. *J Neurosci* 2005, **25**(34):7805–7812.
81. Wu ZH, Zhao SP, Chu LX, Ye HJ: Pioglitazone reduces tumor necrosis factor-alpha serum concentration and mRNA expression of adipose tissue in hypercholesterolemic rabbits. *Int J Cardiol* 2010, **138**(2):151–156.
82. Ghosh S, Patel N, Rahn D, McAllister J, Sadeghi S, Horwitz G, Berry D, Wang KX, Swerdlow RH: The thiazolidinedione pioglitazone alters mitochondrial function in human neuron-like cells. *Mol Pharmacol* 2007, **71**(6):1695–1702.
83. Boris M, Kaiser CC, Goldblatt A, Elice MW, Edelson SM, Adams JB, Feinstein DL: Effect of pioglitazone treatment on behavioral symptoms in autistic children. *J Neuroinflammation* 2007, **4**:3.

84. Campbell DB, Sutcliffe JS, Ebert PJ, Militerni R, Bravaccio C, Trillo S, Elia M, Schneider C, Melmed R, Sacco R, Persico AM, Levitt P: A genetic variant that disrupts MET transcription is associated with autism. *Proc Natl Acad Sci USA* 2006, **103**(45):16834–16839.
85. Sampath S, Bhat S, Gupta S, O'Connor A, West AB, Arking DE, Chakravarti A: Defining the Contribution of CNTNAP2 to Autism Susceptibility. *PLoS One* 2013, **8**(10):ate of Pubaton: 17 Ot 2013.
86. Carter CS: Neuroendocrine perspectives on social attachment and love. *Psychoneuroendocrinology* 1998, **23**(8):779–818.
87. Insel TR, O'Brien DJ, Leckman JF: Oxytocin, vasopressin, and autism: is there a connection? *Biol Psychiatry* 1999, **45**(2):145–157.
88. Ferguson JN, Young LJ, Hearn EF, Matzuk MM, Insel TR, Winslow JT: Social amnesia in mice lacking the oxytocin gene. *Nat Genet* 2000, **25**(3):284–288.
89. Domes G, Heinrichs M, Michel A, Berger C, Herpertz SC: Oxytocin improves "mind-reading" in humans. *Biol Psychiatry* 2007, **61**(6):731–733.
90. Guastella AJ, Mitchell PB, Dadds MR: Oxytocin increases gaze to the eye region of human faces. *Biol Psychiatry* 2008, **63**(1):3–5.
91. Savaskan E, Ehrhardt R, Schulz A, Walter M, Schachinger H: Post-learning intranasal oxytocin modulates human memory for facial identity. *Psychoneuroendocrinology* 2008, **33**(3):368–374.
92. Modahl C, Green L, Fein D, Morris M, Waterhouse L, Feinstein C, Levin H: Plasma oxytocin levels in autistic children. *Biol Psychiatry* 1998, **43**(4):270–277.
93. Wu S, Jia M, Ruan Y, Liu J, Guo Y, Shuang M, Gong X, Zhang Y, Yang X, Zhang D: Positive association of the oxytocin receptor gene (OXTR) with autism in the Chinese Han population. *Biol Psychiatry* 2005, **58**(1):74–77.
94. Yirmiya N, Rosenberg C, Levi S, Salomon S, Shulman C, Nemanov L, Dina C, Ebstein RP: Association between the arginine vasopressin 1a receptor (AVPR1a) gene and autism in a family-based study: mediation by socialization skills. *Mol Psychiatry* 2006, **11**(5):488–494.
95. Jacob S, Brune CW, Carter CS, Leventhal BL, Lord C, Cook EH, Jr: Association of the oxytocin receptor gene (OXTR) in Caucasian children and adolescents with autism. *Neurosci Lett* 2007, **417**(1):6–9.
96. Kim SJ, Young LJ, Gonen D, Veenstra-VanderWeele J, Courchesne R, Courchesne E, Lord C, Leventhal BL, Cook EH,Jr, Insel TR: Transmission disequilibrium testing of arginine vasopressin receptor 1A (AVPR1A) polymorphisms in autism. *Mol Psychiatry* 2002, **7**(5):503–507.
97. Wassink TH, Piven J, Vieland VJ, Pietila J, Goedken RJ, Folstein SE, Sheffield VC: Examination of AVPR1a as an autism susceptibility gene. *Mol Psychiatry* 2004, **9**(10):968–972.

98. Ylisaukko-oja T, Alarcon M, Cantor RM, Auranen M, Vanhala R, Kempas E, von Wendt L, Jarvela I, Geschwind DH, Peltonen L: Search for autism loci by combined analysis of Autism Genetic Resource Exchange and Finnish families. *Ann Neurol* 2006, **59**(1):145–155.
99. Hollander E, Novotny S, Hanratty M, Yaffe R, DeCaria CM, Aronowitz BR, Mosovich S: Oxytocin infusion reduces repetitive behaviors in adults with autistic and Asperger's disorders. *Neuropsychopharmacology* 2003, **28**(1): 193–198.
100. Hollander E, Bartz J, Chaplin W, Phillips A, Sumner J, Soorya L, Anagnostou E, Wasserman S: Oxytocin increases retention of social cognition in autism. *Biol Psychiatry* 2007, **61**(4):498–503.
101. Anagnostou E, Soorya L, Chaplin W, Bartz J, Halpern D, Wasserman S, Wang AT, Pepa L, Tanel N, Kushki A, Hollander E: Intranasal oxytocin versus placebo in the treatment of adults with autism spectrum disorders: A randomized controlled trial. *Mol Autism* 2012, **3**(1).
102. Aman MG, McDougle CJ, Scahill L, Handen B, Arnold LE, Johnson C, Stigler KA, Bearss K, Butter E, Swiezy NB, Sukhodolsky DD, Ramadan Y, Pozdol SL, Nikolov R, Lecavalier L, Kohn AE, Koenig K, Hollway JA, Korzekwa P, Gavaletz A, Mulick JA, Hall KL, Dziura J, Ritz L, Trollinger S, Yu S, Vitiello B, Wagner A, Research Units on Pediatric Psychopharmacology Autism,Network: Medication and parent training in children with pervasive developmental disorders and serious behavior problems: results from a randomized clinical trial. *J Am Acad Child Adolesc Psychiatry* 2009, **48**(12):1143–1154.

Chapter 20

Medical Co-Morbidities in Autism: Clues to Underlying Biological Mechanisms and/or Diagnostic Subtypes?

Margaret L. Bauman*

Abstract

Since its original description in 1943, autism has been largely characterized on the basis of the clinical symptoms of atypical social skills, delayed and disordered language, and repetitive and stereotypic behaviors. It is estimated that approximately 50% of autistic persons are nonverbal, many have associated sensory processing dysfunction, and many exhibit disruptive behaviors that, until recently, have been commonly attributed to being "just part of the autism". In addition, many children on the autism spectrum can be difficult to physically examine within a traditional office practice. Thus, the evaluation and possible significance of associated medical conditions in this disorder has been largely overlooked. With advances in clinical care, the presence of a number of medical conditions have been identified, which when diagnosed and treated, result in improved behavior, developmental outcomes and quality of life. In addition, there is now the realization that by attending more closely to co-morbid medical conditions, it may be possible to begin to identify

*Correspondence to: Department of Anatomy and Neurobiology, Boston University School of Medicine, 72 East Concord Street, Boston, MA02118. Email: mbauman@bu.edu

specific clinical and biologic subtypes within the autism spectrum as well as potentially diagnostic biomarkers and more focused treatment approaches.

Keywords: Behavior; Medical evaluation; Treatment; Clinical subtypes; Developmental outcomes; Quality of life.

Introduction

Autism is a behaviorally defined disorder first described by Leo Kanner in 1943.[1] In his original description, Kanner noted that his patients exhibited the common features of "autistic aloneness" and "obsessive insistence on preservation of sameness". Over the years, the defining clinical features of the disorder have shifted from more psychodynamic explanations toward the concept that the cause of autism may be "inborn". By 1960, with the publication of the DSM-III, disordered language and communication and odd responses to the environment occurring during the first 30 months of life were added as core features. With the publication of the DSM-IV in 1994,[2] the definition of autism became more flexible with the increased recognition that autism represented a spectrum with varying degrees of symptom severity, cognitive abilities and behavioral manifestations, thus broadening the definition of the disorder. The recent release of the DSM-V in 2013[3] has sought to provide more rigorous diagnostic criteria which have been reduced to: (1) persistent impairment in reciprocal social communication and social interaction, and (2) restrictive repetitive patterns of behavior. However, diagnosis also requires a notation of the presence or absence of associated cognitive disability, medical or genetic conditions and/or language impairment.

Although the DSM-V provides an effort to tighten the definition of autism, there remains the contention of both researchers and clinicians that the disorder is heterogeneous in its etiology, neurobiology and clinical presentation, a contention now represented by the generally accepted term, Autism Spectrum Disorder (ASD). Clinically, variable symptoms include levels of cognition, with recent data suggesting that less than half of those affected with ASD have a significant associated intellectual deficit.[4] Measurements of head circumference in early childhood has documented that approximately 20% of autistic children demonstrate a decided macrocephaly of greater than 98 percentile for age and sex.[5]

Until relatively recently, the major focus of both clinical and basic science research related to ASD has been focused on the neurobiology of the brain and many significant advances have been reported as a result. However, over the past several years, there has been a growing appreciation that other organ systems may be playing a significant role in ASD and that many of these systems may be contributors to some of the clinical features of the disorder. Further, dysfunction in these systems may be intimately involved in the biological mechanisms that form the basis of the ASDs. Thus, a more directed evaluation of co-morbid medical conditions in ASD could help to identify phenotypes and genetic clusters of ASD individuals, thereby potentially allowing for the identification of subtypes, and through these subsets, better identify more specific etiologic and biological mechanisms which can ultimately lead to more focused therapies and interventions.

This chapter will address several co-morbid neurological and medical disorders which are being increasingly investigated in relationship to their impact on the clinical features of ASD, what is now known about their neurobiological mechanisms, and how they may inform the field.

Seizures

For many years, it has been recognized that a significant number of individuals with ASD develop epilepsy at some point during their lifetime. This relationship was the focus of a recent forum sponsored by the National Institute of Neurological Disorders and Stroke (NINDS), with the support of a number of other stakeholders and organizations, and the proceedings of this forum were published in the October 2013 issue of the journal *Neurology*.[6] This workshop considered the relationship between autism, epilepsy and intellectual disability (ID). The highest risk of epilepsy in ASD was found in those with ID at a rate of 21%, while the risk in ASD without ID was 8%. Further, studies have indicated that seizures beginning after the age of 12 years and continuing into adulthood may be more likely in ASD than other populations at risk for epilepsy.

Peak risk periods for seizures are said to occur during early childhood and adolescence.[7] Although complex partial seizures are most frequently reported, all types of seizures can occur in ASD including major motor, myoclonic and febrile seizures. Available data suggests that the majority of seizures that occur in ASD occur after 12 years of age with 30% occurring by 20 years of age.[8] While the predilection for seizures in ASD can be associated with ID, as noted above, other factors such as dysmorphic features, motor impairment as well as a long list of syndromes such as Tuberous Sclerosis, Fragile X syndrome, and Rett syndrome can also be important factors.

It has been suggested that the co-occurrence of seizures and ASD may reflect a shared underlying neurobiology and it has been hypothesized that the same pathophysiologic mechanisms may lead to abnormal synaptic plasticity and excitatory/inhibitory imbalance in the developing brain.[6] These abnormalities have been found in syndrome disorders, in rare copy number variants (CNVs) and rare mutations associated with ASD. However, few of the CNVs or mutations reported to be risk factors for ASD are specific to both epilepsy and ASD and all overlap with ID.[9] These factors will need to be taken into consideration when pursuing future genetic studies.

Environmental factors may also play a role in the etiology of ASD, ID and epilepsy. Risk factors that have been reported have included premature birth, low birth weight, certain maternal medications taken during pregnancy and advanced maternal and paternal age. Immunological contributors have also been reported to be potential risk factors.

ASD, epilepsy and ID have been reported to have some overlapping neuropathological and neurophysiologic features, including malformations of cortical development, cellular disorganization, heterotopias, and dysplasia, reflecting abnormalities of neuronal migration and neurogenesis.[6,10] Kang and Barnes (2013) hypothesize that defects in GABAergic signaling could account for the co-morbidity of ASD and seizures. Neurobiological studies in postmortem brain in subjects with ASD as well as in those with epilepsy, have demonstrated alterations in GABAa receptor subunits and GABAergic circuitry.[11–13] Further, both imaging and neuropathological studies have suggested that there may be a shared altered brain development that underlies both ASD and epilepsy, at least in some cases.[10] Future research will need to consider these and other possible neurobiologic mechanisms that more specifically define and characterize ASD individuals with and without seizures.

Gastrointestinal Disorders

At the present time, the true prevalence of gastrointestinal (GI) disorders in ASD is unknown with estimated rates ranging from 9% to 70% or higher, data varying widely depending on patient cohorts and symptoms required for inclusion in a particular study.[14] Although many ASD subjects will present easily recognized symptoms of vomiting, diarrhea, abdominal pain or constipation, others, most especially those who are non-verbal, may not demonstrate obvious clues. Currently, there are no reliable signs or symptoms that can consistently aid the primary care physician to distinguish between disruptive behaviors which may be related to pain and discomfort and those which have psychogenic or emotional origins.

Evidenced-based guidelines for the evaluation of GI disorders in autism are not currently available. However, in 2010, as the result of an extensive review of the literature and combined clinical and research expertise, a group of experienced gastroenterologists published a consensus paper outlining best practices for the diagnosis and treatment of GI disorders in children with autism.[15] The consensus paper focused on disorders such as chronic constipation, diarrhea, vomiting, gastritis, esophagitis, gastroesophageal reflux disease (GERD), colitis, celiac disease, inflammatory bowel disease and Crohn's disease. In addition, the group highlighted a number of disruptive behaviors which, in some cases, have been later found to be related to GI disorders and discomfort and which could be potential clinical "red flags" for diagnosing GI disturbances. Some of these behaviors include, facial grimacing and neck stretching, constant eating and drinking, chewing on non-edibles including shirt sleeves, unusual gulping, chest tapping, putting pressure on the abdomen, unexplained aggression and self-injury and sleep disturbances.[16]

It is well-known that, in typically developing children, disruptive behaviors often occur when a child is not feeling well. There is little reason to believe that it should be any different for children on the autism spectrum. Likewise, it is known that when individuals are unwell, their ability to actively participate in or do well in daily activities, learning environments or in social situations is impaired. By identifying and treating medical condition, such as GI disorders, there is a reasonable likelihood that the treated individual will do better educationally, socially and developmentally.

Studies which are of growing interest in ASD include investigations pertaining to intestinal malabsorption and permeability, gastrointestinal dysmotility, lactose intolerance, the role of food allergies and food sensitivities, the importance of shared neurotransmitter systems between the gut and the brain, and the effect of various dietary interventions. In addition, there is an added interest in the bacterial content of the gut and how this may impact function and general health. Active investigations into these and other factors involving the gut should substantially improve and expand our understanding of potential biologic mechanisms which may underlie this disorder.

It is not yet known whether GI disorders are more common in the ASD population when compared with neurotypical individuals. Similarly, it is not known whether GI disorders in ASD should be treated by the same interventions used for typically developing persons, nor is it known whether the GI disturbances, when present, are inherent to a subset of ASD individuals or are present as an associated phenomenon. There is some evidence to suggest that, in some cases, those ASD persons with GI disorders may represent a subset or part of a subset of individuals on the autism spectrum. In 2009, Campbell *et al.* reported that a disruption of MET gene signaling could contribute to an increased risk for ASD that includes familial GI dysfunction.[17] MET is a pleiotropic gene that is known to be important for brain development as well as GI repair and motility. This study suggests that by including studies of medical disorders, such as GI disorders, as part of the phenotypic characterization of ASD individuals, we may be able to begin to define meaningful subsets within the spectrum of this complex and heterogeneous disorder.

Sleep Disorders

Disturbed sleep patterns are extremely common among individuals with autism, with prevalence rates varying from 40%–80%.[18] Sleep patterns most commonly reported by parents include delayed sleep onset, impaired sleep maintenance, and poor sleep duration.[19] However, other sleep disorders have also been documented in ASD including abnormal patterns of circadian rhythm, atypical sleep-related breathing, parasomnias (nightmares, screaming), and sleep-related movement disorders.[20] While some of these patterns, most especially nighttime awakenings, can be seen in typically developing children, these events can often be more prolonged in the ASD child.

The cause of the high prevalence of sleep disturbances in ASD individuals is largely unknown. Hypotheses include: (1) factors in the home or environment that are not conducive to good sleep, (2) psychological or behavioral disorders associated with core features of ASD, or (3) biological or genetic abnormalities that alter brain structure and/or chemistry. It is possible that any one of these potential factors, or a combination of factors could contribute to these sleep disturbances.[20]

Of the possible psychological factors that could potentially contribute to sleep disturbances in ASD, anxiety, depression, mood disorders, stress and attention deficit hyperactivity disorder each or together could play a significant role. Alternatively certain medical conditions including obstructive apnea, sleep disordered breathing, respiratory compromise secondary to enlarged tonsils and adenoids, as well as gastrointestinal disorders such as GERD could be contributory. It is also recognized that medications often used in the management of associated conditions in ASD, including serotonin reuptake inhibitors, stimulants, some anticonvulsants, corticosteroids and asthma treatments could play a role.

Of the potential biological factors, the role of melatonin has become the subject of growing interest. Melatonin is a neurohormone produced from serotonin in the pineal gland and secreted during the nighttime hours. It is believed to play a role in the organization of circadian rhythm, including sleep-wake cycles, and the regulation of core body temperature cycles. It has been hypothesized that sleep problems in ASD may be related to circadian rhythm dysregulation which may cause the nervous system to fluctuate causing faulty transmission of entrainment cues. Thus, based on what is known about the functions of melatonin, it has been speculated that sleep disorders in ASD may be related to genetic abnormalities associated with melatonin synthesis and its role in modulating synaptic changes.[20]

Sleep disorders can have wide-ranging effects on daytime behaviors, developmental progress and quality of life for the child and his family. Effective interventions therefore become a high priority. Intervention commonly includes efforts to provide predictable bedtime routines and correct problem behaviors, often using behavioral approaches. Pharmacological interventions may also be considered. There is growing evidence to support the use of melatonin which is considered a nutritional supplement and can therefore be obtained without a prescription. Available data suggests that this product can be effective in promoting sleep onset and appears to have few if any long-term negative side effects. However, to date, melatonin has not been carefully studied with regard to dosage, safety and efficacy and further studies are needed. Similarly, there is limited data to support the use of other pharmaceutical products for promoting and sustaining sleep in ASD individuals. Medications, including clonidine, trazodone, fluvoxamine, mirtazapine or clomipramine each have been used successfully in some cases as have some anticonvulsants. None of these products have been rigorously studied with regard to their effectiveness in treating sleep disorders in ASD and further studies are warranted. Studies such as these could potentially result in significant insights into some of the underlying biological mechanisms which may be at play in this disorder, and may further help to delineate subsets within the spectrum.

Metabolic Disorders

Among patients with neurodevelopmental disorders, identifiable metabolic abnormalities are, in general, considered a relative rarity with an estimated prevalence of 1%–2.5%.[21] At the present time, the prevalence of metabolic disorders among the autism population specifically is unknown.

Mitochondrial disorders

Over the past decade, there has been an increasing interest in and growing evidence for the potential role of mitochondrial dysfunction in non-syndromic autism. At this time, estimates of the prevalence of mitochondrial dysfunction in ASD vary widely. In 2005, Oliveira *et al.* published the results of a large population-based study, based on a survey of school-aged ASD children, and noted that 7.2% of their subjects met criteria for a mitochondrial respiratory chain disorder.[22] However, a more recent study has suggested that mitochondrial dysfunction may be present in as many as 80% of ASD children.[23]

Mitochondria are distinct organelles that are known to perform four central functions in a cell that are relevant to the pathophysiology of disease. These functions include: (1) the provision of most of the energy to a cell through the formation of adenosine triphosphate (ATP) as the result of the oxidation of glucose and fatty acids,

(2) they generate and regulate reactive oxygen species, (3) they buffer cytosolic calcium, and (4) they regulate apoptosis through the mitochondrial permeability transition pore.[24] Mitochondria provide much of the energy to a cell, and mitochondrial dysfunction therefore has a greater effect on tissues with high-energy demands such as the brain, muscle, heart, gastrointestinal tract, and the endocrine system. However, any organ system can be involved.

Mitochondrial disorders are heterogeneous and multiple organ systems can be involved. Mitochondrial dysfunction can be caused by genetic defects in mitochondrial DNA (mtDNA) or in the nuclear DNA (nDNA) gene that encodes a mitochondrial protein or structural RNAs. Mitochondrial disorders are believed to be relatively rare, with those related to mtDNA estimated to have a frequency of 1 in 5000, and known pathogenic mtDNA mutations detected in cord blood in 1 in 200 live births. The frequency of nDNA mutations affecting mitochondrial genes may be more common than mtDNA mutations.[24]

Clinically distinguishing ASD children with and without mitochondrial disorders can be challenging, and there is as yet no consistently reliable screening test for these disorders. Diagnostic criteria have been outlined by Morava *et al.* (2006), and have been widely used by both clinicians and researchers.[25] These criteria assign point values for clinical, imaging, metabolic and morphological factors, and the probability of a mitochondrial disorder is based on a total score. The need for a muscle biopsy to confirm a diagnosis of a mitochondrial disorder can be a confounding factor since many families will not consent to this invasive procedure. The absence of a biopsy can negatively impact the certainty of the diagnosis. In 2008, Weissman *et al.* reported on a group of 25 ASD subjects, each having a biopsy-proven mitochondrial disorder.[26] The authors reviewed past history, clinical and laboratory features and developmental trajectory, and concluded that ASD children with and without mitochondrial disorders can be distinguished from each other based a number of symptomatic and laboratory "red flags", including poor physical endurance, delayed motor milestones, repeated regressions after the age of 3 years, and the involvement of multiple organ systems. Other features included abnormalities of blood lactate, pyruvate and plasma alanine. The authors concluded that, given detailed clinical and biochemical evaluations, ASD children with co-occurring mitochondrial disorder could be distinguished from those with essential autism. Further, the authors raised the possibility that ASD with co-occurring mitochondrial dysfunction might represent a subset of individuals within the autism spectrum.

At this time, there continues to be a growing interest in the role of mitochondrial dysfunction in ASD. Whether this dysfunction is primary (causative) or secondary (an associated phenomenon) remains uncertain. The significance of abnormal mitochondrial function in autism will be an important focus for future research.

Other potentially significant metabolic disorders

Given the growing interest in mitochondrial disorders, clinicians and investigators working in the field of autism are expanding research into other potential metabolic disorders that may be associated with autistic symptoms. Space does not allow for a detailed summary of many of these disorders but several disorders will be mentioned here, many important because their identification could, in some cases, lead to meaningful treatments and clinical improvement.

Smith-Lemli-Opitz syndrome (SLOS)

Smith-Lemli-Opitz syndrome (SLOS) has been frequently been associated with symptoms with autism.[27] SLOS is an autosomal recessive disorder associated with an inborn error of cholesterol synthesis caused by mutations of the 7-dehydrocholesterol reductase gene (DHCR7) located on chromosome 11q12-13. Cholesterol is essential for neuroreactive steroid production, growth of myelin membranes as well as normal and embryonic and fetal development. Further, cholesterol is also believed to modulate oxytocin receptors, G-protein coupling of the serotonin-1A receptor and ligand activity. It is suggested that a deficit of cholesterol may negatively impact these biological mechanisms, thus leading to the symptoms of autism observed in this disorder. It has been estimated that 5–75% of SLOS individuals may meet criteria for ASD.[28] The diagnosis of SLOS can be confirmed by the presence of an elevated plasma level of 7-dehydrocholesterol relative to the cholesterol level. Identification of this disorder is important given that it is one cause of autism that is partially treatable with cholesterol supplementation.

Cerebral folate deficiency (CFD) syndrome

Cerebral folate deficiency (CFD) syndrome is believed to be caused by impaired transport of folate across the blood-brain barrier. Six studies have reported ASD in a subset of children with CFD, most of the subjects described being low functioning with significant neurological abnormalities. However, in a recent study of 93 ASD children who showed no significant neurological abnormalities, Frye and Rossignol (2013) found that 75.3% demonstrated the presence of folate receptor autoantibodies (FRAs) which typically interfere with folate transport across the blood brain barrier.[29] The FRA positive children who were treated with leucovorin calcium demonstrated significant improvement in language and communication, attention and stereotypical behavior. Given the apparent substantial prevalence of FRA in ASD children, the authors have recommended that FRA testing be considered in all ASD children.

X-linked inborn error of carnitine biosynthesis

X-linked inborn error of carnitine biosynthesis has recently been hypothesized to be a risk factor for non-dysmorphic autism. Celestino-Soper *et al.* (2012) recently reported a deletion of exon 2 of the trimethyllysine hydroxylase epsilon gene (TMHLE) in a proband with autism.[30] TMLHE is known to map to the X chromosome and encodes the first enzyme in carnitine biosynthesis. TMLHE deficiency was found to be 2.82 times more frequent in male-male multiplex ASD probands than in controls or in simplex families. These findings suggest that dysregulation of carnitine metabolism may be an important factor in non-dysmorphic autism and that the carnitine pathway may be significant as it may contribute to some of the symptoms of autism. Research in this area may lead to possible prevention and therapy.

These disorders, as well as an increasing number of other syndromes have been identified in association with ASD. For example, phenylketonuria has long been identified as a disorder associated with autistic features in some cases. Creatine deficiency,[31] urea cycle disorder,[32] Wilson's Disease and Lesch-Nyhan syndrome[33] have become topics of increasing interest. This growing interest in a number of metabolic disturbances may reflect the fact that, if and when identified, many of these disorders have been found to be treatable and many have shown a reduction of autistic symptomatology as a result. In addition, the identification of these syndromes may begin to categorize the ASDs into subsets which may help to better understand the underlying biological mechanisms that form the basis of some of the core features of ASD.

Discussion

Since its original description in 1943, autism has been largely characterized by a number of behavioral, cognitive, communication/language and social atypicalities, which for some years, were believed to be psychiatric in origin, but for the past twenty-five years, has been identified as a neurodevelopmental disorder of the brain. Evidence indicates that this disorder, in most cases, begins before birth. Autism is now recognized as being heterogeneous clinically, etiologically and biologically, resulting in the now commonly used term, "Autism Spectrum Disorders" (ASDs).

Most of the research related to ASD has been focused on studies of the brain and associated neurological symptoms. However, over the past 5–10 years, it has become increasingly apparent that organ systems outside of the brain are often involved in many cases. In addition, there has been growing and convincing evidence that pain and discomfort, often secondary to any number of medical conditions, can result in disruptive, repetitive and often aggressive behaviors. This phenomenon has been described by Carr and Owen-Deschryver (2007) who reported problem behaviors associated with pain in minimally verbal developmentally disabled individuals.[34] It is estimated that approximately 50% of ASD individuals are non-verbal and therefore unable to report illness or discomfort. In addition, many have sensory processing inefficiencies that limit their ability to identify the location of their distress. Thus, it should be no surprise that the only mechanisms that some ASD persons have to express their discomfort is through behavioral mannerisms.

As noted above, there are a number of medical conditions which have been documented as occurring fairly frequently in ASD. Until recently, these medical disorders have received little attention, probably for a variety of reasons. First, ASD individuals do not present with symptoms that are easily recognized by the average primary care physician in dealing with typically developing children. Further, ASD children can be difficult to examine due to limited verbal/communication skills which is compounded by poor cooperation and office behavior. It is now apparent that medical disorders in ASD can present in the form of disruptive behaviors, which until recently, have been attributed to "simply being part of the autism". While reliable prevalence data does not yet exist, disorders of the gastrointestinal tract appear to be fairly common. However, other potentially pain-related disorders, such as infections, sore throat, dental discomfort, sinusitis, urinary track disorders, or fractures have yet to be explored. Nonetheless, it has become apparent that the presence of challenging behaviors in ASD children should lead to exploration of potential underlying medical conditions before assuming that these symptoms are simply "behavioral" requiring behavioral intervention and/or psychopharmacological medications. Further, there is a growing body of evidence that treating underlying medical conditions leads to an improved sense of well-being, more effective participation in educational and therapeutic programs, and improved quality of life for the patient and his/her family.

Of additional importance is the inclusion of identifiable medical conditions as part of future imaging, genetic, clinical and biomarker studies in ASD. It is now recognized that other organ systems outside of the brain share many neurotransmitters with the brain. It is known, for example that serotonin, long a neurotransmitter associated with autism, is generated in the gut. Identification of associated medical conditions may lead to the isolation of significant subsets within the autism spectrum, which will then allow us a more directed biological investigation of these groups, hopefully leading to a more specific understanding of the underlying mechanisms related to these groups, potential diagnostic biomarkers and more focused treatment approaches.

Future Directions

Given the complexity of the ASDs, it will be critical that future studies, including genetics, imaging, psychopharmacological clinical trials, immunological, and other clinical projects, should include efforts to evaluate and describe any medical conditions that are part of the subject's past or present status. Laboratory studies looking at potential biomarkers should be pursued by looking at a number of laboratory studies relevant to a particular clinical group and based on the presence or absence of defined medical conditions. These studies may involve longitudinal studies in the same population since behavioral, developmental and medical patterns may change over time. Developmental trajectories could be important clues and should be documented in detail when possible. Lastly, evidence of disruptive behaviors or unexplained changes in behavior should be actively investigated to rule out any underlying medical causes.

Executive Summary

- ASD is a heterogeneous and complex disorder, with varied causes and clinical presentations.
- Medical conditions associated with ASD have been poorly studied.

- Evaluation of and defining medical conditions in ASD can lead to the identification of potential treatable conditions which can result in improved developmental outcomes and improved behaviors.
- Evaluation of co-morbid medical conditions can potentially help to define clinical and biological subsets within the autism spectrum.
- Evaluation of co-morbid medical conditions in ASD could help to define potential diagnostic biomarkers and possibly more focused treatment approaches.

References

1. Kanner L: Autistic disturbances of affective contact. *Acta Paedopsychiatr* 1968, **35**(4):100–136.
2. American Psychiatric Association: *Diagnostic and Statistical Manual of Mental Disorders, 4th edition:* 4th ed. Washington, DC: American Psychiatric Association; 1994.
3. American Psychiatric Association: *Diagnostic and Statistical Manual of Mental Disorders, 5th edition:* Arlington, VA: American Psychiatric Association; 2013.
4. Newschaffer CJ, Croen LA, Daniels J, Giarelli E, Grether JK, Levy SE, Mandell DS, Miller LA, Pinto-Martin J, Reaven J, Reynolds AM, Rice CE, Schendel D, Windham GC: The epidemiology of autism spectrum disorders. In *Annu Rev Public Health.* Edited by Anonymous 2007:235–258. [Anonymous, *vol 28.*]
5. Lainhart JE, Bigler ED, Bocian M, Coon H, Dinh E, Dawson G, Deutsch CK, Dunn M, Estes A, Tager-Flusberg H, Folstein S, Hepburn S, Hyman S, McMahon W, Minshew N, Munson J, Osann K, Ozonoff S, Rodier P, Rogers S, Sigman M, Spence MA, Stodgell CJ, Volkmar F: Head circumference and height in autism: a study by the collaborative program of excellence in autism. *Am J Med Genet A* 2006, **140**(21):2257–2274.
6. Tuchman R, Hirtz D, Mamounas LA: NINDS epilepsy and autism spectrum disorders workshop report. *Neurology* 2013, **81**(18):1630–1636.
7. Volkmar FR, Nelson DS: Seizure disorders in autism. *J Am Acad Child Adolesc Psychiatry* 1990, **29**(1):127–129.
8. Rossi PG, Parmeggiani A, Bach V, Santucci M, Visconti P: EEG features and epilepsy in patients with autism. *Brain Dev* 1995, **17**(3):169–174.
9. Mefford HC, Batshaw ML, Hoffman EP: Genomics, intellectual disability, and autism. *N Engl J Med* 2012, **366**(8):733–743.
10. Kang TQ, Barnes G: A common susceptibility factor of both autism and epilepsy: functional deficiency of GABA A receptors. *J Autism Dev Disord* 2013, **43**:68–79.

11. Fatemi SH, Aldinger KA, Ashwood P, Bauman ML, Blaha CD, Blatt GJ, Chauhan A, Chauhan V, Dager SR, Dickson PE, Estes AM, Goldowitz D, Heck DH, Kemper TL, King BH, Martin LA, Millen KJ, Mittleman G, Mosconi MW, Persico AM, Sweeney JA, Webb SJ, Welsh JP: Consensus paper: pathological role of the cerebellum in Autism. *Cerebellum* 2012, **11**(3):777–807.
12. Oblak AL, Gibbs TT, Blatt GJ: Reduced GABAA receptors and benzodiazepine binding sites in the posterior cingulate cortex and fusiform gyrus in autism. *Brain Res* 2011, **1380**:218–228.
13. Oblak A, Gibbs TT, Blatt GJ: Decreased GABAa receptors and benzodiazepine binding sites in the anterior cingulate cortex in autism. *Autism Res* 2009, **2**(4):205–219.
14. Buie T, Campbell DB, Fuchs III GJ, Furuta GT, Levy J, Van De Water J, Whitaker AH, Atkins D, Bauman ML, Beaud AL, Carr EG, Gershon MD, Hyman SL, Jirapinyo P, Jyonouchi H, Kooros K, Kushak R, Levitt P, Levy SE, Lewis JD, Murray KF, Natowicz MR, Sabra A, Wershil BK, Weston SC, Zeltzer L, Winter H: Evaluation, diagnosis, and treatment of gastrointestinal disorders in individuals with ASDs: A consensus report. *Pediatrics* 2010, **125**(SUPPL. 1):S1-S18.
15. Buie T, Fuchs III GJ, Furuta GT, Kooros K, Levy J, Lewis JD, Wershil BK, Winter H: Recommendations for evaluation and treatment of common gastrointestinal problems in children with ASDs. *Pediatrics* 2010, **125**(SUPPL. 1):S19-S29.
16. Buie T: Clinical presentation of gastrointestinal disorders in non-verbal children with autism. Unpublished data, personal communication, 2013.
17. Campbell DB, Buie TM, Winter H, Bauman M, Sutcliffe JS, Perrin JM, Levitt P: Distinct genetic risk based on association of MET in families with co-occurring autism and gastrointestinal conditions. *Pediatrics* 2009, **123**(3): 1018–1024.
18. Johnson KP, Giannotti F, Cortesi F: Sleep patterns in autism spectrum disorders. *Child Adolesc Psychiatr Clin N Am* 2009, **18**(4):917–928.
19. Malow BA, Marzec ML, McGrew SG, Wang L, Henderson LM, Stone WL: Characterizing sleep in children with autism spectrum disorders: a multidimensional approach. *Sleep* 2006, **29**(12):1563–1571.
20. Richdale AL, Schreck KA: Sleep problems in autism spectrum disorders: Prevalence, nature, & possible biopsychosocial aetiologies. *Sleep Med Rev* 2009, **13**(6):403–411.
21. van Karnebeek CDM, Jansweijer MCE, Leenders AGE, Offringa M, Hennekam RCM: Diagnostic investigations in individuals with mental retardation: a systematic literature review of their usefulness. *Eur J Hum Genet* 2005, **13**(1):6–25.
22. Oliveira G, Diogo L, Grazina M, Garcia P, Ataide A, Marques C, Miguel T, Borges L, Vicente AM, Oliveira CR: Mitochondrial dysfunction in autism spectrum disorders: a population-based study. *Dev Med Child Neurol* 2005, **47**(3):185–189.

23. Giulivi C, Zhang Y-, Omanska-Klusek A, Ross-Inta C, Wong S, Hertz-Picciotto I, Tassone F, Pessah IN: Mitochondrial dysfunction in autism. *JAMA — J Am Med Asso* 2010, **304**(21):2389–2396.
24. Wallace DC, Fan W, Procaccio V: Mitochondrial energetics and therapeutics. In *Annu Rev Pathol Mech*. Edited by Anonymous 2010:297–348. [Anonymous, *vol 5*.]
25. Morava E, Van Den Heuvel L, Hol F, De Vries MC, Hogeveen M, Rodenburg RJ, Smeitink JAM: Mitochondrial disease criteria: diagnostic applications in children. *Neurology* 2006, **67**(10):1823–1826.
26. Weissman JR, Kelley RI, Bauman ML, Cohen BH, Murray KF, Mitchell RL, Kern RL, Natowicz MR: Mitochondrial disease in autism spectrum disorder patients: a cohort analysis. *PLoS One* 2008, **3**(11).
27. Bukelis I, Porter FD, Zimmerman AW, Tierney E: Smith-Lemli-Opitz syndrome and autism spectrum disorder. *Am J Psychiatry* 2007, **164**(11):1655–1661.
28. Aneja A, Tierney E: Autism: the role of cholesterol in treatment. *Int Rev Psychiatry* 2008, **20**(2):165–170.
29. Frye RE, Rossignol DA: Mitochondrial dysfunction can connect the diverse medical symptoms associated with autism spectrum disorders. *Pediatr Res* 2011, **69**(5 PART 2):41R-47R.
30. Celestino-Soper PBS, Violante S, Crawford EL, Luo R, Lionel AC, Delaby E, Cai G, Sadikovic B, Lee K, Lo C, Gao K, Person RE, Moss TJ, German JR, Huang N, Shinawi M, Treadwell-Deering D, Szatmari P, Roberts W, Fernandez B, Schroer RJ, Stevenson RE, Buxbaum JD, Betancur C, Scherer SW, Sanders SJ, Geschwind DH, Sutcliffe JS, Hurles ME, Wanders RJA, Shaw CA, Leal SM, Cook Jr. EH, Goin-Kochel RP, Vaz FM, Beaud AL: A common X-linked inborn error of carnitine biosynthesis may be a risk factor for non-dysmorphic autism. *Proc Natl Acad Sci USA* 2012, **109**(21):7974–7981.
31. Nasrallah F, Feki M, Kaabachi N: Creatine and creatine deficiency syndromes: biochemical and clinical aspects. *Pediatr Neurol* 2010, **42**(3):163–171.
32. Görker I, Tüzün Ü: Autistic-like findings associated with a urea cycle disorder in a 4-year-old girl. *J Psychiatry Neurosci* 2005, **30**(2):133–135.
33. Trauner DA: Cognitive and behavioral manifestations of metabolic and chromosomal disorders. In *Pediatric Behavioral Neurology*. Edited by Frank Y. New York, NY: CRC Press; 1996:251–268.
34. Carr EG, Owen-DeSchryver JS: Physical illness, pain, and problem behavior in minimally verbal people with developmental disabilities. *J Autism Dev Disord* 2007, **37**(3):413–424.

Chapter 21

Translational Implications of a Whole-Body Approach to Brain Health in Autism: How Transduction between Metabolism and Electrophysiology Points to Mechanisms for Neuroplasticity

Martha R. Herbert*

Abstract

Autism has been viewed as a highly heritable neurobiological condition of mysterious but presumably genetic origin, which involves lifelong neurocognitive, perceptual and emotional deficits. This conceptual framing has led to a focus on searching for underlying genetic causes of differences in the autistic brain, particularly in anatomical structure, that are presumed to be hardwired into the system.

More recently, it is becoming clear that genes alone do not create autism. The more inclusive emerging view is that genes and environment interact to influence epigenetics, cell signaling and physiology.

*Correspondence to: TRANSCEND Research Program, Massachusetts General Hospital, Harvard Medical School, Higher Synthesis Health, Boston, MA 02129. Email: marthaherbertmd@gmail.com

This formulation goes beyond the idea that a preconceptional or prenatal gene-environment interaction definitively causes a person's autism, and instead emphasizes that the interplay of all of these levels develops over time contingent upon exposures, experiences and lifestyle choices, and that the behaviors are actively produced by living cells that function differently, rather than simply being caused by static brain wiring that is different from the start.

From this vantage point it is important to look freshly at how we think about the brain, and what it is that the brain *does* to create autistic behaviors. A multi-scaled, whole-body, dynamical approach to the processes of signal generation in the brain and how these processes are shaped by ongoing dynamic interplays of multiple modulators offers previously unappreciated opportunities for improvement of brain function.

Keywords: Autism spectrum disorders; Brain; Oscillation; Multi-scale biology; Systems biology; Dynamical systems; Magnetic resonance spectroscopy; Diffusion tensor imaging; EEG; Neuromodulation; Metabolism; Mitochondrial dysfunction; Connectivity; Coherence; Neuroplasticity.

Introduction

Autism (which we will call Autism Spectrum Conditions (ASCs) rather than Autism Spectrum Disorders (ASDs) due to the neutrality of the word "condition" as compared to the judgment of deficit implied in the word "disorder" that is now in dispute both culturally and scientifically) has been viewed as a highly heritable neurobiological condition, or set of conditions, of mysterious but presumably genetic origin, involving lifelong neurocognitive, perceptual and emotional deficits. This framing has led to a focus on searching for underlying genetic causes of differences in brain, particularly in anatomical structure, that are presumed to be hardwired into the system. Researchers have also focused on developing more refined methods to characterize the nature of autism deficits to provide rigorous correlates for the genetic and brain research. Also prominent in this approach is an emphasis on pharmaceutical treatment of gene-related targets.

However, in the emerging "gene-environment-epigenetic-cell signaling-physiology" framework of dynamical multi-scale systems biology, maintaining a simple bottom-up deterministic approach is giving way to multi-scale, interactive, dynamical understandings of ASCs. *This in turn opens the way to entry points into intervention from multiple levels.*

Our framing of the role of the brain in creating and sustaining autism needs to be brought up to date with this systems approach. Therefore, in what follows we will:

- review the reasons for broader, more multisystem and more dynamical understandings of ASCs;
- look at what we already have observed about the brain from this fresh perspective, and
- offer a dynamical, process-oriented, electrophysiologically centered model of the biology that generates the behavioral outputs we label as "autism".

The approach elaborated here emphasizes the sources of brain function plasticity that can derive from whole-body interventions that can modulate brain health. The overall point is that when we are able to transition from the belief that autism is a hardwired trait to the model that much or sometimes all of autism may be related to states that are dynamic and changeable, we will be able to make much more rapid and effective progress toward minimizing suffering and maximizing potential in this growing group of people with huge challenges but also great gifts.

Beyond Hardwired: Autism Not Born but Made

Although autism has been considered a lifelong genetically-caused neurological condition, evidence and clinical observations are pointing to a different framing of autism, namely that autism may not be totally or even mainly hardwired from genetics, but may emerge from a combination of interacting exposures and stressors, some or many of which can be dialed back. Here are some of the considerations pointing toward this reframing of autism:

- ***Numbers of reported cases are increasing markedly,*** and analyses of these trends show that this cannot be ascribed purely to increasing awareness, earlier age at diagnosis or more lenient diagnostic criteria[1,2] — and since you cannot have a genetic epidemic, at least some of these new cases must arise from "environment" — i.e., changes in our world and how we live in it.
- ***We are learning that there is more to autism than genes, brain and behavior,*** because problems becoming more commonly identified in ASCs include a range of medical issues and whole-body/systemic physiological abnormalities that go beyond the disorders of early brain development previously thought to be the causes and core of autism.[3,4]
- ***We are learning that there are more contributors to behavior than brain.*** More broadly, science beyond just autism is revealing that shifts in gut, immune, endocrine and other body systems as well as microbial ecology can change the way the brain functions.[5–12]
- ***Regression really does occur.*** Whereas it used to be thought that parents who claimed that their children had regressed into autism were naïve new parents who had missed earlier signs of abnormality, regression of previously essentially healthy infants and toddlers into autism is now solidly documented; this suggests that autism may develop — and contingently so, depending on the accumulation of exposures and circumstances — rather than be predetermined and hardwired prenatally or even from conception.[13–17]
- ***Regression and medical features may be related,*** given the common association between these phenomena. Children who regress into autism often have a history of medical problems such as recurrent infections, gastrointestinal issues, food sensitivities, sleep disturbances, rashes and irritability.[18–20]
- ***Autism is not fixed but rather is variable.*** More careful observation of individuals with autism has revealed substantial variability of severity *within* as well as between individuals — over short and

longer terms. Sometimes a good day can be turned into a bad day quickly by a food or chemical exposure. Sometimes there is improvement that is transient and goes away — improvement during fever,[21,22] clear-fluid diet before medical procedures, during treatment with some antibiotics.[23] Some children with autism who receive short courses of steroids for acute problems like asthma attacks speak much more fluently and are more interactive for the duration of the treatment. Sometimes improvement persists, even to the point of loss of diagnosis, which has now been documented in a significant minority of individuals with autism. All of this variability undermines the idea that we can treat the level of severity on any particular day as genetically hardwired. This will be discussed more extensively below.

- ***Autism can no longer be considered simply a set of "deficits,"*** since many of the aspects of human functioning previously considered deficient or absent are now known to be present. Cognitive deficiency turns out to have been greatly overestimated, in part from testing people with ASCs with tests not suited for their cognitive styles.[24,25] Certain areas of perception are stronger and more discerning in people with ASCs than in neurotypical people.[26,27] Contrary to stereotypes, there is certainly no lack of emotion, though there may be issues with its expression. Lack of speech does not at all equate with lack of intelligence or even lack of the ability to read or write; more and more people with non-verbal ASCs or with verbal apraxia (difficulties in orally producing speech) and/or motor coordination or proprioceptive challenges are able to type, some even to the point of writing books.[24,26,28]
- ***The different aspects of autism are interconnected.*** Although not everyone has the same combination of symptoms, that does not make each symptom strictly discrete and separated from the others. One piece of support for this is the often widespread impact of systems-oriented treatments and interventions, which although commonly observed needs better published documentation. Though conventional medicine thinks that treatments should target specific symptoms, various kinds of treatments targeting

resiliency more generally — whether improving diet, reducing toxicity or improving neuro-movement coordination — lead to improvement in domains not specifically targeted by the treatments. This suggests that autism is not simply a collection of discrete, independently determined symptoms, but also — or perhaps instead — involves an overall compromise in adaptive function and brain coordination that affects many neurofunctional domains at once.[29] If the latter is the case, it makes sense that improving resiliency (with the idea of improving the brain's capacity to coordinate) can have a constructive impact on many domains at once, or at least over time. This phenomenology will be explored in further detail below in the section on autism as an "emergent property" rather than a discrete entity.

To summarize the implications of the above points in a sentence, it appears that rather than being uniformly predestined and lifelong, autism can be created, can emerge, and thus can also be ameliorated or even resolved if the underlying factors driving its emergence can be overcome or dialed back.

What are These New Observations Telling Us About How Autism "Works"?

Each of the observations above indicates interactions between biological and behavioral levels of functioning.

- Increasing numbers of cases suggests that the way the brain produces behaviors is being altered by environmental influences.
- Multisystem medical issues in autism, which have been labeled as "co-morbidities" as if they were there by coincidence, may actually be integrally related to the overall set of issues that create the "autism."
- Regression into autism suggests that the overall system can be shifted so that the way the brain functions is changed over time.
- The common association of regression with medical problems suggests that these brain changes occur in relation to stressors that are changing the underlying biological "settings" of the brain.

- Day-to-day variability in autism suggests that brain function is not totally hardwired but that its ups and downs may be influenced by environment — for example, brain inflammation may be worse with chemical or allergen exposure (or even sleep deprivation), or better on an allergen-free diet or reduced stress and feeling of safety.
- Transient and persistent improvements and loss of diagnosis suggest that ***autism is a functional issue, and not necessarily a hard-wired trait*** — and that the functional problems are related to ***obstruction*** of the expression of potential, rather than an underlying deficit of potential. The association of improvement with removal of noxious triggers, such as allergens or toxic exposures, is further support for the suggestion that the physiology that modulates brain "settings" may be altered by day-to-day biological influences.
- The observation that many symptoms may improve in coordination with each other suggests that conceptualizing symptoms as discrete entities created by distinct genetic influences may not be biologically accurate.

New Directions for Research and Treatment: Whole-Body Approach to Brain

If autism is something that develops rather than being predestined by genes, a whole new set of questions arise to drive research and treatment. These questions center around the ways different systems of brain and body relate to each other — how they maintain each other, help each other, or hurt each other. This is important because we would like to identify treatments and interventions that have the strongest and most widely distributed positive impacts possible — we want *the most leverage possible* out of what we do. To get this leverage, we need to focus on clarifying critical and intrinsically translational questions such as:

- How does regression happen? What are the changes in brain and body? What triggers drive these changes?
- How do problems with medical and systems physiology throughout the body affect the brain?

- What are the health problems of brain cells and tissue that underlie or contribute to what we label as "autism"?
- In particular, how does compromise of the biological health status of the brain cells and tissues affect the electrical signaling activities of the brain?
- How does remission or recovery happen? What are the changes in brain and body? Are they like playing a movie of regression in reverse? Or are the processes different?

How a Whole Body Approach Reframes the Role of the Brain in Autism

The issue I will focus on most centrally is the implications all of these observations and questions have on our understanding of the brain in autism. ***If*** we simply assume that autism is "hardwired" into the brain, ***then*** we look for the genes that "create" the wiring diagram, and seek treatments to ***compensate*** for the situation, believing still that at its core autism is fixed and unchangeable. It does not change things that much to shift the focus just to epigenetics, if the term is used in a generic, nonspecific way while the belief remains that the outcome is essentially hardwired (even though epigenetics is in its essence *not* hardwired but responsive to physiological and environmental stressors — or supports).

But ***if*** we notice the ways in which the brain is ***not*** fully hardwired, then we start to ***wonder and seek for*** where its flexibility and plasticity might come from. This allows us to be open to noticing and enhancing ways to maximize the expression of brain potential.

I propose that maximizing brain potential is best accomplished through a whole-body approach to the brain. My rationale is that many sources of transformation may arise from the way improvements in the health status of some levels of the brain and body ***transduce into — are transformed into*** — improved function at other levels of the system, particularly the brain.

Multi-Scale Interacting Aspects of Brain Biology and Function

Here are some of the reasons that the "genes-brain-behavior" model is an oversimplification of a remarkably more complex situation.

Problems in the body that have parallels in the brain

The challenges that the brain faces are in major part related to the problems faced by molecules, cells and tissues in other parts of the body. By now abundant research is showing that biochemical, metabolic, or energy-producing processes may work inefficiently or in an unbalanced way in people with autism spectrum conditions (as well as in a variety of other chronic conditions). Cells, tissues and organs may have difficulty with detoxification of noxious substances, or with getting rid of their garbage (such as misfolded proteins).[30] Cell membranes or internal structures like the cellular skeleton (cytoskeleton), receptors and channels in the membranes and various other cellular components may have problems with their structure or function.[31–38] These problems can include alteration of or injury to the lipid/fat composition of membranes that may lead to stiffness or frailty that impedes proper function of receptors and channels that live in the membranes, or that may lead to leakiness that compromises the separations between inside and outside of the cell, or between compartments inside the cell, that membranes are supposed to provide.[39,40] On a larger scale, people's bodies contain barriers such as the gut-blood barrier or the blood-brain barrier whose structures or functions may be compromised in ASCs (or other chronic conditions).[41–45] Cellular and tissue function, energy and resilience may be altered, depleted or hampered by inflammation or oxidative stress.[46,47] Atypical populations of organisms in the microbiome, including missing helpful organisms and/or the presence of harmful organisms, may produce substances that circulate and change the activities of cells in other parts of the body.[48–53] All of these "whole body" problems may impact the cells and tissues in the brain as well as in the rest of the body.

But a critical point for truly understanding the ***autism itself***, and not just the contributors, is that the functions of the brain are not just physical and biological like the rest of the body. If autism were purely whole-body (e.g., immune, GI) and not whole body-brain, then we would have a lot of brilliant people walking around with allergies and gastric reflux but not with brain problems. We need to understand what happens en route from body to brain to make it ***autism***.

Brain: Transduction between Physical and Informational Levels

The brain is an organ that has both organ-type functions and information-type functions. While the rest of the body also engages in energy and information activities, these activities are taken to a much higher and more complex level in the brain. However, in my opinion, brain research in autism has largely been characterized by either lack of insight or confusion regarding how these levels interact.

- On one side we have cognitive neuroscientists who look for genetic correlates of differences in information processing patterns or localization of brain activity without taking account of cellular, metabolic and tissue problems in the underlying brain tissue that prominently include inflammation, mitochondrial dysfunction or oxidative stress.
- On the other side we have tissue pathophysiology-oriented researchers who look at the cellular and biochemical changes in great detail, but do not think through how these changes actually alter the *information processing activities* of the brain so it produces the behaviors we label as "autistic."

When colleagues make the argument that a whole-body approach to autism (i.e. addressing medical and physiological problems and not just using behavioral and psychopharmacological interventions) improves the functioning and quality of life of people with autism, what we are taking for granted without actually explaining it is that ***whole-body interventions help the brain process information better by generating better health for the cells and tissues that do the information processing.***

Therefore, I believe that it is time to explicate this linkage. To really transform treatment strategies in autism, I believe it is imperative that we bridge — translate between — these two levels. In what follows, I will look at older observations from the newer perspective of a whole-body approach to the brain.

Brain as a Physical Organ: The Poorly Explored Relationships between Anatomy and Pathophysiology

A variety of anatomical observations have been made in the brains of people with autism, both microscopically in brain tissue from people with autism who have died, and macroscopically using brain imaging technologies, such as magnetic resonance imaging (MRI), positron emission tomography (PET) or single photon emission computed tomography (SPECT).

The observations show intriguing differences between people diagnosed with autism and people with other diagnoses or with neurotypical development. However, the ways these observations have been interpreted may often be more revealing about the assumptions people have brought to this research than about what may actually be going on. From a whole body-brain vantage point, many of these observations need to be reconsidered freshly and their prior interpretations re-thought, keeping in mind our growing understanding of dynamical pathophysiology. The following sections cite some examples.

Microscopic findings

Changes in cell size and density over time

At the microscopic level, it has been known for decades that cells in the brain in ASCs may have different shapes or sizes than in neurotypicals. For example, Bauman and Kemper have reported that neurons may be larger in children with autism compared with those in neurotypical children and smaller in adults with autism than in neurotypical adults.[54] Why might this be? Although these investigators had originally proposed that autism was created prenatally because the limbic system, which showed smaller and more tightly packed cells in their studies, forms in the brain at around the 32nd week of gestation,[55] their observation that this change in size and density occurred far into the lifetime and long past birth led them to suggest that some subtle but persistent changes might be going on in these cells, though they did not specify what these changes might be. Could processes such as inflammation or oxidative stress be slowly eating away at the robustness and resilience of these cells physically — and also functionally?

Differences in density of the fine dendritic fibers emanating from neurons

Other microscopic observations may also be related to the impacts on cells of persistent irritation from chronic tissue pathophysiology. The fibers from neurons, called dendrites, may be of different density — sometimes thicker, sometimes thinner.[56] The SHANK3 mutation is associated with more dendritic fibers, ironically juxtaposed with less neuronal "firing power" — perhaps a compensation for other factors leading to poorer synaptic activity.[57,58] Dendrite density can also be altered by seizures and epilepsy, which greatly irritate neurons. Zinc deficiency, quite common in autism,[59] can dysregulate the synaptic proSAP/Shank scaffold and might contribute to the development of ASCs.[60] Toxicants such as polychlorinated biphenyl compounds (PCBs) may lead to an overabundance of dendritic fibers by hijacking and confusing signals that regulate cellular development.[61,62] So altered physiology, possibly from environmental exposures, as well as genes can lead to these microscopic changes.

Beyond neurons: Glial cells

Up until recently, brain researchers in autism focused almost exclusively on neurons. Until the past year or two, hardly any papers or even abstracts were written by autism researchers about glial cells, a group of cells including astrocytes (or astroglial cells), microglial cells and oligodendroglial cells, which together outnumber neurons in the brain by 4–10:1, and which perform many vital metabolic, immune and signaling functions.[63] These cells may become "activated" (which among other things makes them bigger) to deal with perceived immunological, toxic, infectious or other cellular emergencies, and produce pro-inflammatory or excitotoxic chemicals in this state that can irritate the tissue. A growing number of studies have documented cellular, gene expression, or immune pattern changes in the brain tissue of people who had autism that are consistent with a state of immune activation or "neuroinflammation."[64–72] Although these changes may change gene expression, they may or may not be caused by genes. But the impact may not require a genetic cause.

Accumulation of protein waste products in the brain

There may be pile-ups of proteins like lipofuscin or amyloid-beta[73,74] that suggest either increase in production of these proteins (e.g. from excitoxocity, seizures or oxidative stress), problems with clearing out problem cellular byproducts, or both. Such aberrant protein accumulations have been identified in several studies of brain tissue from people with autism. The process of clearing waste products from the brain can occur through various process such as autophagy, which may be regulated or impeded by factors including oxidative stress that have been demonstrated in many individuals with autism. It is also dependent upon transport of these cellular waste products or "garbage" out of the brain through fluid channels recently dubbed "glymphatics" (because they are little fluid channels in the brain shaped by glial cells).[75–79] Little is known about fluid drainage in the brains of people with autism, though a recent paper in *Science* reported that sleep drives metabolite clearance from the adult brain,[80] with glial shrinkage during sleep contributing to a 60% increase in interstitial space, greatly increasing convective fluxes and amyloid clearance during sleep, a function with which most people with ASCs have difficulty. In neurodegenerative diseases like Alzheimer's where abnormal substances like beta-amyloid pile up, we are learning that these glymphatic fluid flow systems function poorly, so that misfolded proteins fail to clear properly from the brain, speeding the development of dementia or other neurodegenerative diseases.[81]

Macroscopic findings

Regional size differences

At the macroscopic level, myriad anatomical brain imaging studies have measured the size of different parts of the brain and have found some parts to be larger or smaller in autism than in neurotypical people.[82,83] Much confusion has been generated by inconsistencies between studies — a region may be larger in autism in one study, smaller in another and no different in a third, particularly if you compare studies performed on younger vs. older people, suggesting that these differences change over time — and that different regions of the brain change in different ways, some getting larger over time and some getting smaller.[84,85] Could these brain volume differences start out as impairments of cellular function and only later change lead to macroscopically measurable changes in sizes or properties of brain structures? This question has received very little reflection in the autism brain research literature. Yet the size of a whole brain region is nothing more than the sum of the sizes of all the cellular, vascular, fluid, and extracellular "stuff" in that region.[84] So it would make sense that changes in the health status of any of these components, and not just genetics or high intensity activity, might be contributors to the differences in overall region size.

Overall brain enlargement

One of the most intriguing findings is that, especially in younger people with autism, the brain is on average larger than in neurotypical people, and this size increase cannot be attributed simply to the size of any one region,[86] though enlargement of white matter appears to contribute more to this overall size increase than enlargement of gray matter.[87–89] There are striking parallels between the phenomena of brain size enlargement and of inflammation, but on the other hand there is very little direct documentation of brain inflammation in young children reliably diagnosed with autism, because they rarely die and have their brains donated to research. Brain imaging of children might be able to contribute here, but it is difficult to detect inflammation more than by inference from magnetic resonance spectroscopy imaging which is hard to perform on young children without sedation; PET studies, mostly done on substantially older individuals in part due to risks from the technology and the need to inject radioactive ligands, are only recently being performed to measure this.[90] Many people who think that brain inflammation is likely to be a central player in ASCs think that the brain volume increase is likely to come from microglial cells, but it would not be inconsistent with their point of view if the brain enlargement also or instead originated from astrocytes or even fluid accumulation in the extracellular matrix. It does not, however, appear to come from a higher density of axons (the fibers that neurons send out to link to other neurons over short or long distances). Instead, the density of these fibers has been measured as lower on many occasions,[91] one of the several arguments against attributing brain volume increase in autism to failure of neuronal pruning early in development.[92]

Whole-body-brain take-home messages

Overall, the relationship between the macroscopic changes in total brain and brain region size and the underlying microscopic changes in cells, fluids and other substances in the brain is very poorly understood, with most inferences, even if plausible, going beyond what can be solidly supported. Even so, here are several things we can say about the brain in individuals with autism from observations to date from a whole-body vantage point:

- The finding of larger brains in autism suggests that the assumption that "bigger is better (or healthier)" may not always be correct, even though the details are not well worked out at present.
- The contributions of cell types other than neurons, and of tissue types other than brain cells (e.g. blood vessels, fluids, extracellular matrix) may be important in explaining what we are seeing — as well as how it got that way. Taking their roles into account may even possibly help us identify health improvement strategies that might give the tissues the resources they need to implement their repair capabilities.
- This phenomenon of widespread brain enlargement in association with autistic symptoms is perplexing to those who assume that the specific behaviors we use to diagnose autism must be localized in specific regions of the brain associated with those brain functions. It is less perplexing to those who think that behaviors emerge from dynamic interactions across large areas of brain. This leads nicely into considering the "other life" of the brain — information processing, and how physiological and informational functioning interact.

Brain as a System for Information Processing

The brain is not just a wet set of metabolizing cells; it is also an information processing organ. Many different types of cells and tissues participate in the metabolism, electrical signaling, energetics and physics of this information processing. Although it is often left unstated, the ability of each of these cell types to perform its roles is dependent upon the integrity of cellular structure and function, the supplies of nutritional precursors for biochemical processes and the maintenance of cellular integrity, and the absence of substances (e.g., toxicants or pro-inflammatory cytokines or other metabolites) or other influences (e.g., radiation that interferes with smooth functioning).

Within the domain of research that focuses on brain information processing — called cognitive neuroscience — it is very easy to: (1) get caught up in thinking about one level of this information processing activity and forget about all the other levels that are going on simultaneously and are really part of the same overall process, and (2) forget about the supports for the underlying tissue substrate that makes this information processing possible. So let us remember here that information processing is the result of activities at many levels, ranging from synapses between neurons — including all the underlying processes that make this possible, through aggregations of signals that travel down fiber tracts. It is also related to waves of electrical activity that are related to glial cell as well as neuronal function. It is further related to modulation of blood flow. While thousands of scientists make careers studying these processes separately, in truth they are all integrated, all facets of getting the brain's job done. And from a whole brain-body health perspective, all individually and collectively depend upon underlying cellular health and integrity.

The following sections summarize some of the major areas that have captured people's attention.

"Structural connectivity" through fiber tracts

We know that neurons talk to each other through synapses and that bundles of neuronal processes, or axons, gather together in fiber tracts that link one part of the brain to another. We have measured differences in density, diffusivity and other properties of fiber bundles between people with autism and neurotypical people, using an MRI technique called diffusion tensor imaging (DTI). It is currently very fashionable in autism cognitive neuroscience research to study structural connectivity in this fashion, and to correlate it with fMRI measures to be discussed next. However, while it is tempting to infer that these structural changes "cause" autism, fiber bundles are not the only way that information is moved around the brain. Furthermore, the changes in these fiber bundles are most likely quite a ways "downstream" of factors that include not just genes, but also and perhaps most fundamentally chronic, persistent irritating pathophysiological changes at the cellular level. In spite of the present emphasis on the technically low-hanging fruit of making these measures in fiber tracts, the underlying cellular disturbances rather than the looseness in the fiber tracts may be more basically "causative" of the functional differences leading to the autistic behaviors (by mechanisms to be further discussed below), with the fiber tract changes being secondary manifestations of these more primary cellular problems.

Specific regions that "light up" during brain activity

People with ASCs show differences from neurotypicals in the regions of their brains that activate (or "light up") in fMRI. Brain region activation appears to be more consistent in relation to the task being performed among neurotypical people, whereas the differences from neurotypicals in people with ASCs in the way their brains perform the same tasks may be "all over the map" (that is, "lighting up" different sets of regions from one person to the next).[93] This suggests that there may be no one "autism pattern" in brain activity. This in turn may mean that there is not really an "autism" but more likely a *process of dysregulation* that can create a wide variety of impacts — and that it is the *process* itself, and not the patterns that get created by the process (that in any case are fairly variable), which leads to what we label as "autism."

Tissue-level contributors to brain activation measured by fMRI

Brain activation as measured by fMRI studies arises from differences in brain blood flow and oxygenation.[94,95] This is driven by processes that involve not only neurons but other cells and tissues including astrocytes, otherwise known as astroglial cells, and the blood vasculature. Up until recently astroglial cells got almost no attention in autism, except for a few studies noting that they were inflamed and that markers for astrocytes like GFAP (glial fibrillary acidic protein) were elevated in postmortem brain specimens, suggesting gliosis or glial activation and reactive cellular injury.[96] Astrocytes have little "foot processes" that surround the brain's capillaries and are part of the Blood-Brain Barrier (BBB). Although we know that there is a tendency for astrocytes to be "activated" and involved in brain inflammation (as briefly reviewed above regarding immune alterations, e.g. the Vargas *et al.* (2005) pathbreaking neuroinflammation paper[64]), we do not have an understanding of how that brain tissue pathophysiology impacts the blood flow changes that we measure in fMRI studies. fMRI researchers rarely discuss inflammatory and other influences upon astroglial cells and how they may impact the patterns of brain activation that fMRI is able to measure, because most autism fMRI researchers are not interested in brain inflammation or other brain tissue phenomena in autism; these "cognitive neuroscience" researchers instead focus mainly at the level of neurocognitive information processing but not much on the underlying cellular processes and cellular health prerequisites that make these information processing activities possible. Such researchers are more likely to jump directly to attempting to correlate their findings with genetics, skipping the vast intermediary domain of brain pathophysiology and how it may interfere with the cell's roles in information processing.

Problems with blood perfusion of brain tissues

There are other ways of measuring brain activity that relate to the amount of blood flow, or perfusion, via different measurements, such as PET and SPECT scans. Many of these studies have shown poorer blood perfusion in people with autism, but the parts of the brain that have poorer blood flow are not particularly consistent from one study to another.[83] We know almost nothing about the relationship between these blood flow changes measured by these technologies and the underlying changes in the cellular health and function at the brain tissue level in ASCs. One reason is that autism researchers using these techniques have generally been more interested in correlating changes with psychological function than with cellular and tissue changes (and such changes in any case have not been solidly on the autism map for all that long, and are also not so easy to measure).

Electrical oscillations or "brain waves"

The brain also generates waves of electrical oscillations that are synchronized or desynchronized with each other in complicated ways, and that oscillate at a great range of frequencies.[97] We are beginning to learn about how these oscillations are different in people with autism as compared to neurotypicals, and that the differences are different depending on the frequency of oscillation. These types of brain wave differences cannot be directly mapped onto the differences we have identified in fiber bundle anatomy or density because they are not confined to travelling along those fiber tracts. Generation of synapses and the brain waves that represent aggregations of synaptic activity rest on underlying support systems, including energy production, blood flow, garbage disposal, and the ability of glial cells to smoothly and precisely support and fine-tune what the neurons do.

Temporal vs. spatial resolution of measurements and learning about autism as a PROCESS

Synaptic discharges between neurons occur in the time frame of milliseconds — thousandths of a second. Brain wave oscillations are also measured in milliseconds. These measurements can be made in human beings by EEG (electroencephalography) or MEG (magnetoencephalography), but NOT by fMRI or PET or SPECT. fMRI has much better *spatial* resolution than PET or SPECT, being much less blurry in its anatomical definitions than those technologies, as well as being non-invasive (generally nothing needs to be injected). But its temporal resolution is poor. The popularity of its use goes hand in glove with the dominant belief that autism is a *fixed trait;* measurement of changes from microsecond to microsecond that we can get by using EEG or MEG will help us better understand the ways that autism is a constantly dynamical *process.*

The *process* consideration is important because if electrical oscillations and synaptic activity are occurring in the millisecond range, some of the underlying molecular signaling, chemistry, energetics and quantum physics processes are occurring at even more rapid rates.[98] This has significant implications for how we think about the multi-scale interactions producing autism and how we might more profitably think about autism research, as will be further discussed below.

Molecular and Cellular Underpinnings of Brain Information Processing

Underlying systems supporting brain function: Differences between cell types

From animal studies, in which needles can be placed to measure activity in specific cellular types or layers of the brain, we know that different types of brain cells generate different types and frequencies of oscillations. But since we cannot stick needles into the brains of living human beings to perform similar measurements, we can make only limited inferences from what we are learning about differences in brain oscillations as measured by EEG or MEG regarding underlying cellular status or health of different cell types in the brain. Yet even though we cannot be certain of many of the fine points of which cells are malfunctioning in what specific ways, from a whole-body-brain vantage point we can still look for signs of deficiency of underlying support systems (e.g., insufficient antioxidants, poor supply of essential fatty acids or insulin resistance), or excessive exposure to substances or processes (e.g. free radicals, toxicants, pro-inflammatory cytokines, radiation) that may interfere with optimal function.[99,100]

Underlying systems supporting brain function: Gap junctions and voltage-gated calcium channels

One further way that information is transmitted in the brain is through calcium waves generated by astroglial cells through their gap junctions. Gap junctions are actually tiny tunnels through the membranes of adjacent cells that create physical continuity between large numbers of astroglial cells. Chemicals pass through these tunnels generating calcium waves that relate to the brain oscillations and the regulation of how much activity is going on in particular regions of the brain, in ways that are poorly understood at present, not only for autism but more generally. Gap junction function can be modulated by other features related to pathophysiology, such as buildup of misfolded proteins or acidity.[101,102] For autism and other conditions where astroglial cells are activated or involved in neuroinflammatory processes, we have little or no data about how this pathophysiology impacts information processing.

Example: The potential impact of brain inflammation on brain signal-to-noise ratio

In the sections just preceding, one of the players that has shown up at multiple levels of the system is the astroglial cell, or astrocyte. Astrocytes play roles in synaptic regulation, propagation of chemically generated signals through the brain, capillary lumen caliber, integrity or compromise of the BBB, metabolic support of neurons and much more. Let us now add to this list the role that astrocytes play in regulating excitatory responses. Glutamate, an excitatory neurotransmitter at the neuronal synapse, is normally mopped up by astrocytes after the synaptic event occurred, but astroglial cells fail to do this efficiently when they are activated or inflamed. This pathophysiology-induced loss of function could very well be a major contributor to brain information processing differences in ASCs, but this potential relationship between cellular dysfunction, biochemistry and information has received hardly any attention.

If glutamate is indeed left hanging around in the synapse because activated astrocytes abdicate their clean-up duties, this glutamate will continue to stimulate neuronal responses.[103] The neurons are in no position to make a distinction between the persistent signals generated by this overload of glutamate (or the other excitotoxic or proinflammatory substances generated by the overall dysregulated milieu) and the information-containing signals from the environment which enter the brain via the sensory system. This "glutamate overload" leads to an increase in "noise" in the system. Moreover these same pathophysiological processes that impede post-synapse glutamate mop-up may also degrade the integrity of the myelin sheath around the axonal processes. Compromising the myelin reduces the insulation around the brain's axonal wiring system, which may further reduce the quality and speed of the signal and further increase the system's "noise" level.

In engineering there is a generic ratio, the "signal-to-noise" ratio (SNR), which indexes the quality of signal in a system. More signal is better; the converse (i.e., less signal) means the system is functioning less well. An example familiar to everyone is cell phone signal: a cell phone with five bars of reception has good signal, but with two bars you get a lot of static (i.e., noise) and not much signal. With that kind of low SNR, the noise drowns out the signal making it very hard to have a useful conversation.

Some electrophysiological studies in autism have shown greater power and less coherence in signaling, which is evidence for a degraded signal-to-noise ratio.[104–107] Although more cross-disciplinary studies are needed, one can already say that this evidence may well reflect the impact of underlying cellular pathophysiology on the capacity of the brain to keep its electrical signaling strong and well-organized.

The above-described scenario of under-functioning astrocytes will also increase the "excitation/inhibition (E/I) ratio."[108] The brain needs to maintain a balance between excitation and inhibition to make the amount of signal "just right." Because this overall collection of pathophysiological processes increases pro-inflammatory and pro-excitatory processes and at the same time interferes with inhibitory processes, it tends to increase the E/I ratio while decreasing the signal-to-noise ratio. So you get more noise and less information. This proclivity to overload could plausibly underlie behavioral manifestations in ASCs like poorer attention span, more meltdowns, and sensory irritability, as well as problems with sleep and vulnerability to seizures. Yet while researchers in autism and in neuroscience more broadly are very interested in this altered E/I ratio, they may still believe that the alterations must emanate from genes, and they very rarely consider potential contributors from other parts of multi-scale biology, particularly including these cell-based environmentally modulated pathophysiological processes that could quite plausibly be contributory, as can be seen from even this brief discussion.

Fast Tracking Progress in Autism Research: From Bottom-up to Middle-out and Multi-scale Approaches

Rather than thinking either bottom-up (genes and molecules "cause" everything) or top-down (which in autism would include the idea that behavioral therapy is the best therapy), I think that the "middle-out" approach along with a multi-scale approach,[109] both of which are increasingly discussed in the physiological and systems biological literature, will advance us more rapidly toward effective ways of helping people with autism.

The "middle-out" point of view is articulated by specialists in physiology and metabolism, most eloquently by the British muscle physiologist Denis Noble in his elegant book *The Music of Life: Biology Beyond the Genome.*[110] It emphasizes that the *action* is at the physiological level. The multi-scale approach emphasizes the simultaneous and complex interactions of activities across all the biological scales discussed above.[111,112] From this point of view, genes bias the way that things happen — they set up strengths and vulnerabilities — but they are by no means the sole contributors, so you cannot truly claim that they "cause" things all by themselves.

This point of view is emerging as well in the literature on just how specifically genes and environment interact. Calcium channel dysfunction provides an example that has received some careful attention in this regard. There are a number of genes altering calcium channel function that are associated with autism, but environmental as well as genetic routes to calcium signaling dysfunction have been identified,[113] which include chemicals such as the polyaromatic hydrocarbons. PCB-95 in particular modulates the calcium-dependent signaling pathway responsible for activity-dependent dendritic growth.[61,62] Stamou and colleagues (2012) argue that once a genetic mutation has been associated with altering a critical signaling pathway and conferring risk for autism, chemicals or other environmental agents can be identified that target the same pathways and also confer ASC risk. These authors have reviewed this strategy of identifying multiple mechanisms converging on common signaling pathways regarding Ca(2+)-dependent mechanisms as well as extracellular signal-regulated kinases (ERK)/phosphatidylinositol-3-kinases (PI3K) and neuroligin-neurexin-SHANK.[114] From this point of view, there may be no particular reason to privilege genetic mutations in their contribution to a disturbance of calcium signaling, since whether this function becomes derailed due to a genetic mutation, a chemical toxicant, or electromagnetic perturbation, the functional effect is comparable because it is more directly due to the way the calcium channel physiology changes.

The case that Stamou *et al.* (2012) make is a good example of middle-out thinking. The core point of what they are saying is that *it is the dysfunction in the calcium channel that is altering function in autism.* The gene itself does not directly alter the brain function, and neither does the environmental trigger. In both cases, it is *the impact on the physiological process* that creates the change in brain (and other) function.

A Further Problem: Heterogeneity — How Can There be So Many Routes to "Autism"?

Calcium channel dysfunction not only provides a good example for middle-out thinking, it also provides a good lead-in to a discussion of heterogeneity in autism. While some individuals with autism have calcium channel dysfunction, others may not. At the level of genetic syndromes, over a hundred distinct syndromes have been identified where autism is a comorbidity;[115] and while there are some common themes across the genetic issues underlying these conditions, there is much we do not understand. Many people like to say that autism is a disorder of the synapse, but this actually is a rather vague statement since the synapse is hugely complicated and involves interactions of enormous numbers of proteins, neurotransmitters, receptors and also cell types (not just neurons but also astrocytes) as well as blood vessels, blood flow and extracellular matrix. There is no particular reason why the synaptic problem even needs to derive from a genetic mutation, if it is the synaptic dysfunction rather than the origin of that dysfunction that is the proximal cause of the brain and behavioral (and physiological) changes underlying the "autism." Scientists now talk about the tripartite synapse (neurons, astrocytes, blood vessels)[116] or even the quadripartite synapse (tripartite plus extracellular matrix).[117] Synaptic dysfunction could come from any of the components of this system.

Even here, with so many things capable of causing synaptic dysfunction, are we not faced with the famous "chicken-egg" problem? That is, when a complex system with many interacting parts changes the dynamics of how it operates, how do you pin down one single "cause" that "comes first"? Or does it even make sense to do that?

Okay, So What "is" the "Autism" Anyway?

After reviewing the incredibly complex multi-scale biology of autism, it is difficult to go back to the old assumption that "autism is a behavioral syndrome deriving from genetic alterations of brain development" without feeling that we are missing not just the boat but a giant ocean liner. Instead one is called to become fascinated with how all of these different types of multi-scale challenges to the organism and in particular to the brain, with different combinations of them from one person with autism to the next, *actively* create the behaviors that meet diagnostic criteria for autism.

Autism as an Emergent Property of a System with Many Shifted "Settings"

As mentioned, some neurocognitive researchers have proposed that at the cognitive level many features of autism emerge from difficulties with complex information processing.[29] Why would the system have such difficulties? It is easier to think about this from a multi-scale systems biology point of view. Cells laboring under inflammation, mitochondrial dysfunction and oxidative stress, or calcium channel dysfunction, will lose an edge of adaptability and fluidity or rapidity of response that may take away from the system's resources to coordinate information in the most complex ways possible. It is even possible that some neurons may be challenged enough to go "offline" so that they can survive, rather than burning themselves out when they do not have enough mitochondrial or astroglial support to generate energy or regenerate their metabolic integrity.

In a thoughtful article on the idea of emergence in autism, Anderson (2008) states that various "autism-related phenomena including intellectual disability, seizures, persistence of primitive reflexes, stereotypies, self-injurious behavior, savant abilities, and morphological abnormalities, among others," could potentially be emergent properties.[118] The author suggests that "consideration of the role of emergence in autistic behavior and related phenomena should complement a reductionist approach and might help illuminate the components and complexities of autism."

The multi-scale systems approach sketched above adds more substance and support to this assertion. This approach supports the idea that a system laboring under various combinations of cellular challenges may produce behaviors we label as autistic simply because that is what the brain "does" when it has to conserve its resources, or when it cannot marshal and coordinate sufficient resources to get things done in the most elegant ways.[119]

One ironic and provocative twist here is that some challenges to the system, such as an increase in glutamate (e.g. from the failure of activated astrocytes to remove it from the synaptic cleft after synaptic discharge) may contribute to heightened sensitivity, intelligence or even heightened perceptual acuity. Indeed, superior function in some individuals with ASCs compared to neurotypicals has been documented in just such domains (sometimes this may manifest as savant skills, sometimes as superior perceptual abilities, and sometimes as great and even sophisticated talents). It is just speculation at this point to attribute such findings at the neuropsychological level to underlying neurobiology; but even so, this is an interesting hypothesis worth testing, since it links disparate and seemingly contradictory but clearly co-existing features in ASCs.

From "what causes autism" to "how is autism caused"

When we move beyond genetic determinism to a multi-scale biology, middle-out approach, we see that much of what autism involves is related not to fixed "autistic states of being" but rather to the consequences of changes in interacting sets of complex *processes* that are working in people with ASCs at different rates, with different equilibria or set points, than they are in neurotypical people. This can include processes as disparate as cellular transport, fluid transport, mitochondrial metabolism, immune responsiveness, gut motility or electrophysiological oscillations — or all of these altered together in concert with each other.

Since there are many factors — including nutrition, activity and environmental exposures — that modulate these processes, it is quite conceivable that the cause of the regressive process leading to autism could be a cascade of events that progressively alter the way these internal processes work.

Similarly, from this perspective it becomes conceivable that improvement in functioning in autism, or even loss of diagnosis, may ensue when changes in these internal processes move in the opposite direction from the way they changed during regression.

From "how is autism caused" to "how does autism work": Dynamical process approach

From this perspective, the goal is not only to advance past a static point of view suggesting that a specific gene, a specific toxin or a specific gene-environment interaction might "cause" autism — i.e. singlehandedly shift a system from not-autistic to autistic, but also to advance into a dynamical framework where we see how the autism emerges from ongoing processes in both brain and the whole multi-scaled physiologically-centered system.

Autism as What Emerges from the Moment-to-Moment Impact on Electrophysiology of Depleted Function of Cells Supporting Brain Electrical Activity

From this point of view, I would like to propose that the features we label as "autistic" emerge on a *moment-to-moment basis* from physiological processes, rather than being a product of some kind of hard-wired inborn differences in brain circuitry. Differences in circuitry would then emerge or exaggerate during development, rather than being fully "programmed in" from the start.

Specifically, since electrophysiological oscillations — aggregate patterns emerging from interrelated activities of individual synapses — underlie brain activity, and since electrophysiology is generated by cells whose physiological status is known to be altered in at least large numbers of people with autism — it therefore follows that the autism may emerge from a vast and inter-individually heterogeneous array of physiological alterations that converge upon the intersection of tissue physiology and electrophysiology to create changes. This physiologically altered system creates or generates "outputs" that we label as autistic behaviors.

Clearly this does not exclude the possibility that a certain number of individuals with autism have genetic alterations that contribute substantially to this situation, but it does help us understand why no such genes have been found in the majority of people with autism, and why ASCs are so heterogeneous as well as variable even *within* single individuals. It also follows that even in individuals with strong genetic vulnerabilities, there could be an additional overlay of environmentally modulated cellular dysfunction that is aggravating the situation but is actionable — that is, reducible through improving cellular function, even if the gene mutation itself is immutable.

Practical Implications for Conducting Research and Clinical Care?

If we can establish empirically that tissue pathophysiology is connected to electrophysiological alterations, we will be able to move the center of gravity in autism research from looking for the genes or environmental factors that "cause" autism in a bottom-up fashion, to *looking for the systems perturbations that we can modulate and correct*. We will no longer invest the bulk of our resources in an endless hunt for autism genes, betting on the assumption that identification of prior causes (genetic or environmental) is the only way to make progress. Instead, we will see that there are many entry points into the system's ***function*** through which that system can be modulated.

Lifestyle interventions for the brain: Public health implications

Particularly important is the ground-zero interface between systems physiology and electrophysiology, because by modulating physiology we may very well be able to make substantial improvements in electrophysiology — and the behaviors it generates.

To the extent that pathophysiological processes like inflammation, oxidative stress and mitochondrial dysfunction collectively influence electrophysiological activity as well as underlying synaptic function, this is good news, because we already have many practical ways to influence these types of dysfunction. Prominent among these are everyday lifestyle alterations including optimal high-nutrient-density diet, exercise and good sleep hygiene. High nutrient density food, full of antioxidants and quality lipids, can be anti-inflammatory, can reduce oxidative stress and can be restorative to mitochondria and organ systems function and even neuroprotective.[120,121] And whereas lack of sleep, stress, and lack of exercise produce pro-inflammatory responses, improving these domains is restorative.[122]

In the face of something as severe as autism can be, it may seem trivial to talk about these everyday interventions. But if you look at the daily lives of individuals with autism, they are often characterized by self-restricted diets sorely lacking in fruits or vegetables and very nutrient poor;[123] they often sleep poorly not to speak of keeping the whole family awake too much; and they often exercise very little and feel stressed all the time. None of this is healthy, and all of it is actionable once we realize that there is a profound physiological and brain-health point to it. Improving lifestyle choices in these domains may decrease depletion and improve support for brain maintenance and repair functions.

What it Will Take to Implement a Public Health, Everyday Epigenetics Approach to Autism

What this really means is that we need a public health approach to autism, emphasizing achievable lifestyle interventions that could improve the quality of life at least somewhat for most people with autism, and improve it a huge amount for a significant subset of people.[124] In fact, given the rising number of people with autism and the rising costs, we really cannot afford NOT to implement such an approach.

Future Directions: Toward True Translational Research

Multi-scale biological research is therefore important not only for what new things it may teach us, but also for the rhetorical purpose of showing that a public health approach with everyday lifestyle changes aimed at epigenetic modification of gene expression is NOT crazy but instead imperative.[124]

Therefore:

1. From a research point of view, we need to integrate brain, behavior and pathophysiology research through coordinated projects employing multi-scale systems biological approaches. One way of doing this is to look at the interventions being performed by innovative clinicians on the systemic physiology to see how it affects brain electrophysiology and metabolism

(via EEG (or MEG) and magnetic resonance spectroscopy, respectively). Generation of this type of data will probably optimally come from prospective multisystem sample collection from individuals undergoing intensive and skilled comprehensive treatment of multiple aspects of their physiology, alongside animal models structured in parallel.[125]

2. By taking on transduction issues in ASCs through multi-scale study designs that truly tackle the interfaces between brain electrophysiology and metabolism, we will gain a systems level and mechanistic characterization of transformation, i.e., the passage from greater to lesser options (regression) or lesser to greater options (recovery). This may allow us to reverse-engineer more effective treatments.[99,100]
3. To do this, we need an infrastructure that captures data from both academic research and community-based clinical environments, and does so in a fashion that allows these differently skilled and differently equipped groups to meaningfully collaborate.[125–127]
4. In this fashion, we can perform "bedside-to-bench and back" translational research that may sooner rather than later transform the lives of thousands or millions of individuals with autism and spill over not just into their families but also into the lives of people with other conditions where compromised pathophysiology is interfering with optimal brain function and quality of life.
5. Importantly, we can transform this information into support for a public education campaign that can be started now and enriched as the research utilizing this approach accumulates.

Conclusion: Multi-Scale Dynamical Biology Gives Hope and Empowerment

In summary, based upon insights from multi-scale biology, and with the understanding that the processes involved in electrophysiological generation of "autism" can be modulated by lifestyle interventions that improve underlying physiology and epigenetic regulation of gene expression, we can make a big difference now with what we already know, and we should mobilize our motivation and infrastructure to make it happen.

Executive Summary

- Autism develops through a dynamic interplay amongst genetic, environmental, epigenetic, metabolic, molecular signaling and physiological levels.
- Genetics may bias the system toward autism weakly or strongly, but even if it does so there may be an overlay of surplus physiological complications that may be actionable; if constructive choices are made, improved function and quality of life may still occur.
- Emphasis in brain research on anatomical features of the brain is not sufficient for helping us understand the developmental or dynamic features of autism.
- Many brain anatomical features may be downstream consequences of chronic, persistent tissue irritation, particularly from inflammation, oxidative stress and mitochondrial dysfunction, rather than purely results of genetic differences.
- The "ground zero" of how the brain comes to generate "autism" probably lies in the interface between tissue-based physiology and electrophysiological oscillations.
- Many potential avenues of intervention exist at this "ground zero" interface, particularly through "bottom-up" or "middle-out" improvement of cellular health status through whole body-brain lifestyle changes, including high density nutrition, improved sleep, and reduction of physiological stressors and noxious exposures.
- From a dynamical systems perspective, "autism" can be reframed as what emerges from the moment-to-moment impact on the brain's generation of electrophysiological oscillations when the cells generating this oscillatory activity are depleted or otherwise dysfunctional.
- Bedside-to-bench translational research that collects data on changes across physiological and electrophysiological systems

from whole-body interventions may help us reverse-engineer the critical transitions of regression in ASCs to more effectively guide people to recovery.

- It is an urgent public health matter that these avenues be pursued because the savings in unnecessary human suffering and economic cost could be huge.
- A whole-body approach to body-brain transduction is a science-based empowering source of hope.

References

1. Hertz-Picciotto I, Delwiche L: The rise in autism and the role of age at diagnosis. *Epidemiology* 2009, **20**(1):84–90.
2. King M, Bearman P: Diagnostic change and the increased prevalence of autism. *Int J Epidemiol* 2009, **38**(5):1224–1234.
3. Kohane IS, McMurry A, Weber G, Macfadden D, Rappaport L, Kunkel L, Bickel J, Wattanasin N, Spence S, Murphy S *et al.*: The co-morbidity burden of children and young adults with autism spectrum disorders. *PLoS One* 2012, **7**(4):e33224.
4. Herbert MR: Autism: a Brain disorder or a disorder that affects the brain? *Clin Neuropsychiatry* 2005, **2**(6):354–379 http://www.marthaherbert.com/Herbert%320CN%320autism%320brain%320or%320affecting%320brain%320final.pdf.
5. Foster JA, McVey Neufeld KA: Gut-brain axis: how the microbiome influences anxiety and depression. *Trends Neurosci* 2013, **36**(5):305–312.
6. Heijtz RD, Wang S, Anuar F, Qian Y, Bjorkholm B, Samuelsson A, Hibberd ML, Forssberg H, Pettersson S: Normal gut microbiota modulates brain development and behavior. *Proc Natl Acad Sci U S A* 2011, **108**(7):3047–3052.
7. de Theije CG, Wu J, da Silva SL, Kamphuis PJ, Garssen J, Korte SM, Kraneveld AD: Pathways underlying the gut-to-brain connection in autism spectrum disorders as future targets for disease management. *Eur J Pharmacol* 2011.
8. Naviaux RK: Metabolic features of the cell danger response. *Mitochondrion* 2013.
9. Ochoa-Reparaz J, Mielcarz DW, Begum-Haque S, Kasper LH: Gut, bugs, and brain: role of commensal bacteria in the control of central nervous system disease. *Ann Neurol* 2011, **69**(2):240–247.
10. Savidge TC, Sofroniew MV, Neunlist M: Starring roles for astroglia in barrier pathologies of gut and brain. *Lab Invest* 2007, **87**(8):731–736.
11. Theoharides TC, Doyle R: Autism, gut-blood-brain barrier, and mast cells. *J Clin Psychopharmacol* 2008, **28**(5):479–483.

12. Bercik P, Verdu EF, Foster JA, Macri J, Potter M, Huang X, Malinowski P, Jackson W, Blennerhassett P, Neufeld KA *et al.*: Chronic gastrointestinal inflammation induces anxiety-like behavior and alters central nervous system biochemistry in mice. *Gastroenterology* 2010, **139**(6):2102–2112 e2101.
13. Parr JR, Le Couteur A, Baird G, Rutter M, Pickles A, Fombonne E, Bailey AJ, International Molecular Genetic Study of Autism Consortium M: Early developmental regression in autism spectrum disorder: evidence from an international multiplex sample. *J Autism Dev Disord* 2011, **41**(3):332–340.
14. Yirmiya N, Charman T: The prodrome of autism: early behavioral and biological signs, regression, peri- and post-natal development and genetics. *J Child Psychol Psychiatry* 2010, **51**(4):432–458.
15. Goldberg WA, Thorsen KL, Osann K, Spence MA: Use of home videotapes to confirm parental reports of regression in autism. *J Autism Dev Disord* 2008, **38**(6):1136–1146.
16. Rogers SJ: Developmental regression in autism spectrum disorders. *Ment Retard Dev Disabil Res Rev* 2004, **10**(2):139–143.
17. Werner E, Dawson G: Validation of the phenomenon of autistic regression using home videotapes. *Arch Gen Psychiatry* 2005, **62**(8):889–895.
18. Jyonouchi H, Sun S, Le H: Proinflammatory and regulatory cytokine production associated with innate and adaptive immune responses in children with autism spectrum disorders and developmental regression. *J Neuroimmunol* 2001, **120**(1–2):170–179.
19. Nordahl CW, Lange N, Li DD, Barnett LA, Lee A, Buonocore MH, Simon TJ, Rogers S, Ozonoff S, Amaral DG: Brain enlargement is associated with regression in preschool-age boys with autism spectrum disorders. *Proc Natl Acad Sci U S A* 2011, **108**(50):20195–20200.
20. Hrdlicka M: EEG abnormalities, epilepsy and regression in autism: a review. *Neuro Endocrinol Lett* 2008, **29**(4):405–409.
21. Curran LK, Newschaffer CJ, Lee LC, Crawford SO, Johnston MV, Zimmerman AW: Behaviors associated with fever in children with autism spectrum disorders. *Pediatrics* 2007, **120**(6):e1386–1392.
22. Mehler MF, Purpura DP: Autism, fever, epigenetics and the locus coeruleus. *Brain Res Rev* 2009, **59**(2):388–392.
23. Sandler RH, Finegold SM, Bolte ER, Buchanan CP, Maxwell AP, Vaisanen ML, Nelson MN, Wexler HM: Short-term benefit from oral vancomycin treatment of regressive-onset autism. *J Child Neurol* 2000, **15**(7):429–435.
24. Dawson M, Soulieres I, Gernsbacher MA, Mottron L: The level and nature of autistic intelligence. *Psychol Sci* 2007, **18**(8):657–662.
25. Soulieres I, Dawson M, Gernsbacher MA, Mottron L: The level and nature of autistic intelligence II: what about Asperger syndrome? *PLoS One* 2011, **6**(9):e25372.

26. Mottron L, Dawson M, Soulieres I, Hubert B, Burack J: Enhanced perceptual functioning in autism: an update, and eight principles of autistic perception. *J Autism Dev Disord* 2006, **36**(1):27–43.
27. Soulieres I, Dawson M, Samson F, Barbeau EB, Sahyoun CP, Strangman GE, Zeffiro TA, Mottron L: Enhanced visual processing contributes to matrix reasoning in autism. *Hum Brain Mapp* 2009, **30**(12):4082–4107.
28. Rossetti Z, Ashby C, Arndt K, Chadwick M, Kasahara M: "I like others to not try to fix me": agency, independence, and autism. *Intellect Dev Disabil* 2008, **46**(5):364–375.
29. Williams DL, Goldstein G, Minshew NJ: Neuropsychologic functioning in children with autism: further evidence for disordered complex information-processing. *Child Neuropsychol* 2006, **12**(4–5):279–298.
30. Gregersen N, Bross P: Protein misfolding and cellular stress: an overview. *Methods Mol Biol* 2010, **648**:3–23.
31. Pescosolido MF, Gamsiz ED, Nagpal S, Morrow EM: Distribution of disease-associated copy number variants across distinct disorders of cognitive development. *J Am Acad Child Adolesc Psychiatry* 2013, **52**(4):414–430 e414.
32. Ziats MN, Rennert OM: Sex-biased gene expression in the developing brain: implications for autism spectrum disorders. *Mol Autism* 2013, **4**(1):10.
33. Gargus JJ: Mitochondrial component of calcium signaling abnormality in autism. In: *Autism: Oxidative Stress, Inflammation, and Immune Abnormalities.* Edited by Chauhan A, Chauhan V, Brown T. Boca Raton, FL: CRC Press; 2009: 207–224.
34. Gargus JJ: Genetic calcium signaling abnormalities in the central nervous system: seizures, migraine, and autism. *Ann N Y Acad Sci* 2009, **1151**:133–156.
35. Splawski I, Yoo DS, Stotz SC, Cherry A, Clapham DE, Keating MT: CACNA1H mutations in autism spectrum disorders. *J Biol Chem* 2006, **281**(31):22085–22091.
36. Smith M, Flodman PL, Gargus JJ, Simon MT, Verrell K, Haas R, Reiner GE, Naviaux R, Osann K, Spence MA *et al.*: Mitochondrial and ion channel gene alterations in autism. *Biochim Biophys Acta* 2012, **1817**(10):1796–1802.
37. Lopez-Rodriguez A, Holmgren M: Restoration of proper trafficking to the cell surface for membrane proteins harboring cysteine mutations. *PLoS One* 2012, **7**(10):e47693.
38. Yao Y, Walsh WJ, McGinnis WR, Pratico D: Altered vascular phenotype in autism: correlation with oxidative stress. *Arch Neurol* 2006, **63**(8):1161–1164.
39. Jacob RF, Mason RP: Lipid peroxidation induces cholesterol domain formation in model membranes. *J Biol Chem* 2005, **280**(47):39380–39387.
40. Pamplona R, Barja G, Portero-Otin M: Membrane fatty acid unsaturation, protection against oxidative stress, and maximum life span: a homeoviscous-longevity adaptation? *Ann N Y Acad Sci* 2002, **959**:475–490.
41. Schengrund CL, Ali-Rahmani F, Ramer JC: Cholesterol, GM1, and autism. *Neurochem Res* 2012, **37**(6):1201–1207.

42. Stolp H, Neuhaus A, Sundramoorthi R, Molnar Z: The long and the short of it: gene and environment interactions during early cortical development and consequences for long-term neurological disease. *Front Psychiatry* 2012, **3**:50.
43. Theoharides TC, Angelidou A, Alysandratos KD, Zhang B, Asadi S, Francis K, Toniato E, Kalogeromitros D: Mast cell activation and autism. *Biochim Biophys Acta* 2012, **1822**(1):34–41.
44. de Magistris L, Familiari V, Pascotto A, Sapone A, Frolli A, Iardino P, Carteni M, De Rosa M, Francavilla R, Riegler G *et al.*: Alterations of the intestinal barrier in patients with autism spectrum disorders and in their first-degree relatives. *J Pediatr Gastroenterol Nutr* 2010, **51**(4):418–424.
45. Liu Z, Li N, Neu J: Tight junctions, leaky intestines, and pediatric diseases. *Acta Paediatr* 2005, **94**(4):386–393.
46. Rossignol DA, Frye RE: Mitochondrial dysfunction in autism spectrum disorders: a systematic review and meta-analysis. *Mol Psychiatry* 2011:1–25.
47. Rossignol DA, Frye RE: A review of research trends in physiological abnormalities in autism spectrum disorders: immune dysregulation, inflammation, oxidative stress, mitochondrial dysfunction and environmental toxicant exposures. *Mol Psychiatry* 2012, **17**(4):389–401.
48. Mulle JG, Sharp WG, Cubells JF: The gut microbiome: a new frontier in autism research. *Curr Psychiatry Rep* 2013, **15**(2):337.
49. Williams BL, Hornig M, Parekh T, Lipkin WI: Application of novel PCR-based methods for detection, quantitation, and phylogenetic characterization of Sutterella species in intestinal biopsy samples from children with autism and gastrointestinal disturbances. *MBio* 2012, **3**(1).
50. Hsiao EY, McBride SW, Chow J, Mazmanian SK, Patterson PH: Modeling an autism risk factor in mice leads to permanent immune dysregulation. *Proc Natl Acad Sci U S A* 2012, **109**(31):12776–12781.
51. Hsiao EY, McBride SW, Hsien S, Sharon G, Hyde ER, McCue T, Codelli JA, Chow J, Reisman SE, Petrosino JF *et al.*: Microbiota modulate behavioral and physiological abnormalities associated with neurodevelopmental disorders. *Cell* 2013.
52. Macfabe DF: Short-chain fatty acid fermentation products of the gut microbiome: implications in autism spectrum disorders. *Microb Ecol Health Dis* 2012, **23**.
53. Kang DW, Park JG, Ilhan ZE, Wallstrom G, Labaer J, Adams JB, Krajmalnik-Brown R: Reduced Incidence of and other fermenters in intestinal microflora of autistic children. *PLoS One* 2013, **8**(7):e68322.
54. Bauman ML, Kemper TL: Neuroanatomic observations of the brain in autism: a review and future directions. *Int J Dev Neurosci* 2005, **23**(2–3):183–187.
55. Bauman M, Kemper TL: Histoanatomic observations of the brain in early infantile autism. *Neurology* 1985, **35**(6):866–874.

56. Raymond GV, Bauman ML, Kemper TL: Hippocampus in autism: a Golgi analysis. *Acta Neuropathol* 1996, **91**(1):117–119.
57. Peca J, Feliciano C, Ting JT, Wang W, Wells MF, Venkatraman TN, Lascola CD, Fu Z, Feng G: Shank3 mutant mice display autistic-like behaviours and striatal dysfunction. *Nature* 2011, **472**(7344):437–442.
58. Herbert MR: SHANK3, the synapse, and autism. *N Engl J Med* 2011, **365**(2):173–175.
59. Yasuda H, Yoshida K, Yasuda Y, Tsutsui T: Infantile zinc deficiency: association with autism spectrum disorders. *Sci Rep* 2011, **1**:129.
60. Grabrucker S, Jannetti L, Eckert M, Gaub S, Chhabra R, Pfaender S, Mangus K, Reddy PP, Rankovic V, Schmeisser MJ *et al.*: Zinc deficiency dysregulates the synaptic ProSAP/Shank scaffold and might contribute to autism spectrum disorders. *Brain* 2013.
61. Wayman GA, Yang D, Bose DD, Lesiak A, Ledoux V, Bruun D, Pessah IN, Lein PJ: PCB-95 promotes dendritic growth via ryanodine receptor-dependent mechanisms. *Environ Health Perspect* 2012, **120**(7):997–1002.
62. Wayman GA, Bose DD, Yang D, Lesiak A, Bruun D, Impey S, Ledoux V, Pessah IN, Lein PJ: PCB-95 modulates the calcium-dependent signaling pathway responsible for activity-dependent dendritic growth. *Environ Health Perspect* 2012, **120**(7):1003–1009.
63. Fields RD: *The Other Brain: From Dementia to Schizophrenia, How New Discoveries about the Brain are Revolutionizing Medicine and Science.* New York: Simon and Schuster; 2009.
64. Vargas DL, Nascimbene C, Krishnan C, Zimmerman AW, Pardo CA: Neuroglial activation and neuroinflammation in the brain of patients with autism. *Ann Neurol* 2005, **57**(1):67–81.
65. Zimmerman AW, Jyonouchi H, Comi AM, Connors SL, Milstien S, Varsou A, Heyes MP: Cerebrospinal fluid and serum markers of inflammation in autism. *Pediatr Neurol* 2005, **33**(3):195–201.
66. Fatemi SH, Folsom TD, Reutiman TJ, Lee S: Expression of astrocytic markers aquaporin 4 and connexin 43 is altered in brains of subjects with autism. *Synapse* 2008, **62**(7):501–507.
67. Voineagu I, Wang X, Johnston P, Lowe JK, Tian Y, Horvath S, Mill J, Cantor RM, Blencowe BJ, Geschwind DH: Transcriptomic analysis of autistic brain reveals convergent molecular pathology. *Nature* 2011, **474**(7351):380–384.
68. Schwarz JM, Bilbo SD: Sex, glia, and development: interactions in health and disease. *Horm Behav* 2012, **62**(3):243–253.
69. Pardo CA, Eberhart CG: The neurobiology of autism. *Brain Pathol* 2007, **17**(4):434–447.
70. Li X, Chauhan A, Sheikh AM, Patil S, Chauhan V, Li XM, Ji L, Brown T, Malik M: Elevated immune response in the brain of autistic patients. *J Neuroimmunol* 2009, **207**(1–2):111–116.

71. Lintas C, Sacco R, Persico AM: Genome-wide expression studies in autism spectrum disorder, Rett syndrome, and Down syndrome. *Neurobiol Dis* 2012, **45**(1):57–68.
72. Jung JY, Kohane IS, Wall DP: Identification of autoimmune gene signatures in autism. *Transl Psychiatry* 2011, **1**:e63.
73. Frackowiak J, Mazur-Kolecka B, Schanen NC, Brown WT, Wegiel J: The link between intraneuronal N-truncated amyloid-beta peptide and oxidatively modified lipids in idiopathic autism and dup(15q11.2-q13)/autism. *Acta Neuropathol Commun* 2013, **1**(1):61.
74. Wegiel J, Frackowiak J, Mazur-Kolecka B, Schanen NC, Cook EH, Jr., Sigman M, Brown WT, Kuchna I, Wegiel J, Nowicki K *et al.*: Abnormal intracellular accumulation and extracellular Abeta deposition in idiopathic and Dup15q11.2-q13 autism spectrum disorders. *PLoS One* 2012, **7**(5):e35414.
75. Yang L, Kress BT, Weber HJ, Thiyagarajan M, Wang B, Deane R, Benveniste H, Iliff JJ, Nedergaard M: Evaluating glymphatic pathway function utilizing clinically relevant intrathecal infusion of CSF tracer. *J Transl Med* 2013, **11**(1):107.
76. Mendelsohn AR, Larrick JW: Sleep facilitates clearance of metabolites from the brain: glymphatic function in aging and neurodegenerative diseases. *Rejuvenation Res* 2013, **16**(6):518–523.
77. Iliff JJ, Wang M, Zeppenfeld DM, Venkataraman A, Plog BA, Liao Y, Deane R, Nedergaard M: Cerebral arterial pulsation drives paravascular csf-interstitial fluid exchange in the murine brain. *J Neurosci* 2013, **33**(46): 18190–18199.
78. Iliff JJ, Nedergaard M: Is there a cerebral lymphatic system? *Stroke* 2013, **44**(6 Suppl 1):S93–95.
79. Iliff JJ, Lee H, Yu M, Feng T, Logan J, Nedergaard M, Benveniste H: Brain-wide pathway for waste clearance captured by contrast-enhanced MRI. *J Clin Invest* 2013, **123**(3):1299–1309.
80. Xie L, Kang H, Xu Q, Chen MJ, Liao Y, Thiyagarajan M, O'Donnell J, Christensen DJ, Nicholson C, Iliff JJ *et al.*: Sleep drives metabolite clearance from the adult brain. *Science* 2013, **342**(6156):373–377.
81. Iliff JJ, Wang M, Liao Y, Plogg BA, Peng W, Gundersen GA, Benveniste H, Vates GE, Deane R, Goldman SA *et al.*: A paravascular pathway facilitates CSF flow through the brain parenchyma and the clearance of interstitial solutes, including amyloid beta. *Sci Transl Med* 2012, **4**(147):147ra111.
82. Amaral DG, Schumann CM, Nordahl CW: Neuroanatomy of autism. *Trends Neurosci* 2008, **31**(3):137–145.
83. Herbert MR: The Neuroanatomy of Autism. In: *The Neuropsychology of Autism.* Edited by Fein DA. New York, NY: Oxford University Press; 2011: 47–76.

84. Caviness VS, Jr., Lange NT, Makris N, Herbert MR, Kennedy DN: MRI-based brain volumetrics: emergence of a developmental brain science. *Brain Dev* 1999, **21**(5):289–295.
85. Caviness VS, Jr., Kennedy DN, Richelme C, Rademacher J, Filipek PA: The human brain age 7–11 years: a volumetric analysis based on magnetic resonance images. *Cereb Cortex* 1996, **6**(5):726–736.
86. Herbert MR: Large brains in autism: the challenge of pervasive abnormality. *Neuroscientist* 2005, **11**(5):417–440.
87. Courchesne E, Redcay E, Kennedy DP: The autistic brain: birth through adulthood. *Curr Opin Neurol* 2004, **17**(4):489–496.
88. Courchesne E, Karns CM, Davis HR, Ziccardi R, Carper RA, Tigue ZD, Chisum HJ, Moses P, Pierce K, Lord C *et al.*: Unusual brain growth patterns in early life in patients with autistic disorder: an MRI study. *Neurology* 2001, **57**(2):245–254.
89. Herbert MR, Ziegler DA, Makris N, Filipek PA, Kemper TL, Normandin JJ, Sanders HA, Kennedy DN, Caviness VSJ: Localization of white matter volume increase in autism and developmental language disorder. *Ann Neurol* 2004, **55**(4):530–540.
90. Suzuki K, Sugihara G, Ouchi Y, Nakamura K, Futatsubashi M, Takebayashi K, Yoshihara Y, Omata K, Matsumoto K, Tsuchiya KJ *et al.*: Microglial activation in young adults with autism spectrum disorder. *JAMA Psychiatry* 2013, **70**(1):49–58.
91. Sundaram SK, Kumar A, Makki MI, Behen ME, Chugani HT, Chugani DC: Diffusion tensor imaging of frontal lobe in autism spectrum disorder. *Cereb Cortex* 2008, **18**(11):2659–2665.
92. Dager SR, Friedman SD, Petropoulos H, Shaw DWW: Imaging evidence for pathological brain development in Autism Spectrum Disorders. Totowa, NJ: Humana Press; 2008.
93. Muller RA, Kleinhans N, Kemmotsu N, Pierce K, Courchesne E: Abnormal variability and distribution of functional maps in autism: an FMRI study of visuomotor learning. *Am J Psychiatry* 2003, **160**(10):1847–1862.
94. Ances BM, Leontiev O, Perthen JE, Liang C, Lansing AE, Buxton RB: Regional differences in the coupling of cerebral blood flow and oxygen metabolism changes in response to activation: implications for BOLD-fMRI. *Neuroimage* 2008, **39**(4):1510–1521.
95. Bandettini PA: The BOLD plot thickens: sign- and layer-dependent hemodynamic changes with activation. *Neuron* 2012, **76**(3):468–469.
96. Laurence JA, Fatemi SH: Glial fibrillary acidic protein is elevated in superior frontal, parietal and cerebellar cortices of autistic subjects. *Cerebellum* 2005, **4**(3):206–210.
97. Buzsaki G: *Rhythms of the Brain*. New York: Oxford University Press; 2006.
98. Kholodenko BN: Cell-signalling dynamics in time and space. *Nat Rev Mol Cell Biol* 2006, **7**(3):165–176.

99. Herbert MR, Sage C: Autism and EMF? Plausibility of a pathophysiological link part II. *Pathophysiology* 2013, **20**(3):211–234.
100. Herbert MR, Sage C: Autism and EMF? Plausibility of a pathophysiological link — Part I. *Pathophysiology* 2013, **20**(3):191–209.
101. Haughey NJ, Mattson MP: Alzheimer's amyloid beta-peptide enhances ATP/gap junction-mediated calcium-wave propagation in astrocytes. *Neuromol Med* 2003, **3**(3):173–180.
102. Rorig B, Klausa G, Sutor B: Intracellular acidification reduced gap junction coupling between immature rat neocortical pyramidal neurones. *J Physiol* 1996, **490** (Pt 1):31–49.
103. Miller AH, Maletic V, Raison CL: Inflammation and its discontents: the role of cytokines in the pathophysiology of major depression. *Biol Psychiatry* 2009, **65**(9):732–741.
104. Isler JR, Martien KM, Grieve PG, Stark RI, Herbert MR: Reduced functional connectivity in visual evoked potentials in children with autism spectrum disorder. *Clin Neurophysiol* 2010.
105. Coben R, Clarke AR, Hudspeth W, Barry RJ: EEG power and coherence in autistic spectrum disorder. *Clin Neurophysiol* 2008, **119**(5):1002–1009.
106. Mathewson KJ, Jetha MK, Drmic IE, Bryson SE, Goldberg JO, Schmidt LA: Regional EEG alpha power, coherence, and behavioral symptomatology in autism spectrum disorder. *Clin Neurophysiol* 2012, **123**(9):1798–1809.
107. Murias M, Webb SJ, Greenson J, Dawson G: Resting state cortical connectivity reflected in EEG coherence in individuals with autism. *Biol Psychiatry* 2007, **62**(3):270–273.
108. Rubenstein JL, Merzenich MM: Model of autism: increased ratio of excitation/inhibition in key neural systems. *Genes Brain Behav* 2003, **2**(5):255–267.
109. Vinga S, Neves AR, Santos H, Brandt BW, Kooijman SA: Subcellular metabolic organization in the context of dynamic energy budget and biochemical systems theories. *Philos Trans R Soc Lond B Biol Sci* 2010, **365**(1557):3429–3442.
110. Noble D: *The Music of Life: Biology Beyond the Genome*. New York: Oxford University Press; 2006.
111. Cristofolini L, Taddei F, Baleani M, Baruffaldi F, Stea S, Viceconti M: Multiscale investigation of the functional properties of the human femur. *Philos Transact A Math Phys Eng Sci* 2008, **366**(1879):3319–3341.
112. de Graaf AA, Freidig AP, De Roos B, Jamshidi N, Heinemann M, Rullmann JA, Hall KD, Adiels M, van Ommen B: Nutritional systems biology modeling: from molecular mechanisms to physiology. *PLoS Comput Biol* 2009, **5**(11):e1000554.
113. Pessah IN, Lein PJ: Evidence for Environmental Susceptibility in Autism: What We Need to Know About Gene *x* Environment Interactions: Humana; 2008.
114. Stamou M, Streifel KM, Goines PE, Lein PJ: Neuronal connectivity as a convergent target of gene x environment interactions that confer risk for Autism Spectrum Disorders. *Neurotoxicol Teratol* 2013, **36**:3–16.

115. Betancur C: Etiological heterogeneity in autism spectrum disorders: more than 100 genetic and genomic disorders and still counting. *Brain Res* 2011, **1380**:42–77.
116. Halassa MM, Fellin T, Haydon PG: The tripartite synapse: roles for gliotransmission in health and disease. *Trends Mol Med* 2007, **13**(2):54–63.
117. Dityatev A, Rusakov DA: Molecular signals of plasticity at the tetrapartite synapse. *Curr Opin Neurobiol* 2011, **21**(2):353–359.
118. Anderson GM: The potential role for emergence in autism. *Autism Res* 2008, **1**(1):18–30.
119. Autism WHY and HOW [www.autismWHYandHOW.org]
120. Virmani A, Pinto L, Binienda Z, Ali S: Food, nutrigenomics, and neurodegeneration — neuroprotection by what you eat! *Mol Neurobiol* 2013, **48**(2):353–362.
121. Kussmann M, Van Bladeren PJ: The Extended Nutrigenomics — Understanding the Interplay between the Genomes of Food, Gut Microbes, and Human Host. *Front Genet* 2011, **2**:21.
122. Herbert MR, Weintraub K: *The Autism Revolution: Whole Body Strategies for Making Life All It Can Be*. New York, NY: Random House with Harvard Health Publications; 2012.
123. Hermsdorff HH, Zulet MA, Puchau B, Martinez JA: Fruit and vegetable consumption and proinflammatory gene expression from peripheral blood mononuclear cells in young adults: a translational study. *Nutr Metab (Lond)* 2010, **7**:42.
124. Herbert MR: Everyday epigenetics: from molecular intervention to public health and lifestyle medicine. *N Am J Med Sci* 2013, **6**(3):167–170 (open access).
125. Herbert MR: Learning from the autism catastrophe: key leverage points. *Altern Ther Health Med* 2008, **14**(6):28–30.
126. Herbert M: Treatment-Guided Research: Helping people now with humility, respect and boldness. *Autism Advocate* 2008, **50**(1):8–14 http://www.autism-society.org/site/DocServer/Pages_18-14_TGRI_.pdf?docID=10460.
127. Editorial: Logitudinal thinking. *Nat Biotechnol* 2013, **31**(3):177.

Chapter 22

Achieving Optimal Outcomes in Autism: Treating Potentially Reversible Conditions Associated with Autism Spectrum Disorder

Richard E. Frye* and Daniel A. Rossignol

Abstract

Despite decades of research, effective medical treatments for autism spectrum disorders are very limited. In fact, the idea that autistic children can be successfully treated to achieve optimal outcomes has only recently become more broadly accepted. Recent studies have demonstrated an association between autism and several metabolic abnormalities, including disorders of mitochondrial, folate, redox, glutathione and tetrahydrobiopterin metabolism. Many of these metabolic abnormalities have the potential to be amenable to novel treatments, thereby opening the door for improvements and potential recovery from the disabilities that define and are associated with autism. In fact, several preliminary clinical studies have suggested positive effects from treatment of these metabolic abnormalities. This chapter reviews the metabolic abnormalities that are potentially amenable to treatment, and the evidence for the effectiveness of treatments of these physiological abnormalities in autism. Although preliminary, this evidence paves the way for achieving optimal outcomes in children with autism.

*Correspondence to: Arkansas Children's Hospital Research Institute, Department of Pediatrics, University of Arkansas for Medical Sciences, Little Rock, AR 72202. Email: REFrye@uams.edu

Keywords: Mitochondrial dysfunction; Antioxidants; Glutathione; Redox metabolism; Folate; Folinic acid; Folate receptor alpha autoantibody; Tetrahydrobiopterin; Nitric oxide; Autism spectrum disorder.

Introduction

Despite decades of research, the knowledge about the causes of and treatments for autism spectrum disorder (ASD) remain limited. Currently, the only well-accepted treatment for ASD symptoms is behavioral therapy that requires full-time engagement of a one-on-one therapist over years. This therapy is not only expensive but is typically not covered by medical insurance, thereby limiting its application to a minority of children with ASD. Currently the only medical treatments that are approved by the Food and Drug Administration (FDA) for ASD are antipsychotic medications that only treat a symptom associated with ASD, irritability, but not a core symptom that defines ASD.

The notion that children can recover from symptoms of ASD has only recently been accepted. In the past, many critics dismissed reports of children who recovered from ASD symptoms as the result of initial misdiagnosis. However, a recent multicenter study, in which the original ASD diagnoses were verified by a blind rater, provides considerable support for the idea that optimal outcomes can be achieved in some ASD children.[1] Moreover, another recent study demonstrated that there are at least six different developmental trajectories for children with ASD. This study analyzed the annual change in ASD symptoms from 6,975 children.[2] In general, for social and communication function, children in each group improved over time, with children who were higher functioning improved more than those who were lower functioning. Interestingly, for both social and communication function, a group of 'bloomers' was found who started out as low functioning and then rapidly improved over several years. These studies demonstrate that children with ASD can improve and achieve optimal outcomes. However, the reason that some children improve while others do not is not known. Defining the relationship between treatments received and attaining optimal outcomes is essential to define the best treatments for children with ASD.

ASD research has primarily concentrated on, and continues to concentrate on, genetic causes of ASD[3] despite the fact that inherited single gene and chromosomal defects account for the etiology of ASD in only a minority of cases.[4] In fact, several recent studies have identified rare *de novo* mutations in ASD children, thereby pointing to acquired mutations and/or mutations secondary to errors in the maintenance of DNA integrity rather that inherited genetic syndromes.[5,6] As research in the field of ASD continues, it is becoming clear that the etiology of most ASD cases involves complicated interactions between genetic predisposition with environmental exposures or triggers. Indeed, a recent study of dizygotic twins estimated that the environment contributes a greater percentage of the risk of developing autistic disorder as compared to genetic factors with the contribution of genetic and environmental factors about equal for the risk of developing the wider ASD diagnosis.[7]

Recent studies have expanded the recognition of the abnormalities associated with ASD, implicating abnormalities in basic physiological processes such as redox metabolism and impaired energy generation systems[3] as well as abnormalities in systems responsible for regulating basic metabolites such as folate,[8,9] tetrahydrobiopterin,[10–12] glutathione,[13–15] cholesterol,[16] carnitine[17–20] and branch chain amino acid.[21] Interestingly, several of these abnormalities seem to affect a broad range of children with ASD as they include children with idiopathic ASD where the etiology is not known, as well as those with known genetic causes of ASD. For example, mitochondrial dysfunction is prevalent in both idiopathic ASD[22] and is associated with genetic syndromes in which ASD is common, such as Rett syndrome,[23–25] PTEN mutations,[26] Phelan-McDermid syndrome,[27] 15q11-q13 duplication syndrome,[28,29] Angelman syndrome,[30] Septo-optic dysplasia[31] and Down syndrome.[32,33]

In this chapter, we discuss the evidence for metabolic abnormalities related to ASD that may be amenable to treatment. We demonstrate evidence that these disorders are actually related to the core physiological processes linked to developmental, cognitive and medical disabilities in ASD by discussing the underlying physiological links and providing evidence that increased severity of these physiological abnormalities is associated with greater disability in children with ASD. We also discuss the evidence that novel treatments that address these underlying metabolic abnormalities can be effective in reducing both core and associated ASD symptoms. Lastly, we point out how several of these metabolic abnormalities, as well as their treatments, are interrelated. Through these examples, we aim to demonstrate that there is a potential for treating metabolic abnormalities associated with ASD in order to optimize developmental, cognitive and behavioral outcomes.

Specific Metabolic Disorders Amenable to Novel Treatments

Mitochondrial metabolism

A recent meta-analysis suggested that approximately 5% of children with ASD meet the criteria for the diagnosis of classic mitochondrial disease (MD), while possibly up to 30% or more of children in the general ASD population exhibit abnormal biomarkers consistent with MD.[22] Since these biomarker studies did not investigate whether or not the children met the criteria for classic MD, it is not known whether this high prevalence of abnormal mitochondrial biomarkers translates to a high rate of MD in ASD. However, it has recently been shown that biomarkers of mitochondrial dysfunction can be confirmed approximately half of the time in children with ASD when repeat testing is performed, and that these repeated elevations in biomarker values are valid markers for MD.[34] Furthermore, it appears that these biomarkers may identify different subpopulations of children with MD.[34] Thus, these findings suggested that the prevalence of abnormal biomarker values of MD remains high in the ASD population.[34] A further confirmation of the potentially high prevalence of mitochondrial dysfunction comes from a recent study which compared electron transport chain (ETC) function in lymphocytes between ASD and typically developing controls.[35] This study found that 80% of the children with ASD clearly demonstrated lower than normal ETC function, or in other words, mitochondrial dysfunction.

The ability to define the proportion of ASD children with MD is complicated by the fact that there are multiple criteria used to define MD. The majority of studies that have looked at the prevalence of classic MD in the general ASD population have predominantly used one biomarker of MD, lactic acid, and the modified Walker criterion to define MD.[36] There are certain significant limitations to using the modified Walker criterion. Specifically, this criterion heavily relies on significant reductions in single ETC complex function and known mitochondrial deoxyribonucleic acid (DNA) mutations.[36] However, a review of all of the known published cases of ASD children with MD demonstrates that only 23% of ASD children diagnosed with MD have a known mitochondrial DNA mutation and several reports have noted that some ASD children with MD[22] have ETC complex over-activity rather than deficiencies.[37,38] Clearly, such cases would be missed by solely relying on the modified Walker criteria. Thus, Frye and Rossignol[39] have suggested using the Morava *et al.* criterion,[40] another commonly used criterion that considers a wide variety of findings to diagnose MD, including clinical, metabolic, imaging and morphological findings. Unlike the modified Walker criterion, the Morava *et al.* criterion[40] considers the number of mitochondrial related symptoms and laboratory findings for a particular patient in order to provide a guideline for performing a further workup for MD, including a muscle biopsy, in order to confirm the diagnosis of MD.

There are several links between mitochondrial dysfunction and the underlying pathophysiological mechanisms associated with abnormal cognitive development in ASD. Mitochondrial dysfunction has been demonstrated in the postmortem ASD brain[15,41–44] and in animal models of ASD.[45] Mitochondrial dysfunction of neural tissues can have many effects on neural transmission and function. Neural synapses are areas of high energy consumption[46] and are therefore especially dependent on mitochondrial function.[47] Mitochondria are concentrated in the dendritic and axonal termini where they play an important role in energy production, calcium homeostasis and synaptic plasticity.[48,49] Mitochondrial dysfunction can lead to reduced synaptic function and neurons that have high firing rates, such as GABAergic interneurons, can be severely compromised.[50] Reduction in GABAergic transmission may be particularly significant in ASD. GABA neurons are essential for the generation and synchronization of high-frequency gamma rhythms — rhythms essential for high-level cortical processing of sensory information and 'binding' which have been shown to be abnormal in some ASD children.[51] During the typical age when autistic regression occurs, there is an overabundance of cortical excitatory glutamatergic neurotransmitters and receptors[52,53] and this occurs at the peak of brain growth.[54] Thus, this is a developmental window of time in which mitochondrial energy production is pivotal.

Mitochondrial dysfunction may very well be related to the disability associated with ASD as children with ASD and mitochondrial dysfunction generally have more severe behavioral and cognitive disabilities than children with ASD without mitochondrial dysfunction.[55–57] In addition to contributing to brain dysfunction, mitochondrial dysfunction may also account for the high prevalence of medical problems associated with ASD as children with ASD and MD appear to have a higher prevalence of medical disorders such as gastrointestinal problems, seizures and gross motor delays as compared to ASD children without MD.[22]

Treatments for mitochondrial disease, including antioxidants, coenzyme Q10, L-carnitine, B-vitamins and the ketogenic diet, have been shown to improve function in some children with ASD.[22] Several studies have reported improvements in core and associated ASD behaviors with L-carnitine treatment,[17,29,58–61] including two double-blind placebo-controlled studies.[62,63] Two double-blind placebo-controlled studies using a multivitamin containing B vitamins, antioxidants, vitamin E, and coenzyme Q10 reported various improvements in ASD symptoms compared to placebo.[64,65] Several other antioxidants,[66] including vitamin C[67] and carnosine,[68] have also been reported to significantly improve autistic behaviors. One study reported improvements in ASD symptoms using NADH and

D-ribose.[69] Finally, one study reported improvements in ASD symptoms in 18 of 30 children with the use of a ketogenic diet.[14] Thus, many treatments that are believed to improve mitochondrial function have been shown to be helpful for some children with ASD, although many of the studies on children with ASD may not have specifically targeted mitochondrial function *per se*, and it is not known whether mitochondrial function improved with these treatments in these children with ASD. Nevertheless, these treatments represent those which have a favorable adverse effects profile and also the potential to significantly improve mitochondrial dysfunction in ASD.

Folate metabolism

Several lines of evidence point to abnormalities in folate metabolism in ASD. Several genetic polymorphisms that have been reported to have a higher prevalence in ASD can significantly affect folate metabolism. Polymorphisms reported in the genes coding for methylenetetrahydrofolate reductase,[13,70–78] dihydrofolate reductase[79] and the reduced folate carrier[13] can be associated with decreased production of 5-methyltetrahydrofolate (5MTHF) and impaired folate transport across the blood brain barrier and into neurons. However, perhaps the most significant abnormality in folate metabolism related to ASD is the folate receptor alpha (FRα) autoantibody. Folate in the form of 5MTHF is transported across the blood-brain barrier by an energy-dependent receptor-mediated system that utilizes the FRα.[80] Autoantibodies can bind to the FRα and greatly impair its function. These autoantibodies have been linked to cerebral folate deficiency (CFD) in some children with ASD[81–87] as well as in some children with Rett syndrome.[88–90] Because 5MTHF transport into the brain is adenosine triphosphate dependent, a wide variety of mitochondrial diseases have also been linked to dysfunction of the FRα transport system.[37,87,90–95] Recently, Frye *et al.*[8] reported that 60% and 44% of 93 children with ASD were positive for the blocking and binding FRα autoantibody, respectively. This high rate of FRα autoantibody positivity was confirmed by Ramaekers *et al.*[96] who compared 75 ASD children to

30 non-autistic controls with developmental delay. The blocking FRα autoantibody was positive in 47% of children with ASD but only 3% of the control children.

Deficits in folate metabolism could explain certain abnormalities associated with ASD. For example, deficits in folate metabolism can result in oxidative stress and methylation defects — two major abnormalities that have been shown systemically[13,97,98] and in the brain[99,100] of some children with ASD. Several reports have linked folate metabolism abnormalities with specific autistic behavior. For example, in children with ASD, methylenetetrahydrofolate reductase gene polymorphisms are associated with unusual body movements, hyperactivity and self-injurious behaviors[76] and the FRα autoantibody titers were found to be correlated with increasing aggressive behavior.[82]

Perhaps the most convincing evidence linking folate metabolism to ASD behavior is the marked response to treatment with reduced forms of folate in children with central folate abnormalities. For example, many children with ASD and CFD have been documented to have a marked improvement in clinical status with treatment using folinic acid — a reduced form of folate that can cross the blood-brain barrier using the reduced folate carrier rather than the FRα transport system. Several case reports[86] and case series[82,84] have described neurological, behavioral and cognitive improvements in children with documented CFD and ASD. One case series of five children with CFD and low-functioning autism with neurological deficits found complete recovery from ASD symptoms in one child and substantial improvements in communication in two other children.[84] In another study of 23 children with low-functioning regressive ASD and CFD, two younger children demonstrated full recovery from autism and neurological symptoms, three older children demonstrated improvements in neurological deficits but not ASD symptoms, and the remainder demonstrated improvements in neurological symptoms and partial improvements in some ASD symptoms, mostly prominently in communication.[82] This suggests that CFD represents a potentially reversible cause of ASD, at least in a subset of children.

Recently, Frye *et al.*[8] demonstrated that ASD children who were positive for at least one of the two FRα autoantibodies, but who did not necessarily have CFD, demonstrated an improvement with folinic acid treatment. Forty-four ASD children positive for at least one FRα autoantibody were given 2 mg/kg/day of folinic acid in two divided doses (maximum 50 mg daily) for an average of four months and parents were asked to rate intervention responses using a modified Clinical Global Impression-Improvement subscale on the dimensions of verbal communication, receptive language, expressive language, non-verbal communication, stereotypical behavior, hyperactivity, mood, attention, and aggression. These ratings were compared to a control group of ASD children who were awaiting test results but did not change any treatment over the same average follow-up period. Children treated with folinic acid, as compared to the control group, were found to have significant improvements in verbal communication, receptive and expressive language, attention, and stereotypical behavior with very few adverse effects. Sixteen ASD children who were FRα autoantibody positive underwent a diagnostic lumbar puncture to determine their cerebrospinal fluid (CSF) 5MTHF level prior to any treatment. Although a significant inverse linear relationship between CSF 5MTHF and the serum blocking FRα autoantibody concentration was found, no child demonstrated a CSF 5MTHF concentration below the lower limit of normal, but all CSF 5MTHF concentrations were below the normative mean CSF 5MTHF concentration. This suggests that children with ASD and the blocking FRα autoantibody may have a low normal CSF 5MTHF concentration consistent with a relative deficiency in central folate, or more accurately, a central folate insufficiency, where the folate level in the brain may be lower than what is required to sustain optimal neuronal function. In fact, some investigators have reported that folinic acid treatment significantly improves neurological symptoms in patients who have low normal CSF 5MTHF concentrations.[101] These findings suggest that treatment with reduced forms of folate like folinic acid may provide significant improvement in core and associated ASD symptoms.

Redox metabolism

Several lines of evidence support the notion that some children with ASD have abnormal redox metabolism. Two case-control studies have reported that redox metabolism of children diagnosed with autism is abnormal compared to unaffected control children.[97,102] This includes a significant reduction in glutathione (GSH), the major intracellular antioxidant and mechanism for detoxification, as well as a significant increase in the oxidized disulfide form of glutathione (GSSG). Several metabolic precursors for glutathione synthesis were also lower in the autistic children suggesting that GSH synthesis was insufficient, at least in a subgroup of ASD children. Additionally, GSH deficits have been demonstrated in the cytosol and isolated mitochondria from lymphoblastoid cell lines derived from children with autistic disorder as well as in fresh peripheral blood mononuclear cells derived from children with ASD.[103] The notion that abnormal GSH metabolism could lead to oxidative damage is consistent with studies which demonstrate oxidative damage to protein and DNA in peripheral blood mononuclear cells and postmortem brain from ASD individuals,[15,104] particularly in cortical regions associated with speech, emotion, and social behavior.[100]

Treatments for oxidative stress have been shown to be of benefit for children with ASD. Methylcobalamin and folinic acid have been reported to significantly increase GSH concentrations in children with ASD and appear to improve certain autistic behaviors in open-label studies.[97,105] Other studies suggest that glutathione metabolism can be improved by vitamins and minerals[106] as well as antioxidants, co-enzyme Q10 and B vitamin supplementation[57] and sapropterin treatment[12] in children with ASD. Both small[107,108] and large[106] randomized control trials and large case series[8] demonstrate that novel treatments for children with ASD which can address oxidative stress are associated with improvements in core ASD symptoms,[8,12,107] sleep and gastrointestinal symptoms[108] and hyperactivity, tantruming and parental impression of general functioning.[106] Interestingly, a recent double-blind placebo-controlled study has demonstrated that N-acetyl-L-cysteine improves irritability and social function in children with ASD compared to placebo.[107] N-acetyl-L-cysteine provides a precursor to GSH, and although GSH was not measured in the latter study, this treatment could have conferred its beneficial effects though an improvement in redox metabolism. Several other antioxidants,[66] including vitamin C[67] and carnosine[68] have also been reported to significantly improve autistic behaviors, suggesting that treatment of oxidative stress could be beneficial for children with ASD. Taken together, these studies suggest that the treatment of oxidative stress may improve core and associated symptoms of ASD and, thus, deserves further study. Furthermore, these treatments are generally regarded as safe with a low adverse effect profile.

Tetrahydrobiopterin metabolism

Tetrahydrobiopterin (BH_4) is a naturally occurring substance that is an essential cofactor for several critical metabolic pathways, including those responsible for the production of monoamine neurotransmitters, the breakdown of phenylalanine, and the production of nitric oxide.[11] BH_4 is also readily oxidized by reactive species.[10] Abnormalities in several BH_4 related metabolic pathways or the products of these pathways have been noted in some individuals with ASD, and the CSF concentration of BH_4 has been reported to be depressed in some individuals with ASD.[10,11,109,110] Clinical trials conducted over the past 25 years have reported encouraging results using sapropterin, a synthetic form of BH_4, to treat children with ASD.[11] Three controlled and several open-label trials have documented improvements in communication, cognitive ability, adaptability, social abilities and verbal expression with sapropterin treatment in ASD, especially in children younger than five years of age and in those who are relatively higher functioning at the beginning of the trial.[11,110–123] In addition, a recent double-blind placebo-controlled study has verified the therapeutic effect of sapropterin in children with autism.[122] Frye has shown that the ratio of serum citrulline-to-methionine is related to the BH_4

concentration in the CSF, suggesting that abnormalities in both oxidative stress and nitric oxide metabolism may be related to central BH_4 deficiency.[10] More recently, Frye *et al.* demonstrated, in an open-label study, that sapropterin treatment improves redox metabolism and fundamentally alters BH_4 metabolism in children with ASD. Interestingly, serum biomarkers of nitric oxide metabolism were found to predict response to sapropterin treatment in children with ASD,[12] thereby suggesting that the therapeutic effect of BH_4 supplementation may have been specific to its effect on nitric oxide metabolism.

Common Molecular Mechanisms are Associated with Novel Treatments

The metabolic systems discussed above have clear interrelationships such that abnormalities in one metabolic system can result in abnormalities in another, sometimes resulting in a vicious downward spiral that results in metabolic dysfunction in several systems. Conversely, many of the novel treatments that appear to be helpful in ASD may improve function in multiple metabolic systems. Here we provide some insight into these overlaps and the effects of novel treatments.

Probably the most well-known connection between metabolic abnormalities is the overlap between mitochondrial dysfunction and oxidative stress. Elevated levels of reactive oxygen species (ROS) can result in mitochondrial dysfunction.[3,22] In turn, dysfunctional mitochondria can create a self-perpetuating cycle of progressive damage that amplifies functional deficits.[57] For example, ETC complex I dysfunction can produce high levels of ROS and reactive nitrogen species (RNS).[124,125] In turn, ROS and RNS can inhibit ETC function, such as complex I activity,[126] causing further increases in ROS and RNS.[127] The brain may be a particularly sensitive place for this vicious cycle between worsening ROS/RNS and mitochondrial dysfunction to occur as GSH levels in the brain are about 50 fold lower than in peripheral tissues, such as hepatocytes, making neurons especially sensitive to increases in ROS/RNS resulting from dysfunctional mitochondria.[128] Additionally, since chronic oxidative stress can result in mitochondrial dysfunction,[129] the increased oxidative stress noted in postmortem ASD brain samples[100] could account for the mitochondrial dysfunction found in similar cortical regions in the postmortem ASD brain.[15,130] Interestingly, many novel treatments for ASD that appear to improve ASD symptoms also can improve mitochondrial dysfunction and RNS/ROS, including multivitamins containing B vitamins, antioxidants, vitamin E, and coenzyme Q10.[64–67]

Disorders of folate metabolism can interact with redox and mitochondrial metabolism to cause pathology, especially in the brain. Folate is a necessary precursor for the production of GSH and, thus, polymorphisms in genes involved in folate metabolism known to be associated with ASD can cause a decreased production and availability of GSH.[13] Decreased GSH can result in increases in ROS/RNS which can damage mitochondria, resulting in a further increase in oxidative stress.[129] Mitochondrial dysfunction can compromise the FRα transport system, resulting in decreased brain folate levels. In addition, redox abnormalities in ASD have been associated with elevation in inflammatory mediators suggesting activation of the immune system.[15] Such immune activation could result in increased production of autoantibodies to the FRα which, in turn, could reduce folate transport across the blood brain barrier into the brain.[8] In the context of high ROS, relatively higher folate levels would be needed to produce GSH, thus making the brain particularly vulnerable if folate levels were being suppressed by a dysfunctional FRα transport system. Decreased folate transport into the central nervous system could exacerbate existing problems with mitochondrial dysfunction and oxidative stress, resulting in particularly devastating outcomes. Treatment with folinic acid may be particularly effective as folinic acid, a reduced form of folate, can use the reduced folate carrier to enter the brain and improve the central folate concentration and in the process, bypass metabolic blocks caused by detrimental polymorphisms in folate genes[8] or autoantibodies to the FRα.

Abnormalities in BH_4 metabolism may be particularly associated with redox abnormalities. BH_4 can act as an antioxidant, being easily oxidized in the setting of increased ROS/RNS. Interestingly, recycling BH_4 from its oxidized state requires folate, which may be lowered when ROS/RNS is increased. Oxidation of BH_4 results in a decrease in the reduced-to-oxidized pterin ratio that is essential for regulating the coupling of nitric oxide synthase. When this ratio is reduced, nitric oxide synthase is uncoupled, resulting in a reaction that produces peroxynitrite, a RNS, rather than nitric oxide, thus resulting in a more oxidized microenvironment.[12] This, in turn, can further reduce the reduced-to-oxidized pterin ratio, further exacerbating the abnormalities. The biomarkers studied by Frye[10] suggest that the nitric oxide synthase reaction is up-regulated when uncoupled, probably in an attempt to increase production of nitric oxide. Thus, in a futile attempt to produce nitric oxide, the body may be producing RNS and further decreasing BH_4.[10,12] Interestingly, sapropterin treatment does not merely raise the available BH_4 concentration but appears to fundamentally change pterin metabolism such that oxidized pterins are markedly reduced.[12] This fundamental change in metabolism most likely recouples nitric oxide synthase and restores nitric oxide production.

Future Prospective

This chapter outlines some specific metabolic abnormalities associated with autism as well as their treatments. We have demonstrated that there is evidence that treatments for these metabolic disorders are associated with improvements in cognitive and behavioral function in some children with ASD. As many of these treatments have undergone only preliminary studies, it is unclear how significant the improvement can be in ASD children who respond to specific treatments and what percentage of children with ASD will respond to such treatments. However, some treatments, such as folinic acid, appear very promising in a significant proportion of children with ASD.

One advantage of the treatments outlined above is that the physiological mechanisms are known and biomarkers are available to identify children who might be most amenable to treatment and to correlate changes in the physiological processes with cognitive and behavioral improvements. Preliminary studies suggest that there are a substantial number of ASD children with physiological abnormalities that might be amenable to treatment. For example, the prevalence of mitochondrial dysfunction varies from 5% to 80%[22,34,35] and the prevalence of the blocking FRα autoantibody has been estimated to be between 47%[96] and 60%.[8] Clearly, further studies will be required to clarify the percentage of the ASD population who manifest specific metabolic abnormalities and to determine what percentage of these subgroups will respond to a particular treatment. One potential advantage of having readily available metabolic biomarkers is that children with these underlying physiological abnormalities can be identified early, potentially in the pre-symptomatic period, when treatments may be most effective.

Interestingly, these physiological abnormalities do not only apply to ASD but are shared by a wide variety of psychiatric, neurodevelopmental and neurodegenerative disorders. For example, mitochondrial dysfunction has been implicated in schizophrenia,[131–134] bipolar disorder,[131–133] depression[132,133] and dementia,[132,133] and oxidative stress has been implicated in schizophrenia,[135,136] depression,[137] bipolar disorder,[138,139] Alzheimer's disease[140] and Parkinson's disease.[140] Thus, the treatments for the metabolic abnormalities outlined above for ASD may have implications for a wide range of medical diseases.

Executive Summary

- Autism spectrum disorder is associated with physiological abnormalities in several metabolic systems, including abnormalities in mitochondrial, redox, folate and tetrahydrobiopterin metabolism.
- Abnormalities in several metabolic systems may be amenable to safe, well-tolerated treatments, some of which have been reported to have positive effects on core and associated symptoms of autism spectrum disorder.

- Many of these physiological abnormalities are interrelated such that abnormalities in one metabolic system may result in abnormalities in others and vice-versa. Alternatively, treatments that address one physiological abnormality may improve abnormalities in other metabolic systems.
- These physiological disorders have biomarkers that may facilitate identifying ASD children who will benefit from these treatments as well as potentially detect children early in life, during the pre-symptomatic period, so treatments can be initiated early.
- Many of these metabolic abnormalities have also been implicated in other psychiatric, neurodevelopmental and neurological disorders, suggesting that these physiological abnormalities may represent the final common pathophysiology for a wide range of diseases.

References

1. Fein D, Barton M, Eigsti IM, Kelley E, Naigles L, Schultz RT, Stevens M, Helt M, Orinstein A, Rosenthal M *et al.*: Optimal outcome in individuals with a history of autism. *J Child Psychol Psychiatry* 2013, **54**(2):195–205.
2. Fountain C, Winter AS, Bearman PS: Six developmental trajectories characterize children with autism. *Pediatrics* 2012, **129**(5):e1112–1120.
3. Rossignol DA, Frye RE: A review of research trends in physiological abnormalities in autism spectrum disorders: immune dysregulation, inflammation, oxidative stress, mitochondrial dysfunction and environmental toxicant exposures. *Mol Psychiatry* 2012, **17**(4):389–401.
4. Schaefer GB, Mendelsohn NJ: Genetics evaluation for the etiologic diagnosis of autism spectrum disorders. *Genet Med* 2008, **10**(1):4–12.
5. Neale BM, Kou Y, Liu L, Ma'ayan A, Samocha KE, Sabo A, Lin CF, Stevens C, Wang LS, Makarov V *et al.*: Patterns and rates of exonic de novo mutations in autism spectrum disorders. *Nature* 2012, **485**(7397):242–245.
6. Sanders SJ, Murtha MT, Gupta AR, Murdoch JD, Raubeson MJ, Willsey AJ, Ercan-Sencicek AG, DiLullo NM, Parikshak NN, Stein JL *et al.*: De novo mutations revealed by whole-exome sequencing are strongly associated with autism. *Nature* 2012, **485**(7397):237–241.
7. Hallmayer J, Cleveland S, Torres A, Phillips J, Cohen B, Torigoe T, Miller J, Fedele A, Collins J, Smith K *et al.*: Genetic heritability and shared environmental factors among twin pairs with autism. *Arch Gen Psychiatry* 2011, **68**(11):1095–1102.

8. Frye RE, Sequeira JM, Quadros EV, James SJ, Rossignol DA: Cerebral folate receptor autoantibodies in autism spectrum disorder. *Mol Psychiatry* 2013, **18**(3):369–381.
9. Rossignol D, Frye RE: Folate receptor alpha autoimmunity and cerebral folate deficiency in autism spectrum disorders. *J Pediatr Biochem* 2012, **2**: 263–271.
10. Frye RE: Central tetrahydrobiopterin concentration in neurodevelopmental disorders. *Front Neurosci* 2010, **4**:52.
11. Frye RE, Huffman LC, Elliott GR: Tetrahydrobiopterin as a novel therapeutic intervention for autism. *Neurotherapeutics* 2010, **7**(3):241–249.
12. Frye RE, DeLatorre R, Taylor HB, Slattery J, Melnyk S, Chowdhury N, James SJ: Metabolic effects of sapropterin treatment in autism spectrum disorder: a preliminary study. *Transl Psychiatry* 2013, **3**:e237.
13. James SJ, Melnyk S, Jernigan S, Cleves MA, Halsted CH, Wong DH, Cutler P, Bock K, Boris M, Bradstreet JJ *et al.*: Metabolic endophenotype and related genotypes are associated with oxidative stress in children with autism. *Am J Med Genet B Neuropsychiatr Genet* 2006, **141B**(8):947–956.
14. Evangeliou A, Vlachonikolis I, Mihailidou H, Spilioti M, Skarpalezou A, Makaronas N, Prokopiou A, Christodoulou P, Liapi-Adamidou G, Helidonis E *et al.*: Application of a ketogenic diet in children with autistic behavior: pilot study. *J Child Neurol* 2003, **18**(2):113–118.
15. Rose S, Melnyk S, Pavliv O, Bai S, Nick TG, Frye RE, James SJ: Evidence of oxidative damage and inflammation associated with low glutathione redox status in the autism brain. *Transl Psychiatry* 2012, **2**:e134.
16. Tierney E, Bukelis I, Thompson RE, Ahmed K, Aneja A, Kratz L, Kelley RI: Abnormalities of cholesterol metabolism in autism spectrum disorders. *Am J Med Genet B Neuropsychiatr Genet* 2006, **141B**(6):666–668.
17. Gargus JJ, Lerner MA: Familial autism with primary carnitine deficiency, sudden death, hypotonia and hypochromic anemia. *Am J Hum Genet* 1997, **61**:A98.
18. Frye RE, Melnyk S, Macfabe DF: Unique acyl-carnitine profiles are potential biomarkers for acquired mitochondrial disease in autism spectrum disorder. *Transl Psychiatry* 2013, **3**:e220.
19. Filipek PA, Juranek J, Nguyen MT, Cummings C, Gargus JJ: Relative carnitine deficiency in autism. *J Autism and Dev Dis* 2004, **34**(6):615–623.
20. Celestino-Soper PB, Violante S, Crawford EL, Luo R, Lionel AC, Delaby E, Cai G, Sadikovic B, Lee K, Lo C *et al.*: A common X-linked inborn error of carnitine biosynthesis may be a risk factor for nondysmorphic autism. *Proc Nat Acad Sci USA* 2012, **109**(21):7974–7981.
21. Novarino G, El-Fishawy P, Kayserili H, Meguid NA, Scott EM, Schroth J, Silhavy JL, Kara M, Khalil RO, Ben-Omran T *et al.*: Mutations in BCKD-kinase lead to a potentially treatable form of autism with epilepsy. *Science* 2012, **338**(6105):394–397.

22. Rossignol DA, Frye RE: Mitochondrial dysfunction in autism spectrum disorders: a systematic review and meta-analysis. *Mol Psychiatry* 2012, **17**(3): 290–314.
23. Grosser E, Hirt U, Janc OA, Menzfeld C, Fischer M, Kempkes B, Vogelgesang S, Manzke TU, Opitz L, Salinas-Riester G *et al.*: Oxidative burden and mitochondrial dysfunction in a mouse model of Rett syndrome. *Neurobiol Dis* 2012, **48**(1):102–114.
24. Gibson JH, Slobedman B, K NH, Williamson SL, Minchenko D, El-Osta A, Stern JL, Christodoulou J: Downstream targets of methyl CpG binding protein 2 and their abnormal expression in the frontal cortex of the human Rett syndrome brain. *BMC Neurosci* 2010, **11**:53.
25. Condie J, Goldstein J, Wainwright MS: Acquired microcephaly, regression of milestones, mitochondrial dysfunction, and episodic rigidity in a 46,XY male with a de novo MECP2 gene mutation. *J Child Neurol* 2010, **25**(5): 633–636.
26. Napoli E, Ross-Inta C, Wong S, Hung C, Fujisawa Y, Sakaguchi D, Angelastro J, Omanska-Klusek A, Schoenfeld R, Giulivi C: Mitochondrial dysfunction in Pten haplo-insufficient mice with social deficits and repetitive behavior: interplay between Pten and p53. *PloS One* 2012, **7**(8):e42504.
27. Frye RE: Mitochondrial disease in 22q13 duplication syndrome. *J Child Neurol* 2012, **27**(7):942–949.
28. Frye RE: 15q11.2-13 duplication, mitochondrial dysfunction, and developmental disorders. *J Child Neurol* 2009, **24**(10):1316–1320.
29. Filipek PA, Juranek J, Smith M, Mays LZ, Ramos ER, Bocian M, Masser-Frye D, Laulhere TM, Modahl C, Spence MA *et al.*: Mitochondrial dysfunction in autistic patients with 15q inverted duplication. *Ann Neurol* 2003, **53**(6): 801–804.
30. Su H, Fan W, Coskun PE, Vesa J, Gold JA, Jiang YH, Potluri P, Procaccio V, Acab A, Weiss JH *et al.*: Mitochondrial dysfunction in CA1 hippocampal neurons of the UBE3A deficient mouse model for Angelman syndrome. *Neurosci Lett* 2011, **487**(2):129–133.
31. Schuelke M, Krude H, Finckh B, Mayatepek E, Janssen A, Schmelz M, Trefz F, Trijbels F, Smeitink J: Septo-optic dysplasia associated with a new mitochondrial cytochrome b mutation. *Ann Neurol* 2002, **51**(3):388–392.
32. Pagano G, Castello G: Oxidative stress and mitochondrial dysfunction in Down syndrome. *Adv Exp Med Biol* 2012, **724**:291–299.
33. Pallardo FV, Lloret A, Lebel M, d'Ischia M, Cogger VC, Le Couteur DG, Gadaleta MN, Castello G, Pagano G: Mitochondrial dysfunction in some oxidative stress-related genetic diseases: Ataxia-Telangiectasia, Down Syndrome, Fanconi Anaemia and Werner Syndrome. *Biogerontology* 2010, **11**(4):401–419.
34. Frye RE: Biomarker of abnormal energy metabolism in children with autism spectrum disorder. *N Am J Med Sci* 2012, **5**:141–147.

35. Giulivi C, Zhang YF, Omanska-Klusek A, Ross-Inta C, Wong S, Hertz-Picciotto I, Tassone F, Pessah IN: Mitochondrial dysfunction in autism. *JAMA* 2010, **304**(21):2389–2396.
36. Bernier FP, Boneh A, Dennett X, Chow CW, Cleary MA, Thorburn DR: Diagnostic criteria for respiratory chain disorders in adults and children. *Neurology* 2002, **59**(9):1406–1411.
37. Frye RE, Naviaux RK: Autistic disorder with complex IV overactivity: A new mitochondrial syndrome. *J Pediatr Neurol* 2011, **9**:427–434.
38. Graf WD, Marin-Garcia J, Gao HG, Pizzo S, Naviaux RK, Markusic D, Barshop BA, Courchesne E, Haas RH: Autism Associated With the Mitochondrial DNA G8363A Transfer RNA^{Lys} Mutation. *J Child Neurol* 2000, **15**:357–361.
39. Frye RE, Rossignol DA: Mitochondrial dysfunction can connect the diverse medical symptoms associated with autism spectrum disorders. *Pediatr Res* 2011, **69**(5 Pt 2):41R–47R.
40. Morava E, van den Heuvel L, Hol F, de Vries MC, Hogeveen M, Rodenburg RJ, Smeitink JA: Mitochondrial disease criteria: diagnostic applications in children. *Neurology* 2006, **67**(10):1823–1826.
41. Chauhan A, Gu F, Essa MM, Wegiel J, Kaur K, Brown WT, Chauhan V: Brain region-specific deficit in mitochondrial electron transport chain complexes in children with autism. *J Neurochem* 2011, **117**(2):209–220.
42. Tang G, Gutierrez Rios P, Kuo SH, Akman HO, Rosoklija G, Tanji K, Dwork A, Schon EA, Dimauro S, Goldman J *et al.*: Mitochondrial abnormalities in temporal lobe of autistic brain. *Neurobiol Dis* 2013, **54**:349–361.
43. Anitha A, Nakamura K, Thanseem I, Yamada K, Iwayama Y, Toyota T, Matsuzaki H, Miyachi T, Yamada S, Tsujii M *et al.*: Brain region-specific altered expression and association of mitochondria-related genes in autism. *Mol Autism* 2012, **3**(1):12.
44. Anitha A, Nakamura K, Thanseem I, Matsuzaki H, Miyachi T, Tsujii M, Iwata Y, Suzuki K, Sugiyama T, Mori N: Downregulation of the expression of mitochondrial electron transport complex genes in autism brains. *Brain Pathol* 2013, **23**(3):294–302.
45. Kriaucionis S, Paterson A, Curtis J, Guy J, Macleod N, Bird A: Gene expression analysis exposes mitochondrial abnormalities in a mouse model of Rett syndrome. *Mol Cell Biol* 2006, **26**(13):5033–5042.
46. Ames A, 3rd: CNS energy metabolism as related to function. *Brain Res Brain Res Rev* 2000, **34**(1–2):42–68.
47. Mattson MP, Liu D: Energetics and oxidative stress in synaptic plasticity and neurodegenerative disorders. *Neuromolecular Med* 2002, **2**(2):215–231.
48. Chen H, Chan DC: Mitochondrial dynamics — fusion, fission, movement, and mitophagy — in neurodegenerative diseases. *Human Mol Genet* 2009, **18**(R2):R169–176.

49. Li Z, Okamoto K, Hayashi Y, Sheng M: The importance of dendritic mitochondria in the morphogenesis and plasticity of spines and synapses. *Cell* 2004, **119**(6):873–887.
50. Anderson MP, Hooker BS, Herbert MR: Bridging from cells to cognition in autism pathophysiology: biological pathways to defective brain function and plasticity *Am J Biochem Biotechnol* 2008, **4**(2):167–176.
51. Orekhova EV, Stroganova TA, Nygren G, Tsetlin MM, Posikera IN, Gillberg C, Elam M: Excess of high frequency electroencephalogram oscillations in boys with autism. *Biol Psychiatry* 2007, **62**(9):1022–1029.
52. Herlenius E, Lagercrantz H: Development of neurotransmitter systems during critical periods. *Exp Neurol* 2004, **190**(Suppl 1):S8–21.
53. Huttenlocher PR, Dabholkar AS: Regional differences in synaptogenesis in human cerebral cortex. *J Comp Neurol* 1997, **387**(2):167–178.
54. Groeschel S, Vollmer B, King MD, Connelly A: Developmental changes in cerebral grey and white matter volume from infancy to adulthood. *Int J Dev Neurosci* 2010, **28**(6):481–489.
55. Minshew NJ, Goldstein G, Dombrowski SM, Panchalingam K, Pettegrew JW: A preliminary 31P MRS study of autism: evidence for undersynthesis and increased degradation of brain membranes. *Biol Psychiatry* 1993, **33**(11–12):762–773.
56. Mostafa GA, El-Gamal HA, El-Wakkad ASE, El-Shorbagy OE, Hamza MM: Polyunsaturated fatty acids, carnitine and lactate as biological markers of brain energy in autistic children. *Int J Child Neuropsychiatry* 2005, **2**(2):179–188.
57. Frye RE, Delatorre R, Taylor H, Slattery J, Melnyk S, Chowdhury N, James SJ: Redox metabolism abnormalities in autistic children associated with mitochondrial disease. *Transl Psychiatry* 2013, **3**:e273.
58. Pastural E, Ritchie S, Lu Y, Jin W, Kavianpour A, Khine Su-Myat K, Heath D, Wood PL, Fisk M, Goodenowe DB: Novel plasma phospholipid biomarkers of autism: mitochondrial dysfunction as a putative causative mechanism. *Prostaglandins Leukot Essent Fatty Acids* 2009, **81**(4):253–264.
59. Gargus JJ, Imtiaz F: Mitochondrial energy-deficient endophenotype in autism. *Am J Biochem Biotechnol* 2008, **4**(2):198–207.
60. Poling JS, Frye RE, Shoffner J, Zimmerman AW: Developmental regression and mitochondrial dysfunction in a child with autism. *J Child Neurol* 2006, **21**(2):170–172.
61. Ezugha H, Goldenthal M, Valencia I, Anderson CE, Legido A, Marks H: 5q14.3 deletion manifesting as mitochondrial disease and autism: case report. *J Child Neurol* 2010, **25**(10):1232–1235.
62. Geier DA, Kern JK, Davis G, King PG, Adams JB, Young JL, Geier MR: A prospective double-blind, randomized clinical trial of levocarnitine to treat autism spectrum disorders. *Med Sci Monit* 2011, **17**(6):PI15–23.
63. Fahmy SF, El-hamamsy MH, Zaki OK, Badary OA: l-Carnitine supplementation improves the behavioral symptoms in autistic children. *Res Autism Spect Dis* 2013, **7**(1):159–166.

64. Adams JB, Holloway C: Pilot study of a moderate dose multivitamin/mineral supplement for children with autistic spectrum disorder. *J Altern Complem Med* 2004, **10**(6):1033–1039.
65. Adams JB, Audhya T, McDonough-Means S, Rubin RA, Quig D, Geis E, Gehn E, Loresto M, Mitchell J, Atwood S *et al.*: Effect of a vitamin/mineral supplement on children and adults with autism. *BMC Pediatrics* 2011, **11**:111.
66. Rossignol DA: Novel and emerging treatments for autism spectrum disorders: a systematic review. *Ann Clin Psychiatry* 2009, **21**(4):213–236.
67. Dolske MC, Spollen J, McKay S, Lancashire E, Tolbert L: A preliminary trial of ascorbic acid as supplemental therapy for autism. *Prog Neuro-Psychoph* 1993, **17**(5):765–774.
68. Chez MG, Buchanan CP, Aimonovitch MC, Becker M, Schaefer K, Black C, Komen J: Double-blind, placebo-controlled study of L-carnosine supplementation in children with autistic spectrum disorders. *J Child Neurol* 2002, **17**(11):833–837.
69. Adams JB, Freedenfeld S, Audhya T, Hamada K: Biochemical Effects of Ribose and NADH Therapy in Children with Autism. *Autism Insights* 2011, **3**:3.
70. Frustaci A, Neri M, Cesario A, Adams JB, Domenici E, Dalla Bernardina B, Bonassi S: Oxidative stress-related biomarkers in autism: systematic review and meta-analyses. *Free Radical Bio Med* 2012, **52**(10):2128–2141.
71. Boris M, Goldblatt A, Galanko J, James SJ: Association of MTHFR gene variants with autism. *J Am Physicians Surgeons* 2004, **9**(4):106–108.
72. Mohammad NS, Jain JM, Chintakindi KP, Singh RP, Naik U, Akella RR: Aberrations in folate metabolic pathway and altered susceptibility to autism. *Psychiatr Genet* 2009, **19**(4):171–176.
73. Guo T, Chen H, Liu B, Ji W, Yang C: Methylenetetrahydrofolate reductase polymorphisms C677T and risk of autism in the Chinese Han population. *Genet Test Mol Biomarkers* 2012, **16**(8):968–973.
74. Schmidt RJ, Tancredi DJ, Ozonoff S, Hansen RL, Hartiala J, Allayee H, Schmidt LC, Tassone F, Hertz-Picciotto I: Maternal periconceptional folic acid intake and risk of autism spectrum disorders and developmental delay in the CHARGE (CHildhood Autism Risks from Genetics and Environment) case-control study. *Am J Clin Nutr* 2012, **96**(1):80–89.
75. Liu X, Solehdin F, Cohen IL, Gonzalez MG, Jenkins EC, Lewis ME, Holden JJ: Population- and family-based studies associate the MTHFR gene with idiopathic autism in simplex families. *J Autism Dev Disord* 2011, **41**(7):938–944.
76. Goin-Kochel RP, Porter AE, Peters SU, Shinawi M, Sahoo T, Beaudet al.: The MTHFR 677C-->T polymorphism and behaviors in children with autism: exploratory genotype-phenotype correlations. *Autism Res* 2009, **2**(2):98–108.
77. Pasca SP, Dronca E, Kaucsar T, Craciun EC, Endreffy E, Ferencz BK, Iftene F, Benga I, Cornean R, Banerjee R *et al.*: One carbon metabolism disturbances

and the C677T MTHFR gene polymorphism in children with autism spectrum disorders. *J Cell Mol Med* 2009, **13**(10):4229–4238.

78. Pu D, Shen Y, Wu J: Association between MTHFR gene polymorphisms and the risk of autism spectrum disorders: a meta-analysis. *Autism Res* 2013.
79. Adams M, Lucock M, Stuart J, Fardell S, Baker K, Ng X: Preliminary evidence for involvement of the folate gene polymorphism 19bp deletion-DHFR in occurrence of autism. *Neurosci Lett* 2007, **422**(1):24–29.
80. Wollack JB, Makori B, Ahlawat S, Koneru R, Picinich SC, Smith A, Goldman ID, Qiu A, Cole PD, Glod J *et al.*: Characterization of folate uptake by choroid plexus epithelial cells in a rat primary culture model. *J Neurochem* 2008, **104**(6):1494–1503.
81. Moretti P, Peters SU, Del Gaudio D, Sahoo T, Hyland K, Bottiglieri T, Hopkin RJ, Peach E, Min SH, Goldman D *et al.*: Brief report: autistic symptoms, developmental regression, mental retardation, epilepsy, and dyskinesias in CNS folate deficiency. *J Autism Dev Dis* 2008, **38**(6):1170–1177.
82. Ramaekers VT, Blau N, Sequeira JM, Nassogne MC, Quadros EV: Folate receptor autoimmunity and cerebral folate deficiency in low-functioning autism with neurological deficits. *Neuropediatrics* 2007, **38**(6):276–281.
83. Ramaekers VT, Sequeira JM, Blau N, Quadros EV: A milk-free diet down-regulates folate receptor autoimmunity in cerebral folate deficiency syndrome. *Dev Med Child Neurol* 2008, **50**(5):346–352.
84. Ramaekers VT, Rothenberg SP, Sequeira JM, Opladen T, Blau N, Quadros EV, Selhub J: Autoantibodies to folate receptors in the cerebral folate deficiency syndrome. *N Engl J Med* 2005, **352**(19):1985–1991.
85. Ramaekers VT, Blau N: Cerebral folate deficiency. *Dev Med Child Neurol* 2004, **46**(12):843–851.
86. Moretti P, Sahoo T, Hyland K, Bottiglieri T, Peters S, del Gaudio D, Roa B, Curry S, Zhu H, Finnell RH *et al.*: Cerebral folate deficiency with developmental delay, autism, and response to folinic acid. *Neurology* 2005, **64**(6):1088–1090.
87. Shoffner J, Hyams L, Langley GN, Cossette S, Mylacraine L, Dale J, Ollis L, Kuoch S, Bennett K, Aliberti A *et al.*: Fever plus mitochondrial disease could be risk factors for autistic regression. *J Child Neurol* 2010, **25**(4):429–434.
88. Ramaekers VT, Hansen SI, Holm J, Opladen T, Senderek J, Hausler M, Heimann G, Fowler B, Maiwald R, Blau N: Reduced folate transport to the CNS in female Rett patients. *Neurology* 2003, **61**(4):506–515.
89. Ramaekers VT, Sequeira JM, Artuch R, Blau N, Temudo T, Ormazabal A, Pineda M, Aracil A, Roelens F, Laccone F *et al.*: Folate receptor autoantibodies and spinal fluid 5-methyltetrahydrofolate deficiency in Rett syndrome. *Neuropediatrics* 2007, **38**(4):179–183.
90. Perez-Duenas B, Ormazabal A, Toma C, Torrico B, Cormand B, Serrano M, Sierra C, De Grandis E, Marfa MP, Garcia-Cazorla A *et al.*: Cerebral folate

deficiency syndromes in childhood: clinical, analytical, and etiologic aspects. *Arch Neurol* 2011, **68**(5):615–621.

91. Allen RJ, DiMauro S, Coulter DL, Papadimitriou A, Rothenberg SP: Kearns-Sayre syndrome with reduced plasma and cerebrospinal fluid folate. *Ann Neurol* 1983, **13**(6):679–682.
92. Pineda M, Ormazabal A, Lopez-Gallardo E, Nascimento A, Solano A, Herrero MD, Vilaseca MA, Briones P, Ibanez L, Montoya J *et al.*: Cerebral folate deficiency and leukoencephalopathy caused by a mitochondrial DNA deletion. *Ann Neurol* 2006, **59**(2):394–398.
93. Ramaekers VT, Weis J, Sequeira JM, Quadros EV, Blau N: Mitochondrial complex I encephalomyopathy and cerebral 5-methyltetrahydrofolate deficiency. *Neuropediatrics* 2007, **38**(4):184–187.
94. Hasselmann O, Blau N, Ramaekers VT, Quadros EV, Sequeira JM, Weissert M: Cerebral folate deficiency and CNS inflammatory markers in Alpers disease. *Mol Genet Metab* 2010, **99**(1):58–61.
95. Garcia-Cazorla A, Quadros EV, Nascimento A, Garcia-Silva MT, Briones P, Montoya J, Ormazabal A, Artuch R, Sequeira JM, Blau N *et al.*: Mitochondrial diseases associated with cerebral folate deficiency. *Neurology* 2008, **70**(16):1360–1362.
96. Ramaekers VT, Quadros EV, Sequeira JM: Role of folate receptor autoantibodies in infantile autism. *Mol Psychiatry* 2013, **18**(3):270–271.
97. James SJ, Cutler P, Melnyk S, Jernigan S, Janak L, Gaylor DW, Neubrander JA: Metabolic biomarkers of increased oxidative stress and impaired methylation capacity in children with autism. *Am J Clin Nutr* 2004, **80**(6): 1611–1617.
98. Melnyk S, Fuchs GJ, Schulz E, Lopez M, Kahler SG, Fussell JJ, Bellando J, Pavliv O, Rose S, Seidel L *et al.*: Metabolic imbalance associated with methylation dysregulation and oxidative damage in children with autism. *J Autism Dev Disord* 2012, **42**(3):367–377.
99. Nagarajan RP, Hogart AR, Gwye Y, Martin MR, LaSalle JM: Reduced MeCP2 expression is frequent in autism frontal cortex and correlates with aberrant MECP2 promoter methylation. *Epigenetics* 2006, **1**(4):e1–11.
100. Sajdel-Sulkowska EM, Xu M, McGinnis W, Koibuchi N: Brain region-specific changes in oxidative stress and neurotrophin levels in autism spectrum disorders (ASD). *Cerebellum* 2011, **10**(1):43–48.
101. Hyland K, Shoffner J, Heales SJ: Cerebral folate deficiency. *J Inherit Metab Dis* 2010, **33**(5):563–570.
102. James SJ, Melnyk S, Jernigan S, Cleves MA, Halsted CH, Wong DH, Cutler P, Bock K, Boris M, Bradstreet JJ *et al.*: Metabolic endophenotype and related genotypes are associated with oxidative stress in children with autism. *Am J Med Genet B Neuropsychiatr Genet* 2006, **141B**(8):947–956.

103. James SJ, Rose S, Melnyk S, Jernigan S, Blossom S, Pavliv O, Gaylor DW: Cellular and mitochondrial glutathione redox imbalance in lymphoblastoid cells derived from children with autism. *FASEB J* 2009, **23**(8):2374–2383.
104. Rose S, Melnyk S, Trusty TA, Pavliv O, Seidel L, Li J, Nick T, James SJ: Intracellular and extracellular redox status and free radical generation in primary immune cells from children with autism. *Autism Res Treat* 2012, **2012**:986519.
105. James SJ, Melnyk S, Fuchs G, Reid T, Jernigan S, Pavliv O, Hubanks A, Gaylor DW: Efficacy of methylcobalamin and folinic acid treatment on glutathione redox status in children with autism. *Am J Clin Nutr* 2009, **89**(1):425–430.
106. Adams JB, Audhya T, McDonough-Means S: Nutritional and metabolic status of children with autism vs. neuro- typical children, and the association with autism severity. *Nutr Metab* 2011, **8**:34.
107. Hardan AY, Fung LK, Libove RA, Obukhanych TV, Nair S, Herzenberg LA, Frazier TW, Tirouvanziam R: A randomized controlled pilot trial of oral N-acetylcysteine in children with autism. *Biol Psychiatry* 2012, **71**(11):956–961.
108. Adams JB, Holloway C: Pilot study of a moderate dose multivitamin/mineral supplement for children with autistic spectrum disorder. *J Altern Complement Med* 2004, **10**(6):1033–1039.
109. Tani Y, Fernell E, Watanabe Y, Kanai T, Langstrom B: Decrease in 6R-5,6,7,8-tetrahydrobiopterin content in cerebrospinal fluid of autistic patients. *Neurosci Lett* 1994, **181**(1–2):169–172.
110. Fernell E, Watanabe Y, Adolfsson I, Tani Y, Bergstrom M, Hartvig P, Lilja A, von Knorring AL, Gillberg C, Langstrom B: Possible effects of tetrahydrobiopterin treatment in six children with autism — clinical and positron emission tomography data: a pilot study. *Dev Med Child Neurol* 1997, **39**(5):313–318.
111. Naruse H, Hayashi T, Takesada M, Nakane A, Yamazaki K: Metabolic changes in aromatic amino acids and monoamines in infantile autism and development of new treatment related to the finding. *No To Hattatsu* 1989, **21**(2):181–189.
112. Naruse H, Hayashi T, Takesada M, Nakane A, Yamazaki K: Therapeutic effect of tetrahydrobiopterin in infantile autism. *Proc J Acad* 1987, **63B**:231–233.
113. Naruse H, Takesada M, Nagahata M, Kazamatsuri H, Y. N, Yamazaki K: An open clinical study of sapropterin hydocloride (R-tetrahydrobiopterin SUN 588) in infantile autism: clinical study using a Rating Scale for Abnormal Behaviors in Children. *Rinsho Iyaku* 1990, **6**:1343–1368.
114. Naruse H, Takesada M, Nagahata M, Kazamatsuri H, Nakane Y, Yamazaki K: An open clinical study of sapropterin hydrochloride (R-tetrahydrobiopterin SUN 0588) in infantile autism — clinical study using a Rating Scale for Abnormal Behaviors in Children. *Rinsho Iyaku* 1990, **6**:1859–1875.
115. Nagahata M, Kazamatsuri H, Naruse H, Yamazaki K, Takesada M, Nakane Y, Kaihara S, Ohashi T: Clinical evaluation of aproterin hydrochloride (R-THBP.

SUN 0588) on infantile autism — a multicenter cooperative study. *Rinsho Iyaku* 1990, **6**:1877–1899.

116. Nakane Y, Asuo T, Shimogawa S, Fujiwara T, Kawabata Y, Kubota J: Clinical efficacy and effects on physical development of long-term treatment of R-tetrahydrobiopterin (R-THBP, SUN 0588) for autism. *Kiso to Rinshou* 1990, **24**:4579–4598.
117. Takesada M, Naruse H, Nagahata M: An open clinical study of aprpterin hydrochloride (R-tetrahydrobiopterin, R-THBP) in infantile autism - clinical effects and long-term follow-up. In: *Neurobiology of Infantile Autism*. Edited by Naruse H, Ornitz E.M. New York: Elsevier Science Publishers; 1992: 355–357.
118. Nakane Y, Naruse H, Hayashi T, Takesada M, Yamazaki K: Clinical effect of R-THBP on Infantile Autism. In: *Neurobiology of Infantile Autism*. Edited by Naruse H, Ornitz E.M. New York: Elsevier Science Publishers; 1992: 337–349.
119. Naruse H, Hayashi T, Takesada M: A preliminary study on clinical effect of tetrahydrobiopterin in infantile autism. In: *Report in 1983 for New Drug Development* Ministry of Health and Welfare; 1984: 71–81.
120. Naruse H, Takesada M, Nakane Y, Yamazaki K, Uchiyama T, Kaihara S, Ohashi T: Clinical evaluation of R-tetrahydrobiopterin (SUN 0588) on infantile autism — a double-blind comparative study using placebo as a control. *Rinsho Iyaku* 1990, **6**:1343–1368.
121. Danfors T, von Knorring AL, Hartvig P, Langstrom B, Moulder R, Stromberg B, Torstenson R, Wester U, Watanabe Y, Eeg-Olofsson O: Tetrahydrobiopterin in the treatment of children with autistic disorder: a double-blind placebo-controlled crossover study. *J Clin Psychopharmacol* 2005, **25**(5):485–489.
122. Klaiman C, Huffman L, Masaki L, Elliott GR: Tetrahydrobiopterin as a treatment for autism spectrum disorders: a double-blind, placebo-controlled trial. *J Child Adolesc Psychopharmacol* 2013, **23**(5):320–328.
123. Frye RE, Taylor H, De la Torre R, James SJ: Improvements in language and adaptive behavior associated with tetrahydrobiopterin treatment is related to nitric oxide metabolism and tetrahydrobiopterin redox status. In: *International Meeting for Autism Research*. Toronto, Canada; 2012.
124. Koopman WJ, Nijtmans LG, Dieteren CE, Roestenberg P, Valsecchi F, Smeitink JA, Willems PH: Mammalian mitochondrial complex I: biogenesis, regulation, and reactive oxygen species generation. *Antioxid Redox Signal* 2010, **12**(12):1431–1470.
125. Distelmaier F, Koopman WJ, van den Heuvel LP, Rodenburg RJ, Mayatepek E, Willems PH, Smeitink JA: Mitochondrial complex I deficiency: from organelle dysfunction to clinical disease. *Brain* 2009, **132**(Pt 4):833–842.
126. Chinta SJ, Andersen JK: Nitrosylation and nitration of mitochondrial complex I in Parkinson's disease. *Free Radic Res* 2011, **45**(1):53–58.

127. Wiesner RJ, Zsurka G, Kunz WS: Mitochondrial DNA damage and the aging process: facts and imaginations. *Free Radic Res* 2006, **40**(12):1284–1294.
128. Pastore A, Federici G, Bertini E, Piemonte F: Analysis of glutathione: implication in redox and detoxification. *Clin Chim Acta* 2003, **333**(1):19–39.
129. Chauhan A, Chauhan V: Oxidative stress in autism. *Pathophysiology* 2006, **13**(3):171–181.
130. Chauhan A, Gu F, Essa MM, Wegiel J, Kaur K, Ted Brown W, Chauhan V: Brain region-specific deficit in mitochondrial electron transport chain complexes in children with autism. *J Neurochem* 2011.
131. Clay HB, Sillivan S, Konradi C: Mitochondrial dysfunction and pathology in bipolar disorder and schizophrenia. *Int J Dev Neurosci* 2011, **29**(3):311–324.
132. Kato M, Nakamura M, Ichiba M, Tomiyasu A, Shimo H, Higuchi I, Ueno S, Sano A: Mitochondrial DNA deletion mutations in patients with neuropsychiatric symptoms. *Neurosci Res* 2011, **69**(4):331–336.
133. Scaglia F: The role of mitochondrial dysfunction in psychiatric disease. *Dev Disabil Res Rev* 2010, **16**(2):136–143.
134. Verge B, Alonso Y, Valero J, Miralles C, Vilella E, Martorell L: Mitochondrial DNA (mtDNA) and schizophrenia. *Eur Psychiatry* 2011, **26**(1):45–56.
135. Prabakaran S, Swatton JE, Ryan MM, Huffaker SJ, Huang JT, Griffin JL, Wayland M, Freeman T, Dudbridge F, Lilley KS *et al.*: Mitochondrial dysfunction in schizophrenia: evidence for compromised brain metabolism and oxidative stress. *Mol Psychiatry* 2004, **9**(7):684–697, 643.
136. Tosic M, Ott J, Barral S, Bovet P, Deppen P, Gheorghita F, Matthey ML, Parnas J, Preisig M, Saraga M *et al.*: Schizophrenia and oxidative stress: glutamate cysteine ligase modifier as a susceptibility gene. *Am J Hum Genet* 2006, **79**(3):586–592.
137. Ozcan ME, Gulec M, Ozerol E, Polat R, Akyol O: Antioxidant enzyme activities and oxidative stress in affective disorders. *Int Clin Psychopharmacol* 2004, **19**(2):89–95.
138. Andreazza AC, Kauer-Sant'anna M, Frey BN, Bond DJ, Kapczinski F, Young LT, Yatham LN: Oxidative stress markers in bipolar disorder: a meta-analysis. *J Affect Disord* 2008, **111**(2–3):135–144.
139. Frey BN, Andreazza AC, Kunz M, Gomes FA, Quevedo J, Salvador M, Goncalves CA, Kapczinski F: Increased oxidative stress and DNA damage in bipolar disorder: a twin-case report. *Prog NeuroPsychopharmacol Biol Psychiatry* 2007, **31**(1):283–285.
140. Emerit J, Edeas M, Bricaire F: Neurodegenerative diseases and oxidative stress. *Biomed Pharmacother* 2004, **58**(1):39–46.

Chapter 23

Transcranial Magnetic Stimulation: Application in Autism Treatment

Manuel F. Casanova* and Estate M. Sokhadze

Abstract

Repetitive transcranial magnetic stimulation (rTMS) is a non-invasive method of altering patterns of brain activity. rTMS has been applied to a wide variety of psychiatric and neurological conditions and is generally regarded as safe and has no lasting side effects. Within the context of autism spectrum disorders (ASD), rTMS has a unique application as a treatment modality. Recent evidence suggests the symptoms of ASD may be related to increased cortical excitability and, specifically, an increased ratio of cortical excitation to inhibition accompanied by deficiencies in long-range cortical connectivity. Locally over-connected neural networks may explain the superior ability of autistic children in isolated tasks (e.g., visual discrimination) while, at the same time, deficiencies in long-range connectivity may explain other features of the disorder (e.g., lack of social reciprocity). An increased ratio of cortical excitation to inhibition and higher-than-normal cortical "noise" may also explain the strong aversive reactions to auditory, tactile, and visual stimuli frequently recorded in autistic individuals as well as a higher incidence of epilepsy. Using specific parameters of stimulation, rTMS has been shown to increase cortical inhibition by selectively activating cortical inhibitory neurons. In a number of investigations, our group evaluated the effects of rTMS on indices of

*Correspondence to: Department of Psychiatry & Behavioral Sciences, University of Louisville, 500 S. Preston Street, Bldg. A., #217, Louisville, KY 40202, USA. Email: manuel.casanova@louisville.edu

selective attention and executive functioning, as well as measures of social awareness, hyperactivity, irritability, and repetitive/stereotyped behavior. Subjects with ASD were assessed at baseline and following rTMS with electroencephalographic (EEG) and event-related potential (ERP) measures of selective attention and executive functioning. Subjects were also assessed for ASD symptomatology using neuropsychological questionnaires. Following rTMS, subjects showed significant improvement in EEG and ERP measures of selective attention and executive functioning, and also showed significant improvement in measures of irritability and repetitive/stereotyped behavior. Our preliminary findings in three experimental studies using 6-, 12-, and 18 session-long rTMS courses in children with autism indicate that rTMS has the potential to become an important therapeutic tool in research and treatment, and may play an important role in improving the quality of life for many individuals with ASD. Further research directions are proposed.

Keywords: TMS; Autism; ERP; EEG; Behavior.

Introduction

Repetitive transcranial magnetic stimulation (rTMS) offers a noninvasive method for altering excitability of the brain. It potentially induces a short-term functional reorganization in the human cortex. The magnitude and the direction of rTMS-induced plasticity depend on the variables of stimulation (intensity, frequency, number of stimuli) and the functional state of the cortex targeted by rTMS. Since effects of rTMS are not limited to the stimulated target cortex but give rise to functional changes in anatomically and functionally interconnected cortical areas, rTMS is a suitable tool to investigate neural plasticity within a distributed functional network.[1,2] The lasting effects of rTMS offer new possibilities to study dynamic aspects of the pathophysiology of a variety of diseases and may have therapeutic potential in some psychiatric disorders. By convention, rTMS in the 0.3 Hz–1.0 Hz frequency range is referred to as "slow", whereas "fast" rTMS refers to frequencies greater than 1 Hz. Hoffman and Cavus, in their review of slow rTMS studies, proposed long-term depression and long-term depotentiation as models for understanding the mechanism of slow rTMS.[3] Neocortical long-term depression and changes in the cortical excitability induced by slow rTMS appear to accumulate in an additive fashion as the number of stimulations is increased over many days. Studies of both slow rTMS and long-term depression suggest additive efficacy when higher numbers of stimulations are administered. The reversal, or depotentiation, of previously enhanced synaptic transmission due to long-term potentiation may be the most relevant model for slow rTMS when used as a therapeutic tool.

rTMS is a simple outpatient procedure lasting approximately 20–30 min. Patients are seated in a comfortable, reclining chair and are fitted with a swim cap to outline the TMS coil position and aid in its placement for each session (Figure 1). Before the procedure

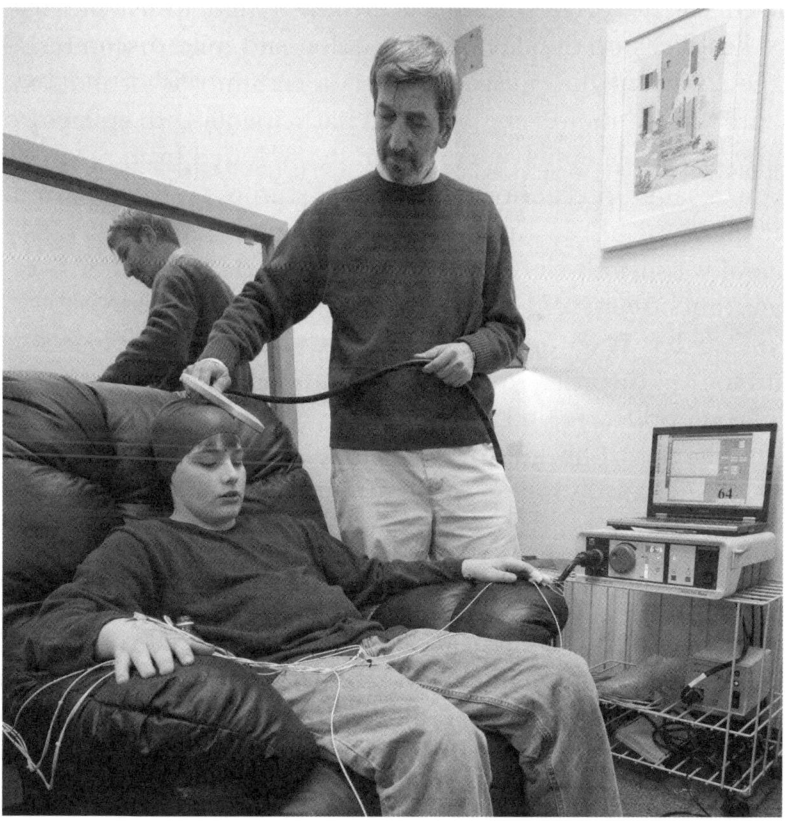

Figure 1. Patient receiving Transcranial Magnetic Stimulation (TMS) treatment with concurrent physiological monitoring.

begins, the "motor threshold" is determined in each patient. Motor threshold (MT) is the intensity of the pulse delivered over the motor cortex that produces a noticeable motor response. Sensors are applied to the hand muscle (i.e., the first dorsal interosseous) opposite the site of stimulation and motor responses are monitored with physiological monitoring tools on a computer. The output of the machine is gradually increased by 5% until a 50 μV deflection on the electromyograph (viz., the monitor) or a visible twitch of the muscle is observed. Once the patient's MT is determined, the coil is moved to the site of stimulation (e.g., the dorsolateral prefrontal cortex) and the pulse intensity is adjusted relative to MT.

TMS is generally regarded as safe, that is, as having no lasting side effects. Reported side effects include a mild, transient tension-type headache on the day of stimulation and mild discomfort due to the sound of the pulses. There is a certain risk of inducing a seizure (Wasserman *et al.*, 1996) and participants with epilepsy or a family history of epilepsy are generally excluded from rTMS studies. As a safety precaution, some rTMS studies adjust the stimulation intensity below the participant's motor threshold. rTMS is generally considered safe for use in pediatric populations, as no significant adverse effects or seizures have been reported.[4,5]

rTMS has been applied to a wide variety of psychiatric and neurological disorders in adult populations. More recently, rTMS has been applied in child and adolescent populations.[6] A number of studies report an improvement in mood after repeated frontal lobe stimulation in both depressed adults and adolescents, and it has been reported that rTMS may improve symptoms of Attention Deficit Hyperactivity Disorder (ADHD).[7,8,9] Furthermore, it has been found that rTMS may improve certain symptoms associated with anxiety disorders, like Post-Traumatic Stress Disorder (PTSD) and Obsessive-Compulsive Disorder (OCD).[10] In Parkinson's disease (PD), most studies have shown beneficial effects of rTMS on clinical symptoms.[11] Currently, only rTMS-based therapy of treatment-resistant major depression has FDA approval; however, it is very likely that in the future rTMS will be approved for the treatment of other mental and neurological disorders.

Inhibition Defects in Autism and the Potential Role of TMS

Recent reviews provide detailed accounts of the current status of rTMS application in autism research and treatment.[12,13] Within the context of autism spectrum disorder (ASD), rTMS has a unique application as a treatment modality. ASD is associated with disturbances in social interaction and communication, restricted and stereotyped behavioral patterns, and frequently abnormal sensory reactions to the environment.[14,15] A wide range of deficits in autism might be understood by an increase in the ratio of cortical excitation to cortical inhibition and increases in local cortical connectivity accompanied by deficiencies in long-range connectivity.[16–18] Locally over-connected neural networks may explain the superior ability of autistic children in isolated tasks (e.g., visual discrimination) while, at the same time, deficiencies in long-range connectivity may explain other features of the disorder (e.g., lack of social reciprocity). An increased ratio of cortical excitation to inhibition and higher-than-normal cortical "noise" may explain the strong aversive reactions to auditory, tactile, and visual stimuli frequently reported in autism.[19]

Autism is associated with cortical cytoarchitectural developmental abnormalities according to the "minicolumnar" hypothesis.[20,21] The reduced neuropil space (i.e., periphery of the minicolumn) reported in autism describes an abnormality of the anatomical compartment where lateral inhibition sharpens the borders of minicolumns and increases their definition.[22–25] The primary source for this inhibitory effect may be derived from axon bundles of double-bouquet cells.[22–24] Double-bouquet cells in the peripheral neuropil space of minicolumns provide a "vertical stream of negative inhibition" surrounding the minicolumnar core.[26,27] Other GABAergic cells in the minicolumn, having collateral projections extending hundreds of microns tangentially, provide lateral inhibition of surrounding minicolumns on a macrocolumnar scale.

The value of each minicolumn's output is insulated to a greater or lesser degree from the activity of its neighbors by GABAergic inhibition in its peripheral neuropil space. This allows for gradations in amplitude of excitatory activity across a minicolumnar field. Rubenstein and Merzenich have posited that reductions in GABAergic inhibitory activity may explain some symptomatology of autism, including increased incidence of seizures and auditory-tactile hypersensitivity.[18] This hypothesis is consistent with findings of reduced minicolumnar peripheral neuropil space in the neocortex of individuals with ASD relative to controls.[20,28] In this model, a reduction in the peripheral neuropil space would result in smaller minicolumns, which would coalesce into discrete, isolated islands of coordinated excitatory activity. Significantly, by puberty, one third of autistic patients will have exhibited at least two unprovoked seizures.[29] Anecdotal case reports have shown that anticonvulsants have ameliorated autistic traits in epileptic patients.[30–34] Anticonvulsants may be of some benefit in autism but at larger doses may induce serious side effects, including stupor and coma. These side effects are due to the non-selective nature of anticonvulsants whose mechanism of action (increasing GABAergic tone) is independent of cell type (e.g., double-bouquet, small and large basket, chandelier). The effects of anticonvulsants stand in contrast to the specificity of slow rTMS. This technique induces electricity in conductors at right angles to an expanding or collapsing magnetic field (law of electromagnetic induction). This effect may be of benefit when selectively attempting to activate the inhibitory cells and fibers surrounding the minicolumn (i.e., peripheral neuropil space) as these anatomical elements have preferential orientation perpendicular to the cortical surface.[16,35,36]

Ogawa *et al.* examined the changes in high frequency oscillations (HFOs) of somato-sensory evoked potentials (SEPs) before and after slow rTMS over the right primary somatosensory cortex (0.5 Hz, 50 pulses, 80% motor threshold intensity).[37] The HFOs, which represent a localized activity of intracortical inhibitory interneurons, were significantly increased after slow rTMS, while the SEPs were not changed. Their results suggest that slow rTMS affects cortical excitability by modulating the activity of the intracortical inhibitory interneurons beyond the time of the stimulation and that rTMS may have therapeutic effects on such disorders. This is in line with our hypothesis that slow rTMS will increase activity of inhibitory cells in minicolumns, which will then enhance the spatial contrast needed to enhance functional discrimination.

The question remains whether rTMS is safe for application in children. Quintana evaluated studies that used TMS in persons younger than 18.[5] The 48 studies reviewed involved a total of 1034 children; 35 of the studies used single-pulse TMS (980 children), 3 studies used paired TMS (20 children), and 7 studies used rTMS (34 children). The TMS studies in persons younger than 18 have been used to examine the maturation and activity of the neurons of various central nervous system tracts, plasticity of neurons in epilepsy, multiple sclerosis, myoclonus, transcallosal inhibition, and motor cortex functioning with no reported seizure risk.[4,38] Repetitive TMS has been applied in children with psychiatric disorders such as ADHD, ADHD with Tourette's, and depression. Although there are a limited number of studies using rTMS in children, these studies did not report significant adverse effects or seizures.

Electrophysiological Studies on Autism

Spectral analysis of the waking electroencephalogram (EEG) in autism have shown increased slow-wave activity, reduced EEG power in frontal and temporal regions, decreased variability, and decreased inter- and intra-hemispheric asymmetries relative to healthy individuals.[39–41] Oscillatory activity in the gamma-band (30 Hz–80 Hz, especially those centered around 40 Hz) of the EEG has been related to *Gestalt* perception and to cognitive functions such as attention, learning, and memory.[42] Gamma activity increase has been most widely associated with top-down attentional processing and object perception, and was shown to be higher in a delayed match-to-sample task with visual, auditory, and audio-visual stimuli, as well as in working memory tasks.[42–50] Human experiments have also found that induced gamma activity correlates with binding.[51] Binding of widely distributed cell assemblies by synchronization of their gamma frequency activity is thought to underlie cohesive stimulus representation in the human brain.[52,53] According to this assumption, changes in gamma EEG activity have been considered indicators of processing of Gestalt-like patterns.[44,54–56]

It has been proposed that some symptoms of autism could result from a reduction in the integration of specialized local networks in the brain caused by a deficit in temporal binding.[17,57] Brown tested adolescents with autism in an experiment that presented either shapes with visual illusions (Kanizsa figures) or random pictures.[58,59] In these perceptual tasks, gamma bursts have been reliably recorded over the posterior attention areas in unimpaired adults at around 300 ms and at around 100 ms post-stimulus.[43,47] Behaviorally, there was no difference between the groups in either reaction time or in accuracy. Whereas both groups showed almost identical activity in the Kanizsa-random condition, they were very different in the Kanizsa-shape condition. When perceiving the shape, the group with autism differed from the control group in three ways: (1) gamma power was higher than in the control group; (2) the group with autism showed a very early burst of gamma activity (80 ms to 120 ms); and (3) the later gamma component in the autistic patients was more powerful and started much earlier. The inability to reduce gamma activity according to Brown would lead to the inability to decide which event requires attention when there are multiple choices.[58] Excessive gamma activity can therefore be linked to a reduction in the ability to focus attention. In autism, powerful and uninhibited gamma activity suggests that none of the circuits in the brain can come to dominance because too many of them are active simultaneously.[60] Abnormalities of gamma oscillatory synchrony in autism can therefore result in significant perceptual and cognitive deficits. The "temporal binding deficits in autism" hypothesis can be tested using a visual attention task with Kanizsa illusory figures as stimuli.[17,57] According to the hypothesis, it is predicted that ASD subjects will exhibit abnormalities of gamma activity over the frontal cortex and visual processing areas during both stimulus encoding and memory rehearsal. Decreased gamma activation in response to task-irrelevant illusory figures following rTMS treatment would support the hypothesis that low frequency rTMS can reduce gamma over-activation patterns.

The event-related potentials (ERP) reflect activation of neural structures in sensory cortex, association cortices, and brain areas related to higher order cognitive processes. Several studies have shown that children and adolescents with autism show abnormalities in ERP.[61,62] The fronto-central N100 component has been found to be shorter in latency and higher in amplitude in an oddball task in children with autism than in normal children.[63,64] Autistic children have been found to differ from typical children mainly with respect to the P300 in standard oddball tasks. Kemner *et al.* have reported an abnormally small occipital P300 in response to visual stimuli.[61,65] Studies using simple visual target detection have found dissociations of frontal (delayed) and posterior P300 (intact) in spatial visual attention tasks.[66] Deficits in central auditory processing in autism, as indexed by ERP, have been described by different authors.[63,67–70] In general, perceptual filtering in autism seems to occur in an all-or-none manner, with little specificity for the location of the stimulus, for the behavioral relevance of the stimulus, or even for the sensory modality of the stimulus. The attention of autistic patients seems founded more on the coarse control of general arousal than on selective activation of specific perceptual systems.

Our own findings of excessive gamma oscillations and ERP responses in visual tasks.[71–74] are in agreement with other studies noting that neural systems in the brains of autistic patients are often inappropriately activated;[75] this may be due to a disruption in the ratio between cortical excitation and inhibition.[18,20,21,28,76] In autism, increased cortical activity made evident by gamma and ERP responses indicate that activity induced by perceptual processes starts earlier and continues longer, because the neural networks subserving cognitive processes involved in combining information processing are not functioning normally. A reduction in the ability to decrease these cortical responses may reflect inhibitory deficits, and may result in the brain of autistic patients being over-activated. Abnormally large cortical responses to sensory stimuli (i.e. signal/sensory amplification) may play an important role in the manifestation of symptoms of ASD (e.g. sensory hypersensitivity, impaired social interaction). Enhanced and weakly differentiated responses to both target and non-target stimuli in sensory specific cortical areas (e.g., visual cortex at occipital EEG sites) and low functional connectivity supports the hypothesis of abnormal regional activation patterns (local over-processing vs. global under-processing).

We hypothesize that contrary to other inhibitory cells (i.e., basket and chandelier) whose projections keep no constant relation to the surface of the cortex, the geometrically consistent orientation of double-bouquet cells and their location at the periphery of the minicolumn (inhibitory surround) makes them the appropriate candidate for induction by a magnetic field applied parallel to cortex. Over a course of treatment, slow rTMS may restore the balance between cortical excitation and cortical inhibition and lead to improved long-range cortical connectivity. Thus far, we have focused on clinical, behavioral, and neuroimaging outcome measures in order to access the effectiveness of rTMS treatment in ASD.

Our Studies in Autism using rTMS

Our studies examined the effects of low frequency (0.5 Hz–1 Hz) rTMS on behavioral, EEG, and ERP outcomes in children and adolescents with autism. We used rTMS over the dorso-lateral prefrontal cortex (DLPFC) on a weekly basis for 6, 12, or 18 weeks in individuals with autism randomly assigned to active-treatment or wait-list groups. We predicted that post-TMS changes in the active-treatment group, as compared to the wait-list group, could be detected during repeated tests using the same functional outcome measures (EEG, ERP etc.) in cognitive tasks. Our prediction was that slow rTMS of the DLPFC will result in an alteration of cortical inhibition through the activation of inhibitory GABAergic interneurons leading to an improvement in the excitatory/inhibitory balance.

In the first of our previous investigations,[77] we measured the EEG gamma band in 8 children with ASD and 5 wait-list participants with ASD during a visual attention task, and then measured the EEG gamma band in the active treatment group after six sessions of slow rTMS to the prefrontal cortex. The study also used 13 age-matched, typically developing children as a control group. We hypothesized that the ASD group would have excess gamma band activity due to a lack of cortical inhibition and that treatment with slow rTMS would help restore inhibitory tone (i.e., reduce excess gamma band activity). We also analyzed clinical and behavioral questionnaires assessing changes in symptoms associated with ASD after rTMS treatment. The visual attention task employed Kanizsa illusory figures, which have been shown to readily produce gamma oscillations during visual tasks.

Subjects were instructed to press a button when they saw the target Kanizsa square and to ignore all other stimuli. Kanizsa stimuli consist of inducer disks of a shape feature and either constitute an illusory figure (square, triangle) or not (collinearity feature). Gamma activity in non-impaired individuals has been found to increase during the presentation of target visual stimuli compared to non-target stimuli. We found that the power of gamma oscillations over the prefrontal cortex was higher in the ASD group and had an earlier onset compared to controls — especially in response to non-target illusory figures (Figure 2). Additionally, there was less

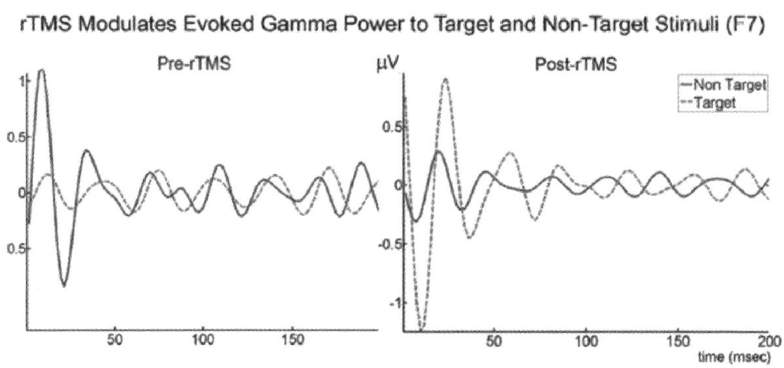

Figure 2. rTMS modulates evoked EEG gamma frequency oscillations to target and non-target illusory figures at frontal site F7. Following a 12 session-long TMS course, early gamma response to non-target decreased, whereas gamma oscillation power to target illusory figures increased.

of a difference in gamma power between target and non-target stimuli in the ASD group, particularly over lateral, frontal, and parietal recording sites. After six sessions of slow rTMS applied to the left prefrontal cortex, the power of gamma oscillations to non-target Kanizsa figures dramatically decreased at frontal and parietal sites on the same side of stimulation, and there was more of a difference between gamma responses to target and non-target stimuli.

The frontal P3a showed a significant decrease in amplitude in response to non-target stimuli without notable changes in response to targets. Repetitive TMS had no effect on the amplitude of the centro-parietal P3b, but decreased P3b latency for non-target stimuli. According to clinical and behavioral evaluations, the ASD group showed a significant improvement on the repetitive behavior scale (RBS), which assesses repetitive and restricted behavior patterns associated with ASD (e.g., stereotyped, self-injurious, compulsive, and restricted range).[78]

In a second study with another pool of participants, we investigated gamma band activity in 16 subjects with ASD in rTMS group and 9 age-matched controls using Kanizsa illusory figures and assessed the effects of 12 sessions of bilateral slow rTMS applied to the prefrontal cortices in the TMS group of the ASD participants.[71,79,80] In individuals with ASD, gamma activity was not discriminative of stimulus type, whereas in controls early gamma power differences between target and non-target stimuli were highly significant. Following rTMS, individuals with ASD showed significant improvement in discriminatory gamma activity between relevant and irrelevant visual stimuli, and there was also a significant reduction in irritability and repetitive behavior as a result of rTMS.

We have also been interested in investigating ERP abnormalities in ASD as ERP provide a neurobiological measure of perceptual and cognitive processing. In a previous paper, we investigated ERP in a three-stimuli, visual task of selective attention in 11 high-functioning children and young adults with ASD and 11 age-matched, typically developing control subjects.[81] Patients with ASD showed significantly amplified and prolonged cortical responses to irrelevant, visual stimuli compared to controls; these

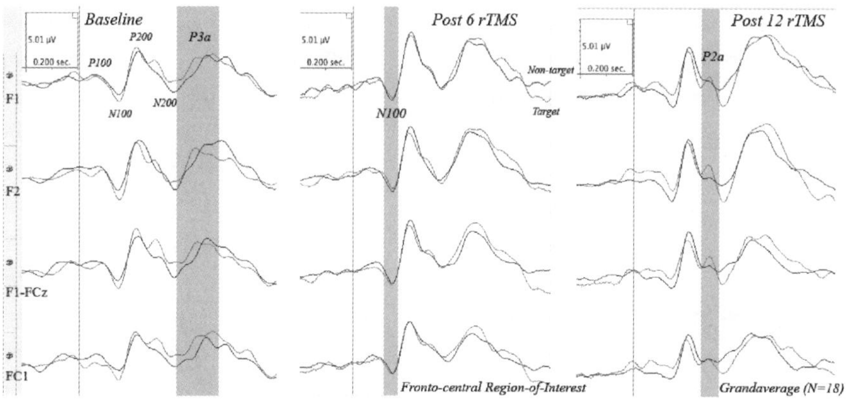

Figure 3. Changes of frontal and fronto-central to target (solid line) and non-target (dashed line) illusory figures at baseline, post 6 and post 12 sessions of low frequency rTMS treatment course (grand average for 18 ASD subjects). ERP components of interest (N100, P2a, and P3a) are highlighted.

results were recently confirmed in a following study assessing ERP responses in 15 subjects with ASD and 15 controls in a similar task using Kanizsa illusory figures.[72] Of special interest for us was evaluating the effects of 12 session-long slow rTMS over the DLPFC on ERP (Figures 3 and 4). The results of our study in 25 subjects with ASD following 12 sessions of bilateral slow rTMS showed significantly improved ERP indices of selective attention, reduced motor response errors, and reductions in both repetitive behavior and irritability.[79,80] We found a significant improvement in both N200 and P300 components as a result of rTMS as well as a significant reduction in response errors. We also found significant reductions in both repetitive behavior and irritability according to clinical behavioral questionnaires as a result of rTMS.

Additionally, in one of our prior ERP studies, we investigated executive functioning in 14 individuals with ASD and 14 age- and IQ-matched controls by evaluating error monitoring and correction.[74] The ASD group showed significant evidence of compromised error detection, evaluation, and correction, which may underlie a general impairment in self-monitoring related to behavioral and/or social disturbances in ASD. We then evaluated the effects of

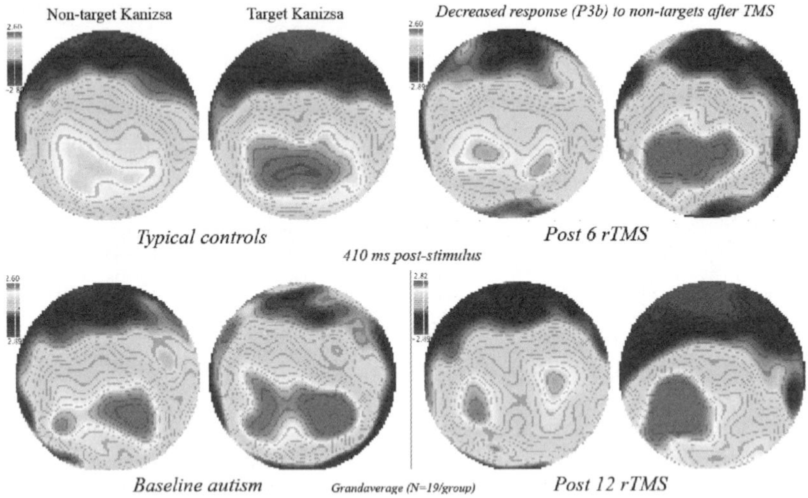

Figure 4. 2D brainwave maps of ERPs (P3b, 410 ms post-stimulus) to target and non-target illusory figures in neurotypical controls, groups of children with autism at the baseline, post 6 TMS and post 12 sessions of rTMS (grand average, $N = 19$ per group). Note the decreased P3b amplitude to non-target stimuli at parietal sites following TMS treatment.

12 sessions of bilateral slow rTMS on error monitoring and correction. The active rTMS group showed significant improvement in error detection and correction compared to a randomized, non-active rTMS group;[80] this may point to improved executive functioning and behavioral performance in ASD as a result of rTMS.

In our most recent study, we compared clinical, behavioral and electrocortical outcomes in two groups of children with autism (TMS, wait-list group, $N = 27$ per group) using 18 weekly sessions of rTMS applied bilaterally over the DLPFC.[13] Post-TMS evaluations showed decreased irritability and hyperactivity on the Aberrant Behavior Checklist (ABC), and decreased stereotypic behaviors on the Repetitive Behavior Scale (RBS-R). Following the rTMS course, we also found decreased amplitude and prolonged latency in the frontal and fronto-central N100, N200, and P3a ERP components to non-targets in the TMS group as compared to the wait-list group (Figure 5). TMS resulted in increase of the frontal

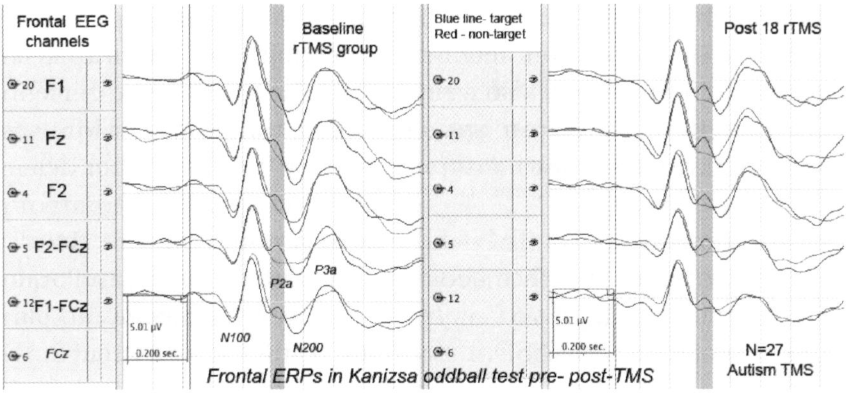

Figure 5. Frontal and fronto-central ERPs in 27 children with ASD before and after 18 sessions of rTMS treatment. Changes of the frontal N100 ERP post-TMS were significant for standards (solid line; $p = 0.031$) and non-target illusory stimuli (dashed line; $p = 0.023$). Both became less negative post-TMS.

P2a difference wave as well. These ERP changes along with increased centro-parietal and parieto-occipital P100 and P300 (P3b) to targets are indicative of more efficient processing of information post-TMS treatment.

Another important finding was decrease of the latency and increase of negativity of error-related negativity (ERN) during commission errors that further confirms improvement in error monitoring and correction function. Facilitated information processing was also manifested in such behavioral response measures as motor response accuracy. Results from our study indicate that rTMS improves executive functioning in ASD as evidenced by normalization of ERP responses, reaction time, and accuracy during tests of executive function, and also by improvements in clinical behavioral evaluations.

Conclusions

Overall, our preliminary findings show promising results for TMS as a treatment modality targeting core symptoms of ASD. Treatment with slow rTMS decreased excess gamma activity and amplified

ERP responses in ASD patients during visual tasks and improved the signal differentiation between processing relevant and irrelevant stimuli. There was also a significant reduction in the percentage of errors in motor responses to target stimuli, and rTMS was associated with a significant improvement in indices of error detection and correction.[71,73,74,77,79,80]

Additionally, slow rTMS dramatically improved the coordinated activity or coherence between different regions of the brain and significantly improved repetitive and restricted behavior patterns associated with ASD.[71,73,77,79] Our results suggest that low-frequency rTMS has the potential to become an important therapeutic tool in ASD treatment and may play an important role in improving the quality of life of many with the disorder.

TMS provides a non-invasive method for both the induction of focal currents and modulation of function in distinct isocortical parcellations. Overall, our studies collectively provide the first exploratory clinical intervention, guided by evidence derived from neuropathological data, which targets core deficits in autism. Other interventions (primarily pharmacological) target accessory symptoms that bear no universality across patients or generalizability within individual patients. Rather than treating end results of cascading interactions, our rTMS studies focus on core pathological features, thereby promising better results for affected patients.

Future Directions

It was not the aim of this chapter to be an exhaustive treatise on the state of research into rTMS application in autism spectrum disorder. Olberman *et al.* recently prepared a review of the current state of art in this area.[12] The neuromodulation approach to autism treatment based on rTMS is a new and rapidly developing field, and substantially more information is needed about many aspects of the rTMS regimen (topography, power, session frequency etc.) as well as specifics of the developmental neuropathology of autism to validate its clinical utility and efficacy.

Here, we mention several areas we consider especially promising for future research:

- More information is needed in order to better define EEG and ERP abnormalities typical of ASD.
- A very important objective for the future of rTMS-based treatment for autism is to conduct randomized clinical trials using double-blind design and sham rTMS as a control condition.
- One of the principal questions to be addressed in future studies is the estimation of how long rTMS-induced positive improvements last (e.g., do they fade out after a certain period of time?), thus emphasizing the need for rigorous follow-up studies.
- It is necessary to understand whether rTMS can be used synergistically in conjunction with other conventional behavioral and pharmaceutical therapies adopted in conventional ASD treatment practice.
- More effort is needed to understand the neurobiological mechanisms underlying core symptoms of autism using different neuroimaging neuroscience techniques, e.g., fMRI, DTI, etc. and their data correlation with clinical instruments of autism symptom severity assessment.
- We also agree with the conclusion of current reviews outlining the need to more carefully design controlled clinical trials in order to evaluate the real potential of rTMS-based treatment of autism, to have it approved by FDA, and to get it accepted by general practitioners and specialists.

With the advances made over the past several years, it is hoped that continued interest will be generated to further rTMS treatment of autism spectrum disorders.

Executive Summary

- Repetitive transcranial magnetic stimulation (rTMS) offers a non-invasive method for altering excitability of the brain.

- The "minicolumnar" hypothesis based on the observed aberrant cortical structure in the autism brain suggests a plausible neurophysiological mechanism for increased neural excitability in ASD.
- Inasmuch as ASD may be a neurological disorder characterized by an imbalance of cortical excitation and inhibition, rTMS holds promise as a novel and evidence-based treatment modality for ASD.
- Electrophysiological studies (EEG and ERP) in autism support the concepts of increased local connectivity and decreased long-range connectivity in the autism brain.
- The location and geometric orientation of certain cortical cells make them appropriate candidates for induction by rTMS.
- Our preliminary and published studies using rTMS to treat children with ASD are highly promising with respect to inducing more normal EEG and ERP responses to sensory stimuli, reducing some of the core symptoms of ASD, and improving ASD-associated behaviors.
- Future research must focus on defining EEG and ERP abnormalities in ASD in order to assess changes towards neurotypical response patterns, optimizing the rTMS treatment protocols, understanding the neurobiological mechanisms underlying the core symptoms of ASD, and conducting carefully designed and controlled clinical studies in order to evaluate the efficacy of rTMS in treating core symptoms of ASD.

References

1. Rossi S, Rossini PM: TMS in cognitive plasticity and the potential for rehabilitation. *Trends Cogn Sci* 2004, **86**:273–279.
2. Ziemann U: TMS induced plasticity in human cortex. *Rev Neurosci* 2004, 15:253–266.
3. Hoffman RE, Cavus I: Slow transcranial magnetic stimulation, long-term depotentiation, and brain hyperexcitability disorders. *Am J Psychiatry* 2002, **159**:1093–1102.
4. Garvey MA, Gilbert DL: Transcranial magnetic stimulation in children. *Eur J Paediatr Neurol* 2004, **8**:7–19.

5. Quintana H: Transcranial magnetic stimulation in persons younger than the age of 18. *J ECT* 2005, 21:88–95.
6. Croarkin PE, Wall CA, Lee J: Applications of transcranial magnetic stimulation (TMS) in child and adolescent psychiatry. *Int Rev Psychiatry* 2011, **23**: 445–453.
7. George MS, Lisanby SH, Avery D, McDonald WM, Durkalski V, Pavlicova M, Anderson B, Nahas Z, Bulow P, Zarkowski P, Holtzheimer PE 3rd, Schwartz T, Sackeim HA: Daily left prefrontal transcranial magnetic stimulation therapy for major depressive disorder: a sham-controlled randomized trial. *Arch Gen Psychiatry* 2010, **67**:507–516.
8. Wall CA, Croarkin PE, Sim LA, Husain MM, Janicak PG, Kozel FA, Emslie GJ, Dowd SM, Sampson SM: Adjunctive use of repetitive transcranial magnetic stimulation in depressed adolescents: a prospective, open pilot study. *J Clin Psychiatry* 2011, **72**:1263–1269.
9. Bloch Y, Harel EV, Aviram S, Govezensky J, Ratzoni G, Levkovitz Y: Positive effects of repetitive transcranial magnetic stimulation on attention in ADHD subjects: a randomized controlled pilot study. *World J Biol Psychiatry* 2010, **11**:755–758.
10. George MS, Belmaker RH: *Transcranial Magnetic Stimulation in Clinical Psychiatry*. Arlington: American Psychiatry Publishing; 2007.
11. Wu AD, Fregni F, Simon DK, Deblieck C, Pascual-Leone A: Noninvasive brain stimulation for Parkinson's disease and dystonia. *Neurotherapeutics*, 2008, 5:345–361.
12. Oberman L, Rotenberg A, Pascual-Leone A: Use of transcranial magnetic stimulation in autism spectrum disorders. *J Autism Dev Disord*, in press.
13. Sokhadze E, El-Baz A, Sears L, Opris I, Casanova MF: Neuromodulation based on rTMS improves electrocortical functional measures of information processing and behavioral responses in autism. *Front Syst Neurosci*, in press.
14. American Psychiatric Association: *Diagnostic and Statistical Manual of Mental Disorders (DSM-IV TR)*. 4th edition, text revision. Washington, D.C.: American Psychiatric Press; 2000.
15. Charman T: Autism spectrum disorders. *Psychiatry* 2008, **7**:331–334.
16. Casanova MF, Buxhoeveden D, and Gomez J: Disruption in the inhibitory architecture of the cell minicolumn: implications for autism. *Neuroscientist* 2003, **9**:496–507.
17. Rippon G, Brock J, Brown C, Boucher J: Disordered connectivity in the autistic brain: Challenges for the 'new psychophysiology'. *Int J Psychophysiol* 2007, 63:164–172.
18. Rubenstein JLR, Merzenich MM: Model of autism: increased ratio of excitation/ inhibition in key neural systems. *Genes Brain Behav* 2003, **2**:255–267.
19. Gillberg C, Billstedt E: Autism and Asperger syndrome: coexistence with other clinical disorders. *Acta Psychiatr Scand* 2000, **102**:321–330.

20. Casanova MF, Buxhoeveden DP, Switala AE, Roy E: Minicolumnar pathology in autism. *Neurology* 2002, **58**:428–432.
21. Casanova MF, van Kooten I, van Engeland H, Heinsen H, Steinbursch HWM, Hof PR, Trippe J, Stone J, Schmitz C: Minicolumnar abnormalities in autism. *Acta Neuropathol* 2006, **112**:287–303.
22. DeFelipe J, Hendry SHC, Hashikawa T, Molinari M, Jones EG: A microcolumnar structure of monkey cerebral cortex revealed by immunocytochemical studies of double bouquet cell axons. *Neuroscience* 1990, **37**:655–673.
23. Favorov OV, Kelly DG: Minicolumnar organization within somatosensory cortical segregates, I: development of afferent connections. *Cereb Cortex* 1994, **4**:408–427.
24. Favorov OV, Kelly DG: Minicolumnar organization within somatosensory cortical segregates, II: emergent functional properties. *Cereb Cortex* 1994, **4**:428–442.
25. DeFelipe J: Chandelier cells and epilepsy. *Brain* 1999, **122**:1807–1822.
26. Mountcastle VB: The columnar organization of the neocortex. *Brain* 1997, **120**:701–722.
27. Mountcastle VB: Introduction. Computation in cortical columns. *Cereb Cortex*, 2003, **13**:2–4.
28. Casanova MF, Buxhoeveden DP, Switala AE, Roy E: Neuronal density and architecture (gray level index) in the brains of autistic patients. *J Child Neurol* 2002, **17**:515–521.
29. Volkmar FR, Nelson DS: Seizure disorders in autism. *J Am Acad Child Adolesc Psychiatry* 1995, **29**:127–129.
30. Plioplys AV: Autism: electroencephalogram abnormalities and clinical improvement with valproic acid. *Arch Pediatr Adolesc Med* 1994, **148**:220–222.
31. Uvebrant P, Bauzienè R: Intractable epilepsy in children: the efficacy of lamotrigine treatment, including non-seizure-related benefits. *Neuropediatrics* 1994, **25**:284–289.
32. Childs JA, Blair JL: Valproic acid treatment of epilepsy in autistic twins. *J Neurosci Nurs* 1997, **29**:244–248.
33. Jambaqué I, Chiron C, Dumas C, Mumford J, Dulac O: Mental and behavioural outcome of infantile epilepsy treated by vigabatrin in tuberous sclerosis patients. *Epilepsy Res* 2000, **38**:151–160.
34. Hollander E, Dolgoff-Kaspar R, Cartwright C, Rawitt R, Novotny S: An open trial of divalproex sodium in autism spectrum disorders. *J Clin Psychiatry* 2001, **62**:530–534.
35. Seldon HL: Structure of human auditory cortex, I: cytoarchitectonics and dendritic distributions. *Brain Res* 1981, **229**:277–294.
36. Seldon HL: Structure of human auditory cortex, II: axon distributions and morphological correlates of speech perception. *Brain Res*, 1981, **229**:295–310.

37. Ogawa A, Ukai S, Shinosaki K, Yamamoto M, Kawaguchi S, Ishii R, Takeda M: Slow repetitive transcranial magnetic stimulation increases somatosensory high-frequency oscillations in humans. *Neurosci Lett* 2004, **358**:193–196.
38. Lin KL, Pascual-Leone A: Transcranial magnetic stimulation and its applications in children. *Chang Gung Med J* 2002, **25**:424–436.
39. Cantor DS, Thatcher RW, Hrybyk M, Kaye H: Computerized EEG analysis of autistic children. *J Autism Dev Disord* 1986, **16**:169–187.
40. Dawson G, Grofer-Klinger L, Panagiotides H, Lewy A, Catelloe P: Subgroups of autistic children based on social behavior display distinct patterns of brain activity. *J Abnorm Child Psychol* 1995, **23**:569–583.
41. Daoust AM, Limoges E, Bolduc C, Mottron L, Godbout R: EEG spectral analysis of wakefulness and REM sleep in high functioning autistic spectrum disorders. *Clin Neurophysiol*, 2004, **115**:1368–1373.
42. Kaiser J: Induced gamma-band activity and human brain function. *Neuroscientist* 2003, **9**:475–484.
43. Tallon-Baudry C, Bertrand O, Peronnet F, Pernier J: Induced gamma-Band activity during the delay of a visual short-term memory task in humans. *J Neurosci* 1998, **18**:4244–4254.
44. Rodriguez E, George N, Lachaux JP, Martinerie J, Renault B, Varela FJ: Perception's shadow: long distance synchronization of human brain activity. *Nature* 1999, **397**:430–433.
45. Gruber T, Muller MM, Keil A, Elbert T: Selective visual-spatial attention alters induced gamma band responses in the human EEG. *Clin Neurophysiol* 1999, **110**:2074–2085.
46. Gruber T, Keil A, Muller MM: Modulation of induced gamma band responses and phase synchrony in a paired associate learning task in the human EEG. *Neurosci Lett*, 2001, **316**:29–32.
47. Herrmann CS, Mecklinger A: Gamma activity in human EEG is related to highspeed memory comparisons during object selective attention. *Vis Cogn* 2001, **8**:593–608.
48. Felli R, Fernandez G, Klaver P, Elger CE, Fries P: Is synchronized neuronal gamma activity relevant for selective attention? *Brain Res Rev* 2003, **42**:265–272.
49. Kaiser J, Hertrich I, Ackermann H, Mathiak K, Lutzenberger W: Hearing lips: gamma-band activity during audiovisual speech perception. *Cereb Cortex* 2005, **15**:646–653.
50. Nakatani C, Ito J, Nikolaev AR, Gong P, Leeuwen CV: Phase synchronization analysis of EEG during attentional blink. *J Cogn Neurosci* 2005, **17**:1969–1979.
51. Gray CM, Singer W: Stimulus-specific neuronal oscillations in orientation columns of cat visual cortex. *Proc Natl Acad Sci USA* 1989, **86**:1698–1702.
52. Kahana MJ: The cognitive correlates of human brain oscillations. *J Neurosci* 2006, **26**:1669–1672.

53. Pavlova M, Birbaumer N, Sokolov A: Attentional modulation of cortical neuromagnetic gamma response to biological movement. *Cereb Cortex*, 2006, **16**:321–327.
54. Keil A, Muller MM, Ray WJ, Gruber T, Elbert T: Human gamma band activity and perception of a Gestalt. *J Neurosci* 1999, **19**:7152–7161.
55. Von Stein A, Rappelsberger P, Sarnthein J, Petsche H: Synchronization between temporal and parietal cortex during multimodal object processing in man. *Cereb Cortex* 1999, **9**:137–150.
56. Herrmann CS, Mecklinger A: Magnetoencephalographic responses to illusory figures: Early evoked gamma is affected by processing of stimulus features. *Int J Psychophysiol* 2000, **38**:265–281.
57. Brock J, Brown CC, Boucher J, Rippon G: The temporal binding deficit hypothesis of autism. *Dev Psychopathol* 2002, **14**:209–224.
58. Brown C: EEG in autism: is there just too much going on in there? In *Recent Developments in Autism Research*. Edited by Casanova MF. New York: Nova Biomedical; 2005:109–126.
59. Kanizsa G: Subjective contours. *Sci Am* 1976, **235**:48–52.
60. Brown C, Gruber T, Boucher J, Rippon G, Brock J. Gamma abnormalities during perception of illusory figures in autism. *Cortex* 2005, **41**:364–376.
61. Kemner C, van der Gaag RJ, Verbaten M, van Engeland H: ERP differences among subtypes of pervasive developmental disorders. *Biol Psychiatry*, 1999, **46**:781–789.
62. Bomba MD, Pang EW: Cortical auditory evoked potentials in autism: a review. *Int J Psychophysiol* 2004, **53**:161–169.
63. Oades RD, Walker MK, Geffen LB, Stern LM: Event-related potentials in autistic and healthy children on an auditory choice reaction time task. *Int J Psychophysiol* 1988, **6**:25–37.
64. Bruneau N, Roux S, Adrien JL, Barthélémy C: Auditory associative cortex dysfunction in children with autism: evidence from late auditory evoked potentials (N1 wave-T complex). *Clin Neurophysiol* 1999, **110**:1927–1934.
65. Kemner C, Verbaten MN, Cuperus JM, Camfferman G, Van Engeland H: Auditory event-related potentials in autistic children and three different control groups. *Biol Psychiatry* 1995, **38**:150–165.
66. Townsend J, Westerfield M, Leaver E, Makeig S, Jung T, Pierce K, Courchesne E: Event-related brain response abnormalities in autism: evidence for impaired cerebello-frontal spatial attention networks. *Brain Res Cogn Brain Res* 2001, **11**:127–145.
67. Courchesne E, Lincoln AJ, Yeung-Courchesne R, Elmasian R, Grillon C: Pathophysiologic findings in nonretarded autism and receptive developmental disorder. *J Autism Dev Disord* 1989, **19**:1–17.
68. Lincoln AJ, Courchesne E, Harms L, Allen M: Contextual probability evaluation in autistic, receptive developmental disorder and control children: event-related potential evidence. *J Autism Dev Disord* 1993, **23**:37–58.

69. Ciesielski KT, Knoght JE, Prince RJ, Harris RJ, Handmaker SD: Event-related potentials in cross-modal divided attention in autism. *Neuropsychologia* 1995, **33**:225–246.
70. Ferri R, Elia M, Agarwal N, Lanuzza B, Musumeci SA, Pennisi G: The mismatch negativity and the P3a components of the auditory event-related potentials in autistic low-functioning subjects. *Clin Neurophysiol*, 2003, **114**:1671–1680.
71. Baruth JM, Casanova MF, El-Baz A, Horrell T, Mathai G, Sears L, Sokhadze E: Low-frequency repetitive transcranial magnetic stimulation (rTMS) modulates evoked-gamma frequency oscillations in autism spectrum disorder (ASD). *J Neurother* 2010, **14**:179–194.
72. Baruth JM, Casanova M, Sears L, Sokhadze E: Early-stage visual processing abnormalities in autism spectrum disorder (ASD). *Transl Neurosci* **1**:177–187.
73. Sokhadze E, Baruth J, Tasman A, Mansoor M, Ramaswamy R, Sears L, Mathai G, El-Baz A, Casanova MF: Low-frequency repetitive transcranial magnetic stimulation (rTMS) affects event-related potential measures of novelty processing in autism. *Appl Psychophysiol Biofeedback* 2010, **35**:147–161.
74. Sokhadze E, Baruth J, El-Baz A, Horrell T, Sokhadze G, Carroll T, Tasman A, Sears L, Casanova MF: Impaired error monitoring and correction function in autism. *J Neurother* 2010, **14**:79–95.
75. Belmonte MK, Yurgelun-Todd DA: Functional anatomy of impaired selective attention and compensatory processing in autism. *Cogn Brain Res* 2003, **17**:651–664.
76. Casanova MF, van Kooten I, Switala AE, van England H, Heinsen H, Steinbuch HWM, Hof PR, Schmitz C: Abnormalities of cortical minicolumnar organization in the prefrontal lobes of autistic patients. *Clin Neurosci Res* 2006, **6**:127–133.
77. Sokhadze E, El-Baz A, Baruth J, Mathai G, Sears L, Casanova MF: Effects of low frequency repetitive transcranial magnetic stimulation (rTMS) on gamma frequency oscillations and event-related potentials during processing of illusory figures in autism. *J Autism Dev Disord* 2009, **39**:619–634.
78. Bodfish JW, Symons FJ, Lewis MH: *The Repetitive Behavior Scale*. Morganton: Western Carolina Center Research Reports; 1999.
79. Casanova MF, Baruth JM, El-Baz A, Tasman A, Sears L, Sokhadze E: Repetitive transcranial magnetic stimulation (rTMS) modulates event-related potential (ERP) indices of attention in autism. *Transl Neurosci* 2012, **3**:170–180.
80. Sokhadze EM, Baruth JM, Sears L, Sokhadze GE, El-Baz AS, Casanova MF: Prefrontal neuromodulation using rTMS improves error monitoring and correction function in autism. *Appl Psychophysiol Biofeedback*, 2012, **37**:91–102.
81. Sokhadze E, Baruth J, Tasman A, Sears L, Mathai G, El-Baz A, Casanova MF: Event-related potential study of novelty processing abnormalities in autism. *Appl Psychophysiol Biofeedback* 2009, **34**:37–51.

Chapter 24

Music Therapy: Personalized Interventions for Individuals with Autism Spectrum Disorder

Petra Kern*

Abstract

The study of autism spectrum disorder (ASD) fascinates various professional disciplines. Music therapy is one of the frontiers in autism research, offering personalized interventions to this specific population and their families. Recent findings provide statistically significant evidence that music therapy interventions are effective for increasing prominent skill areas in young children with ASD, while documentation of additional areas and interventions for a broader age range is emerging. This chapter offers an overview of the impact of music as therapy as well as current music therapy practices, reviews research evidence pertinent to intervention outcomes for individuals with ASD, highlights evolving research and initiatives that are on the horizon, and provides future directions for building on its potential to be an invaluable part of treatment.

Keywords: Autism spectrum disorder; Music therapy; Personalized interventions; Intervention studies; Future developments.

*Correspondence to: Music Therapy Consulting, 618 De La Vista Avenue, Santa Barbara, CA 93103. Email: petrakern@musictherapy.biz.

Introduction

Autism Spectrum Disorder (ASD) is a neurodevelopmental disorder, presently affecting 1 in 88 individuals in the United States across all races, ethnic and socioeconomic groups.[1] Various professional groups, organizations, and individuals investigate its possible causes as well as effective medical treatments and interventions that may improve personal independence and social responsibility of individuals with ASD. As there is currently no known cure for ASD, identifying effective interventions and best fit for each individual's abilities and needs is critical for achieving unique goals that may lead to independent living and well-being.[2]

Music therapy is one of the 22 emerging practices identified by the National Autism Center[3] that addresses and considers the following two core symptoms within each personalized session: (1) persistent deficits in social communication and social interaction across multiple contexts, and (2) restricted, repetitive patterns of behavior, interests, or activities.[4] Music therapy interventions are informed by research evidence and incorporate many of the 11 identified ASD-specific evidence-based practices in each session.[3,5] Yet, research on the effectiveness of music therapy interventions is limited, but gradually expanding.

This chapter highlights why music matters, survey research on current music therapy practices for serving individuals with ASD, historical information, systematic reviews and emerging research, and provides an outlook on prospective developments in the field. A selection of resources and organizational initiatives offers opportunities for more in-depth exploration of the benefits and applications of personalized music therapy interventions for individuals with ASD.

Why Music Matters

Music is deeply ingrained in every culture and history of humanity. People of all nations, ethnicities, socioeconomic backgrounds, and age value and respond to aspects of music in daily life. Infants may fall asleep when listening to lullabies, children may engage in dancing when singing a song, teenagers may socialize while playing in a band, and adults may find joy in attending a concert. Musical enjoyment and involvement in various forms is possible for everyone.[6]

Individuals with disabilities such as ASD demonstrate positive responses and gratitude for music as well. In fact, since the first documentation of ASD by Kanner in 1943, researchers have discussed advanced musical memory, responsiveness, and aptitudes within this population. More recent studies revealed that several high-functioning children with ASD have heightened sensitivity of musical pitch and timbre compared to their typically developing peers, yet similar skills in perceiving musical structure, contour, and emotions.[7,8] However, only a very small percentage of individuals with ASD actually are musical savants.[9] While neurologic mechanisms involved in processing music and musical aptitudes might inform music therapy interventions, having advanced music skills is not necessary for clients to benefit from music therapy services.[10]

Music is engaging and accessible for many people, including those with ASD. It is a remarkable motivator for encouraging and supporting learning and daily functioning across the lifespan. Music therapists understand the therapeutic effects of music and are able to personalize each intervention by reflecting cultural aspects, client's preferences, and individual abilities.[6,11]

Music Therapy Services for ASD

Professional health discipline

Music therapy is a well-established health profession that provides assessment, interventions, and consultative services to individuals with ASD and their families for achieving non-musical therapeutic goals. Music is used as a stimulus to elicit and practice skills identified by the client, his or her family, and the intervention team. Each session is highly personalized to meet the idiosyncratic nature of the disorder.[11,12] Music therapists are board-certified professionals and adhere to a code of ethics and professional standards of practice.[13,14] Recognized as a related service under the Individuals with Disabilities Education Act (IDEA), music therapy is an integral treatment option for individuals with ASD.[15]

Music therapy practice

Based on a recent survey study conducted in the United States by Kern, Rivera, Chandler, and Humpal,[5] music therapists are seeing slightly more individuals with ASD on their caseloads than a decade ago. Currently, music therapists mostly serve those with ASD who are children and adolescents, followed by infants and young children. Most music therapists provide services in public schools (K-12), family homes, private practice, or in community settings. Within school-based settings, music therapists primarily offer group activities, whereas in other work settings individual sessions are more common. Nevertheless, music therapy services continue to be delivered primarily in segregated versus inclusive environments. Most music therapy sessions typically are provided once per week for 30 minutes with an average of 1–3 years of service duration. Music therapists often are part of an interdisciplinary team and collaborate and consult with parents, educators, speech-language pathologists, occupational therapists, and others for treatment planning and implementation.

Therapeutic process

The survey study further revealed that most music therapists assess clients to identify their present level of functioning for intervention planning and to monitor the client's progress for determining the type and intensity of services.[5] Few music therapists are presently involved in the screening and diagnostic evaluation process. As there is no known validated music therapy assessment tool specific to ASD, music therapists employ self-created assessment tools and music-therapy or work-space specific assessment tools.

Only a few utilize validated ASD-specific assessment tools in their clinical practice. The top three goal areas targeted when planning music therapy interventions for individuals with ASD are communication skills, social skills, and emotional skills. Music therapy clients typically achieve personalized intervention goals within 4–6 months. According to research-based practice, most music therapists apply a behavioral approach to music therapy when working with clients on the autism spectrum. Music therapy techniques utilized to achieve therapeutic goals are singing and vocalization, instrument play, movement and dance, free and thematic improvisation, songwriting and compositions, listening to live music/prerecorded music, and combining music with media. The majority of music therapists document clients' progress after each session as well as in summarized regular reports.

Music Therapy Intervention Studies

Building on the past: Anecdotal reports and narrative literature reviews

Music therapy has a long-standing tradition of offering personalized interventions to individuals with ASD worldwide. In the 1960s and 1970s, music therapy pioneers from the United States,[16] Argentina,[17] and the United Kingdom[18] described in numerous anecdotal case reports the positive impact of using music for skill improvement in key areas of ASD. During this early period, pioneers already had begun to report the unusual musical talents and interest in music of children with ASD as well as the importance of using "child-directed" techniques to elicit positive outcomes.

A historical overview by Reschke-Hernández illuminates the evolution of music therapy research and treatments for children with ASD from 1940–2009.[19] This comprehensive article offers an in-depth understanding of how changing autism diagnoses, public laws, organizational developments, or research and educational shifts may have influenced the direction of music therapy research and interventions for children with ASD over time. One of these changes is the "evidence-based practice" movement, which led to the establishment of quality indicators for identifying rigorous research as well as criteria for defining different levels of evidence-based practice. Subsequently, many previous studies do not meet today's scientific standards despite the numerous research reports that have shown positive effects of music therapy interventions on increased communication, socialization, and decreased problem behaviors of those with ASD. Deliberate narrative reviews of the literature related to music intervention for individuals with ASD by Accordino, Comer and Heller[20] and Simpson and Keen[21] confirm the emerging literature, but there is limited empirical support of music therapy interventions with individuals with ASD. Hence, improving the scientific rigor of intervention studies and building the level of evidence to support the positive impact of music therapy intervention for this population remains a strong aim of the discipline.

Learning from the data: Three systematic reviews

The existing research literature presents three systematic reviews related to music-based interventions for individuals with ASD from early childhood to adolescence.[10,22,23] While the systematic reviews were limited by the availability of published literature meeting the studies' criteria, researchers reported treatment benefits from small to large effect sizes. Overall, a positive direction related to an array of outcomes and prospective research is noted in all systematic reviews.

In a meta-analysis by Whipple,[23] nine out of 29 intervention studies with a total of 76 subjects were identified that compared the use of music to non-music conditions with children and adolescents ages 2 to 21 years by analyzing 12 dependent variables (categorized by cognitive skills, communication, and social behavior). Due to the limited literature base at that time, this systematic review included seven studies that had been refereed, but were unpublished master's theses or doctoral dissertations; 50% of the interventionists were not music therapists — factors that are critiqued today. However, results suggested that all music-based interventions were effective (overall effect size of $d = 0.77$; $p < 0.001$) regardless of the age of the client, outcome addressed, discipline of the music provider, theoretical approach, or session type. The investigator concluded, "All effects were in a positive direction, indicating benefits of the use of music in intervention".[23]

A Cochrane review by Gold, Wigram, and Elefant,[22] noted that, out of 52 assessment and intervention studies, only three small randomized controlled trials/controlled clinical trials with a total of 24 subjects met the study's inclusion criteria for comparing music therapy interventions to "placebo" therapy, no treatment, or standard care for children with ASD ages 2 to 9 years. As in the previous systematic review, one of the three studies included in the Cochrane report was an unpublished master's thesis; criteria for inclusion seemed to focus on randomized controlled trials not readily available at that time; and application of the inclusion criteria for controlled clinical trials seemed to have major flaws — details that indicate an update of the Cochrane review is needed. Nevertheless, results note that music therapy interventions had a significant effect on increasing gestural communicative skills (SMD 0.50; medium effect size) and verbal communication skills (SMD 0.36; small to medium effect size) compared to the non-music therapy conditions, but not for reducing behavioral challenges (SMD –0.24; small effect size) in children with ASD. Due to the short-term effect of brief music therapy interventions (i.e., sessions conducted daily for only one week), the studies have limited clinical application for children with ASD. However, the authors concluded, "...the findings indicate that music therapy may help children with autistic spectrum disorder to improve their communicative skills."[22]

The most recent meta-analysis conducted by Whipple[10] identified eight intervention studies with a total of 91 subjects with ASD aged five years or younger that utilized music as an independent variable compared with a no music condition, provided treatment by a certified music therapist, reported quantitative results sufficiently, and were published in peer-reviewed journals. The scientific rigor of the identified studies was evaluated based on the *Scientific Merit Rating* from the NAC[3] and the *Evidenced-Based Practice Rating* including *quality indicators* by Reichow, Volkmar, and Cicchetti[24] — two rating systems meeting today's scientific standards in studying ASD. Results indicated that music therapy interventions with young children with ASD are very effective (overall effect size of $d = 0.76$; $p < 0.0001$) for improving communication (i.e., verbal production and independent responses), interpersonal skills (i.e., eye contact, joint attention, imitation, turn taking, social acknowledgement, and peer interactions), personal responsibility (i.e., cleaning up, hand washing, and toileting), and play (i.e., engaging in meaningful activities). Figure 1 lists the eight selected studies and provides a summary of one. The researcher concluded, "Based on the results of the current meta-analysis, music therapy may be considered an extremely effective treatment for young children with ASD for developing communication, interpersonal, personal responsibility, and play skills."[10]

Moving forward: Emerging research and initiatives

Researchers have continued to study the effectiveness of music therapy interventions for individuals with ASD. More recent studies include a broader age range (3–21 years) and apply more rigorous study designs with larger sample sizes. Topics extend beyond those of previous studies by addressing communication, social skills, joint attention, auditory processing, sensory-motor skills, and family-centered models among others.

For example, Kalas examined the effects of simple versus complex music on joint attention of thirty children with ASD (ages 4 to 6).[25] All children attended three simple and three complex 10-minute individual music conditions over the course of three weeks. Results suggested that music with clear and predictable patterns was more effective in eliciting responses to joint attention in children with severe ASD whereas more complex and variable music was more effective for children with mild/moderate ASD.

Background: For children with ASD enrolled in an inclusive childcare program, playground time can be a major challenge due to the large, undefined space, the lack of predictable and structured play routines, the fast pace of play, and play styles inherent in outdoor play.

Objectives: The purpose of this study was to increase peer interactions and meaningful play of four boys with ASD (ages 3-5) on an inclusive childcare playground by adding an outdoor music center, providing personalized song interventions prepared by a board certified music therapist and implemented by early childhood educators engaging peers.

Method: The effects of the environmental modification, personalized interventions, and collaborative consultative approach were evaluated by using a multiple baseline design across the four subjects. Categories of interaction behaviors were coded by analyzing 10-minute videotaped segments using a 15-second momentary time sampling recording procedure. In addition fidelity data on early childhood educators and peers following the intervention protocol was collected.

Results: The environmental modification (Music Hut) attracted the subjects and peers alike and therefore provided opportunities for play and involvement with peers. Yet, the personalized song interventions resulted in meaningful music play and increased peer interactions on the playground. In addition, the collaborative consultative approach was successful in empowering early childhood educators and peers to implement the interventions with high fidelity ratings.

Conclusions: Music therapy interventions can successfully address play and interpersonal skills in children with ASD, be implemented in natural and inclusive environments, and enable early childhood educators and peers to use music for learning.

Figure 1. Example of a music therapy intervention study (Kern & Aldridge, 2006).

Note: The eight selected studies by Whipple (2012) meeting NAC (2009) standards and four target areas for treatment of children with ASD include Finnigan & Starr[30]; Kern & Aldridge[31]; Kern, Wakeford & Aldridge[32]; Kern, Wolery & Aldridge[33]; Kim, Wigram & Gold[34]; Lim[35]; Lim & Draper[36] and Wimpory.[37]

Schwartzberg and Silverman[26] used a randomized controlled trial to examine the effects of music-based social stories on comprehension and generalization skills of 30 individuals with ASD (ages 9 to 21). All subjects attended one of three experimental music therapy groups or non-music control groups at one-week summer camps. While results indicated that differences between groups were not significant post treatment, anecdotal reports from staff encouraged the use of music-based social stories to support learning of social skills in a motivational way.

Thompson studied the impact of a family-based music therapy model on social engagement of 23 children with ASD (3–6 years) in their family home by comparing a 16-week family-based music therapy program added to their standard care (treatment group) to standard care alone (control group)[27]. Results indicated a significant effect on children's level of engagement during music therapy sessions (p <0.001) as well as improvements in social interactions in their home environment and community (p <0.001). In addition, qualitative data suggested that parent-child relationships improved.

Several other randomized controlled intervention studies (as described on websites and conference programs) are in process: (1) the effects of music therapy groups on the social skills of children with autism spectrum disorders,[28] (2) improvisational music therapy's effectiveness for children with autism spectrum disorders,[29] and (3) the effects of singing on immediate and long-term reading comprehension of children with autism spectrum disorder.[26]

It remains unclear if the recently peer-reviewed and forthcoming studies do meet the standards of the *Scientific Merit Rating*,[3] *Evidenced-Based Practice Rating*[24] or future study inclusion criteria of systematic reviews. Yet, as is apparent by the initiatives set forth by several individuals and organizations, the music therapy profession is determined to improve its scientific merit and to reach evidence-based practice level. In 2008, AMTA established the *Music Therapy and Autism Task Force/Think Tank* (now known as the *AMTA Strategic Priority on Music Therapy and ASD*). This group of music therapy experts is working on strategies to improve awareness and recognition of music therapy services for individuals with ASD, enhance research knowledge and clinical practice, provide advanced training for music therapists on ASD, and engage in information dissemination.[12] As a result, AMTA has published information, collaborates with *Autism Speaks* and the *Autism Society of America*, and is a featured provider on the *National Autism Network*. Table 1 presents a selection of resources created by leaders of the AMTA initiative and individuals who provide music therapy services for this population.

Table 1. Selected resources related to music therapy and ASD

Resources
Fact Sheet
• Music Therapy as a Treatment Modality for Autism Spectrum Disorders (AMTA, 2012) http://www.musictherapy.org/assets/1/7/MT_Autism_2012.pdf
Bibliographies
• Autism Spectrum Disorders: Music Therapy Research and Evidence-Based Practice Support (AMTA, 2010) http://www.musictherapy.org/assets/1/7/bib_autism10.pdf
• Music and Autism Research (Coast Music Therapy, 2013) http://www.coastmusictherapy.com/how-music-helps/autism-research/
Research Snapshots
• *imagine*: early childhood music therapy (de la vista publisher, 2013) http://imagine.musictherapy.biz
Books
• Kern, P., & Humpal, M. (Eds.) (2012). *Early childhood music therapy and autism spectrum disorders: Developing potential in young children and their families.* London and Philadelphia: Jessica Kingsley Publishers.
• Lim, H. A. (2011). *Developmental speech language training through music for children with autism spectrum disorders.* London and Philadelphia: Jessica Kingsley Publishers.
Blogs
• MT Research Blog: Bringing Current Research to Music Therapy Clinicians http://www.musictherapyresearchblog.com/category/develop/autism/
• FAQs: Autism.com: A Resource for Practical Ideas http://faqautism.com/

Outlook on Future Research and Developments

Considering the existing intervention research pertinent to music therapy and the rapid scientific developments germane to ASD, researchers will need to stay well-informed and to continue facing the challenge of studying a highly idiosyncratic group of clients that requires personalized interventions for each individual. While adequate sample size, scientific rigor, documentation of additional goal areas, generalization and maintenance of treatment for this population have been critiqued for a long time, several authors have provided solid recommendations for improving the existing shortcomings in the literature.[10,19–21,23] These recommendations are reflected and complemented to some extent in the following paragraphs, which may inform research and practice standards in the near future.

Sample size and scientific rigor

Although there is a tendency of more experimental single case and group designs (comparing music therapy interventions to non-music conditions) being published, researchers need to continue increasing the sample size and scientific rigor. Collaborations with national autism centers, autism networks, or carefully planned multisite studies may assist in attaining an increased sample size. Sound study designs such as single case experimental designs may account for the subtle differences of music therapy intervention while group experimental designs may allow for statistical analysis that test the significance of any improvements gained by individuals with ASD during music therapy interventions. Both study designs should be considered for establishing evidence-based practice status.[24]

For increasing scientific rigor, researchers may focus on (1) clear criteria for subject inclusion such as tighter age ranges and primary diagnosis (vs. various disabilities) and (2) a detailed study report addressing all primary and secondary quality indicators (especially experimental controls, use of validated psychometric tools, music therapy intervention protocols, fidelity ratings, and generalization and maintenance). As of today, there is a lack of robust generalization and maintenance data supporting skill generalization to other life situations as well as long-term effects of music therapy interventions in individuals with ASD. Researchers may need to look into monitoring changes outside of music therapy sessions and complete long-term follow-up studies after formal sessions have ceased. However, finding funding sources and maintaining access to subjects over a longer period of time may be ongoing challenges in the years to come.

Goal areas and aspects of interventions

There is already strong evidence supporting the use of music therapy for increasing four out of the fourteen treatment targets outlined in the NAC's *Standards Report*[3] (i.e., communication, interpersonal, personal responsibility, and play skills — in children under five years). Nevertheless, more empirical research is needed to confirm the positive effects of music therapy interventions for (1) increasing academic skills, higher cognitive functions, learning readiness, motor skills, placement skills, and self-regulation, and (2) decreased behaviors related to general symptoms of ASD, problem behaviors, restricted repetitive nonfunctional patterns of behavior, and sensory or emotional regulation. As indicated by the recently published survey study,[5] music therapy practitioners already address many of the remaining outcome categories in daily practice; more scientific attention needs to be given to these topics in the future to insure recognition of their efficacy.

Additionally, researchers should explore the efficacy of music therapy techniques (e.g., singing and vocalization, instrument play, movement and dance) or approaches to music therapy (e.g., behavioral music therapy, Nordoff-Robbins Music Therapy, Neurologic Music Therapy) in achieving treatment outcomes for individuals with ASD and document them adequately in the literature. The intensity of interventions (i.e., dosage) is yet another research topic that deserves consideration. As repetition is of utmost importance for ensuring that learning occurs in this population, it remains unclear if weekly 30-minute sessions provided by most music therapy practitioners[5] are sufficient for achieving desirable treatment outcomes. Nonetheless, conducting coherent research that will support the validity and therapeutic value of daily practice may remain a challenge, as it often is influenced by various unpredictable factors.

Context of intervention and collaborations

Although, most music therapists provide intervention services in various settings (e.g., schools, homes), research on ecological validity and inclusion programming is sparse. Natural environments supporting inclusion and skill generalization of individuals with ASD are mandated by U.S. Public Law 94-142. Music therapy practitioners and researchers alike will need to examine effective inclusion practices and supports beyond traditional settings and service delivery models. Much re-thinking of previous theories, approaches, and conventions will be required to assure appropriate access, participation, and supports for individuals with ASD, their families, and collaborators.

Music therapists working with individuals with ASD most likely are part of an interdisciplinary intervention team. The effects of collaborating with other disciplines (e.g., educators, speech-language pathologists, occupational therapists) and including other-mediated interventions based on music need more evaluation. Cost-effectiveness of direct music therapy interventions versus consultative services for this population will be questioned and will need to be evaluated. To date, music therapy services for individuals with ASD mainly are funded by private pay, IDEA/Special Education, State/County Funding, and grants.[5] Future funding of music therapy interventions may continue to be a major concern for both practitioners and parents of individuals with ASD, although securing funding might be less of an issue when sophisticated research, validated within and outside the field of music therapy, becomes readily available and recognized.

Research to practice

The ultimate purpose of music therapy research is to benefit the clients with ASD who are receiving interventions. Therefore, it is essential that research is translated to clinical practice. First, this requires disseminating knowledge widely (i.e., to music therapy consumers, practitioners, and the general public) via various formats (e.g., research synthesis points, case vignettes demonstrating applications to day-to-day practice, or websites devoted to music therapy and ASD). Secondly, advanced training on skills, vital for music therapy practitioners to employ while working with individuals with ASD, should be offered (e.g., online courses, workshops, webinars). Finally, a system to guide practitioners in implementing evidence-based practice (e.g., mentor coaching, peer coaching, or reflective consultation) and monitoring intervention fidelity onsite (e.g., checklists, self-reflection) will need to be in place. Sustainable, outcome-based professional development may assist in ongoing and meaningful implementation of research to practice that optimizes effective delivery of services for individuals with ASD.

Currently, accessing research-based knowledge pertaining to music therapy is challenging for practitioners and the general public. Professional development opportunities related to music therapy and ASD are sparse. Technical support for implementation guidance and monitoring is lacking. Yet, it should be noted that the field of music therapy is small compared to other health professions. Therefore, disseminating knowledge, professional development, and systematic supports for practitioners may take more time than is required for disciplines with more members.

Conclusion

The increased incidence rate of ASD requires the immediate attention and support of individuals and families dealing with this neurodevelopmental disorder. As of yet there is neither a defined cause nor cure; therefore, investigating novel treatments that best fit the idiosyncratic nature of ASD is laudable. Music therapy is one of the emerging practices that provides personalized interventions based on each client's unique abilities and needs. Given that most individuals with ASD respond positively to music, music therapy can function as a stimulus to elicit and practice skills necessary to achieve personal independence and social responsibility in life. Future research will shed more light on the effectiveness of music therapy interventions; outcomes will assist clients, families, and professionals to make more informed decisions on how to best meet the needs of this population.

Executive Summary

- The increased incidence rate of autism spectrum disorder (ASD) calls for effective interventions that best fit each individual's interest, abilities, and needs.
- Music is a preferred medium for many individuals with ASD and can function as a stimulus to elicit and practice skills necessary to live an independent and social life.
- Music therapy is an established health profession that provides evidence-based and personalized interventions for individuals with ASD and their families.
- Since the late 1960s, numerous anecdotal case reports have described the positive impact of personalized music therapy interventions for this population across the world.
- Two systematic reviews report treatment benefits with small to medium effect size, while the most recent meta-analysis reveals a medium to large effect related to music therapy interventions in increasing communication, interpersonal skills, personal responsibility, and play in young children with ASD.
- Emerging research on the effects of music therapy interventions for individuals with ASD is applying more rigorous study designs and larger sample sizes to meet today's scientific standards.
- Researchers studying the effects of music therapy interventions need to continue strengthening the scientific rigor, study remaining goal areas and specific aspects of interventions, and expand research questions to environmental context and professional collaborations.
- Available and future research outcomes should be disseminated more efficiently, advanced professional development opportunities should be provided, and a system guiding and monitoring practitioners in implementing evidence-based practice should be considered.
- More research validated within and outside the field of music therapy will make personalized music therapy interventions more accessible for individuals with ASD.

Acknowledgements

The author is grateful to Marcia Humpal (AMTA Strategic Priority on Music Therapy and ASD, Silver Spring, MD, USA) and Dr. Jennifer Whipple (Charleston Southern University, Charleston, SC, USA) for their valuable comments and thougtful review of this book chapter.

References

1. Autism Spectrum Disorders (ASDs) [http://www.cdc.gov/ncbddd/autism/index.html].
2. National Research Council: Committee on educational interventions for children with autism: *Educating Children with Autism:* Washington, DC: National Academy Press; 2001.
3. National Autism Center: National Standards Project Findings and Conclusions: Addressing the need for evidence-based practive guidelines for Autism Spectrum Disorders: Randolph, MA: National Autism Center; 2009.
4. American Psychiatric Association: *Diagnostic and Statistical Manual of Mental Disorders,* 5th edition: Arlington, VA: American Psychiatric Association; 2013.
5. Kern P, Rivera NR, Chandler A, Humpal M: Music therapy services for individuals with autism spectrum disorder: A survey of clinical practices and training needs. *J Music Therapy* 2013, **50**(4):274–303.
6. Gfeller KE: Music: A human phenomenon and therapeutic tool. In *An Introduction to Music Therapy and Practice.* 3rd edition. Edited by Davis WB, Gfeller KE, Thaut MH. Silver Spring, MD: American Music Therapy Association; 2008:41–78.
7. Heaton P: Pitch memory, labelling and disembedding in autism. *J Child Psychol Psychiatry Allied Disciplines* 2003, **44**(4):543–551.
8. Heaton P: Interval and contour processing in autism. *J Autism Dev Disord* 2005, **35**(6):787–793.
9. Treffert DA: *Islands of Genius: The Bountiful Mind of the Autism, Acquired, and Sudden Savant:* London, England and Philadelphia, PA: Jessical Kingsley Publishers; 2012.
10. Whipple J: Music therapy as an effective treatment with autism spectrum disorders in early childhood: A meta-analysis. In *Early Childhood Music Therapy and Autism Spectrum Disorders: Developing Potential in Young Children and Their Families.* Edited by Kern P, Humpal M. London, England and Philadelphia, PA: Jessica Kingsley Publishers; 2012:pp.59–76.
11. Kern P, Humpal M: Early childhood music therapy and autism spectrum disorders: Developing potential in young children and their families. 2012.

12. American Music Therapy Association, Task Force on Music Therapy and Autism: Autism Strategic Priority Outcomes. 2008.
13. The Standards of Clinical Practice [http://www.musictherapy.org/about/standards/].
14. CBMTScope of Practice [http://www.cbmt.org/].
15. Simpson J: Inclusion of music therapy in early intervention programs. *Imagine* 2011, **2**(1):14–15.
16. Nordoff P, Robbins C: Improvised music as therapy for autistic children. In *Music in Therapy.* Edited by Gaston ET. New York, NY: MacMillan; 1968: 191–193.
17. Benenzon RO: Music therapy in infantile autism. *Brit J Music Therapy* 1976, **7**:10–17.
18. Alvin J, Warwick A: *Music Therapy for the Autistic Child*: Oxford, England: Oxford University Press; 1978.
19. Reschke-Hernandez AE: History of music therapy treatment interventions of children with autism. *J Music Therapy* 2011, **48**(2):169–207.
20. Accordino R, Comer R, Heller WB: Searching for music's potential: A critical examination of research on music therapy with individuals with autism. *Res Autism Spectrum Disorders* 2007, **1**(1):101–115.
21. Simpson K, Keen D: Music interventions for children with autism: Narrative review of the literature. *J Autism Dev Disord* 2011, **41**(11):1507–1514.
22. Gold C, Wigram T, Elefant C: Music therapy for autistic spectrum disorder. Cochrane database of systematic reviews (Online) 2006, (2).
23. Whipple J: Music in intervention for children and adolescents with autism: A meta-analysis. *J Music Ther* 2004, **41**(2):90–106.
24. Reichow B, Volkmar FR, Cicchetti DV: Development of the evaluative method for evaluating and determining evidence-based practices in autism. *J Autism Dev Disord* 2008, **38**(7):1311–1319.
25. Kalas A: Joint attention responses of children with autism spectrum disorder to simple versus complex music. *J Music Ther* 2012, **49**(4):430–452.
26. Schwartzberg ET, Silverman MJ: Effects of singing on immediate and long-term reading comprehension of children with Autism Spectrum Disorder: A randomized effectiveness study. 2013.
27. Thompson GA, McFerran KS, Gold C: Family-centered music therapy to promote social engagement in young children with severe autism spectrum disorder: a randomized controlled study. *Child Care Health Dev* 2013. First published online November 22, 2013. DOI:10.1111/cch.12121.
28. The effects of music therapy groups on the social skills of children with autism [http://www.musictherapy.org/research/fultz/].
29. Geretsegger M, Holck U, Gold C: Randomised controlled trial of improvisational music therapy's effectiveness for children with autism spectrum disorders (TIME-A): Study protocol. *BMC Pediatrics* 2012, 12.

30. Finnigan E, Starr E: Increasing social responsiveness in a child with autism: A comparison of music and non-music interventions. *Autism* 2010, **14**(4): 321–348.
31. Kern P, Aldridge D: Using embedded music therapy interventions to support outdoor play of young children with autism in an inclusive community-based child care program. *J Music Ther* 2006, **43**(4):270–294.
32. Kern P, Wakeford L, Aldridge D: Improving the performance of a young child with autism during self-care tasks using embedded song interventions: A case study. *Music Therapy Perspectives* 2007, **25**(1):43–51.
33. Kern P, Wolery M, Aldridge D: Use of songs to promote independence in morning greeting routines for young children with autism. *J Autism Dev Disord* 2007, **37**(7):1264–1271.
34. Kim J, Wigram T, Gold C: The effects of improvisational music therapy on joint attention behaviors in autistic children: A randomized controlled study. *J Autism Dev Disord* 2008, **38**(9):1758-1766.
35. Lim HA: Effect of "developmental speech and language training through music" on speech production in children with autism spectrum disorders. *J Music Ther* 2010, **47**(1):2–26.
36. Lim HA, Draper E: The effects of music therapy incorporated with Applied Behavior Analysis verbal Behavior approach for children with autism spectrum disorders. *J Music Ther* 2011, **48**(4):532–550.
37. Wimpory D, Chadwick P, Nash S: Brief report: Musical interaction therapy for children with autism: An evaluative case study with two-year follow-up. *J Autism Dev Disord* 1995, **25**(5):541–552.

Chapter 25

The Contributions of Art Therapy in Treatment, Assessment, and Research with People Who Have Autism Spectrum Disorders

Donna Betts,* Rebecca Harmer and Gabriela Schmulevich

Abstract

The contributions of art therapy to treatment, assessment, and research with people who have autism spectrum disorders are presented. An overview of art therapy introduces the reader to the applications and benefits of art therapy, which includes the field's contributions to the knowledge base of autism in the areas of treatment, assessment, and research. The ways in which art therapy can ameliorate symptoms of autism in treatment settings such as schools is described, and applications of art therapy for families and caregivers are presented. Art therapy is discussed as a practical means of assessing symptoms of autism, and the Face Stimulus Assessment is described with support of illustrated case examples. Finally, emerging and expanding areas of art therapy and autism research are delineated, with a focus on the promising intersection of neuroscience and art therapy.

Keywords: Art therapy; Autism; Neuroscience; Assessment; Art therapy research; Face Stimulus Assessment.

*Correspondence to: Art Therapy Program, The George Washington University, Alexandria, VA 22314. Email: dbetts@gwu.edu

Introduction

All individuals with autism can benefit from learning how to express their thoughts, feelings, and interests in a creative, hands-on way, whether to ease and enhance communication, externalize feelings of anxiety, or simply realize their potential as imaginative, productive human beings.[1]

The present chapter provides an overview of art therapy, which includes the field's contributions to the knowledge base of autism in the areas of treatment, assessment, and research. The current status of autism research in art therapy is explored, with a focus on the emerging and expanding areas for investigation, such as neuroscience.

Art Therapy

For those unfamiliar with art therapy, it is broadly defined as follows by the American Art Therapy Association[2]:

> Art therapy is a mental health profession in which clients, facilitated by the art therapist, use art media, the creative process, and the resulting artwork to explore their feelings, reconcile emotional conflicts, foster self-awareness, manage behavior and addictions, develop social skills, improve reality orientation, reduce anxiety, and increase self-esteem. A goal in art therapy is to improve or restore a client's functioning and his or her sense of personal well-being. Art therapy practice requires knowledge of visual art (drawing, painting, sculpture, and other art forms) and the creative process, as well as of human development, psychological, and counseling theories and techniques.

Art therapy is practiced by master's level, credentialed art therapists in a wide variety of clinical and community settings with diverse client populations in individual, couples, family, and group therapy formats.[3]

Art Therapy and Autism

Approximately twenty-five percent of art therapists work with clients who have autism.[4] Furthermore, autism is identified as a top research priority in the profession of art therapy.[5] Art therapy can be a primary or adjunctive treatment for people with autism, depending upon the setting. Services for children and adolescents are typically provided in schools, and people with autism of any age can access art therapy in private clinics, community-based programs, and some hospitals.

Art therapy approaches in schools can target several goals and objectives specific to children with autism. These objectives are implicated in four areas of physiological and psychosocial development: (1) *cognitive growth,* which addresses the objectives of expression of thought, flexibility, attending/focusing, communication, and sensory integration; (2) *emotional regulation,* which targets the objectives of frustration tolerance/delayed gratification, self-monitoring/evaluating, control, fantasy versus reality, sublimation, ego development, self-image, expression and cathartic release of emotions; (3) *adaptive behavioral styles,* which addresses the objective areas of independence, maturation, socialization, positive peer interaction, group identity, motivation, initiative and responsibility; and, (4) *physical development,* which targets the objectives of figure-ground/parts-to-whole, directionality, sequencing, fine-motor coordination.

The wide range of symptoms experienced by people with autism makes them very unique in presentation, so treatments must be tailored to a range of varying needs.[1,6] It is especially important to offer a safe, predictable, and stable environment. By doing so, the art therapist establishes psychological continuity and a stable holding environment. Martin[7] explored the difficulties faced in treating ASD, and described the advantages of art therapy for this population. Martin suggested that the art therapist should be a significant part of the treatment team. She also noted that art therapy provides a space where the person with autism can develop skills, achieve growth, and address complex topics visually while enjoying the creative process.

The "spectrum" nature of autism has presented several challenges in trying to determine the efficacy of treatments for the disorder. Because the severity of symptoms of individuals with ASD are so diverse, individual responsiveness to treatments are varied. Similarly, although autism is considered a lifelong condition emerging in childhood, little research has been conducted on the manifestations of core deficits of autism in adolescence and adulthood. By studying the symptoms of adolescents and adults with autism, Seltzer *et al.*[8] sought to determine the stability of autism diagnoses and the extent to which symptoms of autism differ in severity over time. Their sample included 405 individuals between the ages of 10 and 53 years. Results indicated that while all participants met the criteria for ASD earlier in childhood, about half (54.8%) would have met ASD criteria "if current scores were used to complete the diagnostic algorithm" (p. 565). Moreover, determining the most effective treatment for an autistic individual becomes even more difficult as manifestations of symptoms can change over time.

While studies on different treatments and interventions have been conducted, researchers still do not have a standardized way to assess the efficacies of such treatments. The following sources of literature reflect research on the treatment of autism, in an attempt to illustrate the need for additional methods of measuring efficacy.

Art Therapy Treatment & Autism

Several authors have written on this topic. Recommended books include Evans and Dubowski[6] and Martin.[7] While people with autism may share common general manifestations, the triggers and causes for these manifestations may vary greatly from one individual to another. It is therefore necessary that more information on treatment efficacy be available so that specific symptoms of individuals with autism can be targeted more efficiently.[9] The present section hones in on the subject of art therapy approaches.

Bowker, D'Angelo, Hicks, and Wells[10] studied parental choices of treatments as well as parental perceptions of change due to treatments. Through an online survey, 970 parents of children with ASD reported on the treatments they currently use, those discontinued, and reasons for discontinuation. Results illustrate that parents are likely to use a variety of evidence-based and non-evidence-based treatments simultaneously for their autistic children, which supports earlier findings by Goin-Kochel, Myers and Mackintosh.[11] This study also added new information regarding parental perceptions of treatment efficacy, in that parents reported that their children were most likely to improve behaviorally and least likely to improve socially from treatment.

In a similar study on parental experience with treatments for autism, Green[12] interviewed 19 parents on three specific treatments including, Applied Behavioral Analysis, Sensory Integration therapy, and a combination of Vitamin B6 and Magnesium. The study found that most parents received information about such treatments from the Internet, other parents, or occupational therapists. In addition, the study shows that ease of implementation, time commitment, and perceived effectiveness were also factors in parental decisions to use certain treatments, regardless of their lack of empirical support. The results from this study should be interpreted with caution, however, as the sample was rather small and parents volunteered to participate in the interview. Nevertheless, this study helps inform the decision making process of parents selecting treatments for children with autism. Both the Bowker *et al.*[10] study and the Green[12] study found that several non-evidence-based treatments are regularly used to treat autism, and that parents get their information from a variety of sources, some of which possess no normative data to validate them. Thus, there is a need for parents and individuals with autism to have more information on the efficacy of different treatments.

The topic of art-making and its benefits in addressing symptoms of autism has been discussed in literature. For instance, Weed[13] explored the role of art for a child with ASD from an art education standpoint. Weed analyzed three case studies of young boys with autism and reported perceived improvement and increased creativity in the boys over the treatment period. Weed found that art played an important and beneficial role in the lives of these boys with ASD, specifically that art may aid in the expansion of creativity and creative thinking processes for individuals with ASD.

Elkis-Abuhoff[14] explored the effectiveness of art therapy for individuals with Asperger's syndrome through a case study of a female adolescent with the disorder. She perceived growth and improvement for the adolescent in the areas of communication and socialization after adding art therapy to her treatment program. Elkis-Abuhoff offered extensive background information about the subject through reports and scores from evaluations such as the Wechsler Adult Intelligence Scale (WAIS-III),[15] the Thematic Apperception Test (TAT),[16] the Kinetic-House-Tree-Person (K-H-T-P),[17] the House-Tree-Person (H-T-P),[18] and the Kinetic Family Drawing (K-F-D).[19] Nine projects created over a 7-month treatment period were examined, including collages, template designs, crayon conversation pieces, coffee filter designs, mirror on tile projects, and masks. This case study offers an in-depth exploration of the artwork of an individual with an ASD and the benefits of art therapy.

Another case study was presented by Emery,[20] who looked at the importance and role of art therapy treatment for a 6-year-old boy with autism. The value and importance of art therapy treatment was measured through the boy's perceived development, growth, and ability to relate with others. Emery found that art therapy played a key role in the development of object constancy. Art therapy also had an important role in the boy's process of self-discovery, and in expanding his awareness of his environment. Emery also discussed the important topic of family involvement in the development of relationships and awareness for autistic children. The role of the family is discussed later in the chapter.

Brancheau[21] reported the benefits of art therapy for a high school student, Marshall. Through the creation of mandalas — drawn circles that clients can embellish with art materials — Marshall learned to reduce his agitation, and attain a sense of increased mindfulness and focus. Over the course of two years of art therapy treatment, Marshall also demonstrated improvements in his social interactions, communication skills, internal locus of control, and sense of self/identity.

In a study by Lu, Peterson, Lacroix, and Rousseau,[22] the use of sand play for stimulating creativity was discussed. The authors explored whether sand play can stimulate creative and symbolic play for children who have ASD. Two primary research questions were examined: (1) how does sand play adapted as a semi-structured creative intervention support the developmental skills of children with autism spectrum disorder in the areas of communication, socialization, and symbolic elaboration?; (2) how do children with ASD with different levels of impairment respond to the sand play intervention? The study took place over a 10-week period and the sample was taken from four different special needs classrooms from a school in Montreal, Canada. Lu *et al.*, found that the children who experienced the sand play treatment demonstrated increased verbal expression, greater social interaction, and increased novel and symbolic play. They suggested that creative interventions could be complementary to behavior and socially based intervention models for children with ASD. This study also offers insight into possible materials and methods that might help stimulate creativity and symbolic thought for children with ASD and their role for autistic individuals.

Art Therapy School-Based Programs

At least three articles are recommended as resources for model school-based art therapy programs that provide services to children and adolescents, including those with autism.[23–25] Isis, Bush, Siegel, and Ventura[23] described a successful program in Miami-Dade County, Florida; Lister, Tanguay, Snow, and D'Amico[24] presented information about a creative arts therapies center for people with developmental disabilities in Montreal, Canada; and, a creative arts therapy program in Jersey City public schools, New Jersey, was discussed by art therapist Cindy Lou Nelson.[25]

This topic has also been addressed in the United Kingdom.[26,27] In 2003, Osborne[26] examined art therapy's suitability as a treatment for children with ASD in Berkshire, UK, and discussed why art therapy was not more prevalent in the school systems. Osborne asserted that art therapy offers a common ground between the teacher and student through non-verbal means, and that it provides a mechanism for visual communication. Despite its benefits, some of the difficulties facing art therapy in the UK school system included funding, resources, and limited ability to establish a therapeutic space. More recently, Takeda[27] reported that art therapy has made strides within UK educational settings. This trend is attributed to the introduction of the National Service Frameworks (NSF) by the Department of Health in 2004, resulting in an increase in therapy services offered by schools in the United Kingdom.

A study by Epp[28] examined a social skills program, "Super Kids," in Ridgefield, Connecticut, which utilized art therapy and group therapy for children with autism ages 11 through 18. Art therapy is described as providing a non-threatening environment where the visual thoughts of autistic individuals can be explored with the support of a credentialed professional art therapist. Epp used pre- and post-treatment measurements from instruments distributed to the parents and teachers of the children. The study reports that the "Super Kids" program that involved group interventions and art therapy showed improvement in problem behaviors as measured by the Social Skills Rating System (SSRS).[29] While this offers possible evidence supporting the effectiveness of art therapy treatment, some limitations should be noted. The sample was one of convenience from an upper middle class community and there was a relatively small sample size.[28] Another includes the complexity of the treatment program. The complexity makes it difficult to accurately determine which treatments or what combinations of treatments are causing the change in a child's perceived behavior. Nonetheless, the study results offer some support as to the benefits of development and introduction of creative therapies for individuals with ASD.

Family Art Therapy and Autism

Family art therapy is an effective treatment for children with autism spectrum disorders and their families.[30] Several authors have investigated the effects of ASD on the family unit,[31–36] and the role of family art therapy for individuals affected by autism.[37–41]

While ASD affects the individual with the disorder, others in the family are often directly affected as well. Norton and Drew[34] established the possible family stressors that may result from raising a child with an ASD, such as issues related to communication, unpredictable behavior, difficulties with change, and financial obligations. Through the use of meta-analysis, Shiechel[36] explored the variables that should be considered in treating families with children who have an ASD. Shiechel found that paternal and maternal psychopathology was different between patients with disabled and non-disabled children. Furthermore, Shiechel concluded that siblings of children with disabilities are significantly affected by the disorder or cognitive disability.

Due to the pervasive nature of the potential stressors previously discussed, it has been shown that some families develop adaptive and maladaptive coping mechanisms. Altiere and Kluge[31] studied the functioning and coping behaviors of parents of children with ASDs. They examined variables such as adaptability, cohesion, and social support within the family unit. Their extensive review of the literature revealed that an enmeshed style was most adaptive for families who have children with ASD. They found that as levels of cohesion increase, there is an increase in the use of positive coping mechanisms by the family unit. This suggests that interventions such as art therapy that deal with cohesion and family dynamics might have an important role in the treatment of families who are directly affected by an ASD.

Based on his review of past and present literature, Hillman[33] suggested that children with ASD and their families are under-researched. According to Hillman, psychologists play an important role in treating families affected by ASD because they can provide information about treatments, can offer assistance in overcoming barriers, and can provide support to family members. Osbourn and Scott[35] offered information regarding the diagnosis, treatment, and role of behavioral health for individuals affected by ASD. They suggested that early interventions, family support, and overall behavioral health are important in the treatment plan for a child diagnosed with autism.

Gabriels[38] observed notable perceived benefits of art therapy for autistic individuals and their family members. Art therapy may offer the autistic child a place to build foundational skills, cognitive skills, and social communication skills. Furthermore, multifamily art therapy groups provide support and coping strategies because of the involvement of members who are experiencing similar hardships. Gabriels explored multiple case studies from family art therapy sessions and suggested directives such as a family coat of arms, art directives that involve familiar materials, and family portraits.

As noted in the previously mentioned studies, the primary method of data collection on perceived efficacy of treatments has been through surveys and interviews with parents of individuals with autism and clinicians. However, parental perception of change may not accurately reflect what the individual with autism is actually experiencing or feeling in response to treatment. Specifically, it is common for autistic individuals to have difficulties with communication[42]; therefore they may not be able to fully convey in words what they are experiencing. Art therapy assessment may provide a solution to this issue, as such an approach can serve as an alternate mode of communication for people with ASDs about their treatment experiences. This topic is discussed forthwith.

Art Therapy Assessment & Autism

When an art therapist is required to compose IEP goals and objectives for a child, or treatment goals for an adult with autism, a standardized art therapy assessment can prove to be a useful tool. It can help: (1) identify how the client views him or herself and his or her environment, (2) determine the client's strengths, (3) facilitate understanding and communication about the client among the treatment team, (4) determine the client's readiness for art therapy services, (5) determine progress, and/or (6) prepare a discharge summary.

The Face Stimulus Assessment (FSA),[43] was developed specifically for the purposes delineated above. Instruments such as the FSA provide the clinician with a standardized way to compare a client's artwork to norms and act as a baseline for comparison of the client's own previous work, for instance to help determine treatment progress. This will be illustrated by some examples, following a discussion of the FSA's development.

The Face Stimulus Assessment (FSA)[43] is a projective drawing assessment used with children and adolescents with disabilities, specifically those who are nonverbal. Materials for the FSA include: a standard packet of 8 Crayola markers; a packet of 8 Crayola Multicultural markers; Stimulus picture #1 (a printout of the 8 ½ × 11″ FSA Template 1 depicting the complete face stimulus); Stimulus picture #2 (a printout of the 8 ½ × 11″ FSA Template 2 depicting the outline of the face and neck); Stimulus picture #3 (optional: one 8 ½ × 11″ sheet of blank white paper). (Note: An e-packet with the copyrighted FSA templates and complete instructions for the FSA's standardized administration procedures is available from the first author of this chapter).

Betts'[43] rationale for using a face as a stimulus in the FSA was based upon Alley,[44] who explored the social and applied aspects of face perception. Specifically, the human face attracts and captures our attention from infancy. In addition, the face is considered a primary body region associated with emotional expression and individual identity. Betts' article presents six case examples to support her development of the FSA. Hamilton[45] furthered the development of the FSA by working with Betts to establish a rating system for this art-based assessment. In Hamilton's study, the FSA was administered to 30 undergraduate students, and was scored using nine adapted scales from the Formal Elements Art Therapy Scale: The Rating Manual (FEATS).[46] Statistically significant findings indicated that this scoring system was effective with drawing #2 of the FSA with a normative adult population, which supports the use of the FSA in treatment and research contexts. The FSA is further described in Betts,[47] which includes two case study examples.

Most recently, Mattson and Betts,[48] present two additional studies that are currently being conducted with the FSA. As previously described in Betts,[49] Soon Ran Kim[50] conducted a large scale FSA study in South Korea. Among 921 potential participants identified from the elementary school system, 296 non-normative and 240 normative student drawings were analyzed. The non-normative sample was comprised of 144 students with a developmental disorder and 152 with a hearing impairment. Results showed that the FSA evaluation method developed in this study was able to discriminate between the normative and non-normative groups. The author concluded that the FSA evaluation method she employed demonstrated success in evaluating the cognitive abilities of the children with developmental disabilities and hearing impairment.

Similarly, Betts, Schmulevich, Hu, Kelly, and Choi[51] implemented the FSA in a study on the medication responsiveness of individuals with autism in the U.S. Preliminary results of the FSA drawing analysis in the Betts *et al.* study have shown the ability of many of the FSA rating scales to reflect the developmental level and cognitive abilities of individuals with autism. These data, derived from 57 subjects, are summarized in Table 1.

The FSA: Case Examples

These two FSA series (Figures 1 and 2) were completed by Frank (pseudonym), a 20-year-old African-American male with Severe Mental Retardation, Pervasive Developmental Disorder NOS, and Expressive Language Disorder. Frank's FSA drawings reflect his ability to recall and acknowledge the presence of facial features. It is instructive to compare Frank's baseline FSA, completed at the beginning of art therapy treatment (Figure 1), to a second FSA he completed, 8 months later. In the third drawing of his first FSA series (Figure 1), Frank made a controlled scribble. Eight months later, he drew a contained circle (Figure 2), exemplifying increased fine motor control and cognitive-perceptual skill.

Figures 3 and 4 are the FSA drawings of Zach (pseudonym), a 20-year-old African-American male, with severe mental retardation

Table 1. Scales associated with specific diagnostic categories as described in the FSA rating manual, 2nd edition.[59] The following information is based on current knowledge: more research is needed to further validate the scales and identify implications of each scale

Scale	Autism Data (American Children & Adolescents) (Betts *et al.*, 2013; avg. Scores, N = 57)	Normative Data (American Adults) (Hamilton, 2008; avg. Scores, N = 30)
Prominence of Color	3	4
Color Fit	2	4.5
Implied Energy	2	4
Logic	2.5	4
Realism	1.5	3
Developmental Level	2	4
Details of Objects and Environment	1	4
Line Quality	3	4
Perseveration	4	5 (none)

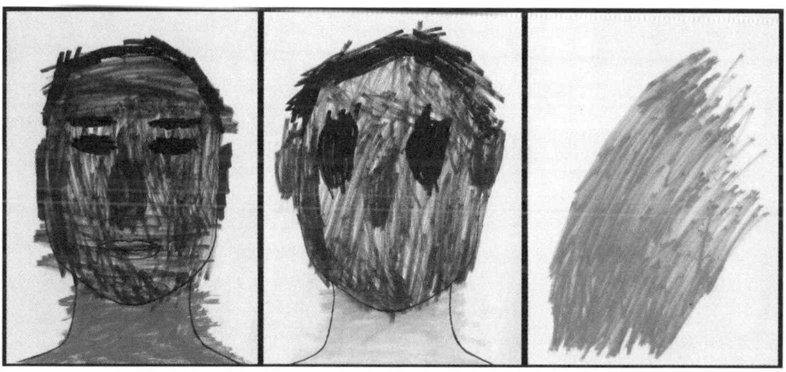

Figure 1. Frank's first Face Stimulus Assessment Series.

and autism. Zach was non-verbal. A difference is noted between his two FSA series. In the baseline series (Figure 3), Zach's third drawing depicts circular shapes, whereas in the follow-up series completed 8 months later (Figure 4), his third drawing shows Zach's

Figure 2. Frank's second Face Stimulus Assessment Series, completed 8 months later.

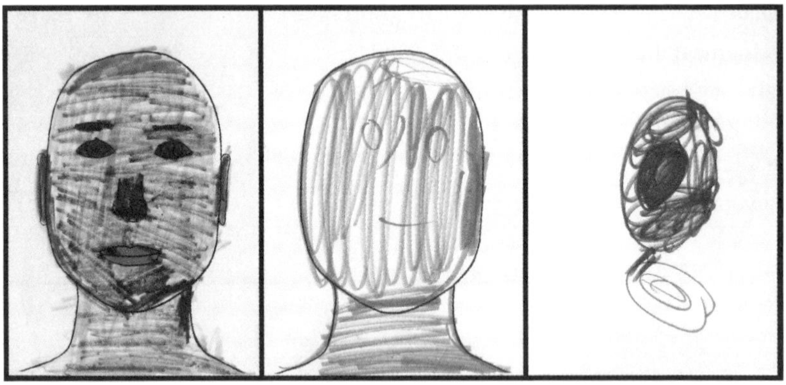

Figure 3. Zach's first Face Stimulus Assessment Series.

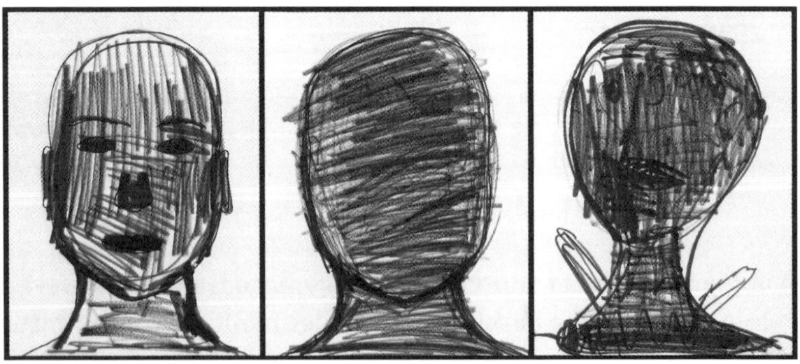

Figure 4. Zach's second Face Stimulus Assessment Series, completed 8 months later.

attempt at reproducing a human form, reflecting increased cognitive-perceptual skill.

These two case vignettes provide a sense of how standardized art therapy assessments can, among other benefits, help determine change over time. Other authors have also explored the topic of art therapy assessments that are applicable for people with ASD. Silver[52] considered how stimulus drawings can be used to express fantasies, cognitive skills, and emotional needs, especially for those who are nonverbal or have language impairments. Silver presented years of her research, including both qualitative and quantitative studies, on the use of stimulus drawings with a broad range of populations. She explained benefits of standardized art therapy assessments, such as how different individuals perceive the same drawing task differently, pointing out that each person's drawings can thus provide insight into their individual experiences and feelings.

Oster and Gould[53] proposed that learning how to use drawings for diagnostic evaluations can offer many opportunities for clinicians to address problems that may not otherwise be discussed or seen through routine verbal examinations. They presented research on the use of drawings in clinical settings as well as applications of art therapy assessments. Several studies are cited showing that drawings can be a primary source for measuring current level of functioning in individuals, as well as a relevant method of expressing emotional and interpersonal problems. In addition, they argued that drawings can be used as retest measures to assess progress or deterioration of functioning over time. The authors noted, however, that there is a lack of information regarding the reliability and validity of such assessments, and underscored the necessity of more research.

The following studies have various limitations in that they lacked standardized implementations of the assessments and that the rating systems used had not been established as valid. Thus, the data from these studies may not be generalizable. These limitations aside, the information presented in the following studies illustrates how art therapy assessments incorporating the human figure or portraits can be well suited for individuals with autism.

For example, Lee and Hobson[54] explored the differences in human figure drawings of autistic individuals compared to non-autistic individuals with learning disabilities. The authors were inspired by previous research demonstrating that children with autism have deficits in social engagement and personal relatedness. Fourteen children with autism were matched with 14 non-autistic children with learning disabilities, and asked to complete both a human figure drawing as well as a house drawing. Drawings were rated using McCarthy's[55] standardized scoring procedures, and results revealed that children with autism showed more differentiation in their drawings of different houses, as compared to their drawings of human figures, whereas the reverse was found for the non-autistic group. This study was limited in that the sample was small and that there was no set time limit for the participants to complete the drawings, but has interesting implications nonetheless.

Lim and Slaughter[56] conducted a study on the human figure drawings of children with Asperger's syndrome. The authors were inspired by previous studies, which had found that the human figure drawings of autistic children were less sophisticated in detail than neurotypical children. Twenty-nine children with Asperger's were matched with 28 neurotypical children on gender, age, and nonverbal IQ. The drawing assessment involved a free drawing, and then drawings of a person, house, and tree. Drawings were rated using Buck's House-Tree-Person scoring system,[18] and Koppitz' human figure scoring system.[57] Results showed that the children with Asperger's produced less sophisticated human figure drawings than typically developing children, in contrast to their house and tree drawings, which were equally as sophisticated as those by the control group.

Martin[58] assessed the portrait drawings of 25 children and adolescents with autism, and 15 neurotypical children. The individuals were asked to draw the researcher's face. During the procedure, 13 drawing characteristics and 17 behaviors were noted as they completed the task. This study was limited by its small sample size, lack of random selection, and lack of control for age. Despite these limitations, the study yielded interesting results. Participants with ASD were more engaged in the drawing process than were the neurotypical children. In addition, portrait drawing was found to be a successful method for ASD individuals to help them engage in relationships as well as a method for developing face recognition and processing skills.

Future research with autism and art therapy assessments should continue to examine what artistic elements are common to people with ASD. Betts has made some progress on this endeavor with the Face Stimulus Assessment.[43] One direction of relevance points to an exploration of *Graphic Equivalents of Diagnostic Criteria for Autistic Disorder*.[59] The concept of "Graphic Equivalents of Symptoms" was introduced by Gantt and Tabone.[46] In their Formal Elements Art Therapy Scale (FEATS) manual, Gantt and Tabone described how the FEATS scales are "logically derived from the core symptoms of a particular disorder" (p. 25). However, Gantt and Tabone cautioned that "readers must not assume that they are statistically correlated in any way. This is a question for future research" (p. 25). Thus, based on Gantt and Tabone's concept, Table 2 shows how the FSA rating scales line up with diagnostic criteria for Autistic Disorder. However, more research needs to be done to determine whether or to what extent the FSA scales might (or might not) actually correlate with specific criteria for autism. Such research would involve confirmation of an individual's diagnostic criteria through a valid and reliable source, such as the Autism Diagnostic Observation Schedule (ADOS),[60] and the Autism Diagnostic Interview-Revised (ADI-R).[61] The ADOS and the ADI-R, used in combination, are seen as the "gold standard" in the diagnostic investigation of autism spectrum disorders (ASD) (p. 88).[62]

Future Directions: Art Therapy Research & Autism

This chapter has presented information on the challenges and opportunities faced by art therapists in working with people who have autism, in the realms of treatment, assessment, and research. If art therapist clinicians and researchers build upon the

Table 2. Graphic equivalents of DSM diagnostic criteria for autistic disorder

DSM IV-TR Diagnostic Categories for 299.00 Autistic Disorder	**DSM IV-TR Diagnostic Criteria for 299.00 Autistic Disorder**	**Observations in the Art Therapy Literature (*Martin, 2009; see Appendix A*)**	**FSA Rating Scales Implicated (*more research needed to substantiate*)**
Impairment in social interactions	a. marked impairment in the use of multiple nonverbal behaviors such as eye-to-eye gaze, facial expression, body postures, and gestures to regulate social interaction b. failure to develop peer relationships appropriate to developmental level c. a lack of spontaneous seeking to share enjoyment, interests, or achievements with other people (e.g., by a lack of showing, bringing, or pointing out objects of interest) d. lack of social or emotional reciprocity	Atypical artistic development Schema stalemate	#2 Color Fit #5 Realism #6 Developmental Level #7 Details of Objects and Environment
Impairment in communication	a. delay in, or total lack of, the development of spoken language (not accompanied by an attempt to compensate through alternative modes of communication such as gesture or mime) b. in individuals with adequate speech, marked impairment in the ability to initiate or sustain a conversation with others c. stereotyped and repetitive use of language or idiosyncratic language d. lack of varied, spontaneous make-believe play or social imitative play appropriate to developmental level	Atypical artistic development Difficulty with imaginative (non-veridical) representations	#2 Color Fit #5 Realism #6 Developmental Level #7 Details of Objects and Environment

(Continued)

Table 2. *(Continued)*

DSM IV-TR Diagnostic Categories for 299.00 Autistic Disorder	DSM IV-TR Diagnostic Criteria for 299.00 Autistic Disorder	Observations in the Art Therapy Literature (*Martin, 2009; see Appendix A*)	FSA Rating Scales Implicated (*more research needed to substantiate*)
Restricted repetitive and stereotyped patterns of behavior, interests, and activities	a. encompassing preoccupation with one or more stereotyped and restricted patterns of interest that is abnormal either in intensity or focus b. apparently inflexible adherence to specific, nonfunctional routines or rituals c. stereotyped and repetitive motor manners (e.g., hand or finger flapping or twisting, or complex whole-body movements) d. persistent preoccupation with parts of objects	Color sequencing Tactile defensiveness High detail Cramming; Impulsive correction/completion; cataloging; graphic perseveration; motor perseveration; schema stalemate	#1 Prominence of Color #3 Implied Energy #4 Logic #7 Details of Objects and Environment #8 Line Quality #9 Perseveration
Mental Retardation	Profound: IQ below 20 Severe: 20–34 Moderate: 35–49 Mild: 50–69 Borderline intellectual functioning: 70–84	Atypical artistic development Impulsive correction/completion	#5 Realism #6 Developmental Level

knowledge base established in the previous literature, then future clients with ASDs will benefit. As such, a brief exploration of future directions and recommendations is warranted.

In terms of challenges, Martin[63] identified several interrelated factors that impede art therapists from achieving a more significant role in the treatment of individuals with autism. She offered some solutions. Art therapists could develop and expand art therapy's research to meet the current expectations of the greater autism community, and pursue an education in autism beyond the scope of art therapy. The present authors also suggest that outreach efforts would improve clients' access to art therapy services. This would include efforts undertaken by the American Art Therapy Association in furthering relationships with autism organizations, such as they have done with the Autism Society of America in the joint creation of an Art Therapy & Autism Toolkit.[64] If art therapy services are sought, a registered and board-certified art therapist should be consulted and hired (see also www.arttherapy.org and www.atcb.org).

Kaiser and Deaver[5] conducted a Delphi study of American art therapy researchers to determine research priorities for the profession. Results identified "quantitative experimental designs that test the effectiveness of art therapy interventions compared to other forms of treatment" as the most important area of research (p. 119). Such studies are exemplified in Lu *et al.*'s[22] action research with 25 subjects, and Saneei and Haghayegh's[65] comparison of 30 participants with autism to 30 normative subjects. The literature would benefit from further quantitative research featuring large subject groups, multi-site or replicated studies, and so forth. In terms of emerging and expanding research priorities for advancing the profession, Kaiser and Deaver's[5] paper cited art therapy and neuroscience as an imperative. People with autism were prioritized as one of the most important client populations to investigate.

In a review of intervention studies on young children with autism, Kassari[66] reported elements that are essential in creating well-designed treatment studies, including randomization/matching procedures and more reliable outcome measures, or measures of change. In addition, the author emphasized the importance of increasing the knowledge on treatment efficacy, by explaining the difficulty in predicting which children need what type and what level of intensity of treatment based on their developmental behaviors. There is a difficulty in comparing the success of different treatments because not all of them are measuring the same variable of change. Most comprehensive studies have relied on pre- and post-measures of change, however Kassari explains that these are rather limited, because change that is occurring over the course of treatment is not being measured.

Neuroscience, Art Therapy and Autism

"Making art stands to give form and meaning to this world in ways that no words can describe" (p. 120).[67] Temple Grandin[32] explained that she, as a person on the spectrum, "thinks in pictures" (p. 15), and is particularly sensitive to sensory stimuli. The possibility that structural differences in the brain might be present in individuals with ASD has been explored through neurological testing. In Grandin's case, her brain is structurally unique in areas that analyze sensory stimuli, and this finding has significant implications for neuroscience research. Furthermore, the new DSM-V[68] has generated much controversy, which has resulted in discussions about the need for future research to focus on neurological and biological underpinnings of disorders.[69]

The burgeoning field of neuroscience offers ample opportunities for advancement of art therapy research, and several authors have begun to explore this promising area.[70–77] Advancements in understanding the brain, especially its implications in autism, provide a ripe area for investigation. Increased understanding of the structure of the brain and its functions, and elucidation of how the process of art-making can influence neural pathways, could lead to improved outcomes of art therapy as an effective treatment for autism.

Executive Summary

- Art therapists are master's level, credentialed and board-certified mental health clinicians who are highly qualified to address the assessment and treatment needs of people with autism.
- Art therapy is widely recognized as beneficial in the treatment of people with autism, particularly in school settings where goals and objectives related to physiological and psychosocial development can be successfully addressed.
- Informing the public and stakeholders, as well as increasing the research base would further enhance awareness of art therapy's benefits for people with autism, and in turn, increase their access to art therapy assessment and treatment services.
- The literature has explored the intersection of art therapy and neuroscience, an area of particular relevance in the further investigation of how art therapy helps people with autism.

References

1. Martin N, Betts DJ: Art therapy approaches to treating autism. In *Cutting Edge Therapies for Autism, 2011–2012.* 2nd ed. Edited by Siri K, Lyons T. New York, NY: Skyhorse Publishing, Inc.; 2011:56–60.
2. American Art Therapy Association: What is art therapy? http://www americanarttherapyassociation org/aata-aboutus html 2013a.
3. American Art Therapy Association: Who benefits from art therapy? http://www americanarttherapyassociation org/aata-aboutus html 2013b.
4. Elkins DE, Deaver SP: American Art Therapy Association, Inc.: 2009 Membership survey report. *J Am Art Ther Assoc* 2010, **27**(3):141–147.
5. Kaiser D, Deaver SP: Establishing a research agenda for art therapy: a Delphi study. Art Therapy: *J Am Art Ther Assoc* 2013, **30**(3):114–121.
6. Evans K, Dubowski J: *Art Therapy with Children on the Autistic Spectrum: Beyond Words*. 2001, Philadelphia, PA (Jessica Kingsley Publishers).
7. Martin N: *Art as an Early Intervention Tool for Children with Autism:* Philadelphia, PA: Jessica Kingsley Publishers; 2009a.
8. Seltzer MM, Krauss MW, Shattuck PT, Orsmond G, Swe A, Lord C: Symptoms of Autism Spectrum Disorders in adolescence and adulthood. *J Autism Dev Disord* 2003, **33**(6):565–581.
9. Siri K, Lyons T: *Cutting-edge Therapies for Autism:* New York, NY: Skyhorse Publishing; 2011.

10. Bowker A, D'Angelo NM, Hicks R, Wells K: Treatments for autism: parental choices and perceptions of change. *J Autism Deve Disord* 2011, **41**(10):1373–1382.
11. Goin-Kochel RP, Myers BJ, Mackintosh VH: Parental reports on the use of treatments and therapies for children with autism spectrum disorders. *Res Autism Spectrum Disord* 2007, **1**(3):195–209.
12. Green V: Parental experience with treatments for autism. *J Dev Physical Disabilities* 2007, **19**:91–101.
13. Weed R: An exploration of identity: art and autism. *Unpublished doctoral dissertation* 2006, Iowa City, Iowa (The University of Iowa).
14. Elkis-Abuhoff DL: Art therapy applied to an adolescent with Asperger's Syndrome. *The Arts in Psychotherapy* 2008, **35**(4):262–270.
15. Wechsler D: *WAIS-III Administration and Scoring Manual:* San Antonio, TX: Psychological Corporation; 1997.
16. Murray HA: *Thematic Apperception Test:* Cambridge, MA: Harvard University Press; 1943.
17. Burns RC: *Kinetic-House-Tree-Person Drawings (K-H-T-P):* New York, NY: Brunner/Mazel; 1987.
18. Buck JN: The HTP technique: a qualitative and quantitative scoring manual. *J Clin Psychology* 1948, **4**:317–396.
19. Burns RC, Kaufman SH: *Kinetic Family Drawings (KFD): An Introduction into Understanding Children through Kinetic Drawing:* New York, NY: Brunner/Mazel; 1970.
20. Emery M: Art therapy as an intervention for autism. *J Am Art Ther Assoc* 2004, **21**(3):143–147.
21. Brancheau D: Art therapy as a treatment choice for autism spectrum disorders. In *Using Art Therapy with Diverse Populations: Crossing Cultures and Abilities.* Edited by Howie P, Prasad S, Kristel J. Philadelphia, PA: Jessica Kingsley Publishers; 2013:143–152.
22. Lu L, Petersen F, Lacroix L, Rousseau C: Stimulating creative play in children with autism through sandplay. *The Arts in Psychotherapy* 2010, **37**(1):56–64.
23. Isis PD, Bush J, Siegel CA, Ventura Y: Empowering students through creativity: art therapy in Miami-Dade County public schools. *J Am Art Ther Assoc* 2010, 27(2):56–61.
24. Lister S, Tanguay D, Snow S, D'Amico M: Development of a creative arts therapies center for people with developmental disabilities. *J Am Art Ther Assoc* 2009, 26(1):34–37.
25. Nelson CL: Meeting the needs of urban students: creative arts therapy in Jersey City public schools. *J Am Art Ther Assoc* 2010, **27**(2):62–68.
26. Osborne J: Art and the child with autism: therapy or education? *Early Child Dev Care* 2003, **173**(4):411–423.
27. Takeda F: The capacity for imagination: Implications for working with children with autism in art therapy. In *Arts Therapies in Schools: Research and*

Practice. Edited by Karkou V. Philadelphia, PA: Jessica Kingsley Publishers; 2009:217–230.

28. Epp K: Outcome-based evaluation of a social skills program using art therapy and group therapy for children on the autism spectrum. *Children Schools* 2008, **30**(1):27–36.
29. Gresham FM, Elliot SN: *Social Skills Rating System Manual:* Circle Pines, MN: American Guidance Service; 1990.
30. Harmer R: The effectiveness of art therapy for families of children with autism. *Unpublished master's project*, 2012, George Washington University (Washington, DC).
31. Altiere M, Kluge S: Family functioning and coping behaviors in parents of children with autism. *J Child Family Studies* 2009, **18**(1):83–92.
32. Grandin T: Visual abilities and sensory differences in a person with Autism. *Bio Psychiatry* 2008, **65**(1):15–16.
33. Hillman J: Supporting and treating families with children on the autistic spectrum: The unique role of the generalist psychologist. *Psychotherapy: Theory, Research, Practice, Training* 2006, **43**(3):349–358.
34. Norton P, Drew C: Autism and potential family stressors. *Am J Family Ther* 1994, **22**(1):67–76.
35. Osbourn P, Scott F, New Mexico State Deptartment of Education: Autism Spectrum Disorders: Guidance on providing supports and services to young children with autism spectrum disorders and their families. *Technical Assistance Manual New Mexico State Department of Education* 2004.
36. Shiechel S: Effects of cognitive disabilities on family dynamics and functioning. *Unpublished doctoral dissertation* 2008, New York, NY (St. John's University).
37. Estrada A, Pinsof W: The effectiveness of family therapies for selected behavioral disorders of childhood. *J Marital Family Ther* 1995, **21**(4):403–440.
38. Gabriels R: Art therapy with children who have autism and their families. In *Handbook of Art Therapy.* Edited by Malchiodi C. New York, NY: Guilford Press; 2003:193–206.
39. Rhodes P: Behavioral and family systems interventions in developmental disability: Towards a contemporary and integrative approach. *J Intellectual Dev Disability* 2003, **28**(1):51–64.
40. Solomon M, Ono M, Timmer S, Goodlin-Jones B: The effectiveness of parent-child interaction therapy for families of children on the autism spectrum. *J Autism Dev Disord* 2008, **38**(9):1767–1776.
41. Stoddart K: Adolescents with Asperger's syndrome: three case studies of individual and family therapy. *Autism* 1999, **3**(3):255.
42. Baron GM, Groden J, Groden G, Lipsitt LP: *Stress and Coping in Autism:* New York, NY: Oxford University Press; 2006.
43. Betts DJ: Developing a projective drawing test: Experiences with the Face Stimulus Assessment (FSA). *J Am Art Ther Assoc* 2003, **20**(2):77–82.

44. Alley TR: Social and applied aspects of face perception: An introduction. In *Social and Applied Aspects of Perceiving Faces.* Edited by Alley TR. Hillsdale, NJ: Lawrence Erlbaum.; 1988:1–8.
45. Hamilton MK: Developing a standardized rating system for the Face Stimulus Assessment (FSA) using 9 scales adapted from the Formal Elements Art Therapy Scale (FEATS). *Unpublished Master's Thesis* 2008, Avila University (Kansas City, MO).
46. Gantt L, Tabone C: *The Formal Elements Art Therapy Scale: The Rating Manual:* Morgantown, WV: Gargoyle Press; 1998.
47. Betts DJ: Introduction to the face stimulus assessment (FSA). In *The Art Therapists' Primer: A Clinical Guide to Writing Assessments, Diagnosis, and Treatment.* Edited by Horovitz EG, Eksten S. Springfield, IL: Charles C Thomas; 2009:77–78.
48. Mattson D, Betts DJ: The face stimulus assessment. In *The Wiley-Blackwell Handbook of Art Therapy.* Edited by Gussak DE, Rosal ML. Hoboken, NJ: Wiley-Blackwell; in press.
49. Betts DJ: Art therapy assessments with diverse populations. In *Using Art Therapy with Diverse Populations: Crossing Cultures and Abilities.* Edited by Howie P, Prasad S, Kristel J. Philadelphia, PA: Jessica Kingsley; 2013a:41–55.
50. Kim SR: A study on development of FSA evaluation standard and its validation. *Unpublished Doctoral Dissertation* 2010, Yeungnam University (Daegu, South Korea).
51. Betts DJ, Schmulevich G, Hu V, Kelly B, Choi J: Diagnostic criteria for autism reflected in graphic features of artwork. *Poster Session Presented at George Washington University Research Days, Washington, DC* 2013, April.
52. Silver R: Art as language: Access to thoughts and feelings through stimulus drawings: Philadelphia, PA: *Psychology Press;* 2001.
53. Oster GD, Gould P: *Using Drawings in Assessment and Therapy: A Guide for Mental Health Professionals:* New York, NY: Brunner-Routledge; 2004.
54. Lee A, Hobson PR: Drawing self and others: How do children with autism differ from those with learning disabilities? *Brit J Dev Psychology* 2006, **24**:547–565.
55. McCarthy D: *McCarthy Scales of Children's Abilities:* New York, NY: Psychological Corporation; 1972.
56. Lim HK, Slaughter V: Brief report: human figure drawings by children with asperger's syndrome. *J Autism Dev Disabilities* 2008, **38**:988–994.
57. Koppitz EM: *Psychological Evaluation of Children's Human-Figure Drawings:* New York, NY: Grune & Stratton; 1968.
58. Martin N: Assessing portrait drawings created by children and adolescents with Autism Spectrum Disorder. *J Am Art Ther Assoc* 2008, **25**(1):15–23.
59. Betts DJ: *The Face Stimulus Assessment (FSA) Rating Manual:* 2nd ed. Washington, DC: Department of Art Therapy, George Washington University; 2013b.

60. Lord C, Rutter M, DiLavore P, Risi S: *Autism Diagnostic Observation Schedule (ADOS):* Los Angeles, CA: Western Psychological Services; 1999.
61. Rutter M, Le Couteur A, Lord C: *Autism Diagnostic Interview-Revised:* Los Angeles, CA: Western Psychological Services; 2003.
62. Kamp-Becker I, Ghahreman M, Heinzel-Gutenbrunner M, Peters M, Remschmidt H, Becker K: Evaluation of the revised algorithm of Autism Diagnostic Observation Schedule (ADOS) in the diagnostic investigation of high-functioning children and adolescents with autism spectrum disorders. *Autism* 2013, **17**(1):87–102.
63. Martin N: Art therapy and autism: overview and recommendations. *J Am Art Ther Assoc* 2009b, **26**(4):187–190.
64. American Art Therapy Association, Autism Society of America: *Art Therapy & Autism Spectrum Disorder: Integrating Creative Interventions (Toolkit):* Alexandria, VA: Authors; 2012.
65. Saneei A, Haghayegh SA: Family drawings of Iranian children with autism and their family members. *The Arts in Psychotherapy* 2011, **38**(5):333–339.
66. Kassari C: Assessing change in early intervention programs for children with autism. *J Autism Dev Disord* 2002, **5**:447–461.
67. Henley D: Annihilation anxiety and fantasy in the art of children with Asperger's Syndrome and others on the autism spectrum. *Am J Art Ther* 2001, **39**:113–121.
68. American Psychiatric Association: *Diagnostic and Statistical Manual of Mental Disorders:* 5th ed. Washington, DC: Author; 2013.
69. Belluck P, Carey B: Psychiatry's guide is out of touch with science, experts say. *The New York Times Retrieved from* http://www.nytimes.com 2013, May 6.
70. Belkofer CM, Konopka LM: Conducting art therapy research using quantitative EEG measures. *J Am Art Ther Assoc* 2008, **25**(2):56–63.
71. Century M, Hustvedt S, Pelli D, Scott J, Wiley KC, Levy EK: Neuroscience and the arts today. *PAJ* 2013, **35**(3):8–23.
72. Gantt L, Tinnin L: Support for a neurobiological view of trauma with implications for art therapy. *The Arts in Psychotherapy* 2009, **36**:148–153.
73. Kaufman SB: The real neuroscience of creativity. http://blogs.scientificamerican.com/beautiful-minds/2013/08/19/the-real-neuroscience-of-creativity/2013, August 19.
74. Lusebrink VB: Art therapy and the brain: An attempt to understand the underlying processes of art expression in therapy. *J Am Art Ther Assoc* 2004, **21**(3):125–135.
75. Lusebrink VB: Assessment and therapeutic application of the Expressive Therapies Continuum: implications for brain structures and functions. *J Am Art Ther Assoc* 2010, **27**(4):168–177.

76. Lusebrink VB, Mārtinsone K, Dzilna-Šilova I: The Expressive Therapies Continuum (ETC): Interdisciplinary bases of the ETC. *Int J Art Ther* 2012, **18**(2):75–85.
77. McNamee CM: Experiences with bilateral art: a retrospective study. *J Am Art Ther Assoc* 2006, **23**(1):7–13.
78. Selfe L: *Normal and Anomalous Representation Drawing Ability in Children.* London: Academic Press, 1983.
79. Craig J, Baron-Cohen S, Scott F: Drawing ability in autism: A window into the imagination. *Israel J Psychiatry* 2001, **38**: 242–253.
80. Rimland B: Savant Capabilities of autistic children and their cognitive implications. In G. Serban (Ed.) *Cognitive Defects in the Development of Mental Illness.* New York, NY: Brunner/Mazel, 1978.

Appendix A: Martin's (2009) characteristics of artwork by children with autism: Definitions of terms

Atypical artistic development: a child's growth is not following the typical trajectory and might have unusual or out-of-order features. This may include skipping the preschematic stage of drawing (moving from unintelligible scribbles to recognizable figures with little or no experimental stage in between), unconventional choice of subjects,[78] or precocious drawing ability (p. 49).

Cataloging: involves creating a visual catalog of interests, often with some relationship to each other (e.g., drawing different models of trains, different species of beetles). Characteristics in the order and organization category seem to be related to perseveration and are often self-soothing and satisfying to perform (pp. 51–52).

Color sequencing: a compulsive need to use colors in a rigid order (p. 51).

Cramming: a tendency to focus on task completion rather than quality of composition. The child may begin drawing with little attention to placement and then run out of room on the paper and "cram" in the remaining features (pp. 49–51).

Difficulty with imaginative (non-veridical) representations: drawing something that does not exist in reality is a challenge for many children with ASD (p. 54).[79]

Graphic perseveration: a persistent theme or topic in the child's art-work (e.g., a tv show character, cars, horses, rainbows). In ASD, this is often at the expense of having other interests, thus reflecting inflexible and compulsive behavior. It might be related to either imagination deficits or the need for order and organization (p. 54).

High detail: children who draw in this manner appear unable to filter out extraneous detail and compulsively include it in their artwork; these children usually develop reputations as artists. This behavior is often described as hyperfocus or "high fidelity attention".[80] The need for high detail may be related to perceptual/sensory issues, perseveration and obsessive-compulsive behavior, or the need for order and organization, particularly if it involves drawing maps or floor plans (p. 51).

Impulsive correction/completion: a low threshold of tolerance for any perceived omission on a drawing or the impulse to self-correct or to make corrections (with or without permission) on others' drawings. For children with ASD, this seems to persist past a developmentally appropriate age and might also be related to obsessive-compulsive feelings or visual-spatial deficits (pp. 50–51).

Schema stalemate: when the development of the child's schemas stalls out. Once the child has determined how to draw "person," "house," and so on, minimal if any changes are made to these schemas without intervention. This is possibly related to lack of experimentation or social deficits (pp. 54–55).

Tactile defensiveness: a sensitivity to touch, including pressure, temperature or textures. This may impact their level of comfort in using certain art materials (p. 57).

Chapter 26

Shifting Paradigms: An Examination of Our Understanding of Adult Autism

Rhonda J. Greenhaw*

Abstract

This chapter explores the current status of adults on the autism spectrum and examines their historical and current status as well as future trends of this population. Discussed is the fact that individuals on the spectrum have been approached as deficient and as needing to be 'fixed,' and that this may not be the best route to habilitation. The current status of adults on the spectrum is reviewed, much of which is troubling in terms of social indicators, but a new *social model* approach to adults on the spectrum is investigated in which the competence of individuals on the spectrum is presumed. This model may hold much promise as an approach for creating inclusive opportunities for adults in employment, education and housing.

Keywords: Autism spectrum disorder; Adult autism; Ableism; Social model of disability; Medical model; Employment of people with disabilities; Self-advocacy; Social impacts of autism; Transition-aged youth; Neurodiversity; Presuming competence.

*Correspondence to: Office of Collaborative Programs and Hussman Center for Adults with Autism, Towson University College of Health Professions, 8000 York Road, Towson, MD 21252. Email: rgreenhaw@towson.edu.

Introduction

In the 1940s, when Leo Kanner first began describing autism in his lab at Johns Hopkins, it was unclear how autism impacted an individual's lifespan.[1] Fast forward to 2010 and an article published in *The Atlantic* provides a glimpse into the life of Donald Gray Triplett, the first child that Kanner diagnosed with autism. Triplett was 77 years old at the time of the article.[2] He had learned to drive a car and found activities that created meaning in his life. Many of Triplett's peers with autism, however, spent their lives in institutions, many misdiagnosed with schizophrenia or other mental health conditions. Fortunately, current life circumstances for many individuals on the autism spectrum have greatly improved.

Today, there are many high-profile examples of individuals with autism who live typical life spans and participate in meaningful activities, perhaps most notably, professor, animal behaviorist, public speaker, advocate and author, Dr. Temple Grandin, who is 65. In addition to Grandin, there is a growing group of very visible individuals on the autism spectrum, such as author and speaker John Elder Robison, now 56 years old, author and advocate Donna Williams (age 50), and Larry Bissonnette (age 55), who starred in the movie *Wretches and Jabberers*,[3] just to name a few. These adults with autism and many advocates on the spectrum across the country are changing our understanding of autism as well as the perception that autism is a just a pediatric issue.

There is a growing awareness in society that individuals with autism can live full lives as members of their communities, but improving life for all adults with autism will require critical attention and increased research on this population. It is now estimated that nearly 50,000 young people on the autism spectrum turn 18 every year and, with the increases in diagnosis described previously, that number will continue to rise.[4] Given the increase in the number of individuals diagnosed with autism, learning more about the current conditions they face as well as the challenges and needs of this population is incredibly important in order to successfully create inclusive educational options, work places, and communities. This chapter will provide an overview of the current issues facing adults on the autism spectrum, examine recent developments in the field of adult autism, and explore future areas for research.

Conditions Impacting Adults on the Spectrum

Although life experiences, education, and interactions with others shape all developing young people, individuals with autism face some unique experiences and challenges as children and adolescents, as described below.

Therapeutic and medicated childhoods

For a growing number of children with autism, their childhood is infused with a barrage of medications, therapies and services. From the time a child is identified as being on the autism spectrum, throughout the rest of their childhood and sometimes their life, they are measured and assessed, and found 'lacking' or having 'deficits'. Currently, scientists have methods of identifying children with autism at increasingly younger ages. Scientists may also be very close to developing a blood test for autism or even an *in utero* test for autism in the near future. Early identification of children and now infants with autism has been seen as vital and necessary in order to provide early intervention services for children with autism. While there are some very effective, respectful practices that use positive behavior supports, some of those practices include the use of aversive treatments and punishments as well as methodologies that focus on training the child with autism to repress their autistic tendencies.

Many adults on the spectrum argue that this approach can be damaging to the child on the spectrum.[5–7] On the Internet, one can find conversations about applied behavior analysis (ABA), for example, in which adults with autism liken these remedies to homosexual conversion therapy, which has been widely discredited.[8] Indeed Ivar Lovaas, who first developed ABA for use with children with autism, also was the principal investigator on the Feminine Boys Project, both with a shared therapeutic goal of rendering the children indistinguishable from their typically developing peers.[9] Gay conversion therapy, as it has been widely called, has been denounced by professional associations such as the American Psychiatric Association, the American Academy of Pediatrics, the American Psychological Association, and other organizations; however, while the practice of using punishments and aversives is becoming less common, it is still widely used for children with autism.[10]

There are other treatments and therapies for children with autism that may diminish their quality of life as well. For example, some children with autism are placed on rigorous and restrictive diets, may be given significant dosages of a myriad of intensive pharmaceuticals, or may participate in highly experimental, untested and potentially dangerous treatments, such as chelation, which was developed to treat life-threatening situations for victims of heavy metal exposure, and which has resulted in the death of at least one child with autism.[11] Aside from overt harmful effects, it is not clear what long-term impact the various prescribed drugs have on the physical, neurological, and psychological development of the child. In addition, even therapies that are less risky can involve many hours of the child's life. It is not uncommon for a child with autism who is receiving therapy to attend school all day and then spend the entire evening in therapy throughout the week. Many adults with autism report that these *therapeutic* experiences and the fixation on curing autism so they would resemble their typical peers were stress-inducing and created feelings of despair and self-hatred that could persist throughout their lifetime.[5,7]

Bullying, abuse and isolation

Many individuals on the autism spectrum experience bullying at a much higher rate than the general population. The Interactive Autism Network conducted a survey of families and found that 63% reported that their child with autism experienced bullying, with middle-school being reported as the time when bullying was most likely to occur.[13] Bullying has impacts on both physical and mental health, and children who are bullied are more likely to experience headaches, stomachaches, and anxiety, and are also more likely to commit suicide.[13] Young adults with autism who experience bullying throughout their school years are certainly impacted by the accumulation of those experiences over the course of their life, and can feel distrust and despair as a result.

Adults with autism are also more likely than the general population to have experienced child abuse, often at the hands of a caregiver, teacher, parent, or other family member.[14] People abused as children can be affected across multiple areas. These areas include increased stress, hyper-arousal, difficulty regulating emotions, increased dependency, increased isolation, and increased impulsivity, among others.[15]

In addition to bullying, young adults with autism are an incredibly isolated group, even more than young people with other intellectual disabilities.[16] For example, nearly 50% of adults on the autism spectrum had no contact with friends, including phone calls and invitations to participate in activities, while only 17% of young adults with intellectual disabilities experienced this level of isolation. This, in addition to the fact that children with autism are more likely to be educated in separate educational settings, contributes to an incredible amount of isolation that the individual with autism will experience over the course of his/her childhood and adolescence, which is only exacerbated after the young person ages out of the school system. Thus, challenges in social communication and understanding associated with autism are significantly compounded by social isolation. In addition, adults with autism also lack the social networking with friends and acquaintances that can provide tremendous assistance during times of crisis, such as when conducting job searches and looking for housing, or just by providing social feedback.

Co-occurring conditions

Co-occurring health and mental health issues can feature prominently in the lives of young adults with autism, creating additional complexities for the individual, families and providers. The majority of children with autism have a co-occurring developmental disorder, with a quarter of children on the spectrum having two or more conditions present.[17] The most prevalent of these conditions include language disorder, intellectual disability, Attention Deficit Hyperactivity Disorder (ADHD), sensory integration, learning disorder, oppositional defiant disorder (ODD), anxiety disorder, emotional disorder, mood disorder, obsessive compulsive disorder (OCD), depression, epilepsy, and encephalopathy.

There is also an emerging awareness among autism self-advocates and researchers that the experiences of isolation, bullying and abuse that often occurs while growing up on the spectrum can result in post-traumatic stress disorder (PTSD) persisting into adulthood.[18] Recent studies indicate that individuals on the autism spectrum may have a 28-fold greater risk for suicide than the general population.[19] These mental health conditions must be taken into consideration when treating, developing programs, or trying to develop understanding for the actual experiences of individuals with autism. Much more research needs to be conducted to learn about how the traumatic experiences of adults with autism and PTSD are manifested in the autism population in order to develop supports that can help heal those affected.

Housing

New research indicates that individuals with autism are less likely than those with learning disabilities, intellectual disabilities, or emotional disturbances, to have ever lived independently, instead living with their families for longer periods or other groups of individuals with disabilities.[20] Indeed, while some individuals with autism may be choosing to remain at home in an environment that is familiar and comfortable, other factors preventing independent living may be a lack of access to supports, such as personal assistants, and financial considerations.

There are different residential solutions for individuals on the spectrum. To understand housing options for people on the spectrum, the state of Pennsylvania conducted a survey of residential models, and determined that there were essentially seven different types of housing utilized by people with autism: remaining at home, family living, renting an apartment or home, purchasing a home, shared housing, intentional communities, and licensed facilities.[21]

President Barak Obama declared 2009 as the *Year of Community Living* for people with disabilities, promising that the U.S. Department of Justice would actively enforce the Supreme Court's decision in *Olmstead v. L.C.*[22] In the Olmstead decision of 1999, the court ruled that Title II of the Americans with Disabilities Act required states to ensure that individuals with disabilities be provided services in the most integrated manner possible. These events represent a major shift in U.S. policy towards people with disabilities, and ultimately will compel states to follow suit.

While most individuals with autism stay in the family home through adulthood, a growing problem exists — many parents of adults with autism are aging, and at some point may become unable to provide care for their adult children on the spectrum. Disability is the one minority status that anyone can acquire at any time, and people are more likely to become disabled as they age. For families with children on the spectrum who are significantly impacted and who require around the clock care and support, this can present a significant dilemma. As a growing number of children on the spectrum enter adulthood, this is an area that requires much more attention. Maintaining the ability of adults with disabilities who desire independent living solutions, while creating supports for individuals who require significant assistance and care and who desire to live with aging parents, will require our country to direct much more resources to the area of housing and residential supports to meet the growing demand.

Education

Transition-aged youth

When someone on the spectrum attends school in K-12 settings, the experience is very often managed by an individualized education plan (IEP). A multi-disciplinary team of educators, therapists, school administrators, parents, and the child develop this plan. The objective of this team is to create services that make the curricula accessible to the child and to facilitate their involvement in school and school-related experiences. Many students with an IEP attend high school until the end of the school year after they turn 21.[23] This is meant to ensure that students with disabilities are given the extra time in schools they may need to complete a diploma or a certificate, and also allows them access to support services that are not mandated at the post-secondary level. An unintended consequence of this policy, however, is that individuals with autism and other disabilities face situations in which their peers go off to college at 17 or 18, while they stay behind in the high school setting for a few more years, which may increase their isolation.

For students 16 and older, the Individuals with Disabilities Education Act (IDEA 2004) mandates that transition services are planned for within IEP meetings.[24] Transition-planning services are designed to develop the child's goals for post-secondary life, whether college, vocational training, or work. Often, however, students with autism matriculate out of the educational experience without a strong plan for post-secondary life. Although mandated, transition to adulthood is often not fully addressed in IEP meetings and plans, and individuals with disabilities, parents, and professionals are recognizing that this is an area that requires much more attention and resources.

New post-secondary academic options

Historically, people with cognitive differences have often been excluded from post-secondary education opportunities. The rate of attendance at a college or university for individuals on the autism spectrum is significantly lower than that of the general population, and even fewer ever earn a degree.[4] This reality is beginning to change, however, as more services for individuals with developmental disabilities such as autism are being created at the college level.

Many colleges and universities around the country have programs that are available to support individuals with autism attain a college experience. A variety of program models exist, including state-government sponsored programs, programs developed and managed by a college or university, and private non-profit or for profit programs. The cost for such services ranges from free programs to those that can cost over $40,000 per semester. The programs available may be a minimal service such as a campus social group for students on the spectrum, to full wrap-around services that provide intensive levels of support for students. These services, ranging from minor assistance in social interaction and communication to 1:1 aides who attend classes with the individual, can provide support for someone on the spectrum for whom the post-secondary experience may otherwise have been out of reach.

The federal government is also taking an interest in supporting the post-secondary academic involvement of individuals with developmental disabilities by providing federal financial aid for certificate programs at many colleges and universities across the country. These programs provide courses in non-traditional subjects such as life and social skills, as well as academic courses. Some programs have on-campus housing, although some of these are in segregated dorms or even in off-campus hotels, rather than in traditional student housing with peers. Students in these programs can now access financial aid from schools that have been approved as comprehensive transition and postsecondary (CTP) programs sites. Community colleges have also begun to offer more support for students on the spectrum, with a growing number of programs offering additional support beyond what is traditionally available through disability support services.

In addition to on-campus options, many individuals on the spectrum are electing to take classes online. Whether traditional or specifically designed for individuals on the spectrum, online degree or certificate programs can be a good option for individuals who have sensory, communication, or behavioral differences. Often, the environment of a college or university is a loud, noisy, confusing, and overwhelming place; for some individuals, the online programs offer another choice that may work better for them.

Post-secondary training options

Outside of academic programs, there are a number of programs that offer post-secondary training for individuals on the spectrum. These programs may address unique needs, such as language development or social interaction, or may provide non-credit bearing courses in a particular field, such as computer-based animation. Some of these programs take place on college campuses, while others are developed and offered by organizations or private businesses. A small percentage of these programs have peers who are not on the spectrum that participate alongside adults with autism. These types of programs provide important opportunities for inclusive experiences that adults on the spectrum need to gain social knowledge and community connection.

Although there are currently not enough resources to fill the growing demand for the post-secondary educational opportunities needed by the population of adults on the spectrum, this is an area that is experiencing tremendous growth as new programs are being developed across the United States and around the world. The increase in post-secondary options is very exciting, and worlds away from the days in which individuals on the spectrum were relegated to uninspiring vocational rehabilitation programs that offered them very little choice about their future direction. The promise of the new crop of post-secondary options should receive more support to address the rapidly expanding need.

Employment

Adults with autism face significant hurdles when it comes to achieving the American dream of a self-supporting job. In 2011, a team of researchers undertook a comprehensive survey to determine the state of adults on the autism spectrum. In the area of employment, the researchers found that more than half of students leaving secondary education did not find work in the two years after school, a rate higher than any other disability group in the nation.[25] Not surprisingly, the researchers found that those from lower socioeconomic backgrounds fared even worse. Other studies have also reported significantly lower rates of employment for adults on the spectrum than the general population, making this a significant issue impacting adults with autism.[26]

There are many reasons individuals on the spectrum face challenges when it comes to gaining full employment in a field of their choosing. These challenges can include stigmatization and fear of employing or interacting with individuals with neurodevelopmental differences that may be exhibited by neurotypical individuals who hold key decision-making positions, difficulty maneuvering through parts of the employment process, challenges inherent to autism such as sensory processing difficulties that certain work environments may exacerbate, communication challenges that may impact areas such as the job interview, and inadequate funding for work-based supports.[27]

Adults with autism, in the past, often participated in work programs called *sheltered workshops*. The idea of a sheltered workshop was to provide individuals who were deemed 'unemployable' with job readiness training, skill development, or just something to do during the day.[28] Sometimes these experiences are paid, but the work is often tedious and low-skilled, and compensation is often irregular and far below the minimum wage. Further, these settings are highly segregated, and can contribute to the feelings of isolation experienced by individuals with disabilities. As a consequence of these and many other issues, several states across the country are moving their support away from sheltered workshops to a system that encourages paid employment in the competitive marketplace.[28]

Individuals on the autism spectrum, in particular, may benefit from systems that move them from the make-work world of sheltered workshops to inclusive employment. In fact, a recent study indicated that individuals with autism fared worse in workshop settings, earning significantly lower wages, and costing more in support dollars in the sheltered employment environment than in a supported employment setting.[29] In addition, the sheltered workshop experience does not allow for individual differences and unique interests, areas that are vital for success for the autism population.

New models of supporting and integrating adults with autism are being developed across the country as adults on the spectrum and their families advocate for more inventive options that take advantage of their abilities and interests and rely less upon restrictive work programs. Bolstered by the research supporting these options and the growing support of state and federal governments, foundations and advocacy groups for finding people with autism paid work in inclusive settings of their choice, creative solutions are being developed for individuals across the spectrum.

Supported employment

Some individuals with autism may benefit from programs characterized as *supported employment*. These programs provide group or individualized support across several areas, including vocation assessments, resume development, job readiness skills and on-site training, as well as on-going support designed to assist the individual to keep the job once hired.[30] Often, the individual in a supported employment program is provided a *job coach* or *employment specialist* who accompanies the individual to the work place to provide assistance and support as the individual learns the tasks required for the position and becomes comfortable with the work setting. Ideally, the job coach will provide less and less support for the individual until they eventually can perform the job independently. A competent employment specialist/job coach can also assist the employee with exercising self-advocacy skills on the job, and facilitate relationships with other employees, supervisors and consumers.

Work-based learning and job trials

Unlike classroom-based training courses, work-based learning and job trials provide opportunities for individuals to learn about the job *in context*. For individuals with autism, learning about the job while performing the work provides many opportunities to enhance skill development. In work-based learning settings, another employee or a job coach can perform the task alongside the person on the spectrum, acting as a mentor and model as they increase their capacity to perform the work.

Another advantage of work-based learning and job trials is that they allow the individual on the spectrum to try the various tasks associated with the position to help determine whether that job and the environment is a good fit for them. This can address situations in which assessments designed to determine job readiness, interest or ability are not capable of providing information about the individual. For example, individuals with autism may have communication differences that render traditional assessment methods meaningless. Work-based learning provides the individual an opportunity to experience the position and provides important feedback that can inform decision-making about careers and jobs.

Work-based learning can also provide an employer with an opportunity to get to know an individual on the autism spectrum, and can alleviate stereotypes, fears, and concerns that employers or co-workers may have about the capacity of the person on the spectrum within a particular job. This may help open doors and create opportunities that may not be available to the individual on the spectrum in the traditional assessment/interview method of acquiring employment.

Customized employment

One of the main differences between competitive employment and sheltered workshops is that, in competitive employment, people are selected to work based upon the needs of the workplace, whereas in sheltered workshops, jobs are based upon the needs of the person.[28] In customized employment, job tasks may be carved out from several different jobs or reassigned. That is, the work may be organized in a unique manner to capitalize on the skills, interests, and strengths of the individual on the spectrum. For example, if a person on the spectrum has a unique talent for creating sophisticated spreadsheets, a customized employment solution might be to carve out or reassign the creation of all spreadsheets for a department to that employee, thus freeing the other employees to focus their attention on other aspects of their work. In another example of customized employment, a non-verbal adult on the spectrum may have difficulty performing all of the functions of a retail worker or may have behavior challenges that make it difficult for him to interact with customers, but he might be hired to organize a storeroom, capitalizing on an interest in organizing and putting things away.

The concept of customized employment and work-based learning for individuals with autism and other disabilities are new concepts in the U.S. workplace, and something that will require more research and investigation into how to most successfully navigate this type of work re-organization. However, the potential benefit to both the employer and the workforce as well as the expected increase in the inclusive experiences of individuals on the spectrum really make a compelling case for more actively supporting the expansion of these types of opportunities for individuals on the spectrum.

Future Trends in Adult Autism

Translational research

According to the National Institutes of Health (2009), translational research is concerned with two areas: taking research conducted in laboratories and applying it to direct human studies, and making basic research applied by informing best practice models that are currently being used in clinical practices.[31] In the field of autism, research on the topography of the autistic brain is occurring at a rapid pace and researchers working in this area are discovering that the brain of someone on the spectrum develops and functions in very different ways than the brain of someone of the general population.[32] Translational research can provide clinicians and researchers with important information that must inform practice with individuals in the autism spectrum.

For adults with autism, translational research has the potential to inform best practice models. Scientific discoveries about brain functioning can be incorporated into developing better assessment tools, academic and work-based accommodations, and methods of training adults on the spectrum. Advances along these lines can improve outcomes for adults on the spectrum and provide important information about how to make society universally accessible.

In addition, research can also inform the development of new pharmaceuticals applications to address the overwhelming sensory issues, treat co-morbidities, and mitigate other difficult manifestations of autism that some individuals on the spectrum experience. In addition, increasing financial support for translational research can save a tremendous amount of money, as it can foster the creation of more effective therapies that can improve outcomes, and reduce reliance on supports, such as personal assistants, for adults on the spectrum who want to live and work independently.

Psychiatry

This year the American Psychiatric Association (APA) was poised to release its fifth Diagnostic and Statistical Manual (DSM-V), the diagnostic book that professionals use to diagnose and categorizes mental health conditions. The association's proposed new language around autism was a much-discussed and heatedly debated topic throughout the autism community.[33] The new version removed Asperger's as a diagnostic category. This was a very impactful conceptual shift to those within the autism population as many individuals diagnosed with Asperger syndrome embraced their diagnosis and felt a strong community connection with others similarly diagnosed.

The release was met with additional broad criticism from Thomas R. Insel, Director of the National Institutes of Mental Health (NIMH).[34] In a strongly worded response regarding the DSM-V classification methods, Insel compelled the scientific community to move away from assessments that had a 'lack of validity' as he termed it, to an understanding "...based on the emerging research data, not on the current symptom-based categories".[35] Insel challenged the scientific community to come up with categories rooted in "...genetic, imaging, physiologic, and cognitive data" and not just in a collection of symptoms.[35]

In addition to the changes proposed by Insel, many individuals on the spectrum are also advocating for change, asserting that autism research become much more participatory.[27] The value that can be gained by conferring with individuals with autism about what they need to best function in the world can also result in measures, assessments and treatment approaches that have more validity as they are informed by individuals on the spectrum who have a direct understanding of the impact of neurological differences.

Embracing the social model of disability

Throughout the disability rights movement, there is considerable interest in moving towards the social model of disability. Many advocates on the autism spectrum agree and believe that the medical model of disability represents an approach to disability that views the affected individual as unwell, broken, or needing to be healed. The social model of disability, originating from the disability community itself, does not view disability as a problem in and of itself, and considers the challenges associated with disability as primarily originating in the manner in which society thinks about and treats those with disabilities.[36]

The social model of disability recognizes the ever-present impact of discrimination or prejudice against individuals with disabilities, sometimes referred to as 'ableism' and 'disableism', and confronts and dismantles the injustices that derive from them. This model of disability has many important implications for adults with autism. For example, this model can foster the generation of new therapies, assessment tools, and data collection methods that consider individuals with autism as agents of their own lives. Interest-based learning, positive behavior supports, and the deconstruction of barriers to inclusive living, such as separate educational settings and institutional living, can create more respectful approaches towards individuals on the autism spectrum.

In the social model of disability, self-determination is a preeminent value. In this model, teaching decision-making skills, self-awareness and self-advocacy become necessary components of education and support programs for adults with autism, and can change the relationship between family, caregivers and the young adult with autism in a positive way.

For autism advocates, the social model of disability has some very important elements, which are discussed below.

The Presumption of Competence: It is very difficult to determine the intelligence of individuals with autism, particularly those with communication differences (Dawson, Soulières, Gernsbacher, Mottron, 2007). Thus, it really should compel people to presume competence when interacting with individuals with autism. This presumption should inform the way we communicate with, provide services to, and think about individuals with disabilities such as autism. In the social model of disability, *infantilization*, when individuals with disabilities are treated as children or as incapable of making their own decisions, is seen as a major barrier for individuals with disabilities, including autism. Therapies, education, and interventions must presume competence in order to be effective and relevant to individuals with disabilities.

Parents, teachers, therapists, and peers bring preconceived notions regarding an individual's lack of abilities into their relationships with the individual with a disability. These constructs do not accurately represent the person with the disability and can serve as barriers in professional practice. For example, a teacher working with a person with autism who does not use verbal language may then presume that the individual cannot understand verbal language, yet individuals with autism often have incredible capacity for receptive understanding. As a result of this presumption, certain classroom materials or lessons are not introduced to the individual, preventing him or her from accessing important academic information.

Neurodiversity and Valuing Multiple Perspectives: The medical model of disability views autism as an aberration from typical development; differences such as autism are viewed as the body and brain being *disordered*. The social model of disability understands disability to be a natural part of human biodiversity. Indeed, biologists have long recognized that species that have diverse representation are more robust. It provides a species the ability to withstand disease and environmental changes. Rather than try to conform individuals with autism to a single standard of being, the principle of neurodiversity embraces the differences that autism manifests, and forwards the notion that these differences are valuable to our species. Individuals that are neurodiverse, then, can provide unique and varied perspectives and those perspectives should be included and valued in all aspects of society.

Self-Determination: The medical model of disability can often struggle to find the balance between habilitation and self-determination (Bannerman, Sheldon, Sherman & Harchik, 1990) whereas the social model of disability asserts the idea that self-determination is a fundamental right of persons with disabilities. Habilitation is the viewpoint that there are areas of deficits within the disabled individual that prevent him or her from being independent, necessitating the design and implementation of treatments to improve the person, even when the individual may object to those treatments (Bannerman *et al.*, 1990). The social model underscores the notion that self-determination, the ability to self-direct all aspects of one's life, is sacrosanct and should guide the creation and delivery of all supports for persons with disabilities. The concept of self-determination is rooted in the understanding that the individual with autism is truly the only one who fully understands their own experiences, needs, and interests; others can be empathic, but cannot fully appreciate the experience of being on the spectrum, and thus, the individual with autism must guide and determine all decisions about the individual's life choices.

Inclusion: The social model of disability promotes the idea that diversity should be valued, and that diverse individuals have the right to participate in society alongside their peers, fully included in all aspects of society. Inclusion is the antithesis of ableism, which leads to the exclusion and differential treatment for people with disabilities.[36] The right to inclusion is premised in the understanding that separate is unequal, and that the right to inclusive experiences at school, at work, and in the community is a human right. This right must be accorded to all. Inclusion also benefits non-disabled individuals and society as a whole, and studies have indicated that children who are educated with their peers with disabilities gain skills in the domains of social empathy and confidence. For adults with autism, access to inclusive settings can decrease loneliness and isolation, and provide important networking and social connection opportunities, and improve social decision-making.

Universal Design: To promote equal access for people with disabilities, an inherent aspect of a social model of disability, universal design must be embraced across all aspects of life. For people with autism, having spaces that are designed to minimize sensory impacts, and having information that is presented in multiple formats to address variances in communication styles, creates opportunities to participate alongside those without autism. Involving people with autism and other disabilities in the design phase of new projects, such as building construction or curriculum development, brings in valuable perspectives that can inform the process and make the end product accessible to a broader group of users. This also can add tremendous value to a project and can save time and money by preventing the need for retroactive adaptations. A good example of this is the billions of dollars that are currently being spent by homeowners in the United States in order to make their homes accessible to the aging baby boomer population. If universal design had been part of the home building process, creating accessible homes could have been done in the design phase, making accessibility features like entrance ramps much more affordable and seamless to incorporate.

Self-Advocacy and Community Advocacy: Founded on the value of self-determination, self-advocacy is an action of primary importance for individuals with disabilities. When practiced by individuals on the autism spectrum, self-advocacy provides opportunities to identify conditions and experiences that create barriers to self-determination, and then to act in some fashion to remove or change those barriers. For example, self-disclosing about one's disability status allows someone on the spectrum to access remedies that emanate from the Americans with Disabilities Act. Access to accommodations, such as being able to take a break when having a sensory impact, can be critical for someone to be able to function in certain environments.

Individual acts of self-advocacy when aggregated across the entire disability community can also benefit the broader disability community. An individual's personal request for accommodations in the workplace, for example, can serve as a teachable moment for an entire group of non-disabled coworkers and can pave the way for other individuals to access accommodations more easily. In addition, self-advocacy is a consequence of the broader disability civil rights struggle that has taken place in the past, and which continues to take place in society today. The Americans with Disabilities Act and other laws and regulations available to protect the rights of individuals with disabilities stem from the personal and community advocacy of the disability rights movement.

Person-directed Living and Person-centered Planning: Just as people without disabilities are free to create their life experiences as they see fit, the social model of disability embraces the principle that people with disabilities should be able to direct their own lives. Person-centered planning is the manifestation of self determination and person-directed living in policy; individuals with disabilities, seen as agents directing their own lives, should then be able to access resources and supports that *they* identify as being vital and helpful. Treatment teams should be led by the individual with a disability, and the team's primary function should be understanding the desires of the individuals whom they support.

Rather than personal idiosyncrasies and differences inherent to autism being seen as deficits, programming and education that teaches self-awareness can assist individuals with autism to learn about what they are interested in and what they need to be happy and self-fulfilled, which can in turn lead to skill acquisition and development. Moving from the medical model to the social model of autism provides a new approach and understanding that embraces diversity and values the uniqueness of the individual on the spectrum.

Trauma-informed programming

People with autism may experience significant trauma as a consequence of abuse, isolation, and stigmatization. In addition, the medical model, when it emphasizes assessment and deficit identification, can result in additional stress and trauma to the individual. Therapies that have focused on compliance-based training, and educational experiences that are restrictive and over-regulated, even when designed to increase habilitation, can lead to prompt dependence and a lack of self-identity.

All of these experiences have an aggregate effect over the course of a person's lifetime, and may contribute to the development of comorbidities such as depression and PTSD. Young adults with autism may also experience trust issues and many have developed coping behaviors to mediate the impact of these experiences. New models of programs that take into account these impacts must be developed for adults on the spectrum. Experiences that allow the individual to be self-directed and behave freely and that provide space for personal expression and validation should be explored more often for people on the spectrum. Care providers, teachers and therapists must also understand that behavior that has historically been viewed as simply a manifestation of autism may instead stem from disabling and traumatic experiences.

Technology and social media and peer-to-peer support

One of the most exciting, dynamic areas of growth that holds incredible promise for adults on the autism spectrum is technology, social media and peer-to-peer support. Technology, such as predictive-typing software, text-to-voice software, apps that provide specific supports, and devices such as the iPad and other high-tech devices, are revolutionizing life for adults on the spectrum. In the area of communication, for example, technology can provide many young adults who may not be able to use verbal communication with a means to connect to the broader world. Communication apps for touch devices can cost thousands of dollars less than previous models, making them much more accessible for a broader range of individuals.

Facilitated communication as a communication support mechanism, in which individuals on the spectrum received physical assistance to type, caused a sensation when research demonstrated that the facilitators were actually the ones communicating, not the individual on the spectrum. This caused many in the field to dismiss supportive typing altogether. However, a growing list of high-profile cases of non-speaking individuals on the spectrum, such as Carly Fleishman, are learning to type independently, causing people to reconsider the value of supported typing as a means of teaching independent typing.[37] Carly Fleishman, who communicated primarily through her behaviors until she was a teenager, was thought to have the mental capacity of a toddler. When she began typing independently, she made it very clear to the world that she had a tremendous amount of intelligence. She is now in college in Toronto, and has written a book with her father about her experiences. Carly, and many others like her, have learned to use technology to access communication and connect to the world, thereby showing the world that assumptions about the intelligence of non-speaking individuals may be largely incorrect.

Social media is another area that is changing conditions for individuals on the spectrum. Facebook, Twitter, YouTube, and other social media sites are providing important forums for individuals with autism to express their ideas, control their own message in the media, and connect to one another. This has provided adult self-advocates the opportunity to monitor and respond to media messages that they disagree with, and to express their own thoughts and educate others about their experiences on the spectrum.

Peer-to-peer support, both from typical peers and other adults on the autism spectrum, is an incredibly rich area for future research and program development. In the past, it was often assumed that individuals on the spectrum were uninterested in socializing; however, there is a growing awareness that adults on the spectrum are often highly interested in connecting with their peers. This creates avenues for education for adults with autism who, often as a consequence of social isolation or sensory processing differences, may be missing important information about social conventions and expectations.

The interest in adult autism is growing, as society recognizes that autism is more than just a pediatric developmental disability. Children with autism grow up to become adults, and programs, educational experiences, therapies, and other supports must recognize that it is inappropriate to simply extend pediatric practices to adult populations. The challenge going forward will be to develop best practices for working with adults on the spectrum that recognize their humanity and their right to self-determination. Autistic adults deserve the same respect that is granted to non-disabled adults. Interest-based programming that embraces the authentic individual, informed by an understanding of the sensory needs and a recognition of the aggregate effect of disabling experiences, can foster personal growth and development, and provide support for the young adult on the spectrum to lead a meaningful life, fully engaged in their community.

Executive Summary

- While children are the focus of many research studies and services provided to individuals with autism, adults make up a rapidly growing majority of individuals on the spectrum.
- Little attention has been paid to the quality of life, needs, and support services that are appropriate to this growing population of adults with autism.

- From childhood, too many individuals with autism have been viewed as "deficient" and as somehow "needing to be fixed" through a myriad of medications and other interventions, which is the primary focus of the prevailing "medical model" of disability.
- Adopting a social model of disability involves first and foremost the presumption of competence of the individual with autism and capitalizes on the individual's strengths. This model is supported by increasing evidence that even individuals without spoken language who were assumed to be intellectually challenged have demonstrated profound thoughts and abilities when provided with the appropriate opportunities to express themselves and to participate in the broader community.
- Issues that need to be addressed regarding adults with autism include how to create inclusive housing, education, and employment opportunities for the population, all of which have been sorely neglected with respect to research on best practices.
- Moving forward, translational research on adults with autism must not only address the above issues, but also investigate strategies to best include these individuals into our communities, in part by using universal design to provide every opportunity for equal access to jobs, recreation, and other opportunities available to the general population.

References

1. Kanner LL: Autistic disturbances of affective contact. *Nervous Child* 1943; **2**: 2217–2250.
2. Donovan J, & Zucher C: Autism's first child. The Atlantic. 2010 (October), online. http://www.theatlantic.com/magazine/archive/2010/10/autisms-first-child/308227/.
3. Biklin D, Wurzburg G (Producers) & Wurzburg G (Director): *Wretches & Jabberers* [Motion Picture]. United States: State of the Art, Inc., 2010.
4. Shattuck PT, Roux AM, Hudson LE, Taylor JL, Maenner MJ, Trani J: Services for adults with an autism spectrum disorder. *Can J Psychiatry* 2012, **57**(5): 284–291.
5. Dawson, M: The misbehaviour of behaviourists: Ethical challenges to the autism-ABA industry. Michelle Dawson 2004. http://www.sentex.net/~nexus23/naa_aba.html.

6. Sinclair, J: Don't mourn for us. In *Autism Network International Newsletter, Our Voice*, 1993, **1**(3) http://www.autreat.com/dont_mourn.html
7. Bascom, J: Quiet hands. Just Stimming 2011. http://juststimming.wordpress.com/2011/10/05/quiet-hands/.
8. Tate, D: Reports from a Resident Alien, comments. Reports from a Resident Alien, 2013. http://chaoticidealism.livejournal.com/113479.html?thread=1108551&
9. Rekers, G, Lovaas, IO: Behavioral treatment of deviant sex-role behaviors in a male child. *J Appl Behav Analysis* 1974, **7**(2):173–190.
10. American Psychological Association: Report of the American Psychological Association Task Force on Appropriate Therapeutic Responses to Sexual Orientation. *Am Psychological Assoc* 2009: 1–138.
11. Associated Press: Boy with autism dies after chelation therapy: 5-year old was receiving controversial treatment in doctor's office. Associated Press 2005, (25 August). http://www.nbcnews.com/id/9074208/ns/health-mental_health/t/boy-autism-dies-after-chelation-therapy/#.Ui9AD7yE67M.
12. Grace EJ: I Was A Self-Loathing FC Skeptic. Tiny Grace Notes (AKA Ask an Autistic) 2012. http://www.tinygracenotes.com/2012/12/i-was-self-loathing-fc-skeptic.html
13. Anderson, C: IAN research report: Bullying and children with ASD. Interactive Autism Network 2012. http://www.iancommunity.org/cs/ian_research_reports/ian_research_report_bullying
14. Child Welfare Information Gateway: The risk and prevention of maltreatment of children with disabilities. *Children's Bureau* 2012. https://www.childwelfare.gov/pubs/prevenres/focus/focus.pdf.
15. United States Department of Health and Human Services Child Welfare Information Gateway: Understanding the effects of maltreatment on brain development. United States Department of Health and Human Services 2009. https://www.childwelfare.gov/pubs/issue_briefs/brain_development/effects.cfm
16. Orsmond GI, Shattuck PT, Cooper BP, Sterzing PR, Anderson KA: Social participation among young adults with an autism spectrum disorder. *J Autism Dev Disord* 2013, **43**(11):2710–2719.
17. Levy SE, Giarelli E, Lee LC, Schieve LA, Kirby RS, Cunniff C, Nicholas J, Reaven J, Rice C: Autism spectrum disorder and co-occurring developmental, psychiatric, and medical conditions among children in multiple populations of the United States. *J Dev Behav Pediatrics* 2010; **31**(4):267–275.
18. Autism National Committee: Can aversives and restraints produce PTSD in people with autism? *The Communicator* 1998.://www.autcom.org/articles/PTSD.html.
19. Mayes SD, Gorman A, J Hillwig-Garcia, Ehsan S: Suicide ideation and attempts in children with autism. *Res Autism Spectrum Disorders*, 2013, **7**(1):109–119.

20. Anderson K, Shattuck P, Cooper B, Roux A, & Wagner M: Prevalence and correlates of postsecondary residential status among young adults with an autism spectrum disorder. *Autism*, 2013. Published online 2013 (30 August). DOI: 10.1177/1362361313481860.
21. Pennsylvania Department of Public Welfare Bureau of Autism Services: Housing options for adults with autism spectrum disorder. Pennsylvania Department of Public Welfare Bureau of Autism Services 2010. http://delco-network.wikispaces.com/file/view/Housing_Options_for_Adults_with_Autism_-_FINAL_-_05-20-10.pdf.
22 United States Department of Justice. Olmstead: Community integration for everyone. ADA Home Page, United States Department of Justice 2013. http://www.ada.gov/olmstead/.
23. Office of Rehabilitation and Special Education Services: Welcome to OSERS: Inclusion, equity, opportunity. Office of Rehabilitation and Special Education Services 2013. http://www2.ed.gov/about/offices/list/osers/osep/index.html
24. Findyouthinfo.gov: How individualized education program (iep) transition planning makes a difference for youth with disabilities. Findyouthinfo.gov, n.d. http://findyouthinfo.gov/youth-briefs/transition-age-youth-disabilities.
25. Shattuck PT, Narendorf SC, Cooper B, Sterzing PR, Wagner M, Taylor JL. Postsecondary education and employment among youth with an autism spectrum disorder. *Pediatrics* 2012, **129**(6):1042–1049.
26. Howlin P, Goode S, Hutton J, Rutter M: Adult outcome for children with autism. *J Child Psychol & Psychiatry* 2004, **45**(2):212–229.
27. Autistic Self-Advocacy Network: Position statement. Autistic Self-Advocacy Network 2013. http://autisticadvocacy.org/policy-advocacy/position-statements/.
28. Migliore A: Sheltered Workshops. In: JH Stone, M Blouin, editors: *International Encyclopedia of Rehabilitation* 2013. http://cirrie.buffalo.edu/encyclopedia/en/article/136/
29. Cimera RE, Wehman P, West M, Burgess S: Do sheltered workshops enhance employment outcomes for adults with autism spectrum disorder? *Autism* 2012, **16**(1):87–94.
30. Wehman P, Lau S, Molinelli A, Brooke V, Thompson K, Moore C, West M: Supported employment for young adults with autism spectrum disorder: Preliminary data. *Research & Practice for Persons with Severe Disabilities* 2012, **37**(3):160–169.
31. National Institutes of Health: Institutional clinical and translational science award. National Institutes of Health 2009. http://grants.nih.gov/grants/guide/rfa-files/RFA-RM-06-002.html.
32. Grandin T, Panek R: *The autistic brain: Thinking across the spectrum*. Boston: Houghton Mifflin Harcourt. Kindle Edition; 2013.

33. American Psychiatric Association: DSM-V: Development. American Psychiatric Association 2012. http://www.dsm5.org/Newsroom/Pages/Commentaries.aspx.
34. Lane, C: The NIMH withdraws support for DSM-5: The latest development is a humiliating blow to the APA. Psychology Today, Side Effects 2013 (May). http://www.psychologytoday.com/blog/side-effects/201305/the-nimh-withdraws-support-dsm-5.
35. Insel, TR: Director's blog: Transforming diagnosis. National Institutes of Mental Health Director's Blog 29 2013 (April). http://www.nimh.nih.gov/about/director/2013/transforming-diagnosis.shtml
36. Carson, G: Social model of disability. Scottish Accessible Information Forum. http://www.ukdpc.net/site/images/library/Social%20Model%20of%20Disability2.pdf.
37. Fleishman A, Fleishman C: *Carly's voice: Breaking through autism.* New York, NY: Simon & Schuster, Inc.; 2012.
38. Dawson M, Soulières I, Gernsbacher M, Mottron L: The level and nature of autistic intelligence. *Psychological Science* (Wiley-Blackwell) 2007, **18**(8):657–662.
39. Dukes C, Smith M: Inclusion: What it is and why it's good news for everyone. In *A Practical Guide to Pre-School Inclusion*. London: SAGE Publications Ltd. 2006:1–10. doi: 10.4135/9781446214190.n1.
40. Grinker R: Unstrange minds: remapping the world of autism. Roy Richard Grinker: New York : Basic Books; 2007.
41. Interactive Autism Network: Research report #3 - September 2007. Interactive Autism Network, 2007. http://www.iancommunity.org/cs/ian_ research_reports/ian_research_report_sep_2007
42. Interagency Autism Coordinating Committee: The 2011 Interagency Autism Coordinating Committee strategic plan for autism spectrum disorder research. *Interagency Autism Coordinating Committee* 2011. http://iacc.hhs.gov/strategic-plan/2011/index.shtml.
43. Seltzer MM, Shattuck P, Abbeduto L, Greenberg JS: Trajectory of development in adolescents and adults with autism. *Mental Retard Dev Disabilities Res Rev* 2004, **10**(4):234–247.
44. Seltzer MM, Krauss MW, Shattuck PT, Orsmond G, Swe A, Lord C: The symptoms of autism spectrum disorders in adolescence and adulthood. *J Autism Dev Disord* 2003, **33**(6):565–581.
45. Winter P: Loud Hands & Loud Voices. In Bascom J., *Loud Hands: Autistic People, Speaking* 2012 (December 8) Kindle Edition.
46. Wohlring G: The politics of ableism. *Development* 2008, **51**:252–258
47. Wolman D: The truth about autism: Scientists reconsider what they think they know. Wired Magazine: 16.03 2008 (25 February). http://www.wired.com/medtech/health/magazine/16-03/ff_autism?currentPage=all

Index

Aberrant Behavior Checklist (ABC), 596
ableism, 670
abnormal histone modifications, 291
abuse, 658
actin polymerization in dendritic spines, 266
acyl-carnitine profiles, 417
adaptive, 431
adaptive behavioral styles, 629
ADARB1, 231, 237
Adenosine deaminase acting on RNA (ADAR) enzymes, 231
adenosine-to-inosine (A-to-I) RNA editing, 229
ADI-R classification, 107
adolescent, 630, 632
adults with autism, 630, 655–657, 661, 677
advanced maternal and paternal age, 500
affected-unaffected siblings, 7
agitation, 632
air pollutant compounds, 344
air pollution exposure, 353
allele-specific methylation, 292
altered methylation, 290, 382
alternative splicing, 156, 158, 161, 165, 210, 384
Alu elements, 379
amantadine, 478
American Art Therapy Association, 646
amniotic fluid testosterone, 457
anatomical brain imaging studies, 528
androgen dysregulation, 459
Angelman syndrome, 287
animal models, 394, 436
animal studies, 395
antidepressant drugs during pregnancy, 345
antioxidant defense, 407, 409, 411
antioxidant pathway, 384
antioxidants, 408, 412, 535, 544, 566
antisense RNAs, 209, 216
antisense RNA therapy, 218
Applied Behavior Analysis (ABA), 657
aptamers, 217
arginine vasopressin (AVP), 484
aripiprazole, 474
art education, 631
art therapy, 628
art therapy assessment, 636
ASD-specific maternal autoantibodies, 383
assisted reproductive technology, 342
ataxias, 212

A to G mutation, 232
A-to-I RNA Editing, 229
ATP production, 410
Autism and Developmental Disabilities Monitoring (ADDM) Network, 336
autism candidate genes, 6, 99, 109, 291, 381
Autism Diagnostic Interview-Revised diagnostic instrument, 7, 189, 254, 453, 643
Autism Diagnostic Observation Schedule, 123, 189, 254, 452, 643
autism pathophysiology, 279
autistic brain, 279
autistic disturbances, 5
autistic regression, 49, 51, 52
autoimmunity, 430
autophagy, 527
aversive treatments, 657

4E-BP2, 252, 259
bacterial content of the gut, 502
BCL2, 290
behavioral phenotypes, 15, 16, 83, 164, 229, 260
behavioral plasticity, 381
beta-alanine, 14
biased sex ratio, 459
bioinformatic approaches, 77, 293
bioinformatics, 100, 101, 109, 110, 112
biological phenotypes, 15
biologic heterogeneity, 3, 5, 474
biomarkers, 3, 5, 9, 11, 109, 125, 148, 162, 175, 178, 193, 210, 212, 438, 475, 498, 511, 560, 568, 570
biomarkers of oxidative stress, 410, 417
biotransformation, 347
Bisphenol A (BPA), 388
blocking elongation, 260
blood, 78, 101, 147–150, 154, 157–160, 165, 381
blood-brain barrier (BBB), 432
brain, 147–149, 152–160, 165, 185, 190, 206, 231, 235, 238, 262, 283, 289, 308, 314, 341, 375, 379, 382, 384, 393, 410, 413, 415, 417, 429, 435, 451–456
brain activation, 532
Brain Drain, 375, 376
brain enlargement, 529
brain function, 455, 669
brain inflammation, 536
brain methylation, 384
brain-reactive maternal antibodies, 432
brain signal-to-noise ratio, 536
brain stem abnormalities, 341
brain waves, 534
branch chain amino acid, 559
broad autism phenotype (BAP), 6, 57
bullying, 658
B vitamin supplementation, 566

calcium channels/signaling, 536–540
calcium homeostasis and synaptic plasticity, 410, 417
Cambridge Fetal Testosterone Project, 457, 459
candidate genes, 6, 58, 82, 87, 99, 109, 192–193, 291, 313, 354, 380
carnitine, 508, 559
case-control study design, 128

case-only designs, 356
categorization of probands, 6
causal inference, 322
cellular metabolism, 382
cerebellar hemisphere, 290
cerebral folate deficiency (CFD) syndrome, 508, 563
cerebral organoid technology, 395
CGG repeats, 256
chemical exposures, 343, 344, 375, 385, 389
Childhood Autism Risks from Genetics and Environment (CHARGE), 357
cholesterol, 507, 559
chromatin modifications, 351
chromatin remodeling, 282
chromosome 5p14.1, 206
circadian dysfunction, 15
circadian gene RORA, 382
circadian rhythm, 14, 151, 503
circadian rhythm dysregulation, 504
circulating IgG levels, 432
classifier accuracy, 130
class prediction analyses, 9, 11
clinical and biologic subtypes, 498
clinical diagnosis, 4, 69, 77, 121
clinical diagnostic criteria, 4
clinical genetic research using WGS, 70
clinical heterogeneity, 3, 5, 25
clinically relevant genetic mutations in ASD, 70, 77
clinical phenotype of FXS, 256
clinical phenotypes, 154, 194, 317
clinical presentation, 498
clinical translation, 69
clinical trials, 215, 264, 267
CNTNAP polymorphisms, 484
CNVs (see copy number variants), 58, 339
co-enzyme Q10, 566
cognitive growth, 629
coherence, 516, 537
college, 662
combined genetic and environmental paradigm, 338
common genetic risk variants, 26, 100
communication skills, 632
community advocacy, 673
comorbidities, 15, 16, 107, 112, 194, 497, 499, 675
Comparative Toxicogenomics Database, 353
complex behaviors, 233
comprehensive transition and postsecondary (CTP) programs, 663
congenital infections, 345
connectivity, 100, 102, 112, 516
context-dependent transcriptional modification, 231
control of mRNA translation, 264
convergent pathways, 26, 34
co-occurring conditions, 659
coping, 635
copy number variants (CNVs), 13, 27, 69, 71, 74, 80, 106, 120, 126, 153, 187, 204, 254, 339, 377, 500
core symptom domains, 474
corpus callosum, 458
covalent modification of histones, 281
CpG dinucleotides, 281, 304, 380
CpG island, 282, 307
creative thinking, 631

critical periods, 341
customized employment, 667, 668

7-dehydrocholesterol reductase gene (DHCR7), 507
daily functioning, 609
Danish biobank, 451, 459
dark matter, 178
D-Cycloserine, 478
decreased antioxidant defense, 411
decreased inhibitory signaling at the synapse, 14
deep sequencing, 238
de novo genetic variants and mutations, 28, 71, 74, 79, 205, 284, 340, 351, 377
de novo model, 284
de novo mutation of protein-coding genes, 205
depression, 675
developmental trajectories, 511, 558
dextromethorphan, 479
diagnosis/prognosis, 3, 70, 83, 121, 229
Diagnostic and Statistical Manual of Mental Disorders, 5th edition (DSM-5), 4, 53, 120, 176
diagnostic subtypes, 497
diagnostic yield, 85
dietary factors, 382
dietary interventions, 502
diethylstilbestrol (DES), 388, 392
differential alternative splicing (DAS), 156, 161
differential exon usage, DEU, 161
differentially methylated regions (DMRs), 290, 291
diffusion tensor imaging (DTI), 531
dihydrofolate reductase, 563
dimensional stratification, 6, 47, 48
direct damage to the DNA, 351
disableism, 670
discordant MZ twins, 290
dizygotic twins, 26, 120, 280, 559
DNA methylation, 31, 102, 112, 188, 281, 290, 292, 304, 305, 308, 317, 351, 375, 378, 382
DNA methyltransferases (DNMTs), 382
Dnmt1, 380
Dnmt3a, 380
dorsolateral prefrontal cortex, 586
double-bouquet cells, 587
Down's syndrome, 214
drawings, 641
drug development, 262
"druggable" pathway, 263
DSM-V, 4, 25, 53, 120, 176, 498
dynamical process systems, 542, 547
dysregulated translation in ASD, 249, 260
dysregulated epigenetic proteins, 288

Early Autism Risk Longitudinal Investigation (EARLI), 357
early intervention services, 657
EEG (electroencephalography), 534, 546
E/I balance, (see excitation to inhibition balance)
eIF4E, 252, 257, 259, 264
eIF4E-dependent translation, 259

eIF4E-eIF4G interaction, 260
eIF4F, 250, 252
eIFs (eukaryotic initiation factors), 250
electrical oscillations, 534
electroencephalogram (EEG), 37, 589
electron transport chain (ETC), 407
electrophysiology, 515
Embedded Figures Test (EFT), 457
EMB theory, (see extreme male brain theory)
emergent property, 520, 541
emotional regulation, 629
Empathizing-Systemizing (E-S) theory, 451, 453
Empathy Quotient (EQ), 453
employment, 664–668, 678
endogenous anticonvulsant molecule, 14
endophenotype, 6, 37, 50, 293, 356
energy metabolism, 407, 408, 413
enrichment of sample population, 128
environmental epidemiology, 347
environmental epigenetics, 288, 351
environmental etiologic work, 337
environmental exposures, 16, 233, 375, 389, 392, 394, 542
environmental risk factors, 103, 294, 341, 375, 408, 500
environmental susceptibility, 359
environmental toxicant/toxins, 149, 288, 306, 343, 376, 382, 384, 417
epidemiology, 338, 352, 357, 359
epigenetic ASD etiologies, 381
epigenetic differences/modifications, 185, 342, 351, 376, 378
epigenetic disruption, 284
epigenetic dysregulation, 279, 284, 290, 291, 295
epigenetic factors, 100, 104, 112, 280
epigenetic inheritance, 292
epigenetic mechanisms, 177, 229, 281, 282, 376
epigenetic regulatory signals, 295
epigenetics, 279, 280, 283, 304, 378, 382, 515
epigenetic signatures, 293, 303
epigenome, 280, 359, 376, 382, 393
epigenomic twin studies, 394
epilepsy, 14, 499
epimutations, 283, 289, 292
ERK1/2, 267
ethnicity/ancestry, 133
event-related potentials (ERP), 591
evidence-based practice, 612
excitation to inhibition (E:I) balance, 255, 476, 500
exome sequencing, 27, 205
exposome, 359
exposure assessment and phenotyping, 354, 358, 360
exposure-disease association, 347
exposure-induced congenital disorders, 338
expression profiling, 8, 149, 153, 158, 181, 185, 186, 290, 293
expression quantitative trait loci (eQTL), 102
Extreme Male Brain (EMB) theory, 451, 453, 454

face perception, 637
face recognition, 643
Face Stimulus Assessment, 637
facilitated communication, 676
family art therapy, 634
family studies, 56
female-containing (FC) ASD cohort, 103
fenobam, 34
fetal alcohol syndrome, 343
fetal androgens, 456
fetal brain, 384, 432, 434
fetal brain antigens, 383, 429, 437
fetal Testosterone (fT) theory, 451, 456, 457, 459
fiber tracts, 531
FMR1, 30, 34, 285, 458, 532, 534
FMRP, 256, 260–263
FMRP target mRNAs, 267
folate, 557, 559, 563
folate metabolism, 288, 352, 382, 563
folate receptor, 383
folate receptor autoantibodies (FRAs), 508, 563
folate transport, 383
folinic acid, 352, 564, 565
Fragile X, 147, 163, 164
fragile X mental retardation 1 gene (FMR1), 30, 34, 255, 285, 458, 532, 534
fragile X mental retardation protein (FMRP), 30, 256, 260–263
Fragile-X Syndrome (FXS), 30, 214, 249, 255, 256, 260, 262, 285
free radicals, 407
FXS-relevant phenotypes, 263
GABA, 476
GABA-B agonists, 479
GABAB receptor 1 agonist R-baclofen, 265
GABAB receptors, 267
GABAergic cells/signaling, 265, 501, 587
gamma activity, 589, 590, 593, 597
gamma power, 590, 594
gap junctions, 536
gastrointestinal (GI) disorders, 15, 70, 107, 439, 484, 501, 518, 562
gender-base stratifications, 105
gender bias, 104–106
gender-specific differential penetrance, 103
gene by environment interaction (GxE), 109, 288, 295, 335, 346, 349, 357, 382, 394, 516
gene dosage, 109
gene expression signatures, 11, 154, 280, 381
gene/gene interactions, 109
gene pathway analysis, 119, 127
generationally cumulative environmental exposures, 376
gene regulatory systems, 111
genes and environment, 335, 515
gene set enrichment, 128
genetic and biological heterogeneity, 3
genetic architecture, 7, 71, 102, 128, 378
genetic causes, 70, 99, 148, 338, 559, 616
genetic classifier, 119, 127, 130, 131, 134
genetic clusters, 499

genetic counseling, 84
genetic heterogeneity, 7, 13, 25, 36, 48, 70, 84, 185
genetic imprinting, 280
genetic linkage analyses of ASD, 13
genetic risk factors, 50, 74, 105, 254, 338, 347, 375, 439
genome-wide association studies (GWAS), 8, 26, 71, 99, 127, 206, 339
genome-wide *de novo* CNV data, 80, 106
genome-wide DNA methylation, 290
genome-wide epigenomic variation, 292
genomic coverage, 85
genomic "dark matter", 175
genomic interactions, 378
genomic technologies, 377
germline, 291
gestational environment, 431
GI disorders (see gastrointestinal disorders)
global DNA methylation analyses, 388, 389, 394
global patterns of epigenomic disruption, 289
glutamate overload, 537
glutamatergic system, 476
glutamic acid decarboxylase protein, 476
glutathione (GSH), 384, 418, 557, 566, 559
glutathione redox imbalance, 407
group 1 metabotropic glutamate receptor (mGluR) theory, 31
gut and the brain, 502
GxE autism epidemiology studies, 350, 352
GxE epidemiological interaction, 350, 352
GxE interactions (see gene by environment interactions)
GxE mechanisms, 348

5-hydroxymethylcytosine, 380
habilitation, 655, 672
heavy metals, 384
heritability, 5, 26, 57, 321, 337
heritability discordance, 135
heritable changes in gene expression, 351
heterogeneity, 5, 37, 48, 107, 186, 194, 280, 291, 474, 539
heterogeneity reduction, 13, 15
heterogeneous clinically, etiologically and biologically, 509
heterogeneous disease, 293, 498
high heritability of autism, 280
high-throughput genome-wide assays, 74, 289, 395
histone complexes, 281
histone lysine 4 trimethylation, 291
histone post-translational modifications, 281, 304
histones, 282
histone tail, 319
hnRNP K-homology (KH) domains, 260
hormone studies, 451
House-Tree-Person, 632, 642
housing, 660, 661, 663, 678
HTR4, 15

human brain function, 458
human brain structure, 457
human epidemiology, 394
human figure drawings, 642
hybrid case-only/case-control designs, 356

idiopathic ASD, 4, 23, 26
IgG antibodies, 432
ILK, 147, 155, 160, 162, 163
immune dysregulation, 429
immune modulators, 479
immune system, 383, 430
imprinted genes/regions, 104, 287, 310, 317, 321, 379
imprinting center (IC), 287
improving communication, 614
inactivation of X chromosomes, 280
inclusion, 672
inclusive educational options, 656
inclusive experiences, 664
inclusive opportunities, 655
incomplete penetrance, 87
increased power in genetic analyses, 13
individual-subject level analysis, 156
Individuals with Disabilities Education Act (IDEA 2004), 662
infantilization, 671
infertility treatment, 342
Infinium 450k array, 382
inflammation, 417, 432, 485, 521, 527, 529, 544
information processing, 530
inherited genetics, 339
inherited mutations, 75, 78, 87
inhibitory neurotransmitter, 14
inhibitory RNA-based therapies, 203
in silico eQTL, 102
integrated-omics approach, 100, 185
integrative approaches, 99–111
interactions among multiple elements, 103
interactome networks, 110
inter- and intra-family heterogeneity, 108
interest-based programming, 677
intergenic and intragenic risk loci, 75
internal locus of control, 632
interpersonal skills, 614
intervention strategies, 359
intra-family heterogeneity, 13, 107
intranasal oxytocin, 484
in utero environment, 342, 375
isolation, 658
isolation of significant subsets, 511

job coach and trials, 666

Kanizsa figures, 590
Kinetic Family Drawing, 632
Kinetic-House-Tree-Person, 632

language and visuospatial ability, 458
language development, 51, 458, 664
language disorders, 47
language delay/impairment, 16, 47–50
later birth order, 393

L-carnitine, 562
Leo Kanner, 5, 498, 656
life and social skills, 663
lipid peroxidation, 407
lithium, 265
lncRNA db, 195
local brain volume, 458
long noncoding RNA (lncRNA), 9, 179, 203, 209, 218
low birth weight, 500
lymphoblastoid cell lines (LCL), 7, 147, 149, 150, 152–154, 158, 165, 181, 185–187, 290

5-methylcytosine (5-mC), 380
5-methyltetrahydrofolate (5MTHF), 563
magnetic resonance, 529, 546
male bias, 452
male-only (MO) ASD cohort, 103
mammalian target of rapamycin (mTOR), 32
mandalas, 632
MAPK/ERK pathway, 252
Markers of Autism Risk in Babies-Learning Early Signs (MARBLES), 357
masculinizing effect of fetal testosterone (fT), 453
maternal age, 342, 393
maternal allergic or autoimmune conditions, 342
maternal autoantibodies, 16, 429, 433, 439, 440
maternal autoantibody related (MAR) autism, 435, 440
maternal immune activation (MIA), 432
maternal immune system, 384, 431
maternal infections, 346
maternally-inherited deletions, 287
maternal medications, 500
maternal metabolic conditions, 342
maternal nutrition, 352
maternal obesity, 393
maternal pregnancy factors, 393
MECP2, 29, 126, 177, 257, 281, 286, 289, 309, 381
medical classification, 82
medical co-morbidities (see co-morbidities)
medical management, 70, 77, 84
medical model of disability, 672
medicated childhoods, 657
medications, 437, 483, 505, 558, 657
MEG (magnetoencephalography), 534, 546
melatonin, 14, 504
memantine, 267, 477
memory formation, 380
metabolic abnormalities, 557
metabolic analyses, 16
metabolic disorders, 505, 507
metabolism, 375, 515, 530
MET gene and polymorphisms, 353, 484, 503
methionine cycle, 382
methionine synthase (MTR), 384
methylation (see DNA methylation)
methylation pathways, 312
methylation profiling, 317

methylation QTL (mQTL), 102
methylcobalamin, 566
methyl-CpG-binding domain (MBD), 29
methyl donors, 382
methylenetetrahydrofolate reductase, 382, 563, 564
methyl-group metabolism, 288
methylmercury, 388
methylome, 292, 375, 380, 382, 393
mGluR5 metabotropic glutamate receptor, 262, 264, 267, 286
mGluR antagonists, 264
mGluR signaling, 477
mGluR theory, 260
microfluidics-based multiplex PCR, 238
microRNA (miRNA), 147, 152, 179, 185, 208
"middle-out" approach, 538, 542
mid-gestation, 432
mid-pregnancy blood specimens, 434
mindfulness, 632
"minicolumnar" hypothesis, 587
minicolumns, 587
minocycline, 266
miRNA (see microRNA)
mitochondria, 384, 544
mitochondrial disease (MD), 384, 413, 505
mitochondrial DNA (mtDNA), 410, 506
mitochondrial dysfunction, 16, 407, 412, 541, 544, 547, 559, 560
mitochondrial ETC complexes, 414
molecular pathways, 23
molecular targets, 351
monogenic disorders, 125
monozygotic twins, 7, 26, 280, 288, 381
motor threshold, 586
mouse models of FXS, 263
mRNA expression profiling (see expression profiling)
mRNA translation, 249, 250, 262, 267
mTOR, 33, 147, 155, 160–165, 251, 257, 262–264
mTORC1, 262, 263
mTOR pathway, 161, 164, 251
"multi-scale" biology, 538, 540, 542, 546
multi-step subject stratification approach, 107
music therapy, 607
mutations in core epigenetic regulatory proteins, 295

Natural Killer (NK) activity, 147, 150, 155, 160, 162, 164
ncRNAs, 179, 182, 191, 203–210
nervous system development, 180
network analysis, 106, 109, 111
neural sexual dimorphism, 455
neurodevelopmental differences and diversity, 655, 665, 672
neuroinflammation, 485, 527
neurological dysfunction, 352
neuromodulation, 516, 598
neuronal methylome, 379
neuronal-specific highly methylated domains (N-HMDs), 380
neuronal splicing factor A2BP1, 36
neuronal stalled polysomes, 253

neuroplasticity (see plasticity)
next generation sequencing, 28, 80, 135, 204, 317
NIEHS Environmental Genome Project, 353
nitric oxide, 568, 570
NLGN3, 36, 150, 339
non-coding RNAs (ncRNA), 13, 88, 101, 112, 175, 179, 203, 281, 283, 288, 292
non-CpG sites, 380
non-genetic influences, 394
non-human primate models, 437
non-syndromic autism, 289, 505
novel therapeutics targeting, 475
nuclear DNA (nDNA), 506
nucleosomes, 281, 282
Nurses Health Study II, 393
nutrition, 375, 394, 418, 431, 504, 530, 542
nutritional supplementation, 418

obligate epimutations, 286, 295
obligatory model, 283
obstetric complications, 342
occipital cortex, 290
omega-3 fatty acids, 481
one-carbon cycle, 384
online degree or certificate programs, 663
optimal outcomes, 557
organochlorine pesticide exposure, 344
organophosphate pesticide exposure, 343, 352
organ systems outside of the brain, 511
overlapping pathways, 378
oxidative stress, 384, 407, 408, 410, 479, 483, 523, 544, 566
oxytocin (OXT), 484
oxytocin receptor gene (OXTR), 289, 313

PAK-specific inhibitors, 266
parental characteristics, 342
parental imprinting, 389
parental medical conditions, 342
parental occupational exposures, 344
partially methylated domains (PMDs), 380
paternal age, 342
paternal diet, 292
paternally-inherited deletions, 287
pathophysiology, 537, 544
pathways, 128, 147, 148, 150, 154, 156, 160–165
peer-to-peer support, 675, 677
periconceptional folate supplementation, 288
peroxisome proliferator-activated receptor (PPAR)-gamma, 482
persistent organic pollutants, 388
personalized medicine, 16, 37, 607
personal responsibility, 614
person-centered planning, 674
person-directed living, 674
pervasive developmental disorder – not otherwise specified (PDD-NOS), 24
pharmaceuticals, 658
pharmacogenomics, 13–15
pharmacological agents, 23
pharmacological clues, 85
pharmacological therapies, 34, 37, 475

pharmacology, 210
phenotype assessment, 356
phenotype definition, 3
phenotype information, 360
phenotypes, 499
phenotypic heterogeneity, 50, 107, 111, 112
phenotypic subgrouping, 8, 12, 107
physical development, 629
physiological and psychosocial development, 629
PI3K, 267
pioglitazone, 482
PIWI-interacting RNAs (piRNAs), 208
plasticity, 28, 125, 181, 233, 251, 253, 266, 409, 475, 485, 500, 515, 562, 584
play skills, 614
polybrominated diphenyl ethers (PBDEs), 343, 388
polychlorinated biphenyls (PCBs), 343, 388, 393
poor fetal growth, 393
population stratification, 120, 130, 133
portrait drawings, 642
positive behavior supports, 657
positron emission tomography (PET), 525
postmortem human brain studies, 235, 290, 293, 381, 394, 501
post-secondary academic options, 662
post-secondary training options, 664
post-translational modifications, 319
post-traumatic stress disorder (PTSD), 660, 675
Prader-Willi, 287
pragmatic disorder, 53
pre- and post-high-throughput sequencing, 70, 71
pregnancy, 432
pregnant rhesus monkeys, 438
premature birth, 500
premutation condition, 256
prenatal alcohol exposure, 343
prenatal exposure, 288, 432
prenatal immune challenges, 430
prenatal vitamins, 382
prescribed medications, 388
presuming competence, 655, 671, 678
prevalence of ASD, 176
processed pseudogenes, 379
projective drawing assessment, 637
protein and DNA oxidation, 407
protein synthesis, 260
pseudogenes, 209
psycho-educational interventions, 475, 485
psychological sex differences, 454
psychopharmacology, 473
PTEN (phosphatase and tensin homolog), 252
pure epimutations, 295
Purkinje cells, 341

15q11-13 region, 287
quantitative trait association analyses, 8, 12

rare *de novo* mutation, 28
rare genetic variation, 27, 81

reactive oxygen species (ROS), 384, 408
Reading the Mind, 484
redox, 557
redox metabolism, 566
reduced folate carrier, 563, 564, 569
regional brain size differences, 528
Repetitive Behavior Scale (RBS-R), 596
repetitive sequences, 379
repetitive transcranial magnetic stimulation (rTMS), 583, 584
reproductive methylomes, 389
residential solutions, 660
restrictive diets, 658
restrictive-repetitive behaviour, 25
Rett syndrome, 4, 29, 125, 176, 257, 286, 309, 559
RGG box, 260
ribosome, 216, 217, 250, 251, 252, 260
riluzole, 477
risk and protective contributors of ASD status, 120
risk for suicide, 660
risperidone, 474
RNA aptamers, 216
RNA editing, 233
RNAi, 216
RNA-seq, 234
RORA, 164, 237, 290, 382, 460

S-adenosylmethionine (SAM), 382
sample stratification, 108
sapropterin treatment, 566
school-based art therapy, 633
segmental duplications, 379
seizures, 499
self-advocacy and self-determination, 671–677
sense of self/identity, 632
sensory processing difficulties, 592, 665, 669, 673
sex bias, 112
sex differences, 377, 451, 453, 459
sex steroids, 451, 460
sexual dimorphism, 458
sexual dimorphism in the human brain, 454, 458
SHANK3, 36, 120, 150, 254, 263, 314, 339, 475, 526
shared genes, 147, 154, 158, 164, 165
shared *in utero* environment, 377
shared neurotransmitter systems, 502
sheltered workshops, 665
short interfering RNAs (siRNAs), 208, 216
siblings, 7, 26, 71, 74, 108, 128, 185, 340, 377, 635
signal transduction pathways, 267
single gene disorders, 258
single nucleotide polymorphisms (SNPs), 127
single photon emission computed tomography (SPECT), 525
site-specific deamination of adenosine to inosine, 231
sleep disorders, 14, 503
small ncRNA, 208
small nucleolar RNA (snoRNA), 179, 208

Smith-Lemli-Opitz Syndrome (SLOS), 507
snoRNABase, 194
snoRNAs, 188, 189
SNP quality control, 132
social-communication, 25
social interaction, 632, 633, 664
social learning, 485
social media, 675, 676
social model, 655, 670, 671, 672, 674, 675, 678
Social Skills Rating System, 634
Spatial Ancestry Analysis (SPA), 133, 134
spectral analysis, 589
splicing, 150, 151, 156, 158, 160, 165
stem cells, 189, 238, 293, 379, 395
stigmatization, 665
stochastic events, 322
stratification of ASD, 5, 100, 106, 107, 475
stress-activated proteins, 384
structural connectivity, 531
structural sexual dimorphism in the brain, 455
Study to Explore Early Development (SEED), 357
subgrouping in autism, 294
subphenotypes of ASD, 3, 37, 194
subtype-dependent case-control association analyses, 12
supported employment, 666
synaptic editing levels, 235
synaptic plasticity (see plasticity)
syndromic autism, 23, 28, 285
synergistic effects, 350
Systemizing Quotient (SQ), 453
systems biology, 100, 517, 541
targeted therapies, 13, 15, 37
targets of ncRNAs, 193
technical/statistical limitations, 295
teratogens, 341
TET enzymes, 381
tetrahydrobiopterin, 557, 559, 567
Thematic Apperception Test, 632
Theory of Mind (ToM), 484
therapeutic agents, 249
therapeutic goals, 609
therapeutic potential, 584
therapeutics, 474
therapeutic strategies, 429
therapeutic targeting of ncRNA, 210
therapeutic targets, 210, 212, 295
timing of the exposure assessment, 354
TMLHE deficiency, 508
toxicants, 343
training adults on the spectrum, 669
Transcranial Magnetic Stimulation (TMS), 583
transcriptionally active, 319
transcriptionally silent, 320
transcriptome analyses (see expression profiling)
transduction, 515, 546, 548
transgenerational effects, 375, 376, 389, 392, 394
transgenerational epidemiological studies, 393
transition-aged youth, 661
transition services, 662
translational control, 249, 251

translation-dependent synaptic plasticity, 267
translation initiation, 250
transmissible regulatory RNAs, 292
transposon-driven repeats, 379
trauma-informed programming, 675
treatment modality, 587, 600
treatment option, 90, 99, 203, 207, 610
trimethyllysine hydroxylase epsilon gene (TMHLE), 508
TSC1 and TSC2, 252
tuberous sclerosis, 32
twin studies, 135, 280, 337, 346, 377, 394, 430

ubiquitin E3A ligase (UBE3A), 287
underlying medical conditions, 510
universal design, 673, 678

variable expressivity, 87
variances in communication styles, 673
verbal expression, 633
vinclozolin, 389
visual communication, 634
visual illusions, 590
vocational rehabilitation programs, 664
vocational training, 662
voltage-gated calcium channels, 536

Wechsler Adult Intelligence Scale, 632
whole-body approach, 515
whole exome sequencing (WES), 6, 72, 74, 103, 204, 292, 377
whole genome sequencing (WGS), 6, 69, 70, 73, 76, 86, 206
windows of susceptibility, 341
work-based accommodations, 669
work-based learning, 666, 667
work-based supports, 665
work places, 656
wrap-around services, 663

X chromosome inactivation, 104, 215, 239, 306, 389
X-linked genes, 75, 104